Graduate Texts in Mathematics

46

M. Loève

Probability Theory II

4th Edition

Springer-Verlag
New York Heidelberg Berlin

M. Loève
Departments of Mathematics and Statistics
University of California at Berkeley
Berkeley, California 94720

AMS Subject Classifications
28–01, 60A05, 60Bxx, 60E05, 60Fxx

Library of Congress Cataloging in Publication Data

Loève, Michel, 1907–
Probability theory.
(Graduate texts in mathematics; 45-46)
Bibliography p.
Includes index.
1. Probabilities. I. Title. II. Series.
QA273.L63 1977 519.2 76–28332

Originally published in the University Series in Higher Mathematics (D. Van Nostrand Company); edited by M. H. Stone, L. Nirenberg, and S. S. Chern.

Printed in the United States of America.

ISBN 0-387-90262-7 Springer-Verlag New York
ISBN 3-540-90262-7 Springer-Verlag Berlin Heidelberg

To Line

and

To the students and teachers
of the School in
the Camp de Drancy

PREFACE TO THE FOURTH EDITION

This fourth edition contains several additions. The main ones concern three closely related topics: Brownian motion, functional limit distributions, and random walks. Besides the power and ingenuity of their methods and the depth and beauty of their results, their importance is fast growing in Analysis as well as in theoretical and applied Probability.

These additions increased the book to an unwieldy size and it had to be split into two volumes.

About half of the first volume is devoted to an elementary introduction, then to mathematical foundations and basic probability concepts and tools. The second half is devoted to a detailed study of Independence which played and continues to play a central role both by itself and as a catalyst.

The main additions consist of a section on convergence of probabilities on metric spaces and a chapter whose first section on domains of attraction completes the study of the Central limit problem, while the second one is devoted to random walks.

About a third of the second volume is devoted to conditioning and properties of sequences of various types of dependence. The other two thirds are devoted to random functions; the last Part on Elements of random analysis is more sophisticated.

The main addition consists of a chapter on Brownian motion and limit distributions.

It is strongly recommended that the reader begin with less involved portions. In particular, the starred ones ought to be left out until they are needed or unless the reader is especially interested in them.

I take this opportunity to thank Mrs. Rubalcava for her beautiful typing of all the editions since the inception of the book. I also wish to thank the editors of Springer-Verlag, New York, for their patience and care.

M. L.

January, 1977
Berkeley, California

PREFACE TO THE THIRD EDITION

This book is intended as a text for graduate students and as a reference for workers in Probability and Statistics. The prerequisite is honest calculus. The material covered in Parts Two to Five inclusive requires about three to four semesters of graduate study. The introductory part may serve as a text for an undergraduate course in elementary probability theory.

The Foundations are presented in:

the Introductory Part on the background of the concepts and problems, treated without advanced mathematical tools;
Part One on the Notions of Measure Theory that every probabilist and statistician requires;
Part Two on General Concepts and Tools of Probability Theory.

Random sequences whose general properties are given in the Foundations are studied in:

Part Three on Independence devoted essentially to sums of independent random variables and their limit properties;
Part Four on Dependence devoted to the operation of conditioning and limit properties of sums of dependent random variables. The last section introduces random functions of second order.

Random functions and processes are discussed in:

Part Five on Elements of random analysis devoted to the basic concepts of random analysis and to the martingale, decomposable, and Markov types of random functions.

Since the primary purpose of the book is didactic, methods are emphasized and the book is subdivided into:

unstarred portions, independent of the remainder; starred portions, which are more involved or more abstract;
complements and details, including illustrations and applications of the material in the text, which consist of propositions with fre-

quent hints; most of these propositions can be found in the articles and books referred to in the Bibliography.

Also, for teaching and reference purposes, it has proved useful to name most of the results.

Numerous historical remarks about results, methods, and the evolution of various fields are an intrinsic part of the text. The purpose is purely didactic: to attract attention to the basic contributions while introducing the ideas explored. Books and memoirs of authors whose contributions are referred to and discussed are cited in the Bibliography, which parallels the text in that it is organized by parts and, within parts, by chapters. Thus the interested student can pursue his study in the original literature.

This work owes much to the reactions of the students on whom it has been tried year after year. However, the book is definitely more concise than the lectures, and the reader will have to be armed permanently with patience, pen, and calculus. Besides, in mathematics, as in any form of poetry, the reader has to be a poet *in posse*.

This third edition differs from the second (1960) in a number of places. Modifications vary all the way from a prefix ("sub" martingale in lieu of "semi"-martingale) to an entire subsection (§36.2). To preserve pagination, some additions to the text proper (especially 9, p. 656) had to be put in the Complements and Details. It is hoped that moreover most of the errors have been eliminated and that readers will be kind enough to inform the author of those which remain.

I take this opportunity to thank those whose comments and criticisms led to corrections and improvements: for the first edition, E. Barankin, S. Bochner, E. Parzen, and H. Robbins; for the second edition, Y. S. Chow, R. Cogburn, J. L. Doob, J. Feldman, B. Jamison, J. Karush, P. A. Meyer, J. W. Pratt, B. A. Sevastianov, J. W. Woll; for the third edition, S. Dharmadhikari, J. Fabius, D. Freedman, A. Maitra, U. V. Prokhorov. My warm thanks go to Cogburn, whose constant help throughout the preparation of the second edition has been invaluable. This edition has been prepared with the partial support of the Office of Naval Research and of the National Science Foundation.

M. L.

April, 1962
Berkeley, California

CONTENTS OF VOLUME II

GRADUATE TEXT IN MATHEMATICS VOL. 46

PART FOUR: DEPENDENCE

CHAPTER VIII: CONDITIONING

CHAPTER IX: FROM INDEPENDENCE TO DEPENDENCE

PART FIVE: ELEMENTS OF RANDOM ANALYSIS

CHAPTER XII: FOUNDATIONS; MARTINGALES AND DECOMPOSABILITY

CHAPTER XIII: BROWNIAN MOTION AND LIMIT DISTRIBUTIONS

CHAPTER XIV: MARKOV PROCESSES

CONTENTS OF VOLUME I

GRADUATE TEXTS IN MATHEMATICS VOL. 45

Part Four

DEPENDENCE

For about two centuries probability theory has been concerned almost exclusively with independence. Yet, very particular forms of dependence appear already in the theory of games of chance. But a first general type of dependence—chains—was introduced only at the beginning of this century by Markov. Another type of dependence—stationarity—appears in ergodic theory, and a related type—second order stationarity—is then introduced in probability theory by Khintchine (1932). Centering at conditional expectations by P. Lévy (1935) gives rise to a new type of dependence—martingales.

At the very core of the study of dependence lies the concept of conditioning—with respect to a function—put in an abstract and rigorous form by Kolmogorov. In this part, the concept of conditioning is introduced in a more general form—with respect to a σ-field—and, as much as possible, the properties of various types of dependence are related to more general results, with emphasis given to the methods.

Chapter VIII

CONDITIONING

§27. CONCEPT OF CONDITIONING

The concept of "conditioning" can be expressed in terms of sub σ-fields of events. Conditional probabilities of events and conditional expectations of r.v.'s "given a σ-field $\mathcal{B}$," to be introduced and investigated in this chapter, are $\mathcal{B}$-measurable functions defined up to an equivalence. If $\mathcal{B}$ is determined by a countable partition of the sure event, then these functions are elementary. In this "elementary case," a constructive approach with a definite intuitive appeal is possible and there are no technical difficulties. In the general case, there is no suitable and rigorous constructive approach, and a descriptive one, requiring more powerful tools, especially the Radon-Nikodym theorem, has to be used.

The R.-N. theorem was obtained in its abstract form in 1930 and the concept of conditional probabilities and of conditional expectations of integrable r.v.'s "given" a measurable function, finite or not, numerical or not, was then put on a rigorous basis by Kolmogorov in 1933.

27.1. Elementary case. Investigation of the elementary case will give us an insight into the ideas involved in the intuitive notion of conditioning and will lead "naturally" to the notions and problems which appear in the general case.

The notion of conditional probability of an event A "given an event B" corresponds to that of frequencies of A in the repeated trials where B occurs; it is one of the oldest probability notions. For every event A, the relation

$$PB \cdot P_B A = PAB$$

defines the *conditional probability* (*c.pr.*) $P_B A$ *of* A *given* B as the ratio PAB/PB, provided B is a nonnull event; if B is null, so is AB, and the foregoing relation leaves P_BA undetermined. In what follows, we assume that, unless otherwise stated, B is nonnull.

The function P_B on the σ-field $\mathcal{A}$ of events, whose values are P_BA, $A \in \mathcal{A}$, is called *conditional pr. given B*. The defining relation shows at once that since P on $\mathcal{A}$ is normed, nonnegative, and σ-additive, so is P_B on $\mathcal{A}$:

$$P_B\Omega = 1, \quad P_B \geqq 0, \quad P_B \sum A_j = \sum P_BA_j.$$

Thus, the conditioning expressed by "given B" means that the initial pr. space $(\Omega, \mathcal{A}, P)$ is replaced by the pr. space $(\Omega, \mathcal{A}, P_B)$. The expectation, if it exists, of a r.v. X on this new pr. space is called *conditional expectation* (*c.exp.*) *given B* and is denoted by E_BX; in symbols

$$E_BX = \int X\,dP_B.$$

Since $P_B = 0$ on $\{AB^c, A \in \mathcal{A}\}$, the right-hand side reduces to $\int_B X\,dP_B$ and, since $P_B = \frac{1}{PB}P$ on $\{AB, A \in \mathcal{A}\}$, it becomes $\frac{1}{PB}\int_B X\,dP$. Therefore, the c.exp. of X given B can be defined directly by

$$PBE_BX = \int_B X\,dP$$

and is determined if B is a nonnull event. In particular,

$$PBE_BI_A = \int_B I_A\,dP = PAB$$

so that the c.pr. P_BA can be defined, thereafter, by

$$P_BA = E_BI_A.$$

Thus, if E_B is the *c.exp. given B*, with values E_BX on the family $\mathcal{E}_B$ of all r.v.'s X whose integral on B exists, the c.pr. P_B becomes the restriction of E_B to the family $I_{\mathcal{A}}$ of indicators of events. Furthermore, properties of P_B become particular cases of the immediate properties of E_B below.

If $X \geqq 0$ then $E_BX \geqq 0$, and if c is a constant then $E_Bc = c$. If the X_j are nonnegative, or if the X_j are integrable and their consecutive sums are uniformly bounded by an integrable r.v., then $E_B \sum X_j = \sum E_BX_j$.

C.exp.'s (hence c.pr.'s) acquire their full meaning when reinterpreted as values of functions, as follows. The number E_BX is no longer assigned to B but to every point of B, and similarly for $E_{B^c}X$, so that we have a two-valued function on Ω, with values $E_B X$ for $\omega \in B$ and $E_{B^c}X$ for $\omega \in B^c$. More generally, let $\{B_j\}$ be a countable partition of Ω and let $\mathcal{B}$ be the minimal σ-field over this partition. Let $\mathcal{E}$ be the family of all r.v.'s X whose expectation EX exists, so that their indefinite integrals, hence c.exp.'s given any nonnull event, exist. Consider the elementary functions

$$E^{\mathcal{B}}X = \sum (E_{B_j}X)I_{B_j}, \quad X \in \mathcal{E}.$$

If some B_j are null, then the corresponding values $E_{B_j}X$ are undetermined, so that $E^{\mathcal{B}}X$ is undetermined on the null event which is the sum of null B_j. Such a possibility, together with the definition of $E_{B_j}X$, leads to the following

Constructive definition. *The elementary function $E^{\mathcal{B}}X$ defined up to an equivalence by*

$$(1) \qquad E^{\mathcal{B}}X = \sum \left(\frac{1}{PB_j}\int_{B_j} X\,dP\right) I_{B_j}, \quad X \in \mathcal{E},$$

is the c.exp. of X given $\mathcal{B}$.

Upon particularizing to indicators, the $\mathcal{B}$-measurable function $P^{\mathcal{B}}A$, defined up to an equivalence by setting

$$P^{\mathcal{B}}A = E^{\mathcal{B}}I_A, \quad A \in \mathcal{A},$$

will be the *c.pr. of A given* $\mathcal{B}$; the contraction of $E^{\mathcal{B}}$ on $I_{\mathcal{A}}$, to be denoted by $P^{\mathcal{B}}$, will be the *c.pr. given* $\mathcal{B}$, and its values are the $\mathcal{B}$-measurable functions $P^{\mathcal{B}}A$, $A \in \mathcal{A}$, defined up to an equivalence.

We say "given (the σ-field) $\mathcal{B}$" and not "given (the partition) $\{B_j\}$," because $E^{\mathcal{B}}X$ determines the c.exp. of X given an arbitrary nonnull event $B \in \mathcal{B}$. In fact, if $\sum'$ denotes the summation over some subclass of $\{B_j\}$, then every event $B \in \mathcal{B}$ is of the form $\sum' B_j$, and we have

$$PBE_BX = \int_{\Sigma' B_j} X\,dP \doteq \sum{}' \int_{B_j} X\,dP = \sum{}' PB_jE_{B_j}X.$$

This relation can also be written as follows: If $P_{\mathcal{B}}$ is the restriction of P to $\mathcal{B}$, defined by

$$P_{\mathcal{B}}B = PB, \quad B \in \mathcal{B},$$

then the right-hand side becomes $\int_B (E^{\mathcal{B}}X)\,dP_{\mathcal{B}}$ while the left-hand side is $\int_B X\,dP$. This leads to the following

DESCRIPTIVE DEFINITION. *The c.exp. $E^{\mathcal{B}}X$ of $X \in \mathcal{E}$ given $\mathcal{B}$ is any $\mathcal{B}$-measurable function whose indefinite integral with respect to $P_{\mathcal{B}}$ is the restriction to $\mathcal{B}$ of the indefinite integral of X with respect to P. Since the indefinite integral with respect to $P_{\mathcal{B}}$ of a $\mathcal{B}$-measurable function determines this function up to an equivalence, this definition means precisely that, for every $X \in \mathcal{E}$, $E^{\mathcal{B}}X$ is defined by*

$$\int_B (E^{\mathcal{B}}X)\, dP_{\mathcal{B}} = \int_B X\, dP, \quad B \in \mathcal{B}.$$

To conclude the discussion of the elementary case, we revert to the initial approach, first defining c.pr.'s and, then, defining c.exp.'s as integrals. According to what precedes, we define $P^{\mathcal{B}}$ on $\mathcal{A}$ either by

$$\int_B (P^{\mathcal{B}}A)\, dP_{\mathcal{B}} = PAB, \quad A \in \mathcal{A}, \quad B \in \mathcal{B}$$

or, equivalently, by

$$P^{\mathcal{B}}A = \sum \frac{PAB_j}{PB_j} I_{B_j}, \quad A \in \mathcal{A},$$

up to an equivalence.

Let B_0 be the null sum of all null B_j and, for every A, *select* $P^{\mathcal{B}}A$ within its equivalence class by taking its values $P_\omega{}^{\mathcal{B}}A$ at $\omega \in B_j$ to be PAB_j/PB_j if B_j is nonnull and PA if B_j is null ($\subset B_0$). Then, for every $\omega \in \Omega$, the function $P_\omega{}^{\mathcal{B}}$ on $\mathcal{A}$, with values $P_\omega{}^{\mathcal{B}}A$, is a probability and we can form integrals with respect to it. Let $X \in \mathcal{E}$ and set

$$E_\omega{}^{\mathcal{B}}X = \int X\, dP_\omega{}^{\mathcal{B}}, \quad \omega \in \Omega.$$

Since, for every $\omega \in B_j$ not contained in the null event B_0, we have

$$\int X\, dP_\omega{}^{\mathcal{B}} = \frac{1}{PB_j} \int_{B_j} X\, dP,$$

it follows that the function on Ω with values $E_\omega{}^{\mathcal{B}}X$ belongs to the equivalence class of $E^{\mathcal{B}}X$. Thus, we can define $E^{\mathcal{B}}X$ to be $P_{\mathcal{B}}$-equivalent to the integral $\int X\, dP^{\mathcal{B}}$ where $P^{\mathcal{B}}$, hence the integral, are functions of $\omega \in \Omega$; in symbols

$$E^{\mathcal{B}}X = \int X\, dP^{\mathcal{B}} \text{ a.s.}$$

27.2. General case. The constructive approach fails at the very start as soon as the "given" σ-fields are not generated by countable partitions. However, the descriptive approach remains possible, thanks to the Radon-Nikodym theorem.

Let $(\Omega, \mathcal{A}, P)$ denote, as usual, the pr. space. Let $\mathcal{B}$, with or without affixes, denote a σ-field contained in $\mathcal{A}$, and let $P_{\mathcal{B}}$ denote the restriction of P to $\mathcal{B}$. Finally, let $\mathcal{E}$ be the family of all $\mathcal{A}$-measurable functions whose integral (hence indefinite integral) exists.

DEFINITION. *The c.exp.* $E^{\mathcal{B}}X$ of $X \in \mathcal{E}$ *given* $\mathcal{B}$ *is a* $\mathcal{B}$*-measurable function, defined up to a* $P_{\mathcal{B}}$*-equivalence by*

(1) $$\int_B (E^{\mathcal{B}}X)\, dP_{\mathcal{B}} = \int_B X\, dP, \quad B \in \mathcal{B}.$$

It follows at once that

1° $E(E^{\mathcal{B}}X) = EX.$

2° *If* $\mathcal{B} = \mathcal{A}$ *or* X *is* $\mathcal{B}$*-measurable, then* $E^{\mathcal{B}}X = X$ *a.s.*

3° $E^{\mathcal{B}}X = E^{\mathcal{B}}X^+ - E^{\mathcal{B}}X^-$ *a.s.*

The definition is justified: the indefinite integral φ of X being σ-additive and P-continuous, its restriction $\varphi_{\mathcal{B}}$ to $\mathcal{B}$ is σ-additive and $P_{\mathcal{B}}$-continuous, the extended Radon-Nikodym theorem applies, and the $\mathcal{B}$-measurable function $E^{\mathcal{B}}X$ defined by (1) exists and is defined up to a $P_{\mathcal{B}}$-equivalence.

If $\varphi_{\mathcal{B}}$ is σ-finite, then the Radon-Nikodym theorem applies, so that, moreover, $E^{\mathcal{B}}X$ is finite except on an arbitrary null event belonging to $\mathcal{B}$. If X is a r.v., then φ is σ-finite, but this does not imply that $\varphi_{\mathcal{B}}$ is σ-finite: take $\varphi(\Omega) = \infty$ and $\mathcal{B} = \{\emptyset, \Omega\}$. However, such a possibility is excluded in the case of integrable r.v.'s for, then, φ and hence $\varphi_{\mathcal{B}}$ are bounded.

We observe that as soon as it is understood that $E^{\mathcal{B}}X$ is, by definition, a $\mathcal{B}$-measurable function, we can replace $P_{\mathcal{B}}$ by P, and properties of $E^{\mathcal{B}}X$ valid, except on a $P_{\mathcal{B}}$-null event, may and will be said "a.s."

The function $E^{\mathcal{B}}$ on $\mathcal{E}$ to the space of $\mathcal{B}$-measurable functions (more precisely, on the space of equivalence classes of $\mathcal{A}$-measurable functions possessing an integral to the space of equivalence classes of $\mathcal{B}$-measurable functions) will be called *c.exp. given* $\mathcal{B}$. $E^{\mathcal{B}}$ can also be considered as a function on $\Omega \times \mathcal{E}$ to $\overline{R} = [-\infty, +\infty]$ with values $E_\omega{}^{\mathcal{B}}X$ for $\omega \in \Omega$, $X \in \mathcal{E}$, the value for all ω belonging to an arbitrary $P_{\mathcal{B}}$-null event being arbitrary.

The restriction of $E^{\mathcal{B}}$ to the family $I_{\mathcal{A}}$ of indicators of events is called *c.pr. given* $\mathcal{B}$ and is denoted by $P^{\mathcal{B}}$; in other words, $P^{\mathcal{B}}$ is a function on $\mathcal{A}$ whose values are $\mathcal{B}$-measurable functions $P^{\mathcal{B}}A$ defined up to an equivalence by $P^{\mathcal{B}}A = E^{\mathcal{B}}I_A$ or, directly, by

$$\int_B (P^{\mathcal{B}}A)\, dP_{\mathcal{B}} = PAB, \quad B \in \mathcal{B}.$$

Extension. It is "natural" to require of the definition of c.exp.'s that $E^{\mathcal{A}}X = X$ a.s., whatever be the measurable function X. Yet, the foregoing definition does not apply to those X whose integrals do not exist. However, it is possible to extend the definition so as to achieve the foregoing requirement, as follows: Write $X = X^+ - X^-$ where, as usual, X^+ and X^- are the positive and negative parts of X, respectively; $E^{\mathcal{B}}X^+$ and $E^{\mathcal{B}}X^-$ always exist but may be infinite. *Define* $E^{\mathcal{B}}X$ by $E^{\mathcal{B}}X = E^{\mathcal{B}}X^+ - E^{\mathcal{B}}X^-$ so that $E^{\mathcal{B}}X$ exists on the set on which the difference is not of the form $+\infty - \infty$ up to a $P_{\mathcal{B}}$-null event. This generalized c.exp. exists a.s. if the event $[X \geqq 0]$ and hence $[X < 0]$ belong to $\mathcal{B}$ (and, in particular, if $\mathcal{B} = \mathcal{A}$), for then $E^{\mathcal{B}}X^+ \times E^{\mathcal{B}}X^- = 0$ a.s. If $\mathcal{B} = \mathcal{A}$, then $E^{\mathcal{A}}X^+ - E^{\mathcal{A}}X^- = X^+ - X^- = X$ a.s., whatever be the r.v. X.

27.3. Conditional expectation given a function. We connect now the foregoing definition with the usual definition of c.exp., but we do not assume, as usually done, that the c.exp.'s are restricted to those of integrable r.v.'s.

Let Y be a function on $(\Omega, \mathcal{A}, P)$ to a measurable space $(\Omega', \mathcal{A}')$ and let $\mathcal{B}_Y \subset \mathcal{A}$ and $\mathcal{B}'_Y \subset \mathcal{A}'$ be the σ-fields *induced* by Y on Ω and Ω' respectively: $\mathcal{B}'_Y$ is the σ-field of all sets of $\mathcal{A}'$ whose inverse images under Y are events ($\in \mathcal{A}$) and $\mathcal{B}_Y$ is the σ-field of these events. Let P_Y and P'_Y be the probabilities *induced* by Y on $\mathcal{B}_Y$ and $\mathcal{B}'_Y$, respectively, defined by

$$P_YB = PB \quad B \in \mathcal{B}_Y; \quad P'_YB' = PB, \quad B' \in \mathcal{B}'_Y, \quad B = Y^{-1}(B').$$

(If Y is measurable, then $\mathcal{B}'_Y = \mathcal{A}'$. If no $\mathcal{A}'$ is given, then we take $\mathcal{A}' = \mathcal{S}(\Omega')$.)

If $\mathcal{B} = \mathcal{B}_Y$ in the definitions of the preceding subsection, then we replace every $\mathcal{B}_Y$ by Y. Thus, we write E^YX instead of $E^{\mathcal{B}_Y}X$, and call it *c.exp. of X given Y*. The reason for this terminology is that, as we shall show now, E^YX is a function of the function Y. We require the following proposition.

a. *For every numerical measurable function g on* $(\Omega', \mathcal{B}'_Y, P'_Y)$

$$\int_{B'} g\, dP'_Y = \int_B g(Y)\, dP_Y, \quad B' \in \mathcal{B}'_Y, \quad B = Y^{-1}(B'),$$

in the sense that, if one of these integrals exists, so does the other, and both are equal.

Proof. The asserted equality is true if $g = I_{A'}$ is the indicator of a set $A' \in \mathcal{B}'_Y$, for, setting $A = Y^{-1}(A')$, we have $g(Y) = I_A$ and, hence,

$$\int_{B'} I_{A'}\, dP'_Y = P'_Y(A'B') = P_Y(AB) = \int_B I_A\, dP_Y.$$

Being true for indicators, the equality is true for simple functions g and, by the monotone convergence theorem, for nonnegative measurable g. The assertion follows upon decomposing a measurable g into its positive and negative parts.

We are now in a position to prove the above-stated property.

A. *The c.exp. of $X \in \mathcal{E}$ given Y is a function of the function Y.*

Proof. If φ is the indefinite integral of X and φ' on $\mathcal{B}'_Y$ is defined by

$$\varphi'(B') = \varphi(B), \quad B' \in \mathcal{B}'_Y, \quad B = Y^{-1}(B'),$$

then φ' is σ-additive and P'_Y-continuous, the extended Radon-Nikodym theorem applies to φ' and P'_Y and defines a measurable function g on $(\Omega', \mathcal{B}'_Y)$ by

$$\int_{B'} g\, dP'_Y = \varphi'(B') = \int_B X\, dP.$$

Since

$$\int_B X\, dP = \int_B (E^Y X)\, dP,$$

it follows, upon applying **a**, that

$$\int_B g(Y)\, dP_Y = \int_{B'} g\, dP'_Y = \int_B (E^Y X)\, dP,$$

so that the indefinite integrals of the $\mathcal{B}_Y$-measurable functions $g(Y)$ and $E^Y X$ are the same, and the assertion is proved.

As defined, $E^Y X$ is a Y-measurable function *on the original space* $(\Omega, \mathcal{A}, P)$. However, the usual interpretation of the c.exp. of X given Y is that it is the function g on Ω' defined by

$$\int_{B'} g\, dP'_Y = \int_B X\, dP, \quad B' \in \mathcal{B}'_Y, \quad B = Y^{-1}(B').$$

We prefer to consider c.exp.'s as functions on the original pr. space. Yet, on account of the foregoing theorem, both interpretations are possible: either $E^Y X$ is considered as a *function of the function* Y with values

$E_\omega{}^Y X$ for $\omega \in \Omega$, or it is considered as a *function of* Y with values $E_y{}^Y X$ for $Y = y$, defined up to a P_Y or P'_Y-equivalence, respectively.

Notation. The following symbols are and will be used according to convenience:

$E^Y X$ or $E(X \mid Y)$ or $E(Y; X)$, $E_y{}^Y X$ or $E(X \mid y)$ or $E(y; X)$;
$P^Y A$ or $P(A \mid Y)$ or $P(Y; A)$, $P_y{}^Y A$ or $P(A \mid y)$ or $P(y; A)$;
and similarly with y replaced by ω and/or Y replaced by $\mathcal{B}$.

***27.4. Relative conditional expectations and sufficient σ-fields.** The Radon-Nikodym theorem applies to σ-finite and μ-continuous signed measures on $\mathcal{A}$ with σ-finite measures μ on $\mathcal{A}$. Therefore, the concept of c.exp. continues to apply if, in what precedes, P is replaced by any such measure μ. But, then, we have to specify that the c.exp.'s are taken *with respect to* μ—they are *relative c.exp.'s*. To simplify, we limit ourselves to finite and P-continuous measures μ. (Yet, we shall see in the next volume that, led by physics, we may have to replace pr.'s by σ-finite measures and, thus, use fully the foregoing conditioning.)

Given the pr. space $(\Omega, \mathcal{A}, P)$, the measures μ are indefinite integrals of nonnegative r.v.'s Z and we say that the relative c.exp.'s are taken *with respect to* Z. *In what follows, the r.v.'s* Z, *with or without affixes, are nonnegative and integrable, and the* μ, *with the same affixes if any, are their indefinite integrals.*

If a r.v. X possesses an integral with respect to μ, then the c.exp. of X given $\mathcal{B}$ *with respect to* Z is a $\mathcal{B}$-measurable function, defined up to a $\mu_{\mathcal{B}}$-equivalence by

$$\int_B (E_Z{}^{\mathcal{B}} X)\, d\mu_{\mathcal{B}} = \int_B X\, d\mu, \quad B \in \mathcal{B}.$$

Since

$$\mu A = \int_A Z\, dP, \quad A \in \mathcal{A},$$

it follows that

$$\mu_{\mathcal{B}} B = \int_B Z\, dP = \int_B (E^{\mathcal{B}} Z)\, dP_{\mathcal{B}}, \quad B \in \mathcal{B},$$

and this definition is equivalent to

$$\int_B (E^{\mathcal{B}} Z)(E_Z{}^{\mathcal{B}} X)\, dP_{\mathcal{B}} = \int_B ZX\, dP = \int_B (E^{\mathcal{B}} ZX)\, dP_{\mathcal{B}}, \quad B \in \mathcal{B},$$

which, in its turn, is equivalent to

(1) $$E^{\mathcal{B}} Z \cdot E_Z{}^{\mathcal{B}} X = E^{\mathcal{B}} ZX \text{ a.s.}$$

$E_Z^{\mathcal{B}}X$ is defined up to a $\mu_{\mathcal{B}}$-equivalence and $P_{\mathcal{B}}A = 0$ entails $\mu_{\mathcal{B}}A = 0$; furthermore,

$$\mu_{\mathcal{B}}[E^{\mathcal{B}}Z = 0] = \int_{[E^{\mathcal{B}}Z=0]} (E^{\mathcal{B}}Z)\, dP_{\mathcal{B}} = 0.$$

Therefore, up to a $\mu_{\mathcal{B}}$-equivalence, $E_Z^{\mathcal{B}}X$ is given by

$$(1') \qquad E_Z^{\mathcal{B}}X = E^{\mathcal{B}}ZX/E^{\mathcal{B}}Z,$$

so that

a. *Relative c.exp.'s are reducible, up to an equivalence, to ratios of ordinary c.exp.'s.*

It may happen that c.exp.'s given $\mathcal{B}$ relative to the r.v.'s of a family $\{Z_t\}$ collapse together: there exists a r.v. Z such that, for every t,

$$(2) \qquad E_{Z_t}^{\mathcal{B}}X = E_Z^{\mathcal{B}}X$$

in the sense that, whenever the left-hand side exists, so does the right-hand side, and both are equal. But these sides are determined up to $(\mu_t)_{\mathcal{B}}$- and $(\mu_{\mathcal{B}})$-equivalences, respectively. Thus, equality might be interpreted in the sense that the $\mu_{\mathcal{B}}$-equivalence class of $E_Z^{\mathcal{B}}X$ belongs to every $(\mu_t)_{\mathcal{B}}$-equivalence class of $E_{Z_t}X$'s. This is certainly true as soon as the equality holds for an element of each class, provided every μ_t is μ-continuous. Then, moreover, whenever $E_{Z_t}^{\mathcal{B}}X$ exists so does $E_Z^{\mathcal{B}}X$. Finally, we are led to the following definition.

Let X be "admissible" for the family $\{Z_t\}$ if its integrals with respect to every u_t exist. A sub σ-field $\mathcal{B}$ of events is *sufficient* (with Z) for the family $\{Z_t\}$ if there exists a Z such that every μ_t is μ-continuous and, for every admissible X, (2) holds up to a $(\mu_t)_{\mathcal{B}}$-equivalence. This concept of sufficient sub σ-fields is slightly more general than the usual concept of "sufficient statistics" which plays a considerable role in statistics. Clearly every σ-field P-equivalent to a sufficient $\mathcal{B}$ is sufficient. Thus, in what follows, we assume that every sufficient σ-field is defined up to a P-equivalence.

The basic result (originating with Neyman and put in its final form by Halmos and Savage—in terms of sufficient statistics) is as follows:

A. FACTORIZATION THEOREM. *The sub σ-field $\mathcal{B}$ of events is sufficient for the family $\{Z_t\}$ if, and only if, there exists a Z such that every $Z_t = g_tZ$ a.s. and every g_t is $\mathcal{B}$-measurable; then every $g_t = E^{\mathcal{B}}Z_t/E^{\mathcal{B}}Z$ up to a $\mu_{\mathcal{B}}$-equivalence.*

We require two properties of c.exp.'s: 1° $E^{\mathcal{B}}$ and $\mathcal{B}$-measurable functions commute (25.2, **3**); 2° if, for Y integrable or nonnegative, and for every indicator I_A of events, $E^{\mathcal{B}}YI_A = E^{\mathcal{B}}Y'I_A$ a.s., then $Y = Y'$ a.s. since, for every $A \in \mathcal{A}$,

$$\int_A Y\,dP = \int YI_A\,dP = \int (E^{\mathcal{B}}YI_A)\,dP_{\mathcal{B}} = \int (E^{\mathcal{B}}Y'I_A)\,dP_{\mathcal{B}}$$

$$= \int Y'I_A\,dP = \int_A Y'\,dP.$$

Proof. $Z_t = g_tZ$ a.s. entails μ-continuity of μ_t and, by (1),

$$E^{\mathcal{B}}Z_t\cdot E_{Z_t}{}^{\mathcal{B}}X = E^{\mathcal{B}}Z_tX = g_tE^{\mathcal{B}}ZX = g_tE^{\mathcal{B}}Z\cdot E_Z{}^{\mathcal{B}}X = E^{\mathcal{B}}Z_t\cdot E_Z{}^{\mathcal{B}}X \text{ a.s.}$$

The sets $[E^{\mathcal{B}}Z_t = 0]$ being $(\mu_t)_{\mathcal{B}}$-null, $E_{Z_t}{}^{\mathcal{B}}X = E_Z{}^{\mathcal{B}}X$ up to $(\mu_t)_{\mathcal{B}}$-equivalence. The set $[E^{\mathcal{B}}Z = 0]$ being $\mu_{\mathcal{B}}$-null, it follows from

$$E^{\mathcal{B}}Z_t = g_tE^{\mathcal{B}}Z \text{ a.s.}$$

that $g_t = E^{\mathcal{B}}Z_t/E^{\mathcal{B}}Z$ up to a $\mu_{\mathcal{B}}$-equivalence.

Conversely, if, for all indicators X, every $E_{Z_t}{}^{\mathcal{B}}X = E_Z{}^{\mathcal{B}}X$ up to a $(\mu_t)_{\mathcal{B}}$-equivalence, then, by (1),

$$E^{\mathcal{B}}Z\cdot E^{\mathcal{B}}Z_tX = E^{\mathcal{B}}Z_tE^{\mathcal{B}}ZE_Z{}^{\mathcal{B}}X = E^{\mathcal{B}}Z_tE^{\mathcal{B}}ZX \text{ a.s.}$$

or

$$E^{\mathcal{B}}(Z_tXE^{\mathcal{B}}Z) = E^{\mathcal{B}}(ZXE^{\mathcal{B}}Z_t) \text{ a.s.,}$$

so that

$$Z_tE^{\mathcal{B}}Z = ZE^{\mathcal{B}}Z_t \text{ a.s.}$$

and, hence, on $B = [E^{\mathcal{B}}Z > 0]$,

$$Z_t = \frac{E^{\mathcal{B}}Z_t}{E^{\mathcal{B}}Z}Z \text{ a.s.} \tag{3}$$

Since μ_t is μ-continuous, from

$$\mu B^c = \int_{B^c} Z\,dP = \int_{B^c} (E^{\mathcal{B}}Z)\,dP = 0,$$

so that $Z = 0$ on B^c except for P-null subsets, it follows that $\mu_tB^c = 0$; hence $Z_t = 0$ on B^c except for P-null subsets. Thus (3) is trivially true on B^c. This completes the proof.

Underlying the concept of sufficient σ-fields with Z is the fact that every μ_t is supposed to be μ-continuous. This alone implies that every

$Z_t = g_t Z$ a.s. where the g_t are measurable. Thus, the whole σ-field $\mathcal{A}$ of events is trivially sufficient with any such Z and, in particular, with $Z = 1$. And every sub σ-field $\mathcal{B}$ of events such that all the g_t are $\mathcal{B}$-measurable is sufficient with such a Z; in particular, the sub σ-field $\mathcal{B}$ induced by the family $\{g_t\}$ is the least fine sufficient with Z. The question arises whether there exists some Z, say Z_0, such that the least fine sufficient σ-field with Z_0 is the least fine of all possible sufficient σ-fields for the family $\{Z_t\}$—the *minimal sufficient* σ-field for the family $\{Z_t\}$. The answer is in the affirmative, as follows:

According to Chapter II: Complements and Details 23, there exists a Z_0 such that

(i) $\mu_0 A = 0 \Leftrightarrow$ every $\mu_t A = 0$;

or, equivalently,

(i′) up to P-null subsets, $Z_0 = 0$ on $A \Leftrightarrow$ every $Z_t = 0$ on A.

Since, on account of (i′), every $E_Z{}^{\mathcal{B}}X$ common to all the equivalence classes $E_{Z_t}{}^{\mathcal{B}}X$ belongs also to the equivalence class of $E_{Z_0}{}^{\mathcal{B}}X$, it follows that every sufficient σ-field $\mathcal{B}$ with Z is also sufficient with Z_0. Therefore, the least fine sufficient σ-field with Z_0 is the minimal one. On account of (i′), the corresponding factorization—every $Z_t = g_t Z_0$ a.s.—is such that $Z_0 = 0$ a.s. $\Rightarrow$ every $Z_t = 0$ a.s. Thus:

B. Minimality criterion. *Write every Z_t in the form $Z_t = g_t Z_0$ a.s., with Z_0 such that every $Z_t = 0$ a.s. $\Rightarrow Z_0 = 0$ a.s.; this is always possible. Then the minimal sufficient σ-field for the family $\{Z_t\}$ is the one induced by the family $\{g_t\}$.*

§28. PROPERTIES OF CONDITIONING

To avoid constant repetitions, it will be assumed in this and the following section that the integrals of all functions which figure under the integration and c.exp. signs exist. We recall that an a.s. relation between $\mathcal{B}$-measurable functions is a $P_{\mathcal{B}}$-equivalence.

28.1. Expectation properties. Loosely speaking, *c.exp.'s have a.s. all properties of expectations.*

Let x_k, c, and c' be numbers.

1. *If $X = c$ a.s. then $E^{\mathcal{B}}X = c$ a.s., and if $X \geqq Y$, a.s. then $E^{\mathcal{B}}X \geqq E^{\mathcal{B}}Y$ a.s.*

$E^{\mathfrak{B}}$ is an a.s. linear operation: $E^{\mathfrak{B}}(cX + c'X') = cE^{\mathfrak{B}}X + c'E^{\mathfrak{B}}X'$ *a.s. In particular,*

$$P^{\mathfrak{B}}\Omega = 1 \text{ a.s.}, \quad P^{\mathfrak{B}}\emptyset = 0 \text{ a.s.}, \quad P^{\mathfrak{B}}A \geqq 0 \text{ a.s.}$$

and

$$E^{\mathfrak{B}}\left(\sum_{k=1}^{n} x_k I_{A_k}\right) = \sum_{k=1}^{n} x_k P^{\mathfrak{B}} A_k \text{ a.s.}$$

These properties follow at once from the definition of c.exp.'s and properties of integrals.

CONDITIONAL INEQUALITIES. *Upon replacing E by $E^{\mathfrak{B}}$, the c_r-, Minkowski and Hölder inequalities, as well as their consequences and the inequalities for convex functions, remain valid, almost surely.*

For, on account of **1**, their proofs remain valid up to a $P_{\mathfrak{B}}$-equivalence (for Hölder's inequality use also 25.2, **3**).

2. CONVERGENCE IN THE rTH MEAN. *If $X_n \xrightarrow{r} X$, then $E^{\mathfrak{B}}X_n \xrightarrow{r} E^{\mathfrak{B}}X$ for $r \geqq 1$.*

MONOTONE CONVERGENCE. *If $0 \leqq X_n \uparrow X$ a.s., then $0 \leqq E^{\mathfrak{B}}X_n \uparrow E^{\mathfrak{B}}X$ a.s. In particular, $P^{\mathfrak{B}} \sum_{k=1}^{\infty} A_k = \sum P^{\mathfrak{B}}A_k$ a.s.*

FATOU-LEBESGUE CONVERGENCE. *Let Y and Z be integrable. If $Y \leqq X_n$ a.s. or $X_n \leqq Z$ a.s., then $E^{\mathfrak{B}} \liminf X_n \leqq \liminf E^{\mathfrak{B}}X_n$ a.s., resp.,* $\limsup E^{\mathfrak{B}}X_n \leqq E^{\mathfrak{B}} \limsup X_n$ *a.s.*

In particular, if $Y \leqq X_n \uparrow X$ a.s., or $Y \leqq X_n \leqq Z$ a.s. and $X_n \xrightarrow{\text{a.s.}} X$, then $E^{\mathfrak{B}}X_n \xrightarrow{\text{a.s.}} E^{\mathfrak{B}}X$.

The first assertion follows by

$$E|E^{\mathfrak{B}}X_n - E^{\mathfrak{B}}X|^r = E|E^{\mathfrak{B}}(X_n - X)|^r$$
$$\leqq E(E^{\mathfrak{B}}|X_n - X|^r) = E|X_n - X|^r \to 0.$$

As for the monotone convergence assertion, since $X_{n+1} \geqq X_n$ a.s. implies $E^{\mathfrak{B}}X_{n+1} \geqq E^{\mathfrak{B}}X_n$ a.s., it follows that $E^{\mathfrak{B}}X_n \uparrow X'$ a.s. where X' is $\mathfrak{B}$-measurable. Therefore, the monotone convergence criterion applies to both sequences X_n and $E^{\mathfrak{B}}X_n$, for every $B \in \mathfrak{B}$,

$$\int_B X'\,dP \uparrow \int_B (E^{\mathfrak{B}}X_n)\,dP_{\mathfrak{B}} = \int_B X_n\,dP \uparrow \int_B X\,dP = \int_B (E^{\mathfrak{B}}X)\,dP_{\mathfrak{B}},$$

and the assertion follows. Upon taking $X_n = \sum_{k=1}^{n} I_{A_k}$ so that $E^{\mathfrak{B}}X_n = \sum_{k=1}^{n} P^{\mathfrak{B}}A_k$ a.s., the particular case is proved.

The Fatou-Lebesgue assertion follows from that of monotone convergence as in the nonconditional case.

28.2. Smoothing properties. Loosely speaking, *the operation* $E^{\mathcal{B}}$ *is a* $\mathcal{B}$-*smoothing.*

1. *On every nonnull atom* $B \in \mathcal{B}$, $E^{\mathcal{B}}X$ *is constant and its value* $E_B X$ *is the average of values of* X *on* B *with respect to* P.

By definition, B is a nonnull atom of $\mathcal{B}$ if $PB > 0$ and B contains no other sets belonging to $\mathcal{B}$ than itself and the empty set.

Proof. The first assertion follows from the fact that $E^{\mathcal{B}}X$ is a $\mathcal{B}$-measurable function defined up to a $P_{\mathcal{B}}$-equivalence and a $\mathcal{B}$-measurable function is constant on atoms of $\mathcal{B}$. Therefore, on every atom B of $\mathcal{B}$,

$$E_B X \cdot PB = \int_B (E^{\mathcal{B}}X)\, dP_{\mathcal{B}} = \int_B X\, dP$$

and, for $PB > 0$,

$$E_B X = \frac{1}{PB}\int_B X\, dP.$$

This proves the second assertion and completes the proof.

Thus, $E^{\mathcal{B}}X$ is a $\mathcal{B}$-smoothed X, in the sense that on atoms of $\mathcal{B}$ which are not atoms of $\mathcal{A}$, $E^{\mathcal{B}}X$ is an "averaged X" and, on the whole, has "fewer values" than X. In particular, if $\mathcal{B}$ is the minimal σ-field over a countable partition $\{B_j\}$ of Ω, so that the B_j are atoms of $\mathcal{B}$, then, as is to be expected,

$$E^{\mathcal{B}}X = \sum (E_{B_j}X) I_{B_j} \text{ a.s.};$$

the right-hand side is a.s. defined since the $E_{B_j}X$ are determined except for null B_j whose countable sum is necessarily null. For the "least fine" or "smallest" of all possible σ-fields $\mathcal{B} \subset \mathcal{A}$, that is, for $\mathcal{B}_0 = \{\emptyset, \Omega\}$, we obtain $E^{\mathcal{B}_0}X = EX$ a.s. The same conclusion holds for every $\mathcal{B}$ independent of the σ-field $\mathcal{B}_X$ of events induced by X:

2. *If* $\mathcal{B}$ *and* $\mathcal{B}_X$ *are independent, then* $E^{\mathcal{B}}X = EX$ *a.s.*

For, X and I_B being independent for every $B \in \mathcal{B}$,

$$\int_B (E^{\mathcal{B}}X)\, dP_{\mathcal{B}} = \int_B X\, dP = E(XI_B) = EX \cdot PB = \int_B (EX)\, dP_{\mathcal{B}}.$$

In particular, since $E^Y X$ denotes $E^{\mathcal{B}_Y}X$ and independence of X and Y means independence of $\mathcal{B}_X$ and $\mathcal{B}_Y$, we have

If X *and* Y *are independent, then* $E^Y X = EX$ *a.s.*, $E^X Y = EY$ *a.s.*

The operation $E^{\mathcal{B}}$ transforms $\mathcal{A}$-measurable functions (whose integrals exist) into $\mathcal{B}$-measurable functions (whose integrals exist); in fact, it transforms classes of P-equivalence into classes of $P_{\mathcal{B}}$-equivalence. In particular, as is to be expected, the operation $E^{\mathcal{B}}$ does not modify classes of $P_{\mathcal{B}}$-equivalence, in the sense that, if X is $\mathcal{B}$-measurable, then $E^{\mathcal{B}}X = X$ a.s.; since, then, for every $B \in \mathcal{B}$,

$$\int_B (E^{\mathcal{B}}X)\, dP_{\mathcal{B}} = \int_B X\, dP = \int_B X\, dP_{\mathcal{B}}.$$

More generally, $E^{\mathcal{B}}$ and $\mathcal{B}$-measurable factors commute, as follows:

3. *If X is $\mathcal{B}$-measurable, then $E^{\mathcal{B}}XY = XE^{\mathcal{B}}Y$ a.s.*

The assertion holds for $X = I_{B'}$ where $B' \in \mathcal{B}$, since, for every $B \in \mathcal{B}$,

$$\int_B (E^{\mathcal{B}}I_{B'}Y)\, dP_{\mathcal{B}} = \int_B I_{B'}Y\, dP = \int_{BB'} (E^{\mathcal{B}}Y)\, dP_{\mathcal{B}} = \int_B (I_{B'}E^{\mathcal{B}}Y)\, dP_{\mathcal{B}}.$$

Therefore, it holds for simple functions X_n:

$$E^{\mathcal{B}}X_nY = X_nE^{\mathcal{B}}Y \text{ a.s.}$$

and, by the monotone convergence theorem for c.exp.'s, it holds for nonnegative functions—take $0 \leq X_n \uparrow X$ and let $n \to \infty$ in the foregoing relation. The assertion follows.

4. *If $\mathcal{B} \subset \mathcal{B}'$, then*

$$E^{\mathcal{B}}(E^{\mathcal{B}'}X) = E^{\mathcal{B}}X = E^{\mathcal{B}'}(E^{\mathcal{B}}X) \text{ a.s.}$$

Since $\mathcal{B} \subset \mathcal{B}'$ implies that $P_{\mathcal{B}}$ is restriction of $P_{\mathcal{B}'}$ to $\mathcal{B}$, we have, for every $B \in \mathcal{B}$,

$$\int_B (E^{\mathcal{B}}(E^{\mathcal{B}'}X))\, dP_{\mathcal{B}} = \int_B (E^{\mathcal{B}'}X)\, dP_{\mathcal{B}'} = \int_B X\, dP = \int_B (E^{\mathcal{B}}X)\, dP_{\mathcal{B}},$$

and the left-hand side equality is proved.

Since $\mathcal{B} \subset \mathcal{B}'$ implies that a $\mathcal{B}$-measurable function is $\mathcal{B}'$-measurable, the right-hand equality follows either from **3** or directly from

$$\int_B (E^{\mathcal{B}'}(E^{\mathcal{B}}X))\, dP_{\mathcal{B}'} = \int_B (E^{\mathcal{B}}X)\, dP = \int_B (E^{\mathcal{B}}X)\, dP_{\mathcal{B}}, \quad B \in \mathcal{B}.$$

Thus, the smoothing $E^{\mathcal{B}}$ can be performed in steps and remains a.s. invariant under "finer" smoothings.

Together, **3** and **4** yield the

A. Basic smoothing property. *If* $\mathcal{B} \subset \mathcal{B}'$ *and* X' *is* $\mathcal{B}'$*-measurable, then*

$$E^{\mathcal{B}}XX' = E^{\mathcal{B}}(X'E^{\mathcal{B}'}X).$$

In particular, denoting by $E^{X',Y}$ the c.exp. given the σ-field $\mathcal{B}_{X',Y}$ of events induced by the couple (X', Y), we have

$$E^{Y}XX' = E^{Y}(X'E^{X',Y}X) \text{ a.s.}$$

***28.3. Concepts of conditional independence and of chains.** Under conditioning, the concept of independence extends as follows:

We say that $\mathcal{B}_1$ and $\mathcal{B}_2$ are *conditionally independent* (*c.ind.*) given $\mathcal{B}$ if, for every $B_1 \in \mathcal{B}_1$ and $B_2 \in \mathcal{B}_2$,

$$P^{\mathcal{B}}B_1B_2 = P^{\mathcal{B}}B_1 \cdot P^{\mathcal{B}}B_2 \text{ a.s.}$$

If $\mathcal{B} = \mathcal{A}$, then this relation becomes $I_{B_1B_2} = I_{B_1}I_{B_2}$ a.s., so that two sub σ-fields of events are always c.ind. given the σ-field $\mathcal{A}$ of events, and the concept of c.ind. given $\mathcal{A}$ is trivial.

If $\mathcal{B} = \mathcal{B}_0 = \{\emptyset, \Omega\}$, then this relation becomes $PB_1B_2 = PB_1 \cdot PB_2$ a.s., so that independence is c.ind. given $\mathcal{B}_0$, the "smallest" of all sub σ-fields of events.

In what follows, we drop the parentheses and commas in writing compound σ-fields.

A. $\mathcal{B}_1$ *and* $\mathcal{B}_2$ *are c.ind. given* $\mathcal{B}$ *if, and only if, for every* $B_2 \in \mathcal{B}_2$,

$$P^{\mathcal{B}\mathcal{B}_1}B_2 = P^{\mathcal{B}}B_2 \textit{ a.s.};$$

the subscripts 1 *and* 2 *can be interchanged.*

Proof. Let $B_1 \in \mathcal{B}_1$ and $B_2 \in \mathcal{B}_2$ be arbitrary. We have to prove that

$$(1) \qquad E^{\mathcal{B}}I_{B_1}I_{B_2} = E^{\mathcal{B}}I_{B_1} \cdot E^{\mathcal{B}}I_{B_2} \text{ a.s.}$$

is equivalent to

$$(2) \qquad E^{\mathcal{B}\mathcal{B}_1}I_{B_2} = E^{\mathcal{B}}I_{B_2} \text{ a.s.}$$

Since, on account of smoothing properties (25.2),

$$E^{\mathcal{B}}I_{B_1}I_{B_2} = E^{\mathcal{B}}(I_{B_1}E^{\mathcal{B}\mathcal{B}_1}I_{B_2}) \text{ a.s.}$$

and

$$E^{\mathcal{B}}I_{B_1} \cdot E^{\mathcal{B}}I_{B_2} = E^{\mathcal{B}}(I_{B_1}E^{\mathcal{B}}I_{B_2}) \text{ a.s.,}$$

it suffices to prove that (2) is equivalent to

(3) $$E^{\mathcal{B}}(I_{B_1}E^{\mathcal{B}\mathcal{B}_1}I_{B_2}) = E^{\mathcal{B}}(I_{B_1}E^{\mathcal{B}}I_{B_2}) \text{ a.s.}$$

Upon multiplying both sides in (2) by I_{B_1} and performing the operation $E^{\mathcal{B}}$, (3) follows.

Conversely, (3) implies that, for every $B \in \mathcal{B}$,

$$\int_B (I_{B_1}E^{\mathcal{B}\mathcal{B}_1}I_{B_2})\, dP = \int_B (I_{B_1}E^{\mathcal{B}}I_{B_2})\, dP$$

or, both c.exp.'s being $\mathcal{B}\mathcal{B}_1$-measurable,

$$\int_{BB_1} (E^{\mathcal{B}\mathcal{B}_1}I_{B_2})\, dP_{\mathcal{B}\mathcal{B}_1} = \int_{BB_1} (E^{\mathcal{B}}I_{B_2})\, dP_{\mathcal{B}\mathcal{B}_1}.$$

Since bounded indefinite integrals coinciding on the class of all sets BB_1 coincide on the σ-field $\mathcal{B}\mathcal{B}_1$, it follows that the integrands $E^{\mathcal{B}\mathcal{B}_1}I_{B_2}$ and $E^{\mathcal{B}}I_{B_2}$ are $P_{\mathcal{B}\mathcal{B}_1}$-equivalent, and the proof is complete.

Upon following literally the pattern used for the investigation of the concept of independence, the concept of c.ind. extends to arbitrary families of σ-fields and hence of r.v.'s, random vectors and random functions, and the investigations of the case of independence (Part III) can be transposed to the case of c.ind.

Furthermore, the concept of c.ind. leads to another generalization of that of independence, as follows: Let $\mathcal{B}_n$ be a sequence of sub σ-fields of events. The $\mathcal{B}_n$ are said to form a *chain*, or to be *chained* (or *chain-dependent* or *Markov-dependent*) if, for arbitrary integers m and n, the σ-fields $\mathcal{B}_1, \cdots, \mathcal{B}_{n-1}$, and $\mathcal{B}_{n+1}, \cdots, \mathcal{B}_{n+m}$ are c.ind. given $\mathcal{B}_n$. In symbols, the $\mathcal{B}_n$ are chained, if, for every m, n, $B_k \in \mathcal{B}_k$,

(1) $$P^{\mathcal{B}_n}B_1 \cdots B_{n-1}B_{n+1} \cdots B_{n+m} = P^{\mathcal{B}_n}B_1 \cdots B_{n-1}P^{\mathcal{B}_n}B_{n+1} \cdots B_{n+m} \text{ a.s.}$$

or, equivalently, on account of **A**,

(2) $$P^{\mathcal{B}_1\mathcal{B}_2\cdots\mathcal{B}_n}B_{n+1} \cdots B_{n+m} = P^{\mathcal{B}_n}B_{n+1} \cdots B_{n+m} \text{ a.s.}$$

or

(3) $$P^{\mathcal{B}_n\mathcal{B}_{n+1}\cdots\mathcal{B}_{n+m}}B_1 \cdots B_{n-1} = P^{\mathcal{B}_n}B_1 \cdots B_{n-1} \text{ a.s.}$$

If $n = 1, 2, \cdots$, is interpreted as the "time," we can say, loosely speaking, that the $\mathcal{B}_n$ form a chain if the "past" and the "future" are a.s. independent when the "present" is given, or, equivalently, the "future" ("past") is a.s. independent of the given "past" ("future"), when the "present" is given. We shall use mostly the defining property (2).

Let $k_1 < \cdots < k_{n-1} < k_n < k_{n+1} < \cdots < k_{n+m}$ be arbitrary integers and apply the operation $E^{\mathcal{B}_{k_1}\cdots\mathcal{B}_{k_n}}$ to both sides of (2) where n and $n + m$ are replaced by k_n and k_{n+m} respectively. It follows, by 24.2, and upon replacing by Ω all events whose subscripts are different from $k_{n+1}, \cdots, k_{n+m}$, that we have the seemingly more general property

$$P^{\mathcal{B}_{k_1}\cdots\mathcal{B}_{k_n}} B_{k_{n+1}} \cdots B_{k_{n+m}} = P^{\mathcal{B}_{k_n}} B_{k_{n+1}} \cdots B_{k_{n+m}} \text{ a.s.}$$

Loosely speaking, whatever be the "future" it depends a.s. only upon the last given "past."

As usual, if $\mathcal{B}_n = \mathcal{B}_{X_n}$ are σ-fields of events induced by r.v.'s (random vectors, random functions) X_n, we replace above $\mathcal{B}_n$ by X_n and speak about the *chain* of r.v.'s (random vectors, random functions) X_n.

§ 29. REGULAR PR. FUNCTIONS

29.1. Regularity and integration. Since c.exp.'s behave at first sight as integrals with respect to c.pr.'s, the question arises whether c.exp.'s can be so defined. More precisely, according to 25.1,

1° Properties of functions $P^{\mathcal{B}}A$ are almost surely those of pr. values:

$$P^{\mathcal{B}}\Omega = 1 \text{ a.s.}, \quad P^{\mathcal{B}}A \geqq 0 \text{ a.s.}, \quad P^{\mathcal{B}} \sum A_j = \sum P^{\mathcal{B}} A_j \text{ a.s.}$$

2° Properties of functions $E^{\mathcal{B}}X$ are almost surely those defining integrals with respect to $P^{\mathcal{B}}$:

$$E^{\mathcal{B}} \sum_{k=1}^{n} x_k I_{A_k} = \sum_{k=1}^{n} x_k P^{\mathcal{B}} A_k \text{ a.s.},$$

$$0 \leqq X_n \uparrow X \text{ implies } E^{\mathcal{B}} X_n \uparrow E^{\mathcal{B}} X \text{ a.s.},$$

$$E^{\mathcal{B}} X = E^{\mathcal{B}} X^{+} - E^{\mathcal{B}} X^{-} \text{ a.s.}$$

Yet, to speak about integrals with respect to the $P_\omega^{\mathcal{B}}$, we have to know that the $P_\omega^{\mathcal{B}}$ are pr.'s for every $\omega \in \Omega$ or, the c.exp.'s being defined up to an equivalence, that at the least the $P_\omega^{\mathcal{B}}$ are pr.'s except for ω belonging to some null event. Thus, we have to assume that $P^{\mathcal{B}}$ is "regular."

A c.pr. $P^{\mathcal{B}}$ is said to be *regular* if, for every $A \in \mathcal{A}$, it is possible to select $P^{\mathcal{B}}A$ within its class of equivalence in such a manner that the $P_\omega^{\mathcal{B}}$ are pr.'s on $\mathcal{A}$ except for points ω belonging to a $P_{\mathcal{B}}$-null event N. A regular pr.f. $P^{\mathcal{B}}$ can be said to be defined up to an equivalence, in the sense that if all the functions $P^{\mathcal{B}}A$ are modified arbitrarily on an arbitrary but fixed $P_{\mathcal{B}}$-null event, the new c.pr. is still regular. In particular, a regular c.pr. $P^{\mathcal{B}}$ can be selected within its equivalence class so that $P_\omega^{\mathcal{B}}$ is a pr. on $\mathcal{A}$ for *every* $\omega \in \Omega$. For example, for every ω belonging to the exceptional $P_{\mathcal{B}}$-null event N set $P_\omega^{\mathcal{B}} = P_N$ where P_N is a pr. on $\mathcal{A}$. Unless otherwise stated, regular c.pr.'s will be so selected. In other words,

a regular c.pr. $P^{\mathcal{B}}$, with values $P^{\mathcal{B}}(\omega; A)$, will be a function on $\Omega \times \mathcal{A}$ with the following properties:

(i) *$P^{\mathcal{B}}(\omega; A)$ is $\mathcal{B}$-measurable in $\omega(\in \Omega)$ for every fixed A and is a pr. in $A(\in \mathcal{A})$ for every fixed ω.*
(ii) *For every $A \in \mathcal{A}$ and $B \in \mathcal{B}$,*

$$\int_B (P^{\mathcal{B}}A)\, dP_{\mathcal{B}} = PAB.$$

In the case of regular c.pr.'s the answer to the question stated at the beginning of this section is, as might be expected, in the affirmative.

A. INTEGRATION THEOREM. *If $P^{\mathcal{B}}$ is a regular c.pr., then*

$$E^{\mathcal{B}}X = \int X\, dP^{\mathcal{B}} \quad a.s.$$

Proof. Since all $P_\omega^{\mathcal{B}}$ are pr.'s on $\mathcal{A}$, we can write

$$P_\omega^{\mathcal{B}}A = \int I_A\, dP_\omega^{\mathcal{B}}, \quad \omega \in \Omega,$$

that is,

$$E^{\mathcal{B}}I_A = P^{\mathcal{B}}A = \int I_A\, dP^{\mathcal{B}}.$$

It follows, on account of relations 2°, that

$$E^{\mathcal{B}} \sum_{k=1}^{n} x_k I_{A_k} = \int \left(\sum_{k=1}^{n} x_k I_{A_k} \right) dP^{\mathcal{B}} \text{ a.s.};$$

$0 \leq X_n \uparrow X$, where the X_n are simple functions, implies that $E^{\mathcal{B}}X =$

$$\lim E^{\mathcal{B}}X_n = \lim \int X_n\, dP^{\mathcal{B}} = \int X\, dP^{\mathcal{B}} \text{ a.s.};$$

$$E^{\mathcal{B}}X = E^{\mathcal{B}}X^+ - E^{\mathcal{B}}X^- = \int X^+\, dP^{\mathcal{B}} - \int X^-\, dP^{\mathcal{B}} = \int X\, dP^{\mathcal{B}} \text{ a.s.};$$

and the assertion is proved.

The basic smoothing property becomes

B. BASIC INTEGRATION PROPERTY. *If* $\mathcal{B} \subset \mathcal{B}' \subset \mathcal{A}$ *and* $P^{\mathcal{B}}$, $P^{\mathcal{B}'}$ *are regular, then, for* $\mathcal{A}$*-measurable functions* X *and* $\mathcal{B}'$*-measurable functions* X',

$$\int XX'\,dP^{\mathcal{B}} = \int X'\,dP^{\mathcal{B}} \int X\,dP^{\mathcal{B}'} \quad a.s.$$

The iterated integrations are to be read, as usual, from right to left. The foregoing relation can be written explicitly as follows: except for ω belonging to a $P_{\mathcal{B}}$-null event

$$\int P^{\mathcal{B}}(\omega; d\omega')X(\omega')X'(\omega') = \int P^{\mathcal{B}}(\omega; d\omega'')X'(\omega'') \int P^{\mathcal{B}'}(\omega''; d\omega')X(\omega').$$

***29.2. Decomposition of regular c.pr.'s given separable σ-fields.** The "elementary" case investigated in 24.1 corresponds to a given σ-field $\mathcal{B}$ generated by a countable disjoint class of events. It can then be assumed, without restricting the generality, that the class is a partition of the form $\sum B_j + B_0$, where every $PB_j > 0$ and B_0 is null but not necessarily empty. The corresponding "elementary" c.pr.'s can be written as

$$P^{\mathcal{B}} = \sum (P_{B_j})I_{B_j} + (P_{B_0})I_{B_0}, \tag{1}$$

where every P_{B_j} is a pr. on $\mathcal{A}$ defined by

$$P_{B_j}A = \frac{PAB_j}{PB_j}, \quad A \in \mathcal{A}, \tag{2}$$

so that $P_{B_j}B_j = 1$, and P_{B_0} is an arbitrary pr. on $\mathcal{A}$ which disappears when B_0 is empty. Thus, an "elementary" c.pr. is regular and can be said to be "decomposed" into a countable set of pr.'s. We intend to show that regular c.pr.'s given separable σ-fields can be decomposed in an analogous manner.

A σ-field $\mathcal{B}$ is *separable* if it is generated by (is minimal over) a countable class of sets.

a. *If a* σ*-field* $\mathcal{B}$ *is separable, then every set* $B \in \mathcal{B}$ *is a sum of atoms* B_t *of* $\mathcal{B}$ *such that* $\sum_{t \in T} B_t = \Omega$ *with* $T \subset R$.

Proof. Let B_j be the generators of $\mathcal{B}$ and let B_t be the nonempty distinct sets of the form $\bigcap \bar{B}_j$ where $\bar{B}_j = B_j$ or B_j^c. Since the set of j's is countable, the power of the set T of t's is at most that of the continuum, so that T can be supposed to lie in R. Since the B_t are disjoint and any $\omega \in \Omega$ belongs to one of them, $\sum_{t \in T} B_t = \Omega$. Since $\mathcal{B}$ is the σ-field generated by the B_j, every B_t belongs to $\mathcal{B}$, and by construction contains no other sets belonging to $\mathcal{B}$ than itself and the empty set and is not empty; further, every $B \in \mathcal{B}$ is a sum of B_t's. The assertion is proved.

The functions $P^{\mathcal{B}}A$, being $\mathcal{B}$-measurable, reduce to constants on atoms of $\mathcal{B}$. In fact, they reduce to constants on possibly larger events, namely, on atoms of the σ-field $\mathcal{B}_P \subset \mathcal{B}$ induced by these functions for A varying over $\mathcal{A}$. The σ-field $\mathcal{B}_P$ is generated by events of the form $[P^{\mathcal{B}}A \in S]$ where S are arbitrary Borel sets in R; and it suffices to take events of the form $[P^{\mathcal{B}}A < r]$ where the r are positive rationals. The atoms of $\mathcal{B}_P$ will be called $P^{\mathcal{B}}$*-atoms* and every event contained in a $P^{\mathcal{B}}$-atom will be called $P^{\mathcal{B}}$*-indecomposable;* for example, atoms of $\mathcal{B}$ are $P^{\mathcal{B}}$-indecomposable.

A. DECOMPOSITION THEOREM. *If $P^{\mathcal{B}}$ is a regular c.pr. and $\mathcal{B}$ contains a separable σ-field $\mathcal{B}'$ whose atoms are $P^{\mathcal{B}}$-indecomposable, then there exists a partition $\Omega = \sum_{t \in T} B_t + N$ with $T \subset R$ and $P_{\mathcal{B}}N = 0$ such that, except on $N \times \mathcal{A}$,*

$$P^{\mathcal{B}} = \sum_{t \in T} (P_{B_t}) I_{B_t},$$

where the P_{B_t} are pr.'s on $\mathcal{A}$ and $P_{B_t}B_t = 1$.

Proof. Let the countable class $\{B_j\}$ generate $\mathcal{B}' \subset \mathcal{B}$. The field generated by the B_j is a countable class and, hence, it may be assumed that $\{B_j\}$ is a field.

Let $\sum_{t \in T'} B_t = \Omega$ be the partition into atoms $B_t \in \mathcal{B}'$, as constructed in **a**. Since, by assumption, these atoms are $P^{\mathcal{B}}$-indecomposable, the functions $P^{\mathcal{B}}A (A \in \mathcal{A})$ reduce to constants $P_{B_t}A$ on B_t. Since $P^{\mathcal{B}}$ is regular, the P_{B_t} are pr.'s on $\mathcal{A}$. It remains to show that, upon lumping together some atoms B_t into a $P_{\mathcal{B}}$-null event, $P_{B_t}B_t = 1$ for the remaining ones.

For all $B \in \mathcal{B}$, the indicators I_B being $\mathcal{B}$-measurable coincide with their c.exp. $P^{\mathcal{B}}B$ given $\mathcal{B}$ except on a $P_{\mathcal{B}}$-null event. Let N_j be the $P_{\mathcal{B}}$-null event on which $P^{\mathcal{B}}B_j \neq I_{B_j}$. Since the functions $P^{\mathcal{B}}B_j$ do not vary on the atoms B_t, N_j is the sum of some B_t and the union $N = \bigcup N_j$ of all those exceptional atoms is $P_{\mathcal{B}}$-null. Fix ω belonging to a remaining atom B_t. Since $P_{B_t}B' (= P_\omega^{\mathcal{B}}B')$ and $I_{B'}(\omega)$ are values of pr.'s on $\mathcal{B}'$ and coincide on the generating field $\{B_j\}$, it follows that $P_{B_t}B' = I_{B'}(\omega)$

for every $B' \in \mathcal{B}'$; in particular, $P_{B_t}B_t = I_{B_t}(\omega) = 1$, and the proof is concluded.

COROLLARY 1. *If $P^{\mathcal{B}}$ is a regular c.pr., and one of the σ-fields $\mathcal{A}$ or $\mathcal{B}$ or $\mathcal{B}_P$ is separable, then the foregoing decomposition holds.*

Proof. Since atoms of $\mathcal{B}$ and of $\mathcal{B}_P(\subset\mathcal{B})$ are $P^{\mathcal{B}}$-indecomposable, we have only to prove the assertion when $\mathcal{A}$ is separable. Thus, let $\{A_j\}$ be the countable class which generates $\mathcal{A}$; the class can be assumed to be a field. The countable class of events of the form $[P^{\mathcal{B}}A_j < r]$, r rational, generates a separable σ-field $\mathcal{B}' \subset \mathcal{B}_P \subset \mathcal{B}$. It suffices to show that its atoms B_t are $P^{\mathcal{B}}$-indecomposable.

The functions $P^{\mathcal{B}}A_j$ reduce to constants $P_{B_t}A_j$ on atoms B_t of $\mathcal{B}'$ and, for $\omega \in B_t$, the pr.'s $P_\omega{}^{\mathcal{B}}$ and P_{B_t} on $\mathcal{A}$ coincide on the field $\{A_j\}$. Therefore, they coincide on $\mathcal{A}$ and, hence, for every $A \in \mathcal{A}$, the functions $P^{\mathcal{B}}A$ reduce to constants $P_{B_t}A$ on atoms B_t. The proof is terminated.

COROLLARY 2. *Under conditions of the decomposition theorem*

$$E^{\mathcal{B}}X = \sum_{t \in T}(E_{B_t}X)I_{B_t}\ a.s.,$$

where

$$E_{B_t}X = \int X\,dP_{B_t},\quad t \in T.$$

Apply the integration theorem 26.1**A**.

In the elementary case, relation (2) can be written

$$P_{B_j}A = \int_A p_{B_j}(\omega')\,dP(\omega'),\quad A \in \mathcal{A}$$

with

$$p_{B_j}(\omega') = \frac{1}{PB_j}I_{B_j}(\omega'),\quad \omega' \in \Omega.$$

Therefore the decomposition (1) becomes, for $\omega \notin B_0$,

$$P^{\mathcal{B}}(\omega; A) = \int_A p^{\mathcal{B}}(\omega, \omega')\,dP$$

with

$$p^{\mathcal{B}}(\omega, \omega') = \sum \frac{1}{PB_j}I_{B_j}(\omega)I_{B_j}(\omega'),\quad \omega \notin B_0,\quad \omega' \in \Omega,$$

and, taking $P_{B_0} = P$, the integral relation holds for all $\omega \in \Omega$, provided we add $I_{B_0}(\omega)$ to $p^{\mathcal{B}}(\omega, \omega')$.

In general, let μ be a σ-finite measure on $\mathcal{A}$. We say that a regular c.pr. $P^{\mathcal{B}}$ is *μ-continuous* if there exists an $\mathcal{A}$-measurable function $p_\mu^{\mathcal{B}}(\omega, \omega')$ in ω' such that, for every $\omega \in \Omega$ and $A \in \mathcal{A}$,

$$P^{\mathcal{B}}(\omega; A) = \int_A p_\mu^{\mathcal{B}}(\omega, \omega')\, d\mu(\omega').$$

The function $p_\mu^{\mathcal{B}}$ will be called the *conditional pr. density* given $\mathcal{B}$ with respect to μ. It can and will be assumed to be nonnegative and finite. Furthermore, μ can and will be assumed to be a pr. on $\mathcal{A}$. If μ is finite, it suffices to set

$$\mu' = \mu/\mu\Omega, \quad p_{\mu'}^{\mathcal{B}} = p_\mu^{\mathcal{B}} \cdot \mu\Omega.$$

If μ is strictly σ-finite and $\sum A_n = \Omega$ is a partition such that every $\mu A_n < \infty$, it suffices to set

$$\mu' A = \sum \mu A A_n / 2^n \mu A_n, \quad A \in \mathcal{A},$$

and

$$p_{\mu'}^{\mathcal{B}}(\omega, \omega') = p_\mu^{\mathcal{B}}(\omega, \omega') \cdot 2^n \mu A_n, \quad \omega \in \Omega, \quad \omega' \in A_n.$$

COROLLARY 3. *Under conditions of the decomposition theorem, if $P^{\mathcal{B}}$ is μ-continuous, then the decomposition is countable; more precisely, the decomposition is*

$$\Omega = \sum B_j + N, \quad \mu B_j > 0, \quad PN = 0.$$

Proof. Since μ is a pr. on $\mathcal{A}$ and hence on $\mathcal{B}$, there exists only a countable class $\{B_j\}$ of non μ-null atoms B_t. On the other hand, if B_t is one of the μ-null atoms then, for any $\omega \in B_t$,

$$1 = P_{B_t} B_t = \int_{B_t} p_\mu^{\mathcal{B}}(\omega, \omega')\, d\mu(\omega') = 0$$

so that B_t must be empty.

§ 30. CONDITIONAL DISTRIBUTIONS

30.1. Definitions and restricted integration. A regular c.pr. $P^{\mathcal{B}}$ restricted to a sub σ-field of events still has the regularity properties: it is $\mathcal{B}$-measurable and it is a pr. on the sub σ-field to which it is restricted. However, the converse is not necessarily true. Thus, in the search for regular c.pr.'s, it will be convenient to begin by investigating the weaker "restricted regularity." In fact, it will prove useful to extend this concept to functions of a point in a measurable space $(\Omega_1, \mathcal{A}_1)$ with points ω_1 and measurable sets A_1 and of a measurable set in a measurable space $(\Omega_2, \mathcal{A}_2)$ with points ω_2 and measurable sets A_2.

We shall not hesitate to proceed to the usual abuse of language, that is, according to convenience and the possible degree of confusion, we shall speak of "the function $h(\omega_1, A_2)$" instead of "the function h on $\Omega_1 \times \mathcal{A}_2$." We say that the function $h(\omega_1, A_2)$ is an *$\mathcal{A}_1$-measurable pr.* if it is $\mathcal{A}_1$-measurable in ω_1 for every fixed A_2 and is a pr. in A_2 for every fixed ω_1. Observe that, whenever there exists a pr. P_1 on $\mathcal{A}_1$, then the function

$$P_{12}(A_1 \times A_2) = \int_{A_1} P_1(d\omega_1)h(\omega_1, A_2)$$

determines, by the extension theorem (for measures) a pr. on the product-measurable space $(\Omega_1 \times \Omega_2, \mathcal{A}_1 \times \mathcal{A}_2)$.

Let X be a family of r.v.'s on the pr. space $(\Omega, \mathcal{A}, P)$. Let $\mathcal{A}_X$ be the σ-field of events induced by X, that is, the σ-field of the inverse images $[X \in S]$ of Borel sets S in the range space $\mathfrak{X}$ of X. If a c.pr. $P^{\mathcal{B}}(\omega, A)$, where A varies only over $\mathcal{A}_X$, is a pr. on $\mathcal{A}_X$ for every fixed $\omega \in \Omega$, we say that it is a *conditional distribution* (*c.d.*) of X given $\mathcal{B}$. Clearly

A function $P^{\mathcal{B}}(\omega, A)$, where A varies over $\mathcal{A}_X$, is a c.d. of X given $\mathcal{B}$ if, and only if,

(CD_1) *$P^{\mathcal{B}}(\omega, A)$ is a $\mathcal{B}$-measurable pr.*

(CD_2) $\int_B P(d\omega)P^{\mathcal{B}}(\omega, A) = PAB$

for every $A \in \mathcal{A}_X$ and every $B \in \mathcal{B}$.

To c.pr.'s $P^{\mathcal{B}}(\omega, A)$ restricted to $\mathcal{A}_X$, we make correspond $\mathcal{B}$-functions $Q^{\mathcal{B}}(\omega, S)$ such that, for every fixed Borel set S in $\mathfrak{X}$,

(C) $Q^{\mathcal{B}}(\omega, S) = P^{\mathcal{B}}(\omega, [X \in S])$ *up to a $P_{\mathcal{B}}$-equivalence.*

If a function $Q^{\mathcal{B}}(\omega, S)$ in (C) is a pr. on the Borel sets S, we say that it is a *mixed c.d.* of X given $\mathcal{B}$. Clearly, if there exists a c.d. of X given $\mathcal{B}$, then there exists a mixed c.d. of X given $\mathcal{B}$ but the converse is not necessarily true.

The importance of c.d.'s and mixed c.d.'s of X is due to the fact that they still have the integration property of regular c.pr.'s, provided the integrand depends only upon X.

A. RESTRICTED INTEGRATION THEOREM. *Let g be a Borel function on the range space $\mathfrak{X}$ of a family X of r.v.'s, such that $Eg(X)$ exists.*

If there exists a c.d. (mixed c.d.) of X given $\mathcal{B}$, *then, except for points* ω *in a* $P_{\mathcal{B}}$*-null set,*

$$E^{\mathcal{B}}(\omega, g(X)) = \int_{\Omega} P^{\mathcal{B}}(\omega, d\omega')g(X(\omega'))\left(= \int_{\mathfrak{X}} Q^{\mathcal{B}}(\omega, dx)g(x)\right).$$

Proof. By definition of a c.d. (mixed c.d.), the asserted equality holds for indicators $g = I_S$. It follows, as usual, that it holds for simple, then for nonnegative, Borel functions, and the theorem follows.

If $Q^{\mathcal{B}}(\omega, S)$ is a mixed c.d. of a random vector $X = (X_1, \cdots, X_n)$, we set

$$F^{\mathcal{B}}(\omega, x) = Q^{\mathcal{B}}(\omega, (-\infty, x)), \quad x = (x_1, \cdots, x_n),$$

and call this function a *conditional distribution function* (*c.d.f.*) of X given $\mathcal{B}$; it is $\mathcal{B}$-measurable in ω and a d.f. in x. Thus, we can form its Fourier-Stieltjes transform

$$f^{\mathcal{B}}(\omega, u) = \int e^{iux}\, dF^{\mathcal{B}}(\omega, x), \quad u = (u_1, \cdots, u_n)$$

where $ux = u_1x_1 + \cdots + u_nx_n$, and shall call this function a *conditional characteristic function* (*c.ch.f.*) of X given $\mathcal{B}$; it is $\mathcal{B}$-measurable in ω and a ch.f. in u.

COROLLARY. *To a c.ch.f.* $f^{\mathcal{B}}(\omega, u)$ *of a random vector X given* $\mathcal{B}$, *there correspond c.exp.'s* $E^{\mathcal{B}}e^{iuX}$ *such that, for every* ω *and every* u,

$$E^{\mathcal{B}}(\omega, e^{iuX}) = f^{\mathcal{B}}(\omega, u).$$

For, we can select the c.exp.'s such that, according to the theorem, the equality holds for every rational point u, and then use the continuity property of ch.f.'s in passing to the limit along rational points.

30.2. Existence. The problem of existence of regular c.pr.'s has been investigated principally by Doob who begins by solving the problem of c.d.'s as follows.

a. EXISTENCE LEMMA. *If there exists a c.d. of a family X of r.v.'s given* $\mathcal{B}$, *then there exists a mixed c.d. of X given* $\mathcal{B}$. *The converse is true when the range of X is a Borel set.*

We recall that the range of X is the set of values $X(\omega)$ as ω varies over Ω.

Proof. We use repeatedly the correspondence relation (C). The direct assertion follows at once by setting, for every $\omega \in \Omega$ and every Borel set S in the range space of X,

$$Q^{\mathcal{B}}(\omega, S) = P^{\mathcal{B}}(\omega, [X \in S]).$$

In general, the converse is not true because a set $A \in \mathcal{A}_X$ may be inverse image of different Borel sets, say, S and S'. However, when the range of X is itself a Borel set S_X, then

$$Q^{\mathcal{B}}(\omega, S_X^c) = P^{\mathcal{B}}(\omega, [X \in S_X^c]) = P^{\mathcal{B}}(\omega, \emptyset) = 0$$

except for points ω of some $P_{\mathcal{B}}$-null set N. Therefore, when there exists a mixed c.d. $Q^{\mathcal{B}}(\omega, S)$, that is, a $\mathcal{B}$-measurable pr., then, SS'^c and S^cS' being in S_X^c, we have

$$Q^{\mathcal{B}}(\omega, S) = Q^{\mathcal{B}}(\omega, S') = Q^{\mathcal{B}}(\omega, SS'), \quad \omega \notin N.$$

It follows that, in (C), we can select a $\mathcal{B}$-measurable pr. by setting, for every $A \in \mathcal{A}_X$,

$$P^{\mathcal{B}}(\omega, A) = Q^{\mathcal{B}}(\omega, S), \quad \omega \notin N$$

$$P^{\mathcal{B}}(\omega, A) = Q^{\mathcal{B}}(\omega_0, S), \quad \omega \in N, \quad \omega_0 \notin N,$$

where S is any image of A. This function is an asserted c.d., and the proof is complete.

A. C.D.'S EXISTENCE THEOREM. *There always exists a mixed c.d. of a countable family X of r.v.'s given $\mathcal{B}$. If the range of X is a Borel set, then there exists a c.d. of X given $\mathcal{B}$.*

Proof. On account of the existence lemma, it suffices to prove that there exists a mixed c.pr. We show first that a c.d.f. exists; the proof is based upon the fact that the countable set of rational points $r = (r_1, \cdots, r_n)$ of an n-dimensional euclidean space is dense in it.

Let x, x' and r, r' denote points and rational points, respectively, of the range space of a random vector $X = (X_1, \cdots, X_n)$. Let $P(\omega, A)$ be a c.pr. given $\mathcal{B}$, and, for every r, set

$$F^{\mathcal{B}}(\omega, r) = P^{\mathcal{B}}(\omega, [X < r]), \quad \omega \in \Omega;$$

the right-hand sides are selected arbitrarily within their $P_{\mathcal{B}}$-equivalence classes and kept fixed. Let N, with or without affixes, denote $P_{\mathcal{B}}$-null sets. On account of a.s. properties of c.pr.'s, we have

$$F^{\mathcal{B}}(\omega, -\infty) = 0, \quad F^{\mathcal{B}}(\omega, +\infty) = 1, \quad \omega \notin N_0$$

$$\Delta_{r'-r} F^{\mathcal{B}}(\omega, r) \geqq 0, \ 1, \quad r < r', \quad \omega \notin N_{rr'}$$

$$F^{\mathcal{B}}(\omega, r) \uparrow F^{\mathcal{B}}(\omega, r') \quad \text{as} \quad r \uparrow r', \quad \omega \notin N_{r'}.$$

The countable union

$$N = N_0 \cup \bigcup_{r<r'} N_{rr'} \cup \bigcup_{r'} N_{r'}$$

of $P_{\mathfrak{B}}$-null sets is $P_{\mathfrak{B}}$-null. For every x, set

$$F^{\mathfrak{B}}(\omega, x) = \lim_{r \uparrow x} F^{\mathfrak{B}}(\omega, r), \quad \omega \not\in N,$$

$$F^{\mathfrak{B}}(\omega, x) = F^{\mathfrak{B}}(\omega_0, x), \quad \omega \in N, \quad \omega_0 \not\in N.$$

For every $\omega \in \Omega$, the function so defined is a d.f. and, by the correspondence theorem (for d.f.'s), the relation

$$Q^{\mathfrak{B}}(\omega, (-\infty, x)) = F^{\mathfrak{B}}(\omega, x)$$

determines a pr. $Q^{\mathfrak{B}}(\omega, S)$ in Borel sets S.

This function is an asserted mixed c.d., provided we prove that this function is $\mathfrak{B}$-measurable in ω and that, for every S,

$$Q^{\mathfrak{B}}(\omega, S) = P^{\mathfrak{B}}(\omega, [X \in S])$$

up to a $P_{\mathfrak{B}}$-equivalence. By construction, the assertion is true for every $S = (-\infty, r)$. Hence, on account of the a.s. properties of c.pr.'s, it is true on the field of all finite sums of intervals S and, by monotone passages to the limit, it is still true on the minimal monotone field over this field, that is, on the σ-field of all Borel sets S.

Now, let $X = (X_1, X_2, \cdots)$ be a countable family of r.v.'s. Once the $F^{\mathfrak{B}}(\omega; r_1, \cdots, r_n)$ are selected, we can select the $F^{\mathfrak{B}}(\omega; r_1, \cdots, r_n, r_{n+1})$ within the defining $P_{\mathfrak{B}}$-equivalence classes so that, for every $\omega \in \Omega$,

$$F^{\mathfrak{B}}(\omega; r_1, \cdots, r_n, r_{n+1}) \to F^{\mathfrak{B}}(\omega; r_1, \cdots, r_n) \quad \text{as} \quad r_{n+1} \to \infty.$$

Then, for every $\omega \in \Omega$, the foregoing construction yields consistent d.f.'s, hence consistent pr.'s, and, proceeding step by step with $n = 1, 2, \cdots$, we obtain a consistent family of pr.'s which, by the consistency theorem (for measures), determines a mixed c.d. $Q^{\mathfrak{B}}(\omega, S)$on the σ-field of all Borel sets S in the range space of X. The theorem is proved.

Sample pr. spaces. As long as we are concerned only with a given family of r.v.'s, we can always take for pr. space, the sample pr. space of the family. To simplify the statements and the notations, we consider a countable family $X = (X_1, X_2, \cdots)$ of r.v.'s (or random vectors or random sequences). Set $R_{k_1 \cdots k_m} = \prod_{j=1}^{m} R_{k_j}$, denote by $S_{k_1 \cdots k_m}$ and

$\mathcal{A}_{k_1\cdots k_m}$ the Borel sets and their σ-field in this real space, and let $P_{k_1\cdots k_m}$ be the distribution of $X_{k_1\cdots k_m} = (X_{k_1}, \cdots, X_{k_m})$ defined by

$$P_{k_1\cdots k_m}(S_{k_1\cdots k_m}) = P[X_{k_1,\cdots,k_m} \in S_{k_1\cdots k_m}].$$

If the same affixes occur inside and outside a bracket, we shall omit either the inside or the outside ones, according to our convenience.

The sample pr. space of X consists of the space $R_1 \times R_2 \times \cdots$, the σ-field of Borel sets in this space, and the distribution of X on this σ-field. We take it for our pr. space $(\Omega, \mathcal{A}, P)$. Then

$$X(x_1, x_2, \cdots) = (x_1, x_2, \cdots)$$

and the range of any X_n coincides with its range space R_n. Therefore, the existence theorem applies and, for every σ-field $\mathcal{B} \subset \mathcal{A}$, there exists a c.d. of any subfamily of X given $\mathcal{B} \subset \mathcal{A}$; in fact, there exists a c.d. of the countable family X given $\mathcal{B}$, that is, a regular c.pr. $P^{\mathcal{B}}$. Thus

B. REGULARITY THEOREM. *C.pr.'s in sample pr. spaces of countable families of r.v.'s can be regularized and c.d.'s of their subfamilies always exist.*

In the remainder of this section we take for pr. space of $X = (X_1, X_2, \cdots)$ its sample pr. space and can and will assume that the c.pr.'s given a measurable function Y on Ω are expressed as functions of Y and are regularized. By applying repeatedly the restricted integration theorem, we obtain

b. *If g is a Borel function on $R_{1\cdots n}$, such that $Eg(X_1, \cdots, X_n)$ exists, then*

$$Eg(X_1, \cdots, X_n)$$

$$= \int g\, dP_{1\cdots n}$$

$$= \int P(dx_1) \int P(x_1; dx_2) \cdots \int P(x_1, \cdots, x_{n-1}; dx_n) g(x_1, \cdots, x_n)$$

and, except for a P_1-null set of points x_1,

$$E^{\mathcal{B}}(x_1; g(X_1, \cdots, X_n))$$

$$= \int P(x_1; dx_{2\cdots n}) g(x_1, \cdots, x_n)$$

$$= \int P(x_1; dx_2) \cdots \int P(x_1, \cdots, x_{n-1}; dx_n) g(x_1, \cdots, x_n).$$

The c.d.'s defining properties separate as follows:

c. *Property*

(CD_1) *$P(x_1, S_2)$ is an $\mathfrak{A}_1$-measurable pr. on $\mathfrak{A}_2$,*

characterizes c.d.'s of X_2 given X_1, and property

(CD_2) $P(S_1 \times S_2) = \int_{S_1} P(dx_1)P(x_1, S_2)$

relates the distributions of X_1 and (X_1, X_2).

Applications. 1° The law of the countable family $(X_1, X_2, \cdots)$ is defined by the distribution of this family which determines and is determined by a consistent family of distributions PS_1, PS_{12}, $\cdots$. Because of the consistency requirement, this family of distributions is superabundant. Conditioning permits us to determine the law by means of a nonsuperabundant family of measurable pr.'s (that is, with no required relations among members). For, by applying repeatedly the above propositions, we find that

The law of the countable family $X_1, X_2, \cdots$ determines and is determined by a family $P(S_1)$, $P(x_1; S_2)$, $P(x_1, x_2; S_3)$, $\cdots$ of c.d.f.'s.

Clearly, we can replace c.d.'s by c.d.f.'s or by c.ch.f.'s.

2° Let $X_1, X_2, \cdots$ be r.v.'s on their sample space $(\Omega, \mathfrak{A}, P)$ with joint d.f.'s $F_{k_1 \cdots k_m}$ and c.d.f.'s $F_{k_1 \cdots k_m}{}^{\mathfrak{B}}$ of $(X_{k_1}, \cdots, X_{k_m})$. We can define conditional independence of the X's given $\mathfrak{B}$ by the property

$$F_{k_1 \cdots k_m}{}^{\mathfrak{B}} = F_{k_1}{}^{\mathfrak{B}} \cdots F_{k_m}{}^{\mathfrak{B}}$$

for arbitrary finite subsets $k_1, \cdots, k_m$ of subscripts. Then

$$F_{k_1 \cdots k_m} = E(F_{k_1}{}^{\mathfrak{B}} \cdots F_{k_m}{}^{\mathfrak{B}})$$

where the expectation is obtained by integrating with respect to $P_{\mathfrak{B}}$.

Conversely, any family $X_1, X_2, \cdots$ is trivially conditionally independent—given $\mathfrak{A}$; we exclude this trivial case.

If the r.v.'s are conditionally independent with common c.d.f. $F^{\mathfrak{B}}$, then

$$F_{k_1 \cdots k_m}(x_1, \cdots, x_m) = E(F^{\mathfrak{B}}(x_1) \cdots F^{\mathfrak{B}}(x_m)).$$

Thus, the joint d.f.'s of any m of the r.v.'s do not depend upon their subscripts but only upon their number m. If the joint d.f.'s have this property, that is, for every finite subset $k_1, \cdots, k_m$

$$G_m = F_{k_1 \cdots k_m},$$

we say that the r.v.'s are *exchangeable*. This concept was introduced by de Finetti and his basic result, in terms of conditional independence and in a somewhat more precise form, is as follows.

The concept of exchangeability is equivalent to that of conditional independence with common c.d.f.

The second concept implying the first one, it suffices to prove the converse. Thus, let $G_m = F_{k_1 \cdots k_m}$ and set, for every $x \in R$,

$$\xi_n(x) = \frac{1}{n}\sum_{j=1}^{n} I_{[X_{k_j} < x]}.$$

Since, as $m, n \to \infty$,

$$E(\xi_m(x) - \xi_n(x))^2 = \frac{|m - n|}{mn}(G_1(x) - G_2(x, x)) \to 0,$$

it follows that there exists a r.v. $\xi(x)$ such that $E(\xi_n(x) - \xi(x))^2 \to 0$, and hence $\xi_n(x) \xrightarrow{P} \xi(x)$. Since the $\xi_n(x)$ are bounded by 1, it follows, by the dominated convergence theorem and a.s. invariance under finite permutations of X's of $B \in \mathcal{B} = \mathcal{B}(\xi(x), x \in R)$, that

$$E(\xi(x_1)\cdots\xi(x_m)I_B) \leftarrow E(\xi_n(x_1)\cdots\xi_n(x_m)I_B) \\ \to P([X_1 < x_1, \cdots, X_m < x_m]I_B).$$

Thus $P^{\mathcal{B}}[X_1 < x_1, \cdots, X_m < x_m] = \xi(x_1)\cdots\xi(x_m)$ a.s. Finally, since the function $\xi_n(\omega, x)$ is a d.f. in x, it follows that the function $\xi(\omega, x)$ has a.s. the properties of a d.f. in x and therefore, in the preceding relation, $\xi(x)$ can be replaced by a c.d.f. (use, for example, the same method as in the proof of the c.d.'s existence theorem).

30.3. Chains; the elementary case. In the case of random vectors the definition of chain is as follows:

A sequence X_n of random vectors is a *chain* if, for every integer n, a c. distribution of X_{n+1} given $X_1, \cdots, X_n$ can (and will) be so selected that it coincides with a c. distribution of X_{n+1} given X_n; in symbols $P_{X_{n+1}}^{X_1,\cdots,X_n} = P_{X_{n+1}}^{X_n}$ or, equivalently, the c. distribution $P(x_1, \cdots, x_n; S_{n+1})$ is independent of the *a priori* arguments $x_1, \cdots, x_{n-1}$. (On account of 27.2**b**, this definition entails chain-dependence as defined in 25.3; apply the second relation with n replaced by $m + 1$, $\alpha_1 = (1, \cdots, n)$, $\alpha_2 = n + 1, \cdots, \alpha_{m+1} = n + m$, and $g = I_{S_{n+1}\times\cdots\times S_{n+m}}$.)

Usually, the chained random vectors have a common range-space; to fix the ideas, consider a chain X_n of r.v.'s. The terminology used is phenomenological. The chain is a "system" X whose "state" at "time" n is X_n and has for values points $x \in R$—the "possible" states. The c. distribution $P_{X_{n+1}}^{X_n}$ is the *one-step transition pr.* at "time" n. By the classical abuse of language, it is represented by the same symbol as its values. This symbol will be $P^{n,n+1}(x; S)$ and is read "pr. of passage from x at time n into S at time $n + 1$." The very language used contains implicitly the assumption of chain-dependence.

If $P^{n,n+1}(x; S) = P(x; S)$ is independent of n, then the chain is said to be *constant* (in time); $P(x; S)$ is called the *transition pr. function* (f.) of the chain and read "pr. of passage from x into S in one step." From a phenomenological point of view, a constant chain represents a "random system" whose "law of evolution" does not vary in time. Let $P_n(S)$ denote the distribution of X_n. Since, for every n,

$$P_{n+1}(S) = \int P_n(dx)P(x; S),$$

it follows, by induction, that, for every pair m, n of integers,

$$P_{m+n}(S) = \int P_m(dx)P^n(x; S)$$

where

$$P^n(x; S) = \int P(x; dx_1) \int P(x_1; dx_2) \cdots \int P(x_{n-2}; dx_{n-1})P(x_{n-1}; S).$$

Clearly, this relation implies and is implied by the relation

$$P^{n+p}(x; S) = \int P^n(x; dx')P^p(x'; S); \quad n, p = 1, 2, \cdots.$$

$P^n(x; S)$ is called the *n-step transition pr.* and read "pr. of passage from x into S in n-steps." Upon applying 27.2**b**, it is easily seen that the n-step transition pr. is a c. distribution of X_{m+n} given $X_m (m = 1, 2, \cdots)$.

Upon applying 27.2**a** and 27.2**A**, we can summarize the basic properties of constant chains, as follows:

A. *A function $P(x; S)$, of points $x \in R$ and Borel sets $S \subset R$, is the transition pr.f. of a constant chain of r.v.'s if, and only if, it is a Borel function in x for every fixed S and a pr. in S for every fixed x.*

The law of a constant chain of r.v.'s X_n, with distributions $P_n(S)$, is determined by the initial distribution $P_1(S)$ and the transition pr.f. $P(x; S)$.

For every pair m, n of integers,

$$P_{m+n}(S) = \int P_m(dx)P^n(x; S),$$

where the function $P^n(x; S)$ is determined by the relation

$$P^{n+p}(x; S) = \int P^n(x; dx')P^p(x'; S); \quad n, p = 1, 2, \cdots.$$

Let $P(x;\ S)$ be a transition pr.f. If there exists an initial distribution $P_1(S)$ such that, for every n, the consecutive distributions $P_n(S)$ coincide with the initial one, the transition pr.f. is said to *possess an invariant distribution* $P_1(S)$ and $P_1(S)$ is said to be *invariant* under the transition pr.f. $P(x;\ S)$; the chain whose law is determined by the invariant distribution $P_1(S)$ and by $P(x;\ S)$ is said to be *stationary*. In symbols, $P_1(S)$ is invariant under $P(x;\ S)$ if, for every n,

$$P_1(S) = \int P_1(dx)P^n(x;\ S).$$

Since, for every n,

$$P_{n+1}(S) = \int P_n(dx)P(x;\ S),$$

it suffices to require that this relation be valid for $n = 1$. It easily follows from 27.2**b** that, if the chain X_n is stationary, then, for every n, the distribution of $(X_m, \cdots, X_{m+n})$ is independent of m.

A transition pr.f. $P(x, S)$ is *elementary* if there exists a countable partition $\sum S_k = \Omega$ such that $P(x, S_k) = P_{jk}$ for all $x \in S_j$; thus it reduces to a *transition pr. matrix* and the only values of initial distributions which matter are of the form $P_j = P_1(S_j)$. We set $P_j^n = P_n(S_j)$ and $P_{jk}^n = P^n(x, S_k)$ for $x \in S_j$ and the basic properties of constant chains become

$$P_{jk}^n \geqq 0, \quad \sum_k P_{jk}^n = 1, \quad P_{jk}^{m+n} = \sum_h P_{jh}^m P_{hk}^n,$$

$$P_j^n \geqq 0, \quad \sum_j P_j^n = 1, \quad P_j^{m+n} = \sum_h P_{jh}^m\, P_{hj}^n.$$

EXPONENTIAL CONVERGENCE. The basic limit problem for constant chains is that of the asymptotic behavior of n-step transition pr.f.'s $P^n(x, S)$. A particularly simple yet a cornerstone case, which in essence goes back to Markov, is the *exponential convergence case:* there exists a set function $\bar{P}(S)$ and positive constants a, b such that, for n sufficiently large, $|P^n(x, S) - \bar{P}(S)| \leqq ae^{-bn}$ whatever be x and S. This implies at once that $\bar{P}(S)$ is a pr.

In what follows we use repeatedly the fact that differences $\varphi(S)$ of two pr.'s vanish for $S = R$ so that $2\varphi(S)$ and $2|\varphi(S)|$ attain the same supremum $\operatorname{Var} \varphi = \int |\varphi(dy)|$ at a positive Hahn decomposition set of $\varphi(S)$ to be denoted by H, with or without affixes.

a. INVARIANCE LEMMA. *$\bar{P}(S)$ is invariant under transition pr.f.'s and $|P_{n+1}(S) - \bar{P}(S)| \leqq ae^{-bn}$ whatever be $P_1(S)$.*

For

$$\left|\int\left\{P^n(x, dy) - \bar{P}(dy)\right\}P^m(y, S)\right| \leqq \int\left| P^n(x, dy) - \bar{P}(dy)\right| \leqq 2ae^{-bn}$$

implies that

$$\bar{P}(S) \leftarrow P^{n+m}(S) = \int\{P^n(x, dy)P^m(y, S) \rightarrow \int \bar{P}(dy)P^m(S),$$

and

$$| P_{n+1}(S) - \bar{P}(S) | \leq \int P_1(dx) \, | P^n(x, S) - \bar{P}(S) | \leqq ae^{-bn}.$$

We introduce now a "measure" of chain dependence which originated with Markov. Let $\Delta_{n,n+m} = \sup_{x,y} \sup_S \{P^n(x, S) - P^{n+m}(y, S)\}$. The (generalized) *Markov measure* is $\Delta_n = \Delta_{n,n}$. Clearly $0 \leqq \Delta_n \leqq 1$, and in the independence case $\Delta_n = 0$ (since x, y disappear) while in the deterministic case $\Delta_n = 1$ (since $P^n(x, S) = I(x, S)$).

b. Basic inequalities:

$$\Delta_{n,n+m} \leqq \Delta_n \quad \textit{and} \quad \Delta_{n+m} \leqq \Delta_n\Delta_m.$$

For

$$| P^n(x, S) - P^{n+m}(y, S) | \leqq \int P^m(y, dz) \, | P^n(x, S) - P^n(z, S) | \leqq \Delta_n$$

and if $\varphi_n(S) = P^n(x, S) - P^n(y, S)$, then

$$\begin{aligned} | \varphi_{n+m}(S) | = \left| \int_{H_n+H_n^c} \varphi_n(dz)P^m(z, S) \right| &\leqq \varphi_n(H_n) \sup_z P^m(z, S) \\ &\quad + \varphi_n(H_n^c) \inf_z P^m(z, S) \\ &= \varphi_n(H)\{\sup P^m(z, S) - \inf P^m(z, S)\} \leqq \Delta_n\Delta_m. \end{aligned}$$

B. Exponential convergence criterion. *Exponential convergence holds if, and only if,* $\Delta_h < 1$ *for some integer h.*

Proof. If exponential convergence holds, then

$$\begin{aligned} | P^n(x, S) - P^n(y, S) | \leqq | P^n(x, S) - \bar{P}(S) | \\ + | P^n(y, S) - \bar{P}(S) | \leqq 2ae^{-bn} \end{aligned}$$

Conversely, if $\Delta_h < 1$ then, by **b**, as $m, n \rightarrow \infty$,

$$|P^n(x, S) - P^{n+m}(y, S) | \leqq \Delta_n \leqq \Delta_h^{[n/h]} \rightarrow 0,$$

hence $\lim P^n(x, S) = \bar{P}(x, S)$ exists and this limit is a function $\bar{P}(S)$ of S only (set $m = 0$), so that $|P^n(x, S) - \bar{P}(S)| \leqq \Delta_h^{[n/h]}$ (let $m \to \infty$).

Let μ be a σ-finite measure. By the Lebesgue decomposition theorem

$$P^h(x, S) = \int p^n(x, y)\mu(dy) + P_s^n(x, S)$$

where $p^n(x, y) \geqq 0$ and $P_s^n(x, S)$ is μ-singular.

MARKOV CASE (GENERALIZED). *If* $\inf\limits_x p^h(x, y) \geqq \delta > 0$ *for all* y *in some* μ*-positive set* S, *then exponential convergence holds.*

For, if H is a Hahn set for the difference of pr.'s in Δ_h, then

$$\begin{aligned} P^h(x, S') - P^h(y, S') &\leqq 1 - \{P^h(x, H^c) + P^h(y, H)\} \\ &\leqq 1 - \{P^h(x, H^cS) + P^h(y, HS)\} \\ &\leqq 1 - \left\{\int_{H^cS} p^h(x, y)\mu(dy) + \int_{HS} p^h(y, z)\mu(dz)\right\} \\ &\leqq 1 - \delta\mu(S) < 1. \end{aligned}$$

c. *Let* $X_1, X_2, \cdots$ *be a sequence of chained r.v.'s in the exponential convergence case and let* Y *be a r.v. bounded by* c *defined on* $X_{n+m}, X_{n+m+1}, \cdots$. *If* $\bar{E}$ *refers to* $\bar{P}$, *then*

$$|\bar{E}Y - E(Y \mid X_n)| \leqq 2ace^{-bm}.$$

For,

$$\begin{aligned} &|\bar{E}Y - E(Y \mid X_n = x)| \\ &\quad = \left|\int E(Y \mid X_{n+m} = y)\{\bar{P}(dy) - P^m(x, dy)\}\right| \leqq 2ace^{-bm}. \end{aligned}$$

This sequence behaves asymptotically as in the case of independence, as follows:

C. EXPONENTIAL CONVERGENCE THEOREM. *In the exponential convergence case with chained r.v.'s* $X_1, X_2, \cdots$, *whatever be* $P_1(S)$,

(i) $\{g(X_1) + \cdots + g(X_n)\}/n \xrightarrow{\text{a.s.}} \int g(x)\bar{P}(dx)$

for every Borel function g for which the integral exists

(ii) *the limit laws of normed sums* $\dfrac{g(X_1) + \cdots + g(X_n)}{b_n} - a_n$, $b_n \to \infty$, *where g is a finite Borel function, are stable and independent of* $P_1(S)$.

Proof:

1° If $P_1 = \bar{P}$ then, by **a**, the sequence $g(X_1), g(X_2), \cdots$ is stationary and also is indecomposable since, for every invariant set C, $I_C(x) = P(x, C) = P^n(x, C) \to \bar{P}(C)$, so that assertion (i) follows by the stationarity theorem. Since the limit on and the indicator of the convergence set of the averages are tail functions on $X_1, X_2, \cdots$, it follows from **a** that (i) holds for any P_1.

2° To prove (ii) we can take $a_n = 0$ on account of the convergence of types theorem. Thus, let $\mathfrak{L}(S_n/b_n) \to \mathfrak{L}(X)$ with ch.f. f, where $S_n = \sum_{k=1}^{n} g(X_k)$. Then, by the same theorem, upon excluding the trivial case of degenerate limit laws, $b_n/b_{n+1} \to 1$ so that, given positive constants, c, c', there exists a sequence $m = m(n)$ such that $b_m/b_n \to c'/c$.

Let $P_1 = \bar{P}$. Then, by **a**, in

$$(1)\quad cS_n/b_n + c(S_{n+p} - S_n)/b_n + c(S_{m+n+p} - S_{n+p})/b_n = cS_{m+n+p}/b_n,$$

the law of the middle term is $\mathfrak{L}(S_p/b_n) \to \mathfrak{L}(0)$ for every fixed p. Thus we can and do select $p = p(n) \uparrow \infty$ such that, for these p, $\mathfrak{L}((S_{n+p} - S_n)/b_n) \to \mathfrak{L}(0)$ and hence, in passing to limit laws, we neglect the corresponding term in (1) while the "distance" p between S_n and $S_{m+n+p} - S_{n+p}$ increases indefinitely. But, by **a** and **c**,

$$\begin{aligned} E(\exp\{iuc(S_{m+n+p} &- S_{n+p})/b_n\} \mid S_n) \\ &= E(\exp\{iuc(S_{m+1+p} - S_{1+p})/b_n\} \mid X_1) \\ &= E(\exp\{iuc(S_{m+1+p} - S_{1+p})/b_n\} + o(1) \\ &= E(\exp\{iuc(b_m/b_n)(S_m/b_m)\}) \to f(c'u), \end{aligned}$$

so that the ch.f. $E(\exp\{iucS_n/b_n\}E(\exp\{iuc(S_{m+n+p} - S_{n+p})/b_n\} \mid S_n))$ of the sum of the extreme terms in the left side of (1) converges to $f(cu)f(c'u)$. It follows, by the convergence of types theorem, that there exists a constant c'' such that the ch.f. of cS_{m+n+p}/b_n converges to $f(c''u) = f(cu)f(c'u)$ so that the limit law is stable.

Let $P_1 \neq \bar{P}$. For every fixed k, we can replace S_n by $S_{n+k} - S_n$ in $\lim_n |\bar{E}(\exp\{iuS_n/b_n\}) - E(\exp\{iuS_n/b_n\} \mid X_1)|$ so that, by **c**, this expression is bounded by $2ae^{-bk} \to 0$ as $k \to \infty$. Therefore, the limit ch.f. given X_1 reduces to the limit ch.f. under $\bar{P}$, so does its expectation, and the proof is terminated

COMPLEMENTS AND DETAILS

1. Let $\mathfrak{B}$ be the σ-field in $\Omega = [0, 1]$ of Borel sets B, with or without affixes, and let λ be the Lebesgue measure on $\mathfrak{B}$. Let $C \subset \Omega$ be a set of outer Lebesgue measure 1 and inner Lebesgue measure 0. Take for pr. space $(\Omega, \mathfrak{A}, P)$, where $\mathfrak{A}$ is the σ-field of all sets of the form $A = B_1C + B_2C^c$ and $PA = \frac{1}{2}\lambda B_1 + \frac{1}{2}\lambda B_2$. Then $PB = \lambda B$, $PC = \frac{1}{2}$, and there is no regular c.pr. given $\mathfrak{B}$.

Chapter IX

FROM INDEPENDENCE TO DEPENDENCE

The problems in and the methods developed for the independence case can be transposed to the general case. This permits us to enlarge the domains of validity of the results obtained in the independence case and also to realize the range of the methods.

In the last section of this chapter appears a different method—of indefinite expectations—which leads to more general results for a.s. convergence and is used extensively in the next chapter.

§31. CENTRAL ASYMPTOTIC PROBLEM

The Central Limit Problem is concerned with convergence of sequences $\mathfrak{L}(X_n)$ of laws of sums $X_n = \sum_{k=1}^{k_n} X_{nk}$ of r.v.'s. In order to investigate this problem in the case of dependent summands, we have to extend it to a *Central Asymptotic Problem* concerned with the comparison of the asymptotic behaviors (as $n \to \infty$) of $\mathfrak{L}(X_n)$ and of suitably chosen laws $\mathfrak{L}(Y_n)$. In fact, already in the case of independent summands, the investigation of the Central Limit Problem was based upon the comparison of laws of sums with suitably chosen infinitely decomposable laws.

The tools we shall require are, naturally enough, extensions of those used in the Central Limit Problem for independent summands. We write

$$H = F - G, \quad h = f - g, \quad \hat{h} = \hat{f} - \hat{g},$$

with the same affixes (if any) throughout, for differences of d.f.'s F, G and corresponding ch.f.'s f, g and integral ch.f.'s $\hat{f}$, $\hat{g}$.

31.1. Comparison of laws. In what follows we state properties which either result at once from those of d.f.'s and ch.f.'s or are obtained by means of identical arguments.

I. *Every function* $H = F - G$ *is bounded by* 1 (*in absolute value*), *is continuous from the left, has a countable discontinuity set, and*

$$H(x+0) = F(x+0) - G(x+0),$$

$$\int dH = \operatorname{Var} F - \operatorname{Var} G, \quad \operatorname{Var} H = \int |\, dH\,| \leqq 2.$$

We write $H_n \xrightarrow{w} H$ (up to additive constants) when $\int g dH_n \to \int g dH$ for all $g \in C_0$—the family of continuous functions on R vanishing at infinity. Note that in the case of d.f.'s, by the weak convergence criterion, this convergence is their weak convergence.

The weak compactness theorem is valid for functions H: every sequence H_n is weakly compact.

We write $H_n \xrightarrow{c} H$ when $H_n \xrightarrow{w} H$ and $\int dH_n \to \int dH$.

The Helly-Bray theorem is not valid for functions H.
Its proof breaks down the moment we use convergence of variations, since $\int dH_n \to \int dH$ does not entail $\int |\, dH_n\,| \to \int |\, dH\,|$.

II. *The functions h and $\hat{h}$ are defined by*

$$h(u) = \int e^{iux}\, dH(x), \quad \hat{h}(u) = \int_0^u h(v)\, dv = \int \frac{e^{iux} - 1}{ix}\, dH(x).$$

h on R is continuous and bounded by 2 but the relation $|\, h\,| \leqq h(0)$ is not valid.

The inversion formula is valid:

$$H(a) - H(b) = \lim_{U \to \infty} \frac{1}{2\pi} \int_{-U}^{+U} \frac{e^{-iua} - e^{-iub}}{-iu}\, h(u)\, du.$$

The weak convergence criterion is valid: $H_n \xrightarrow{w} H$ up to additive constants if, and only if, $\hat{h}_n \to \hat{h}$.

The continuity theorem is valid: if $h_n \to k$ continuous at $u = 0$, then $H_n \xrightarrow{c} H$ up to additive constants and $h = k$.

However, *the converse is not valid*, for the proof given for d.f.'s breaks down when the Helly-Bray theorem—which is no longer valid—is to be applied; for example, if $h_n(u) = e^{-inu} - e^{+inu}$, then the sequence h_n does not converge on R, while $H_n(x) = 1$ for $|x| < n$ and $=0$ for $|x| > n$ so that $H_n \xrightarrow{c} 1$.

III. EXPANSION OF h. *If S is a Borel set and $0 < \delta \leqq 1$ then, provided the integrals exist,*

$$|h(u)| \leqq \left|\int dH\right| + \sum_{j=l+1}^{m} \frac{|u|^j}{j!}\left|\int_S x^j\, dH\right| + c_{m\delta}|u|^{m+\delta}\int_S |x|^{m+\delta}|\, dH|$$
$$+ \sum_{j\leqq l} \frac{|u|^j}{j!}\left|\int_{S^c} x^j\, dH\right| + c_{l\gamma}|u|^{l+\gamma}\int_{S^c} |x|^{l+\gamma}|\, dH|$$

where, if $l \geqq 1$, then $0 < \gamma \leqq 1$, and, if $l = 0$, then $0 \leqq \gamma \leqq 1$; the c's depend only on their subscripts.

If the right side is infinite, the inequality is trivially true. If the right side is finite, it follows from

$$h(u) = \int e^{iux}\, dH(x) = \int dH + \int_S (e^{iux} - 1)\, dH + \int_{S^c} (e^{iux} - 1)\, dH;$$

use limited expansions of e^{iux} of order $l + \gamma$ and $m + \delta$; and for $l + \gamma = 0$ use $|e^{iux} - 1| \leqq 2$.

We can now proceed to the comparison of sequences of laws. Two sequences $\mathfrak{L}(X_n)$ and $\mathfrak{L}(Y_n)$ are said to be *weakly equivalent*, and we write $\mathfrak{L}(X_n) \overset{w}{\sim} \mathfrak{L}(Y_n)$, if the two sequences have the same weak limit laws for same subsequences of subscripts; in other words, if $\mathfrak{L}(X_{n'}) \xrightarrow{w} \mathfrak{L}$, then $\mathfrak{L}(Y_{n'}) \xrightarrow{w} \mathfrak{L}$ and conversely. We observe that $\mathfrak{L}(X_{n'}) \xrightarrow{w} \mathfrak{L}$ means that $F_{n'} \xrightarrow{w} F$ up to additive constants. We define complete equivalence $\mathfrak{L}(X_n) \overset{c}{\sim} \mathfrak{L}(Y_n)$ by replacing in what precedes "weakly" by "completely."

In what follows we use repeatedly properties I and II without further comment.

A. WEAK EQUIVALENCE CRITERION. *$\mathfrak{L}(X_n) \overset{w}{\sim} \mathfrak{L}(Y_n)$ if, and only if, $F_n - G_n \xrightarrow{w} 0$ up to additive constants or $f_n - g_n \to 0$.*

Proof. It suffices to consider $F_n - G_n$; the assertion with $f_n - g_n$ follows.

Let $F_n - G_n \xrightarrow{w} 0$ up to additive constants. The weakly compact sequence F_n contains subsequences $F_{n'} \xrightarrow{w} F$ to which correspond subsequences $G_{n'} = F_{n'} - (F_{n'} - G_{n'}) \xrightarrow{w} F$ up to additive constants. It follows that $\mathfrak{L}(X_n) \overset{w}{\sim} \mathfrak{L}(Y_n)$.

Conversely, let $\mathfrak{L}(X_n) \overset{w}{\sim} \mathfrak{L}(Y_n)$. The weakly compact sequence $F_n - G_n$ contains subsequences $F_{n'} - G_{n'} \xrightarrow{w}$ some H and the weakly compact sequence $F_{n'}$ contains subsequences $F_{n''} \xrightarrow{w}$ some F. By hypothesis $G_{n''} \xrightarrow{w} F$ up to additive constants and, hence, $F_{n''} - G_{n''} \xrightarrow{w} H = F - F = 0$ up to additive constants. It follows that the weakly compact sequence $F_n - G_n \xrightarrow{w} 0$ up to additive constants. The proof is concluded.

B. Complete equivalence criterion. *Let the sequences $\mathfrak{L}(X_n)$ or $\mathfrak{L}(Y_n)$ be completely compact. Then $\mathfrak{L}(X_n) \overset{c}{\sim} \mathfrak{L}(Y_n)$ if, and only if, $F_n - G_n \xrightarrow{c} 0$ up to additive constants or $f_n - g_n \to 0$.*

Proof. Since $f_n - g_n \to 0$ implies that $F_n - G_n \xrightarrow{c} 0$ up to additive constants, it suffices to prove the "if" assertion with $F_n - G_n$ and the "only if" assertion with $f_n - g_n$.

If $F_n - G_n \xrightarrow{c} 0$ up to additive constants, then to every completely convergent subsequence $F_{n'} \xrightarrow{c}$ some F there corresponds the subsequence $G_{n'} \xrightarrow{c} F$ up to additive constants, and conversely. It follows that $\mathfrak{L}(X_n) \overset{c}{\sim} \mathfrak{L}(Y_n)$.

If $\mathfrak{L}(X_n) \overset{c}{\sim} \mathfrak{L}(Y_n)$, then one of the sequences f_n or g_n being completely compact in the sense of convergence to continuous functions, the same is true of both sequences and, hence, of the sequence $f_n - g_n$. If $f_{n'} - g_{n'} \to h$, then the sequence $f_{n'}$ contains a subsequence $f_{n''} \to$ some f and, by hypothesis, $g_{n''} \to f$. Therefore, $f_{n''} - g_{n''} \to h = 0$—unique limit element of the completely compact sequence $f_n - g_n$. It follows that $f_n - g_n \to 0$, and the proof is concluded.

Remark. In the proof of the "if" assertion we made use of the complete compactness of $\mathfrak{L}(X_n)$ only to assert that $F_{n'} \xrightarrow{c}$ some F. Let us make the natural convention that, when neither of the sequences $\mathfrak{L}(X_n)$ and $\mathfrak{L}(Y_n)$ has a complete limit element, then $\mathfrak{L}(X_n) \overset{c}{\sim} \mathfrak{L}(Y_n)$. Thus, $F_n - G_n \xrightarrow{c} 0$ up to additive constants implies that if the sequence $\mathfrak{L}(X_n)$ has no complete limit element the same is true of the sequence $\mathfrak{L}(Y_n)$, and conversely. In other words, with the foregoing convention the assumption of complete compactness is unnecessary for the "if" assertion:

If $F_n - G_n \xrightarrow{c} 0$ *up to additive constants, or* $f_n - g_n \to 0$, *then* $\mathcal{L}(X_n) \overset{c}{\sim} \mathcal{L}(Y_n)$.

We shall frequently center the X_n at some suitably chosen conditional expectations ξ_n. We observe that

COROLLARY. *If* $\xi_n \xrightarrow{P} 0$, *then* $\mathcal{L}(X_n - \xi_n) \overset{c}{\sim} \mathcal{L}(X_n)$.

This follows from the law-equivalence lemma.

31.2. Comparison of summands. Let

$$X_n = \sum_k X_{nk}, \quad Y_n = \sum_k Y_{nk}, \quad k = 1, \cdots, k_n.$$

$$Z_{nk} = X_{n0} + \cdots + X_{n,k-1} + Y_{n,k+1} + \cdots + Y_{n,k_n+1},$$
$$X_{n0} = Y_{n,k_n+1} = 0.$$

To X and Y with or without affixes there correspond their d.f.'s F and G and ch.f.'s f and g with same affixes if any; *primes will denote conditioning by* Z_{nk}, *unless otherwise stated*; for example,

$$F'_{nk} = P[X_{nk} < x \mid Z_{nk}], \quad f'_{nk}(u) = E'e^{iuX_{nk}} = E(e^{iuX_{nk}} \mid Z_{nk}).$$

For every fixed value of Z_{nk}, the selected conditional d.f.'s and ch.f.'s have all the properties of d.f.'s and ch.f.'s, and all properties of differences $H = F - G$, $h = f - g$ given in the preceding subsection are valid for the conditional differences $H' = F' - G'$, $h' = f' - g'$.

We intend to compare the sequences $\mathcal{L}(X_n)$ and $\mathcal{L}(Y_n)$ through the summands X_{nk} and Y_{nk}. (Let us observe that it is frequently convenient to compare suitably selected partial sums, each partial sum to be considered as a single summand.) We are at liberty to introduce any suitable dependence between the sets $\{X_{nk}\}$ and $\{Y_{nk}\}$, provided the laws of each of these sets are not modified and, in fact, provided the sequences $\mathcal{L}(X_n)$ and $\mathcal{L}(Y_n)$ remain the same.

A. COMPARISON THEOREM. $\mathcal{L}(X_n) \overset{c}{\sim} \mathcal{L}(Y_n)$

if

$$\sum_k E|f'_{nk} - g'_{nk}| \to 0$$

or if, S being a Borel set fixed or not (depending on n and/or k or not),

(i) $$\sum_k E\left|\int x^j(dF'_{nk} - dG'_{nk})\right| \to 0, \quad j \leq l,$$

(ii) $$\sum_k E\int_{S^c} |x|^{l+\gamma} |dF'_{nk} - dG'_{nk}| \to 0,$$

(iii) $$\sum_k E\left|\int_S x^j(dF'_{nk} - dG'_{nk})\right| \to 0, \quad j = l+1, \cdots, m,$$

(iv) $$\sum_k E\int_S |x|^{m+\delta}|dF'_{nk} - dG'_{nk}| \to 0, \quad 0 < \delta \leqq 1 \text{ fixed.}$$

If $l = 0$ *condition* (i) *disappears and* $0 \leqq \gamma \leqq 1$; *if* $l \geqq 1$, *then* $0 < \gamma \leqq 1$.

Proof. The first assertion follows by the complete convergence criterion from the inequality

$$|f_n - g_n| \leqq \sum_k E|f'_{nk} - g'_{nk}|$$

given by

$$\left|E\left(e^{iu\sum_k X_{nk}} - e^{iu\sum_k Y_{nk}}\right)\right| = \left|E\sum_k (e^{iuX_{nk}} - e^{iuY_{nk}})e^{iuZ_{nk}}\right|$$
$$\leqq \sum_k E\left|E'(e^{iuX_{nk}} - e^{iuY_{nk}})\right|.$$

The second assertion follows then from the expansion 28.1 III, and the theorem is proved.

It is important to observe that the theorem, and hence all which follows, remain valid with "finer" conditioning than by Z_{nk}. In other words, we can condition by any collection of X_{nk}'s and Y_{nk}'s of which Z_{nk} is a function. In particular, we can condition by the random vectors $Z'_{nk} = (X_{n0}, \cdots, X_{n,k-1}, Y_{n,k+1}, \cdots, Y_{n,k_n+1})$ or $Z''_{nk} = (X_{n0} + \cdots + X_{n,k-1}, Y_{n,k+1} + \cdots + Y_{n,k_n+1})$.

FIRST APPROACH. To $X_n = \sum_k X_{nk}$ we make correspond $Y_n = X^*_n = \sum_k X^*_{nk}$ where the summands X^*_{nk} are independent, and independent of the X_{nk}, and $\mathcal{L}(X^*_{nk}) = \mathcal{L}(X_{nk})$. Loosely speaking, $\mathcal{L}(X^*_n)$ is obtained from $\mathcal{L}(X_n)$ by suppressing the dependence between summands. If $\mathcal{L}(X_n) \overset{c}{\sim} \mathcal{L}(X^*_n)$, we say that the summands X_{nk} are *asymptotically independent.* The foregoing equivalence and comparison theorems yield conditions for asymptotic independence upon replacing g' by f and G' by F. We can thus transform the results of the investigation of the Central Limit Problem in the case of independence. Furthermore, we use the conditioning by the vector Z''_{nk}. It is easily seen that, because of the independence assumption, it reduces to conditioning by $X_{n0} + \cdots + X_{n,k-1}$. As an example, let us give a first extension of Liapounov's theorem.

Let $X_{nk} = X_k/s_n$ with $EX_k = 0$, $s_n^2 = \sum_k \sigma^2 X_k$, $k \leqq n$. The conditioning is by $X_1 + \cdots + X_{k-1}$.

B. UNDER LIAPOUNOV'S CONDITION

$$\frac{1}{s_n^{2+\delta}} \sum_k E|\, X_k\,|^{2+\delta} \to 0,$$

if

$$\frac{1}{s_n} \sum_k E|\, E'X_k\,| \to 0 \quad \textit{and} \quad \frac{1}{s_n^2} \sum_k E|\, E'X_k^2 - EX_k^2\,| \to 0,$$

then $\mathfrak{L}(\sum_k X_k/s_n) \to \mathfrak{N}(0, 1)$.

It suffices to apply the comparison theorem with $l = 0$, $m = 2$, and $S = R$ to X_{nk} and $Y_{nk} = X^*_{nk}$, use the inequality

$$E\int |\, x\,|^{2+\delta} |\, dF'_{nk} - dF_{nk}\,| \leqq EE'|\, X_{nk}\,|^{2+\delta} + E|\, X_{nk}\,|^{2+\delta} = 2E|\, X_{nk}\,|^{2+\delta},$$

and apply Liapounov's theorem to the X^*_{nk}.

SECOND APPROACH. We can obtain *directly* results which even in the case of independence are more general than those we obtained for the Central Limit Problem (since they pertain to the more general Central Asymptotic Problem):

For every fixed n, the comparison summands Y_{nk} are selected so that

—*the Y_{nk} are independent and $\mathfrak{L}(Y_n)$ belongs to the family of limit laws we seek to obtain*

—*the sets $\{Y_{nk}\}$ and $\{X_{nk}\}$ are independent.*

As an example, let us prove a Lindeberg type of normal convergence. Let the X_{nk} be centered at their conditional expectations so that $EX_{nk} = EE'X_{nk} = 0$, and set $\sigma_{nk}^2 = EX_{nk}^2$, $\sigma'^2_{nk} = E'X_{nk}^2$.

C. UNDER LINDEBERG'S CONDITION:

(i) $$\sum_k \int_{|x| \geqq \epsilon} x^2\, dF_{nk} \to 0 \quad \textit{for every} \quad \epsilon > 0, \quad \textit{and}$$

(ii) $$\sigma_n^2 = \sum_k \sigma_{nk}^2 \leqq \sigma^2 < \infty \textit{ for every } n,$$

if

(iii) $$\sum_k E|\, \sigma'^2_{nk} - \sigma_{nk}^2\,| \to 0,$$

then $\mathfrak{L}(X_n) \overset{c}{\sim} \mathfrak{N}(0, \sigma_n^2)$.

Proof. Since, by (i), as $n \to \infty$ and then $\epsilon \to 0$,

$$\max_k \sigma_{nk}^2 \leqq \epsilon^2 + \sum_k \int_{|x| \geqq \epsilon} x^2 \, dF_{nk} \to 0,$$

it follows, by (ii), that

$$\text{(1)} \qquad \sum \sigma_{nk}^3 \leqq \sigma^2 \max_k \sigma_{nk} \to 0.$$

Take the summands Y_{nk} to be mutually independent and normal $\mathfrak{N}(0, \sigma_{nk}^2)$, and take $S = (-\epsilon, +\epsilon)$, $l = \gamma = 1$, $m = 2$, $\delta = 1$. The comparison conditions for $j = 1$ and 2 are fulfilled, the first because of the centering and the second because of (iii) and because the condition for $l + \gamma$ is fulfilled by (i) and (1) since

$$\sum_k E \int_{|x| \geqq \epsilon} x^2 | \, dF'_{nk} - dG'_{nk} \, | \leqq \sum_k \int_{|x| \geqq \epsilon} x^2 \, dF_{nk} + c\epsilon^{-1} \sum_k \sigma_{nk}^3 \to 0.$$

Finally, the condition for $m + \delta$ is fulfilled by (ii) and (1), since

$$\sum_k E \int_{|x| < \epsilon} | \, x \, |^{2+\delta} | \, dF'_{nk} - dG_{nk} \, | \leqq 2\epsilon^\delta \sigma^2$$

and $\epsilon > 0$ is arbitrarily small. The theorem is proved.

The reader may proceed in a similar fashion and obtain or extend other results of the case of independence.

***31.3. Weighted prob. laws.** The second approach outlined in the preceding subsection yields the same prob. laws as in the case of independence. However, as we shall see, under similar but less restrictive conditions, disappearance of independence brings forth not the same prob. laws but their "weighted averages." The conditional law of a r.v. X given a sub σ-field $\mathcal{B}$ of events is defined by the conditional d.f. $F^{\mathcal{B}}$ or the conditional ch.f. $f^{\mathcal{B}}$. The d.f. F and the ch.f. f are then given by

$$F = EF^{\mathcal{B}}, \quad f = Ef^{\mathcal{B}}.$$

If the conditioning σ-field $\mathcal{B}$ is induced by some measurable function V not necessarily finite, nor even necessarily numerical, then, denoting by W the pr. distribution of V, we write

$$F = \int F^v \, dW(v), \quad f = \int f^v \, dW(v).$$

We say that W is the *weight function* of the *parameter* V and F (or f) represents the *weighted law* over the family F^v (or f^v) of laws.

Examples

1° A weighted law over the family of degenerate laws is of the form

$$f_W(u) = \int e^{iu\alpha}\, dW(\alpha)$$

where W on R is a d.f. In other words, if w is the ch.f. corresponding to W, we simply have $f_W = w$. It follows that, if the only "weighted" parameter is the shift-parameter, that is, the family of laws consists of $f_\alpha(u) = e^{iu\alpha}f(u)$, $\alpha \in R$, then

$$f_W(u) = \int e^{iu\alpha}f(u)\, dW(\alpha) = w(u)f(u),$$

and the "weighting" over the family reduces to the composition of a law with that represented by f. In other words, the weighting of the shift-parameter alone reduces to the composition of two laws and presents no new interest.

2° The limit laws which emerged in the development of pr. theory are the normal, Poisson, and, more generally, the infinitely decomposable laws. The corresponding weighted laws are

weighted normal:

$$f_W(u) = \int \exp\left[iu\alpha - \frac{\sigma^2 u^2}{2}\right] dW(\alpha, \sigma^2)$$

weighted Poisson:

$$f_W(u) = \int_0^\infty \exp\,[\lambda(e^{iu} - 1)]\, dW(\lambda)$$

weighted infinitely decomposable:

$$f_W(u) = \int \exp\left[iu\alpha + \int g(x, u)\, d\Psi(x)\right] dW(\alpha, \Psi)$$

where $g(x, u) = \left(e^{iux} - 1 - \dfrac{iux}{1 + x^2}\right)\dfrac{1 + x^2}{x^2}$ and the functions Ψ on R are nondecreasing, continuous from the left, and of bounded variation.

If W degenerates at some element (α, σ^2) or (λ) or (α, Ψ), then we get back the corresponding nonweighted laws. A systematic investigation, with restrictions on α, say α constant (since any law is a weighted degenerate), would be of great interest. We say only a few words about the weighted symmetric stable laws.

A *weighted symmetric stable* is defined by

$$f_W(u) = \int_0^\infty \exp\,[-c|\,u\,|^\gamma/2]\,dW(c), \quad 0 < \gamma \leqq 2.$$

It is a Laplace-Stieltjes transform in $|\,u\,|^\gamma$. Hence, on account of the known properties of such transforms,

a. *There is a one-to-one correspondence between a weighted symmetric stable and the weight d.f. defined up to additive constants. In particular, a weighted symmetric stable reduces to a symmetric stable if, and only if, the weight function is a degenerate d.f. of a r.v.*

Furthermore, if $W_n \xrightarrow{w} W$ up to additive constants, then, by the extended Helly-Bray lemma and the fact that we can set $W_n(x) = 0$ for $x < 0$,

$$f_{W_n}(u) = \int_0^\infty \exp\,[-c|\,u\,|^\gamma/2]\,dW_n(c)$$
$$\to \int_0^\infty \exp\,[-c|\,u\,|^\gamma/2]\,dW(c) = f_W(u).$$

Conversely, let $f_{W_n} \to g$. By the weak compactness theorem, there is a d.f. W and a subsequence $W_{n'} \xrightarrow{w} W$, so that, by what precedes, $g(u) = \int_0^\infty \exp\,[-c|\,u\,|^\gamma/2]\,dW(c)$. But by **a**, g determines W up to additive constants. Hence $W_n \xrightarrow{w} W$ up to additive constants and $g = f_W$. Thus

b. *The limit elements of a sequence of weighted symmetric stable are weighted symmetric stable with same exponent.*

Weighted stable laws appear in the case of sequences of exchangeable r.v.'s since, by 27.2, 2°, they are conditionally independent given a sub σ-field and 23.4 applies under this conditioning. In a different guise, weighted laws appear in the third approach where

The conditional laws $\mathfrak{L}'(Y_{nk})$ will be of the limit type obtained under similar conditions in the independence case.

We use the following notation:

$$\alpha'_{nk}(\epsilon) = \int_{|x|<\epsilon} x\,dF'_{nk}, \quad \alpha'_n(\epsilon) = \sum_k \alpha'_{nk}(\epsilon),$$

$$\sigma'^2_{nk}(\epsilon) = \int_{|x|<\epsilon} x^2\,dF'_{nk} - \left(\int_{|x|<\epsilon} x\,dF'_{nk}\right)^2, \quad \sigma'^2_n(\epsilon) = \sum_k \sigma'^2_{nk}(\epsilon);$$

we drop the primes if F' is replaced by F, and drop ϵ if $\epsilon = +\infty$.

Before we attack the extension of the more general i.d. case, let us give, as an example, the extension of the historically important Liapounov's theorem.

A. *Let the X_{nk} be centered at their conditional expectations. If*

$$\sum_k E|X_{nk}|^{2+\delta} \to 0$$

for a $\delta > 0$, *then* $\mathcal{L}(X_n) \overset{c}{\sim} \mathcal{L}(\sum_k Y_{nk})$ *with* $\mathcal{L}'(Y_{nk}) = \mathfrak{N}(0, \sigma'^2_{nk})$.

Proof. Take the Y_{nk} to be conditionally normal $\mathfrak{N}(0, \sigma'^2_{nk})$, so that the law of Y_{nk} is the weighted symmetric normal $E\mathfrak{N}(0, \sigma'^2_{nk})$, and apply the comparison theorem with $S = R$, $l = 0$, $m = 2$, and $\delta \leqq 1$. The comparison conditions for $j = 1, 2$ are fulfilled, since the corresponding sums vanish, the first because of the centering and the second because of $E'Y_{nk}^2 = E'X_{nk}^2$. The condition with $m + \delta$ is fulfilled, since

$$E|Y_{nk}|^{2+\delta} = EE'|Y_{nk}|^{2+\delta} = cE\sigma'^{2+\delta}_{nk} \leqq cEE'|X_{nk}|^{2+\delta}$$
$$= cE|X_{nk}|^{2+\delta}$$

and hence

$$\sum_k E\int |x|^{2+\delta}|dF'_{nk} - dG'_{nk}| \leqq (1 + c)\sum_k E|X_{nk}|^{2+\delta} \to 0.$$

The theorem is proved.

REMARK. If we add the hypothesis that the d.f. W_n of σ'^2_n converges weakly to W, then the sequence $\mathcal{L}(X_n)$ converges to a weighted symmetric normal. This limit law is that of a r.v. if, and only if, $W_n \overset{c}{\to} W$, and then it is normal if, and only if, W degenerates. Similar considerations apply to what follows.

We pass now to the limit weighted i.d. laws. We require the notion of *conditional uniform asymptotic negligibility*, for short *uan'*, defined by

$$\max_k P'[|X_{nk}| \geqq \eta] \overset{P}{\to} 0 \quad \text{for every} \quad \eta > 0.$$

In the case of independent summands, the uan' condition reduces to the uan condition and in the general case implies it, since, by the dominated convergence theorem,

$$\max_k P[|X_{nk}| \geqq \eta] \leqq E(\max_k P'[|X_{nk}| \geqq \eta]) \to 0.$$

c. *Under uan' condition, for every* $\epsilon > 0$

(i) $$\max_k \int_{|x|<\epsilon} |x|^s \, dF'_{nk} \xrightarrow{P} 0 \quad \text{for} \quad s > 0;$$

(ii) $$\max_k \int_{|x|<\epsilon} |x - \alpha'_{nk}(\epsilon)|^s \, dF'_{nk} \xrightarrow{P} 0 \quad \text{for} \quad s \geqq 1;$$

(iii) $$\max_k \gamma'_{nk}(u) = \max_k \int (e^{iu(x-\alpha'_{nk}(\epsilon))} - 1) \, dF'_{nk} \xrightarrow{P} 0.$$

Proof. Let $0 < \eta < \epsilon$. Since

$$\int_{|x|<\epsilon} |x|^s \, dF'_{nk} \leqq \eta^s + \epsilon^s P'[|X_{nk}| \geqq \eta]$$

assertion (i) follows by taking $\max_k$ and letting $n \to \infty$ and then $\eta \to 0$. Assertion (ii) follows, on account of (i), from

$$\int_{|x|<\epsilon} |x - \alpha'_{nk}(\epsilon)|^s \, dF'_{nk} \leqq 2^{s-1} \int_{|x|<\epsilon} |x|^s \, dF'_{nk} + 2^{s-1} |\alpha'_{nk}(\epsilon)|^s$$

and
$$|\alpha'_{nk}(\epsilon)|^s \leqq \epsilon^{s-1} |\alpha'_{nk}(\epsilon)|.$$

Finally, assertion (iii) follows, on account of (ii), from

$$|\gamma'_{nk}(u)| \leqq 2 \int_{|x|\geqq\epsilon} dF'_{nk} + |u| \int_{|x|<\epsilon} |x - \alpha'_{nk}(\epsilon)| \, dF'_{nk}.$$

Let
$$\log g'_{nk}(u) = iu\alpha'_{nk}(\epsilon) + \gamma'_{nk}(u).$$

d. $\sum_k E|f'_{nk}(u) - g'_{nk}(u)| \leqq c \sum_k E|\gamma'_{nk}(u)|^2.$

This follows (upon dropping the subscripts n, k) by $|\gamma'| \leqq 2$, hence $e^{|\gamma'|} \leqq e^2$, from

$$|f'(u) - g'(u)| = |e^{-iu\alpha'(\epsilon)} f'(u) - e^{\gamma'(u)}| = |1 + \gamma'(u) - e^{\gamma'(u)}|$$
$$\leqq \tfrac{1}{2} |\gamma'(u)|^2 e^{|\gamma'(u)|} \leqq c |\gamma'(u)|^2.$$

B. *Under uan', if for every* n

(i) $$\sum_k \int_{|x|\geqq\epsilon} dF_{nk}(x) \leqq c' < \infty$$

and

(ii) $$\sum_k E\sigma'^2_{nk}(\epsilon) \leqq c'' < \infty,$$

then $\mathcal{L}(\sum_k X_{nk}) \overset{c}{\sim} \mathcal{L}(\sum_k Y_{nk})$ *where the summands* Y_{nk} *are conditionally i.d. with ch.f.'s* g'_{nk}.

Proof. On account of (i), condition (ii) is equivalent to

$$\text{(iii)} \qquad \sum_k E\int_{|x|<\epsilon} (x - a'_{nk}(\epsilon))^2\, dF'_{nk}(x) \leqq c''' < \infty,$$

since

$$\sum_k E\left|\int_{|x|<\epsilon} (x - a'_{nk}(\epsilon))^2\, dF'_{nk}(x) - \sigma'^2_{nk}(\epsilon)\right|$$

$$= \sum_k Ea'^2_{nk}(\epsilon)\int_{|x|\geqq\epsilon} dF'_{nk}(x) \leqq \epsilon^2 c'.$$

Therefore, upon substituting on $(-\epsilon, +\epsilon)$ the limited expansion of order 2 of the integrand in $\gamma'_{nk}(u)$,

$$\sum_k E|\gamma'_{nk}(u)| \leqq (2 + \epsilon|u|)\sum_k \int_{|x|\geqq\epsilon} dF_{nk}$$

$$+ \frac{u^2}{2}\sum_k E\int_{|x|<\epsilon} (x - a'_{nk}(\epsilon)^2\, dF'_{nk}$$

$$\leqq (2 + \epsilon|u|)c' + \frac{u^2}{2}c''$$

so that, by **c**,

$$\sum_k E|\gamma'_{nk}(u)|^2 \leqq \max_k |\gamma'_{nk}(u)| \sum_k E|\gamma'_{nk}(u)| \overset{P}{\to} 0.$$

But the left-hand side sum is a number and hence converges to 0. Thus, by **d**,

$$\sum_k E|f'_{nk}(u) - g'_{nk}(u)| \to 0,$$

the comparison theorem in terms of ch.f.'s applies and, hence, $\mathcal{L}(\sum_k X_{nk}) \overset{c}{\sim} \mathcal{L}(\sum_k Y_{nk})$, where the summands Y_{nk} are conditionally i.d. with ch.f. g'_{nk} and mutually independent. The theorem follows.

REMARK. In the case of independence it can be shown that, under uan condition, (i) and (ii) hold when the sequences $\mathcal{L}(X_n)$ or $\mathcal{L}(Y_n)$ are completely compact so that, then, $\mathcal{L}(X_n) \overset{c}{\sim} \mathcal{L}(Y_n)$. This extends the Central Limit theorem. The proof is left to the reader.

RANDOM VECTORS. The extension to random vectors X_{nk} can be obtained as usual either by reinterpreting the symbols used or by making correspond to the random vectors X_{nk} the r.v.'s vX_{nk}—scalar products of X_{nk} and of an undetermined sure vector v.

Random number of r.v.'s. Let the number ν_n of summands in the nth sum $\sum_{k=1}^{\nu_n} X_{nk}$ be a r.v. Set

$$p_n(r) = P[\nu_n = r], \quad P_n(s) = \sum_{r \geqq s} p_n(r)$$

and denote by E_r the conditional expectation given $\nu_n = r$. Assume that the expressions below exist and are finite—they certainly do if $E\nu_n < \infty$. Then

$$\begin{aligned} f_n(u) - g_n(u) &= E(\exp[iu \sum_k X_{nk}] - \exp[iu \sum_k Y_{nk}]) \\ &= \sum_{r=1}^{\infty} p_n(r) E_r(\exp[iu \sum_{k=1}^{r} X_{nk}] - \exp[iu \sum_{k=1}^{r} Y_{nk}]). \end{aligned}$$

But, when all the expectations are conditioned by $\nu_n = r$, then the comparison theorem applies. Hence,

$$E_r(\exp[iu \sum_1^r X_{nk}] - \exp[iu \sum_1^r Y_{nk}]) \leqq \sum_{k=1}^{r} E_r |f'_{nk}(u) - g'_{nk}(u)|$$

and

$$\begin{aligned} |f_n - g_n| &\leqq \sum_{r=1}^{\infty} p_n(r) \{ \sum_{k=1}^{r} E_r |f'_{nk} - g'_{nk}| \} \\ &= \sum_{k=1}^{\infty} \{ \sum_{r \geqq k} p_n(r) E_r \} |f'_{nk} - g'_{nk}|. \end{aligned}$$

Write E_{nk} for the operator $\sum_{r \geqq k} p_n(r) E_r$. The relation becomes

$$|f_n - g_n| \leqq \sum_{k=1}^{\infty} E_{nk} |f'_{nk} - g'_{nk}|,$$

and, hence,

C. *When the number of summands is random, the results obtained by using the comparison theorem remain valid provided* $\sum_k E$ *is replaced by* $\sum_r p_n(r) \sum_{k=1}^{r} E_r$ *or by* $\sum_k E_{nk}$.

If ν_n is independent of the X_{nk} and Y_{nk}, then $E_r = E$ and hence $E_{nk} = P_n(k)E$, and it suffices to multiply F'_{nk} and G'_{nk} by $P_n(k)$. If, moreover, ν_n degenerates at k_n, then $P_n(k) = 1$ or 0, according as $k \leqq k_n$ or $k > k_n$, and, as is to be expected, we fall back on sure number k_n of summands.

§ 32. CENTERINGS, MARTINGALES, AND A.S. CONVERGENCE

32.1. Centerings. Conditions for a.s. convergence (and a.s. stability) of sums of independent r.v.'s were obtained by means of centerings at expectations or at medians. The methods continue to apply to the general case, provided the centering quantities are conditioned and, thus, become themselves r.v.'s. Furthermore, as has to be expected, the conditions so obtained will be sufficient but no more necessary. Since the proofs run parallel to those in the case of independence, we shall be content with essentials and shall leave the complete transcription of Chapter V to the reader.

Centering at conditional medians. We say that a r.v. $\mu^{\mathcal{B}}X$ is a *conditional median* of X given $\mathcal{B}$, where $\mathcal{B}$ is a sub σ-field of events, if

$$P^{\mathcal{B}}[X - \mu^{\mathcal{B}}X \geqq 0] \geqq \tfrac{1}{2} \leqq P^{\mathcal{B}}[X - \mu^{\mathcal{B}}X \leqq 0] \text{ a.s.}$$

When independence is not assumed, the proof of inequality 17.1**C** breaks down at the point where PA_kB_k is replaced by PA_kPB_k. Yet, if we observe that $PA_kB_k = E\{I_{A_k}P(B_k \mid S_1, \cdots, S_k)\}$ and replace medians $\mu(S_k - S_n)$ by conditional medians $\mu(S_k - S_n \mid S_1, \cdots, S_k)$, then the proof remains valid. Thus

A. EXTENDED P. LÉVY INEQUALITY. *If the sums S_k are centered at conditional medians $\mu(S_k - S_n \mid S_1, \cdots, S_k)$, then*

$$P[\max_{k \leqq n} \mid S_k \mid \geqq \epsilon] \leqq 2P[\mid S_n \mid \geqq \epsilon].$$

The propositions in 17.2 which result from P. Lévy's inequality continue to hold with similar modifications. Let us state the most important one.

B. CONVERGENCE THEOREM. *If the sequence of sums $S_n \xrightarrow{P} S$, then there exists a sequence ξ_n of conditional medians of suitably selected partial sums such that $\xi_n \xrightarrow{P} 0$ and $S_n - \xi_n \xrightarrow{\text{a.s.}} S$.*

REMARK. Propositions much more similar to those of the case of independence are obtainable by means of centerings at conditional expectations and, as we shall see in the next subsection, such centerings provide an important dependence model—of "martingales," which is a "natural" generalization of that of consecutive sums of independent r.v.'s centered at expectations. Yet the power of the centerings at medians accompanying symmetrizations in the case of independence leads one to think that it would be of interest to investigate in detail the dependence model that such centerings provide.

Centering at conditional expectations. We suppose that the r.v.'s X_n are integrable so that c.exp.'s $\xi_n = E(X_n \mid X_1, \cdots, X_{n-1})$ exist and are finite; for $n = 1$ the conditioning disappears and $\xi_1 = EX_1$ a.s. We have $E\xi_n = EX_n$ and

$$E(\xi_n \mid X_1, \cdots, X_{n-1}) = \xi_n \text{ a.s.}$$

Therefore, for $m < n$,

$$E\{(X_m - \xi_m)(X_n - \xi_n) \mid X_1, \cdots, X_{n-1}\}$$
$$= (X_m - \xi_m)E(X_n - \xi_n \mid X_1, \cdots, X_{n-1}) = 0 \text{ a.s.,}$$

so that

$$E(X_m - \xi_m)(X_n - \xi_n) = 0$$

and, hence,

$$E\left\{\sum_{k=1}^{n} (X_k - \xi_k)\right\}^2 = \sum_{k=1}^{n} E(X_k - \xi_k)^2.$$

We say that the r.v.'s X_n of a sequence are *centered at c.exp.'s given the predecessors*, if $\xi_n = 0$ a.s. (Such centerings were first systematically used by P. Lévy.) Thus

a. EXTENDED BIENAYMÉ EQUALITY. *If the r.v.'s X_n of a sequence are centered at c.exp.'s given the predecessors, then they are centered at exp.'s and*

$$\sigma^2 S_n = \sum_{k=1}^{n} \sigma^2 X_k.$$

In fact, more is true. If $\xi_n = 0$ a.s. and $A_{n-1} \in \mathcal{B}(X_1, \cdots, X_{n-1})$ is an event defined in terms of $X_1, \cdots, X_{n-1}$, then

$$E(S_{n-1}I_{A_{n-1}} X_n \mid X_1, \cdots, X_{n-1}) = S_{n-1}I_{A_{n-1}} E(X_n \mid X_1, \cdots, X_{n-1})$$
$$= 0 \text{ a.s.}$$

and, hence,

$$E(S_{n-1}I_{A_{n-1}} X_n) = 0.$$

Because of this orthogonality property, the proof of the right-hand side of Kolmogorov's inequality remains valid word for word. Thus

C. EXTENDED KOLMOGOROV INEQUALITY. *If the r.v.'s X_k, $k = 1, \cdots n$, are centered at c.exp.'s given the predecessors, then*

$$P[\max_{k \leq n} \mid S_k \mid \geq \epsilon] \leq \frac{1}{\epsilon^2} \sum_{k=1}^{n} \sigma^2 X_k.$$

The propositions in 16.3 which result from Kolmogorov's inequality and those in 16.4 hold with similar modifications. Let us state the most important ones.

D. CONVERGENCE THEOREM. *If the series* $\sum \sigma^2 X_n$ *converges and the series* $\sum \xi_n$ *converges a.s., then the series* $\sum X_n$ *converges a.s.*

More generally, if for some positive constant c the series $\sum P[|X_n| \geqq c]$ *and* $\sum E\{X_n^c - E(X_n^c \mid X_1, \cdots, X_{n-1})\}^2$ *converge and the series* $\sum E(X_n^c \mid X_1, \cdots, X_{n-1})$ *converges a.s., then the series* $\sum X_n$ *converges a.s.*

E. STABILITY THEOREM. *If* $\sum \frac{\sigma^2 X_n}{b_n^2} < \infty$ *with* $b_n \uparrow \infty$, *then*

$$\frac{1}{b_n} \sum_{k=1}^{n} \{X_k - E(X_k \mid X_1, \cdots, X_{k-1})\} \xrightarrow{\text{a.s.}} 0.$$

Let X be a r.v. and let x vary on $[0, +\infty)$. *If* $E|X|^r < \infty, r < 2$, $P[|X_n| \geqq x] \leqq P[|X| \geqq x]$ *or* $P\{|X_n| \geqq x \mid X_1, \cdots, X_{n-1}\} \leqq P\{|X| \geqq x \mid X_1, \cdots, X_{n-1}\}$ *a.s., according as* $r \neq 1$ *or* $r = 1$, *then*

$$\frac{1}{n^{1/r}} \sum_{k=1}^{n} (X_k - \eta_k) \xrightarrow{\text{a.s.}} 0$$

with $\eta_k = 0$ *or* $E(X_k \mid X_1, \cdots, X_{k-1})$ *according as* $0 < r < 1$ *or* $1 \leqq r < 2$.

Let the r.v.'s X_n of a sequence be centered at c.exp.'s given the predecessors. Since $S_1 = X_1, \cdots, S_n = X_1 + \cdots + X_n$ determine and are determined by $X_1, \cdots, X_n$, it follows that

$$\begin{aligned} E(S_n \mid S_1, \cdots, S_{n-1}) &= E(S_{n-1} + X_n \mid S_1, \cdots, S_{n-1}) \\ &= S_{n-1} + E(X_n \mid X_1, \cdots, X_{n-1}) = S_{n-1} \text{ a.s.} \end{aligned}$$

This property of the sequence S_n is called a "martingale" property. Conversely, if a sequence S_n has the martingale property, then setting $X_n = S_n - S_{n-1} (S_0 = 0)$, we have

$$\begin{aligned} E(X_n \mid X_1, \cdots, X_{n-1}) &= E(S_n - S_{n-1} \mid S_1, \cdots, S_{n-1}) \\ &= S_{n-1} - S_{n-1} = 0 \text{ a.s.} \end{aligned}$$

Thus, the martingale property characterizes consecutive sums of r.v.'s centered at c.exp.'s given the predecessors. Since we are interested in a.s. properties of such sums, it is "natural" to investigate them directly without writing them as sums.

32.2. Martingales: generalities. A possible interpretation of a "fair game" is as follows: Let X_t represent the debt or fortune of a gambler at time s. The game is fair if the gambler's expected fortune at time t, given the past up to the time $s < t$, equals his fortune at time s. To this interpretation corresponds the concept of martingale. It has been introduced and investigated in the form of consecutive sums by P. Lévy, then studied by Ville and systematically explored by Doob—to whom most of the results are due—and, finally, extended to "advantageous games" or submartingales by Doob and, in a different formulation, by Andersen and Jessen.

In this section we assume that, unless otherwise stated, *the expectations of the r.v.'s under consideration exist*, and denote by

$$\mathcal{B}_n = \mathcal{B}(X_1, \cdots, X_n), \quad \mathcal{C}_n = \mathcal{B}(X_n, X_{n+1}, \cdots),$$

$$\mathcal{B} = \mathcal{C}_1 = \mathcal{B}(X_1, X_2, \cdots), \quad \mathcal{C} = \bigcap \mathcal{C}_n,$$

the sub σ-fields of events induced by the families of r.v.'s $(X_1, \cdots, X_n)$, $(X_n, X_{n+1}, \cdots)$, $(X_1, X_2, \cdots)$ and the tail of the sequence $\{X_n\}$, respectively.

Definitions. Let $\{X_t, t \in T\}$ be a family of r.v.'s on a set T ordered by the relation "$\prec$," and let $\mathcal{B}_t = \mathcal{B}\{X_{t'}, t' \prec t\}$ be the sub σ-field of events induced by the subfamily of all the $X_{t'}$ with $t' \prec t$.

The family is said to be a *martingale* if, for every pair $s \prec t$,

$$X_s = E^{\mathcal{B}_s} X_t \text{ a.s., equivalently, } \int_{B_s} X_s = \int_{B_s} X_t, \quad B_s \in \mathcal{B}_s.$$

The martingale is said to be *closed* on the left or on the right according as it has a first or a last member (it may have neither or both).

If in the foregoing definitions "$=$" is replaced by "$\leqq$," the family is said to be a *submartingale*. If the inequality sign is reversed, it is a *supermartingale*. Changing the X_t into $-X_t$ interchanges "sub" and "super."

Note that the X_t being r.v.'s, the above c.exp.'s are a.s. finite for martingales while their negative parts are a.s. finite for submartingales.

We intend to investigate submartingales $\{X_n, n = 1, 2, \cdots\}$. The subscripts are ordered either by the relation "$\leqq$" and then we have a *submartingale sequence* $X_1, X_2, \cdots$ (closed on the left by X_1), or by the relation "$\geqq$" and then we have a *submartingale reversed sequence* $\cdots X_2, X_1$ (closed on the right by X_1). Because of the basic smoothing property of c.exp.'s the foregoing definitions reduce as follows. The r.v.'s $X_n, n = 1, 2, \cdots$ form

a martingale sequence, if $X_n = E(X_{n+1} \mid X_1, \cdots, X_n)$ a.s.,
a closed (*by* X) *martingale sequence*, if $X_n = E(X_{n+1} \mid X_1, \cdots, X_n)$, $X_n = E(X \mid X_1, \cdots, X_n)$ a.s.,
a martingale reversed sequence, if $X_{n+1} = E(X_n \mid X_{n+1}, X_{n+2}, \cdots)$ a.s.
a closed (*by* X) *martingale reversed sequence*, if $X_{n+1} = E(X_n \mid X_{n+1}, X_{n+2}, \cdots, X)$ a.s., $X = E(X_n \mid X)$ a.s.

For example, in the first case, $\mathcal{B}_m \subset \mathcal{B}_{m+1} \subset \cdots \subset \mathcal{B}_{n-1}$ for $m < n$ and, by the basic smoothing property, we have

$$E^{\mathcal{B}_m} X_n = E^{\mathcal{B}_m} E^{\mathcal{B}_{m+1}} \cdots E^{\mathcal{B}_{n-1}} X_n = X_m \text{ a.s.};$$

similarly in the other cases.

If, in the foregoing relations, "=" is replaced by "$\leqq$," martingales become *submartingales.s.*

Examples

1° Let $X_n = \sum_{k=1}^{n} Y_k$, $n = 1, 2, \cdots$. If the r.v.'s Y_k are independent with $EY_k = 0$, or dependent with $E(Y_k \mid Y_1, \cdots, Y_{k-1}) = 0$ a.s. for $k > 1$, then, according to 29.1, the X_n form a martingale sequence.

2° Let $\mathcal{A}_1, \mathcal{A}_2, \cdots$ be a sequence of sub σ-fields of events and let every X_n be $\mathcal{A}_n$-measurable.

If $\mathcal{A}_1 \subset \mathcal{A}_2 \subset \cdots$ and every $X_n = E^{\mathcal{A}_n} X_{n+1}$ a.s., then the X_n form a martingale sequence, since $\mathcal{B}_n \subset \mathcal{A}_n$ and, by the smoothing property,

$$E^{\mathcal{B}_n} X_{n+1} = E^{\mathcal{B}_n} E^{\mathcal{A}_n} X_{n+1} = X_n \text{ a.s.}$$

Similarly, if $\mathcal{A}_1 \subset \mathcal{A}_2 \subset \cdots$ and every $X_n = E^{\mathcal{A}_n} X$, then the X_n form a martingale sequence closed on the right by X. For example, for any r.v. X and random sequence Y_n, the sequence $E(X \mid Y_1, \cdots, Y_n)$ is such a martingale.

Similarly, if $\mathcal{A}_1 \supset \mathcal{A}_2 \supset \cdots$ and every $X_n = E^{\mathcal{A}_n} X$ a.s., then the X_n form a martingale reversed sequence. For example, for any r.v. X and random sequence Y_n, the reversed sequence $\cdots E(X \mid Y_n, Y_{n+1}, \cdots) \cdots E(X \mid Y_2, Y_3, \cdots), E(X \mid Y_1, Y_2, \cdots)$ is a martingale.

Decomposition of submartingales. To simplify, we assume that the r.v.'s below are integrable, and leave to the reader the discussion of the case when their expectations exist but are not necessarily finite.

1° Let $X_1, X_2, \cdots$ be a sequence of r.v.'s and set $X''_1 = 0$

$$X_n = X'_n + X''_n, \quad X''_n = \sum_{k=2}^{n} \{E(X_k \mid X_1, \cdots, X_{k-1}) - X_{k-1}\}.$$

It follows that

$$E(X'_{n+1} \mid X_1, \cdots, X_n) = X'_n \text{ a.s.},$$

and hence

$$E(X'_{n+1} \mid X'_1, \cdots, X'_n, = X'_n \text{ a.s.}$$

Thus the sequence $X'_1, X'_2, \cdots$ is a martingale. In particular, if the sequence $X_1, X_2, \cdots$ is a submartingale, then every summand in X''_n is a.s. nonnegative. Therefore, a submartingale sequence X_n is decomposable into a martingale sequence X'_n and an a.s. nonnegative and nondecreasing sequence X''_n; more precisely,

$$X_n = X'_n + X''_n, \quad E(X'_{n+1} \mid X_1, \cdots, X_n) = X'_n \text{ a.s.}, \quad 0 \leqq X''_n \uparrow \text{ a.s.}$$

and, hence,

$$EX_n = EX'_n + EX''_n, \quad E|X'_n| \leqq E|X_n| + EX''_n, \quad 0 \leqq EX''_n \uparrow.$$

Let $\sup E|X_n| < \infty$. Then, it follows that X'_n and X''_n are integrable and $\sup E|X'_n| < \infty$, $\sup EX''_n < \infty$. Thus $0 \leqq X''_n \uparrow X''$ a.s. finite, and the study of the convergence of the submartingale sequence reduces to that of the martingale sequence X'_n with $\sup E|X'_n| < \infty$. Moreover, the limits, if any, differ by an integrable r.v.

2° Similarly, let the reversed sequence $\cdots X_2, X_1$ be a submartingale with EX_1 finite and set

$$X_n = X'_n + X''_n, \quad X''_n = \sum_{k=n}^{\infty} \{E(X_k| X_{k+1}, X_{k+2}, \cdots) - X_{k+1}\}.$$

The summands of the infinite sum are a.s. nonnegative, so that a.s. $0 \leqq X''_n \downarrow$ with $EX''_1 = EX_1 - \lim EX_n$ (the limit exists since, clearly, $EX_1 \geqq EX_2 \geqq \cdots$). Let $\lim EX_n > -\infty$ so that EX''_1 is finite. Then X''_1 is a.s. finite, $0 \leqq X''_n \downarrow 0$ a.s., and the study of the convergence of the submartingale reversed sequence reduces to that of the martingale reversed sequence $\cdots X'_2, X'_1$ with $E|X'_1| < \infty$; moreover, the limit, if any, is the same.

The interpretation of a martingale as a sequence of fortunes of a gambler raises the question whether in the long run ($n \to \infty$) his fortune was or becomes stabilized, that is, whether there is convergence—in some sense. To answer the question we require a few inequalities.

a. *Let g be convex and continuous on R with $g(+\infty) = +\infty$. If EX exists and $E^{\mathfrak{B}}X > -\infty$ a.s., then $g(E^{\mathfrak{B}}X) \leqq E^{\mathfrak{B}}g(X)$ a.s.*

For, if $E^{\mathfrak{B}}X < \infty$ a.s. the conditional convexity inequality applies; otherwise apply it to $X_n = XI_{[X<n]} + nI_{[X \geqq n]}$ and let $n \to \infty$.

A. SUBMARTINGALE INEQUALITIES. *Let the r.v.'s X_j form a countable submartingale. Then*

(i) *the r.v.'s X_j^+ form a submartingale; and if the $X_j \geqq 0$ a.s. or the X_j form a martingale, then, for every $r \geqq 1$, the $|X_j|^r$ form a submartingale.*

(ii) *if the r.v. Y closes on the right the submartingale, then, for every $c > 0$,*

$$cP[\sup X_j > c] \leqq \int_{[\sup X_j > c]} Y;$$

and if the $X_j \geqq 0$ a.s. or the X_j form a martingale then, for every $r \geqq 1$,

$$c^r P[\sup |X_j| > c] \leqq \int_{[\sup |X_j| > c]} |Y|^r.$$

Proof. (i) follows from **a** by taking, respectively, $g(x) = x^+$, or $g(x) = 0$ for $x < 0$ and $g(x) = x^r$ for $x \geqq 0$, or $g(x) = |x|^r$.

To prove (ii), set $A_j = [X_j > c$, the predecessors $\leqq c]$, so that $B = [\sup X_j > c] = \sum A_j$ and, since Y closes the submartingale,

$$\int_B Y = \sum \int_{A_j} Y = \sum \int_{A_j} E(Y \mid X_j \text{ and the predecessors})$$

$$\geqq \sum \int_{A_j} X_j \geqq \sum c\, PA_j = cPB,$$

so that, the first inequality is proved and the second follows on account of (i).

32.3. Martingales: convergence and closure. The limit properties of submartingales are summarized in the convergence theorem below. The proof is based on an ingenious inequality due to Doob.

Let x_k, $k = 1, \cdots, n$, be finite numbers. The number h of *crossings* from the left of the interval $[a, b]$ is the number of times that, starting with x_1 and proceeding to x_n, we pass from the left of the interval to its right. More precisely, let

$$x_{k_1} \leqq a, \quad x_{k_2} \geqq b, \quad x_{k_3} \leqq a, \quad x_{k_4} \geqq b, \cdots$$

where k_1 is the first subscript k, if any, such that $x_{k_1} \leqq a$, then k_2 is the first subscript $k > k_1$, if any, such that $x_{k_2} \geqq b$, and so on. If k_{j_0} is the last subscript so obtained, set $k_j = n + 1$ for $j_0 < j \leqq n$; if there is none, then $k_1 = \cdots = k_n = n + 1$. Thus, to every $k > k_1$, if any, there corresponds an integer j determined by the values of $x_1, \cdots, x_{k-1}$ and such that $k_j < k \leqq k_{j+1}$. For $k > 1$, if $k \leqq k_1$ set $i_k = 0$, and if $k > k_1$ set $i_k = 0$ or 1 according as the corresponding j is odd or even. When $k_2 \leqq n$, the number of crossings is the largest integer h such that $k_{2h} \leqq n$; when $k_2 > n$, the number of crossings $h = 0$.

Let $h > 0$. If $k_{2h+1} \leqq n$, then

$$\sum_{k=2}^{n} i_k(x_k - x_{k-1}) = (x_{k_3} - x_{k_2}) + \cdots + (x_{k_{2h+1}} - x_{k_{2h}}) \leqq (a - b)h.$$

If $k_{2h+1} > n$, then

$$\sum_{k=2}^{n} i_k(x_k - x_{k-1}) = (x_{k_3} - x_{k_2}) + \cdots + (x_{k_{2h-1}} - x_{k_{2h-2}}) + (x_n - x_{k_{2h}})$$
$$\leqq (a - b)h + (x_n - a).$$

Let $h = 0$. Then the left-hand sum is null, and the first inequality is trivially true.

Thus, in either case

$$\sum_{k=2}^{n} i_k(x_k - x_{k-1}) \leqq (a - b)h + (x_n - a)^+.$$

Now, to r.v.'s $X_1, \cdots, X_n$ we make correspond a r.v. H_n and r.v.'s $I_k(k > 1)$ determined by $X_1, \cdots, X_{k-1}$. We define them by $H_n(\omega) = h$, $I_k(\omega) = i_k$ for $x_1 = X_1(\omega), \cdots, x_n = X_n(\omega)$, $\omega \in \Omega$, where h and i_k are the numbers introduced above. The inequality established above becomes

$$\sum_{k=2}^{n} I_k(X_k - X_{k-1}) \leqq (a - b)H_n + (X_n - a)^+$$

and, by taking expectations assumed finite, we have

$$\sum_{k=2}^{n} \int_{[I_k=1]} (X_k - X_{k-1}) \leqq (a - b)EH_n + E(X_n - a)^+.$$

If $X_1, \cdots, X_n$ is an integrable submartingale, then every left-hand integral is nonnegative and, hence,

$$(b - a)EH_n \leqq E(X_n - a)^+ = \sup_{k \leq n} E(X_k - a)^+.$$

If $E(X_n - a)^+ = \infty$, the inequality is trivially true. If $E(X_n - a)^+ < \infty$, note that H_n is also the number of crossings of $[0, b - a]$ by the integrable submartingale $(X_1 \dot{-} a)^+, \cdots, (X_n - a)^+$. It follows that the inequality is always true.

Similarly, but proceeding from x_n to x_1 instead of from x_1 to x_n, if $X_n, \cdots, X_1$ is a submartingale, then

$$(b-a)EH_n \leqq E(X_1 - a)^+ = \sup_{k \leqq n} E(X_k - a)^+.$$

To summarize (see also 36.2)

a. *If $X_1, \cdots, X_n$ or $X_n, \cdots, X_1$ is a submartingale, then*

$$(b-a)EH_n \leqq \sup_{k \leqq n} E(X_k - a)^+.$$

We are now in a position to prove the basic

A. Submartingales convergence theorem. *Let the r.v.'s X_n form a submartingale sequence or reversed sequence.*

(i) *If* $\sup EX_n^+ < \infty$, *then* $X_n \xrightarrow{\text{a.s.}} X < \infty$ *with* $EX \leqq \sup EX_n^+$ *and* $E|X| \leqq \sup E|X_n|$.

(ii) $X_n \xrightarrow{r} X$ *where* $r \geqq 1$ *if, and only if, the* $|X_n|^r$ *are uniformly integrable, and then* $X_n \xrightarrow{\text{a.s.}} X$.

Proof. 1° Since

$$[X_n \not\to] = \bigcup_{a,b} A_{a,b} \quad \text{with} \quad A_{a,b} = [\liminf X_n < a < b < \limsup X_n],$$

where a, b vary over the denumerable set of all rationals, the divergence set is null if, and only if, every set $A_{a,b}$ is null.

We apply the foregoing lemma to an arbitrary set $A_{a,b}$. Since $H_n \uparrow H = \infty$ on $A_{a,b}$, this set is null, provided $P[H = \infty] = 0$; it will be so, whether the submartingale is a sequence or a reversed sequence, provided

$$EH = \sup EH_n \leqq \sup E(X_n - a)^+/(b-a) < \infty.$$

Therefore, $\sup EX_n^+ < \infty$ and hence $\sup E(X_n - a)^+ < \infty$, for every $a \in R$, imply that $X_n \xrightarrow{\text{a.s.}}$ some X finite or not, and, by the Fatou-Lebesgue theorem, $E|X| \leqq \sup E|X_n|$.

It follows that $X_n^+ \xrightarrow{\text{a.s.}} X^+$ and, by the same theorem, $EX^+ \leqq \sup EX_n^+ < \infty$. Thus, X^+ is integrable hence a.s. finite. Therefore, upon modifying if necessary X^+ hence X on a null set, we can take X^+ to be finite so that $X \leqq X^+ < \infty$. Also, EX exists and

$$EX \leqq EX^+ \leqq \sup EX_n^+.$$

The first assertion is proved.

2° If $X_n \xrightarrow{r} X$ for some $r \geqq 1$, then, by the L_r-convergence theorem, the $|X_n|^r$ are uniformly integrable. Conversely, let the $|X_n|^r$ be uniformly integrable for some $r \geqq 1$. Then the $E|X_n|^r$, *a fortiori* the $E|X_n| \leqq E^{1/r}|X_n|^r$, are uniformly bounded. Therefore, by (i), $X_n \xrightarrow{a.s.} X$ and, by the L_r-convergence theorem, $X_n \xrightarrow{r} X$. The second assertion is proved.

The foregoing convergence theorem yields

B. SUBMARTINGALES CLOSURE THEOREM. *Let* $r \geqq 1$.

(i) *Let* $\{X_n\}$ *be a martingale or a nonnegative submartingale sequence or reversed sequence. If* $Y \in L_r$ *closes it on the right, then* $X_n \underset{a.s.}{\xrightarrow{r}} X$. *If* $\sup E|X_n|^r < \infty$ *with* $r > 1$, *then such a* Y *exists.*

(ii) *Let* $\{X_n\}$ *be a (sub)martingale sequence or reversed sequence. If* $X_n \xrightarrow{r} X$, *then* $X_n \xrightarrow{a.s.} X$ *and* X *closes on the right, respectively, on the left the (sub)martingale; in fact,* X *is the nearest of the closing r.v.'s.*

Proof. 1° Let $Y \in L_r$ close $\{X_n\}$. Set $B_n = [|X_n| > c]$ so that $B = [\sup|X_n| > c = \bigcup B_n$, and use 29.2**A**. Since $c^r PB \leqq \int_B |Y|^r$

$\leqq E|Y|^r < \infty$ and $\int_{B_n} |X_n|^r \leqq \int_B |Y|^r$, it follows that, as $c \to \infty$,

$PB \to 0$, hence $\int_B |Y|^r \to 0$, and the $|X_n|^r$ are uniformly integrable.

Thus **A** applies, and $X_n \underset{a.s.}{\xrightarrow{r}} X$.

If $\sup E|X_n|^r < \infty$ with $r > 1$, then, by 9.4**C**, Cor. 2, and by **A**, $X_n \xrightarrow{1} X$. Since $E|X|^r \leqq \sup E|X_n|^r$ and, by (ii), X closes $\{X_n\}$, (i) is proved, provided we prove (ii);

2° Let the assumptions of (ii) hold. Then $X_n \xrightarrow{r} X$ implies that $X_n \xrightarrow{1} X$ and also, by **A**, that $X_n \xrightarrow{a.s.} X$. Thus we can pass to the limit under the integration sign, as follows:

In the submartingale sequence case, we have, for every $B_n \in \mathcal{B}_n$,

$$\int_{B_n} X_n \leqq \int_{B_n} X_{n+m},$$

and, by letting $m \to \infty$, we obtain

$$\int_{B_n} X_n \leqq \int_{B_n} X = \int_{B_n} E^{\mathcal{B}_n} X.$$

Therefore, $X_n \leqq E^{\mathcal{B}_n} X$ a.s., that is, the submartingale sequence is closed on the right by X. If a r.v. Y also closes the sequence on the right, that is, for every $B_n \in \mathcal{B}_n$,

$$\int_{B_n} X_{n+m} \leqq \int_{B_n} Y,$$

then, by letting $m \to \infty$, we obtain

$$\int_{B_n} X \leqq \int_{B_n} Y.$$

Therefore, on every $\mathcal{B}_n$ and hence on $\bigcup \mathcal{B}_n$, the indefinite integral of $Y - X$ (it exists since X is integrable) is a σ-finite measure and, by the extension theorem, determines a σ-finite measure on $\mathcal{B}$. Thus, the indefinite integral of $Y - X$ on $\mathcal{B}$ is nonnegative, that is, for every $B \in \mathcal{B}$,

$$\int_B X \leqq \int_B Y = \int_B E^{\mathcal{B}} Y.$$

Since X is equivalent to a $\mathcal{B}$-measurable function, it follows that $X \leqq E^{\mathcal{B}} Y$ a.s. and hence the submartingale $X_1, X_2, \cdots, X$ is closed on the right by Y; that is, X is the "nearest" of the closing r.v.'s.

Similarly, in the case of a submartingale reversed sequence, for every $C \in \mathcal{B}(X)$ ($\subset \mathcal{C}$ since X is $\mathcal{C}$-measurable), as $m \to \infty$,

$$\int_C X \leftarrow \int_C X_{n+m} \leqq \int_C X_n = \int_C E(X_n \mid X),$$

so that X is a closing r.v. on the left and if Y is another closing r.v., then for $C \in \mathcal{B}(Y)$

$$\int_C Y \leqq \int_C X_n \to \int_C X = \int_C E(X \mid Y)$$

so that Y closes on the left the submartingale $X, \cdots, X_2, X_1$. Finally, for martingales all foregoing inequalities become equalities. The proof is terminated.

Various cases. Let us put together the properties of the various types of martingales and submartingales which are contained in 29.2**A** and 29.3**A** and **B**. In what follows $r \geqq 1$.

I MARTINGALE SEQUENCE $X_1, X_2, \cdots$

Inequalities:

$$EX_1 = EX_2 = \cdots; \quad EX_1^+ \leqq EX_2^+ \leqq \cdots; \quad E|X_1|^r \leqq E|X_2|^r \leqq \cdots.$$

Convergence. *If* $\lim EX_n^+ < \infty$ *or* $\lim EX_n^- < \infty$, *then* $X_n \xrightarrow{\text{a.s.}} X < +\infty$ or $> -\infty$.

Closure. *The martingale is closed on the right by a r.v.* $Y \in L_r$ *if, and only if, the* $|X_n|^r$ *are uniformly integrable; then* $X_n \underset{r}{\xrightarrow{\text{a.s.}}} X$ *and* X *is the nearest of the closing r.v.'s. In particular, the martingale is closed by a r.v.* $\in L_r$ *when* $\lim E|X_n|^r < \infty$ *with* $r > 1$.

II SUBMARTINGALE SEQUENCE $X_1, X_2, \cdots$

Inequalities:

$$EX_1 \leqq EX_2 \leqq \cdots; \quad EX_1^+ \leqq EX_2^+ \leqq \cdots;$$
$$X_n \geqq 0 \text{ a.s.} \Rightarrow EX_1^r \leqq EX_2^r \leqq \cdots.$$

Convergence. *If* $\lim EX_n^+ < \infty$, *then* $X_n \xrightarrow{\text{a.s.}} X < \infty$. *If*

$$\sup E|X_n| < \infty,$$

in particular if either every $X_n \leqq 0$ a.s. *or every* $X_n \geqq 0$ a.s. *and* $\lim |EX_n| < \infty$, *then* $X_n \xrightarrow{\text{a.s.}} X$ *finite.*

Closure. *If the* $|X_n|^r$ *are uniformly integrable, then* $X_n \xrightarrow{\text{a.s.}} X \in L_r$ *and* X *is the nearest of closing r.v.'s. If every* $X_n \geqq 0$ a.s., *then the* X_n^r *are uniformly integrable, if, and only if, there is a closing on the right r.v.* $Y \in L_r$, *and there is one when* $\lim EX_n^r < \infty$ *with* $r > 1$.

III MARTINGALE REVERSED SEQUENCE $\cdots, X_2, X_1$

Inequalities:

$$\cdots = EX_2 = EX_1; \quad \cdots \leqq EX_2^+ \leqq EX_1^+; \quad \cdots \leqq E|X_2|^r \leqq E|X_1|^r.$$

Convergence. *If* $EX_1^+ < \infty$ *or* $EX_1^- < \infty$, *then* $X_n \xrightarrow{\text{a.s.}} X < \infty$ *or* $> -\infty$, *respectively.*

Closure. *If* $E|X_1|^r < \infty$, *then* $X_n \underset{r}{\xrightarrow{\text{a.s.}}} X \in L_r$ *and* X *is the nearest of the closing r.v.'s.*

IV SUBMARTINGALE REVERSED SEQUENCE $\cdots, X_2, X_1$.

Inequalities:

$$\cdots \leqq EX_2 \leqq EX_1; \quad \cdots \leqq EX_2^+ \leqq EX_1^+;$$

$$\text{every } X_n \geqq 0 \text{ a.s.} \Rightarrow \cdots \leqq EX_2^r \leqq EX_1^r.$$

Convergence. *If* $EX_1^+ < \infty$, *then* $X_n \xrightarrow{\text{a.s.}} X < \infty$.

Closure. *If the* $|X_n|^r$ *are uniformly integrable, then* $X_n \underset{r}{\xrightarrow{\text{a.s.}}} X \in L_r$ *and* X *is the nearest of the closing r.v.'s. In particular,* $X_n \underset{1}{\xrightarrow{\text{a.s.}}} X$ *if and only if* $\sup E|X_n| < \infty$, *equivalently,* $E|X_1| < \infty$, $\lim EX_n > -\infty$. (see 36.1**c**).

REMARK. By using the decomposition of submartingales given in 29.2, we can deduce their properties from those of martingales:

1° Let $X_1, X_2, \cdots$ be a submartingale sequence with $\sup E|X_n| < \infty$. Then $X_n = X'_n + X''_n$ where $X'_1, X'_2, \cdots$ is a martingale sequence with $\sup E|X'_n| < \infty$, and $0 \leqq X''_n \uparrow X''$ finite a.s. Therefore, $X'_n \xrightarrow{\text{a.s.}} X'$ finite and $X_n \xrightarrow{\text{a.s.}} X = X' + X''$ finite.

2° Let $\cdots X_2, X_1$ be a submartingale reversed sequence with $E|X_1| < \infty$ and $\lim EX_n > -\infty$. Then $X_n = X'_n + X''_n$ where $\cdots X'_2, X'_1$ is a martingale reversed sequence with $E|X'_1| < \infty$ and $0 \leqq X''_n \downarrow 0$ a.s., $EX''_1 < \infty$. Therefore $X'_n \underset{1}{\xrightarrow{\text{a.s.}}} X'$ and $X_n \underset{1}{\xrightarrow{\text{a.s.}}} X'$.

32.4. Applications. We use now the properties of martingales in order to extend various properties obtained in the case of independence. In general, we shall revert to P. Lévy's form of martingale sequences $X_n = \sum_{k=1}^{n} Y_k$ with $E(Y_{k+1} \mid Y_1, \cdots, Y_k) = 0$ a.s.; then $\mathcal{B}_n = \mathcal{B}(X_1, \cdots, X_n) = \mathcal{B}(Y_1, \cdots, Y_n)$, and we set $\mathcal{B}_0 = \{\emptyset, \Omega\}$.

We shall have use for a truncation of subscripts, first introduced by P. Lévy and which transforms martingales into martingales. Let ν be an integer-valued measurable function, finite or not, and such that the events $[\nu > n]$ are defined on the first n terms of a sequence $Y_1, Y_2, \cdots$ of r.v.'s, that is, $[\nu > n] \in \mathcal{B}_n$. We set $Y'_n = Y_n I_{[\nu \geqq n]}$, so that $\mathcal{B}(Y'_1, \cdots, Y'_n) \subset \mathcal{B}_n$ and $E(Y'_{n+1} \mid Y_1, \cdots, Y_n) = I_{[\nu \geqq n+1]}\, E(Y_{n+1} \mid Y_1, \cdots, Y_n)$. Thus, if every $E(Y_{n+1} \mid Y_1, \cdots, Y_n) = 0$ a.s., then $E(Y'_{n+1} \mid Y_1, \cdots, Y_n) = 0$ a.s., and the martingale sequence $X_n = \sum_{k=1}^{n} Y_k$ is transformed by the above "ν-truncation" into the martingale sequence $X'_n = \sum_{k=1}^{n} Y'_k$. Observe that, the $\mathcal{B}_n$ being closed under countable operations, we have

$$[\nu > n] \in \mathcal{B}_n \Leftrightarrow [\nu \leqq n] \in \mathcal{B}_n \Leftrightarrow [\nu = n] \in \mathcal{B}_n, \quad n = 1, 2, \cdots.$$

I. ZERO-ONE LAWS. The zero-one laws of the case of independence extend as follows:

Let $Y, Y_1, Y_2, \cdots$ be r.v.'s and apply 29.3**A**. The sequence $Z_n = E(Y \mid Y_1, \cdots, Y_n)$ is a martingale closed on the right by Y whose expectation is assumed to exist, say, $EY^+ < \infty$. Since every $EZ_n{}^+ \leqq EY^+$ and hence $\sup EZ_n{}^+ < \infty$, it follows that $Z_n \xrightarrow{\text{a.s.}} Z < \infty$. If $E|Y|^r < \infty$ for some $r \geqq 1$, then $Z_n \underset{r}{\xrightarrow{\text{a.s.}}} Z = E(Y \mid Y_1, Y_2, \cdots)$. If, moreover, Y is defined on the Y_n, then $Z_n \underset{r}{\xrightarrow{\text{a.s.}}} Y$. To summarize

A. *If $EY < \infty$, then $E(Y \mid Y_1, \cdots, Y_n) \xrightarrow{\text{a.s.}} Z < \infty$. If $E|Y|^r < \infty$ for some $r \geqq 1$, then $E(Y \mid Y_1, \cdots, Y_n) \underset{r}{\xrightarrow{\text{a.s.}}} E(Y \mid Y_1, Y_2, \cdots)$, which reduces a.s. to Y when Y is defined on the Y_n.*

We specialize now these properties. If $Y = I_B$ where the event B is defined on the Y_n, whence B is a "property" of the sequence, then $P(B \mid Y_1, \cdots, Y_n) \xrightarrow{\text{a.s.}} I_B$. (P. Lévy.)
In more intuitive terms

The sequence $P(B \mid Y_1, \cdots, Y_n)$ of c.pr.'s of a property B of the sequence $Y_1, Y_2, \cdots$, given the first n terms of the sequence, converges a.s. to 1 or to 0 according as the sequence has or has not this property.

In particular, if $P(B \mid Y_1, \cdots, Y_n) = PB$ a.s. for every value of n (or for a sequence of values of n), then $PB = I_B$ a.s. (Kolmogorov.)
In more intuitive terms

a. *If the c.pr. of a property of a sequence of r.v.'s, given any finite number of its terms, degenerates into a constant, then, a.s., the property is either sure or impossible.*

Also Borel's zero-one law extends as follows: Let $B_1, B_2, \cdots$ be a sequence of events and set $\mathcal{B}_n = \mathcal{B}(I_{B_1}, \cdots, I_{B_n})$. The two events $[\sum I_{B_n} < \infty]$ and $[\sum P^{\mathcal{B}_{n-1}} B_n < \infty]$ are equivalent (P. Lévy). In more intuitive terms

b. *The number of occurrences of the events B_n is a.s. finite or infinite according as the series of their c.pr.'s $\sum P^{\mathcal{B}_{n-1}} B_n$ is a.s. finite or infinite.*

Proof. The sequence $X_n = \sum_{k=1}^{n} Y_k$ where $Y_k = I_{B_k} - P^{\mathcal{B}_{k-1}} B_k$ (hence $|Y_k| \leqq 1$ a.s.) is a martingale. Let $a > 0$ be a finite number and define

ν by $[\nu = n] = [\sup_{k<n} X_k \leqq a,\ X_n > a]$, $[\nu = \infty] = [\sup X_n \leqq a]$. The ν-truncated sequence $X'_n = \sum_{k=1}^{n} Y'_k$ is a martingale bounded above by $a + 1$, and hence $X'_n \xrightarrow{\text{a.s.}} X'$ finite.

Since $X_n = X'_n$ on $[\sup X_n < a]$ and $a > 0$ is an arbitrary finite number, it follows that $X_n \to X$ finite on $[\sup X_n < \infty]$ except for a null event. Since changing the X_n into $-X_n$ preserves the martingale property, it follows that $X_n \to X$ finite on $[\inf X_n > -\infty]$ except for a null event.

But $0 \leqq \sum_{k=1}^{n} I_{B_k} \uparrow$ and $0 \leqq \sum_{k=1}^{n} P^{\mathfrak{B}_{n-1}} B_k \uparrow$ a.s. so that both sequences have a.s. a limit, finite or not. If one of them is finite and the other is infinite, then $\sup X_n(\omega) = +\infty$ or $\inf X_n(\omega) = -\infty$. Thus both limits are a.s. simultaneously either finite or infinite. The assertion is proved.

II. CONVERGENCE OF SERIES. Let $X_n = \sum_{k=1}^{n} Y_k$ form a martingale sequence. According to 29.2**A**

$$P[\sup_{k \leqq n} | X_k | > c] \leqq E| X_n |^r / c^r, \quad r \geqq 1.$$

For $r = 2$, the summands are orthogonal, $EX_n^2 = \sum_{k=1}^{n} EY_k^2$, the inequality reduces to the extended Kolmogorov inequality and yields the results of 29.1. The martingales convergence properties yield more, but the assumptions to be made are not easily expressed in terms of the summands. However, the ν-truncation method yields a direct extension of the convergence property established in the case of independent and uniformly bounded summands, as follows (P. Lévy):

If the summands of the martingale sequence $X_n = \sum_{k=1}^{n} Y_k$ are uniformly bounded ($| Y_k | \leqq c < \infty$), then $X_n \xrightarrow{\text{a.s.}} X$ finite if, and only if, the series $\sum E^{\mathfrak{B}_{n-1}} Y_n^2$ is a.s. finite.

Proof. Let $a > 0$ be a finite number and define ν by $[\nu = n] = [\sup_{k<n} | X_k | \leqq a,\ | X_n | > a]$, $[\nu = \infty] = [\sup | X_n | \leqq a]$. The sequence $X'_n = \sum_{k=1}^{n} Y'_k$ of ν-truncated summands is a martingale bounded by $a + c < \infty$, and hence $X'_n \xrightarrow{r} X'$ for every $r \geqq 1$. In particular, for $r = 2$, we have $EX'^2 = \sum EY'^2_n < \infty$, so that the series $\sum E^{\mathfrak{B}_{n-1}} Y'^2_n$, whose expectation EX'^2 is finite, is a.s. finite. Since $E^{\mathcal{B}_{n-1}}Y'^2_n = E^{\mathcal{B}_{n-1}}Y_n^2$ on $[\nu \geqq n]$ and vanishes on $[\nu < n]$, it follows that the series $\sum E^{\mathcal{B}_{n-1}}Y_n^2$ is a.s. finite on $[\nu = \infty] = [\sup |X_n| \leqq a]$. Since a is arbitrary, this series is a.s. finite on $[X_n \to X \text{ finite}]$.

Conversely, define ν by

$$[\nu = n] = \left[\sum_{k=1}^{n-1} E^{\mathcal{B}_{k-1}}Y_k^2 \leqq a, \quad \sum_{k=1}^{n} E^{\mathcal{B}_{k-1}}Y_k^2 > a\right],$$

$$[\nu = \infty] = [\sum E^{\mathcal{B}_{n-1}}Y_n^2 \leqq a].$$

The corresponding sequence of ν-truncated sums is a martingale bounded by a. Upon taking the expectations, it follows that

$$E^2|X'_n| \leqq EX'^2_n \leqq \sum EY'^2_n \leqq a + c^2,$$

so that, by 29.3**A**, $X'_n \xrightarrow{\text{a.s.}} X'$ finite; and, as above, $X_n \xrightarrow{\text{a.s.}} X$ finite on $[\sum P^{\mathcal{B}_{n-1}}Y_n^2 < \infty]$. The proof is terminated.

III. Strong laws of large numbers. Let $X_n = \sum_{k=1}^{n} Y_k$ where the Y_k are "conditionally exchangeable" with respect to addition (implied by ordinary exchangeability), that is, for every n and $k \leqq n$,

$$E(Y_k \mid X_n, X_{n+1}, \cdots) = E(Y_1 \mid X_n, X_{n+1}, \cdots) \text{ a.s.}$$

According to 29.3**B**, if $E|X_1|^r < \infty$ for some $r \geqq 1$, then

$$E(Y_1 \mid X_n, X_{n+1}, \cdots) \underset{r}{\xrightarrow{\text{a.s.}}} E^{\mathcal{C}}Y_1.$$

Therefore

$$\frac{X_n}{n} = E\left(\frac{X_n}{n}\middle| X_n, X_{n+1}, \cdots\right) = \frac{1}{n}\sum_{k=1}^{n} E(Y_k \mid X_n, X_{n+1}, \cdots)$$

$$= E(Y_1 \mid X_n, X_{n+1}, \cdots) \underset{r}{\xrightarrow{\text{a.s.}}} E^{\mathcal{C}}Y_1.$$

To summarize

If $X_n = \sum_{k=1}^{n} Y_k$ *where the* Y_k *are exchangeable and* $E|Y_1|^r < \infty$ *for some* $r \geqq 1$, *then* $\dfrac{X_n}{n} \underset{r}{\xrightarrow{\text{a.s.}}} E^{\mathcal{C}}Y_1$.

IV. Independence. The foregoing results, in fact all the results of this chapter, were obtained under the guidance but not by the use of similar results in the case of independence. Thus we have new proofs of the latter under the supplementary assumptions of independence.

First consider results in I. Proposition **a** reduces to the zero-one law for tail events on sequences Y_n of independent r.v.'s; since any tail event $B \in \mathcal{C}_n = \mathcal{B}(Y_{n+1}, Y_{n+2}, \cdots)$ whatever be n, and $\mathcal{B}_n$ and $\mathcal{C}_n$ are independent σ-fields, it follows that $P^{\mathcal{B}_n}B = PB$ a.s. whatever be n. Similarly, proposition **b** reduces to the Borel zero-one criterion, since then $P^{\mathcal{B}_{n-1}}B_n = PB_n$ a.s.

Now, let $X_n = \sum_{k=1}^{n} Y_k$ where the summands are independent r.v.'s centered at expectations. Then, in II, the inequality extends Kolmogorov's inequality for $r = 2$ to any $r \geqq 1$, while the proposition proved there reduces to the fact that, when the summands are uniformly bounded, the series $\sum Y_n$ converges a.s. if, and only if, it converges in q.m. ($r = 2$). As for III, it yields Kolmogorov's strong law of large numbers, since then the tail σ-field is $\{\emptyset, \Omega\}$ a.s. and the limit is a tail function.

To summarize, as had to be expected, the results in I, II, and III provide the basic convergence properties of the case of independence, but nothing new. However, if we do not limit ourselves to results expressed in terms of summands, we get more (Marcinkiewicz).

A. *Let* $X_n = \sum_{k=1}^{n} Y_k$ *be consecutive sums of independent r.v.'s centered at expectations, and let* $r \geqq 1$. *Then* $X_n \xrightarrow{\text{a.s.}} X \in L_r$ *if, and only if,* $X_n \xrightarrow{r} X$.

Proof. If $X_n \xrightarrow{r} X$, then, by 29.3**B**, $X_n \xrightarrow{\text{a.s.}} X \in L_r$. Conversely, let $X_n \xrightarrow{\text{a.s.}} X \in L_r$. Since $r \geqq 1$, the r.v. X is integrable. Since, for every p, $X_{n+p} - X_n$ is independent of $X_1, \cdots, X_n$, it follows that $X - X_n$ is independent of $X_1, \cdots, X_n$, so that

$$E^{\mathcal{B}_n}X = E^{\mathcal{B}_n}(X - X_n) + E^{\mathcal{B}_n}X_n = E(X - X_n) + X_n = EX + X_n \text{ a.s.}$$

But $X_n \xrightarrow{\text{a.s.}} X$, while, by **IA**, $E^{\mathcal{B}_n}X \xrightarrow{\text{a.s.}} X$. Therefore, $EX = 0$ so that $E^{\mathcal{B}_n}X = X_n$ a.s. and the martingale sequence X_n is closed by $X \in L_r$. Thus, theorem 29.3**B** applies and, hence, $X_n \xrightarrow{r} X$. The proof is terminated.

***32.5. Indefinite expectations and a.s. convergence.** Convergence properties of martingales and, hence, all the applications of the preceding subsection are but particular cases of a convergence theorem that we establish now.

We consider sequences X_n of r.v.'s whose indefinite expectations φ_n defined by

$$\varphi_n(A) = \int_A X_n, \quad A \in \mathcal{A},$$

exist. We recall that $\mathcal{B} = \mathcal{B}(X_1, X_2, \cdots)$ is the minimal σ-field over the field $\mathcal{B}_0 = \bigcup \mathcal{B}(X_1, \cdots, X_n)$, and $\mathcal{C} = \bigcap \mathcal{B}(X_n, X_{n+1}, \cdots)$ is the tail σ-field of the sequence X_n. We introduce the following hypothesis.

(H): *There exists a set function φ on $\mathcal{C}$ such that, as $n \to \infty$ and then $m \to \infty$,*

$$\sum_{k=m}^{n} \varphi_k(B_{mk}C) \to \varphi(C'C)$$

whatever be the disjoint in k events $B_{mk} \in \mathcal{B}\{X_m, \cdots, X_k\}$ such that $\sum_{k=m}^{n} B_{mk} \to C'$ and whatever be the tail events C and C'.

If the foregoing events B_{mk} are replaced by disjoint in k events $B_{kn} \in \mathcal{B}\{X_k, \cdots, X_n\}$, the so modified hypothesis will be denoted by (H′).

a. Basic inequalities. *Under* (H) *or* (H′),

$$\varphi(\underline{C}_a C) \leqq aP(\underline{C}_a C) \quad \text{and} \quad \varphi(\overline{C}_b C) \geqq bP(\overline{C}_b C)$$

whatever be the tail event C and whatever be the finite numbers a, b in the tail events

$$\underline{C}_a = [\liminf X_n < a], \quad \overline{C}_b = [\limsup X_n > b].$$

Proof. Let $a < a_m \downarrow a$ as $m \to \infty$ and set

$$B_{mm} = [X_m < a_m], \quad B_{mk} = [X_m \geqq a_m, \cdots, X_{k-1} \geqq a_m, X_k < a_m],$$

so that, as $n \to \infty$ and then $m \to \infty$,

$$\sum_{k=m}^{n} B_{mk} = [\inf_{m \leqq k \leqq n} X_k < a_m] \to \underline{C}_a.$$

Since, for every event C,

$$\sum_{k=m}^{n} \varphi_k(B_{mk}C) = \sum_{k=m}^{n} \int_{B_{mk}C} X_k \leqq a_m P\left(\sum_{k=m}^{n} B_{mk}C\right),$$

it follows, upon letting $n \to \infty$ and then $m \to \infty$, that (H) entails

$$\varphi(\underline{C}_a C) \leqq aP(C_a C).$$

The same inequality is entailed by (H′) upon setting $B_{kn} = [X_n \geqq a_m, \cdots, X_{k+1} \geqq a_m, X_k < a_m]$, and the first asserted inequality is proved. The second one follows by changing the X_n into $-X_n$ and a into $-b$. The proof is complete.

A. BASIC CONVERGENCE THEOREM. *Under* (H) *or* (H′), $X_n \xrightarrow{\text{a.s.}} X$ *a.s. finite above or below according as* φ *is bounded above or below. If, moreover,* φ *on* $\mathcal{C}$ *is* σ*-additive and* σ*-finite, then* $X_n \xrightarrow{\text{a.s.}} X = \dfrac{d\varphi}{dP_{\mathcal{C}}}$ *a.s. finite.*

$\dfrac{d\varphi}{dP_{\mathcal{C}}}$ denotes the $\mathcal{C}$-measurable function whose indefinite integral is the $P_{\mathcal{C}}$-continuous part φ_c of φ, and $P_{\mathcal{C}}$ is the restriction of P to $\mathcal{C}$.

Proof. Since the set $D = [\liminf X_n \neq \limsup X_n]$ of divergence of the sequence X_n can be written as a denumerable union $D = \bigcup C_{ab}$ where

$$C_{ab} = [\liminf X_n < a < b < \limsup X_n]$$

and $a, b (a < b)$ vary over all rationals, it suffices to prove that every event C_{ab} is null. But, by taking $C = C_{ab}$ in the basic inequality so that

$$\underline{C}_a C_{ab} = \overline{C}_b C_{ab} = C_{ab},$$

it follows that

$$bPC_{ab} \leqq \varphi(C_{ab}) \leqq aPC_{ab},$$

and, since $b > a$, we have $PC_{ab} = 0$. Thus, if $X = \lim X_n$ on D^c and we set, say, $X = 0$ on D, then $X_n \xrightarrow{\text{a.s.}} X$ where X is a $\mathcal{C}$-measurable function—not necessarily finite. If $\varphi \leqq c < \infty$, then, taking $C = \Omega$ in the second basic inequality, we have, for every finite $b > 0$,

$$P[X = +\infty] = P[\limsup X_n = +\infty] \leqq P\overline{C}_b \leqq c/b \to 0 \quad \text{as} \quad b \to \infty,$$

so that $X < +\infty$ a.s. Similarly when φ is bounded below, and the first assertion is proved.

Let now φ be σ-additive and σ-finite. By taking, if necessary, a denumerable partition of Ω into events of φ-finite measure, it suffices as usual to prove the last assertion for φ on $\mathcal{C}$ σ-additive and finite, hence bounded. Then X is a.s. finite, and we can take X to be finite by including the null event $[X = \pm\infty]$ in the null event D and setting, say, $X = 0$ on D.

Let

$$C_{nk} = \left[\frac{k}{2^n} \leqq X < \frac{k+1}{2^n}\right], \quad k = 0, \pm 1, \pm 2, \cdots$$

and set

$$X'_n = \sum_{k=-\infty}^{+\infty} \frac{k}{2^n} I_{C_{nk}}$$

so that

$$X'_n \leqq X < X'_n + \frac{1}{2^n} \quad \text{on} \quad D^c.$$

On the other hand, the basic inequalities with C replaced by $CC_{nk}D^c$ and a,b replaced by $\frac{k+1}{2^n}, \frac{k}{2^n}$, respectively, become

$$\frac{k}{2^n} P(CC_{nk}D^c) \leqq \varphi(CC_{nk}D^c) \leqq \frac{k+1}{2^n} P(CC_{nk}D^c)$$

so that, by summing over k, we obtain

$$\int_{CD^c} X'_n \leqq \varphi(CD^c) \leqq \int_{CD^c} X'_n + \frac{1}{2^n}.$$

It follows that

$$\varphi(CD^c) - \frac{1}{2^n} \leqq \int_{CD^c} X \leqq \varphi(CD^c) + \frac{1}{2^n}$$

and, letting $n \to \infty$,

$$\int_C X = \int_{CD^c} X = \varphi(CD^c) = \varphi_c(C), \quad C \in \mathcal{C}.$$

Since X is $\mathcal{C}$-measurable, the last assertion is proved.

Variant. It may happen that φ is defined on the σ-field $\mathcal{B}$ of the whole sequence X_n and at the same time (H) or (H′) continues to hold when arbitrary $C \in \mathcal{C}$ are replaced by arbitrary $B \in \mathcal{B}$. The so modified hypotheses will be denoted by (H_0) and (H'_0), respectively. The same proofs continue to apply with B's instead of C's and we obtain

a$_0$. *Under* (H_0) *or* (H'_0),

$$\varphi(\underline{C}_a B) \leqq aP(\underline{C}_a B), \quad \varphi(\overline{C}_b B) \geqq bP(\overline{C}_b B)$$

whatever be $B \in \mathcal{B}$ *and the finite numbers* a, b.

A$_0$. *Under* (H_0) *or* (H'_0), $X_n \xrightarrow{\text{a.s.}} X$ *a.s. finite above or below according as* φ *is bounded above or below. If, moreover,* φ *on* $\mathcal{B}$ *is* σ*-additive and* σ*-finite, then* $X_n \xrightarrow{\text{a.s.}} X = \frac{d\varphi}{dP_{\mathcal{B}}}$ *a.s. finite.*

COROLLARY 1. *Let* φ *on* $\mathcal{B}_0$ *be a* σ*-finite signed measure, so that it determines its* σ*-additive and* σ*-finite extension* φ *on* $\mathcal{B}$.

If (H) *or* (H′) *hold when* $C \in \mathcal{C}$ *are replaced by* $B_0 \in \mathcal{B}_0$, *then* $X_n \xrightarrow{\text{a.s.}} X = \dfrac{d\varphi}{dP_{\mathcal{B}}}$ *a.s. finite.*

Proof. The basic inequalities hold with $C \in \mathcal{C}$ replaced by $B_0 \in \mathcal{B}_0$. Therefore, by continuity of P and φ, they hold with $B \in \mathcal{B}$ and the above variant applies.

COROLLARY 2. *Let* φ *on* $\mathcal{B}_0$ *be a* σ*-finite signed measure. If, given any* $\epsilon > 0$, *for k sufficiently large*

$$| \varphi_k(B) - \varphi(B) | \leqq \epsilon PB$$

whatever be $B \in \mathcal{B}(X_k, X_{k+1}, \cdots)(B \in \mathcal{B}(X_1, \cdots, X_k))$, *then* $X_n \xrightarrow{\text{a.s.}} X = \dfrac{d\varphi}{dP_{\mathcal{C}}}\left(\dfrac{d\varphi}{dP_{\mathcal{B}}}\right)$.

Proof. In the first case, hypothesis (H′) holds, since φ extends to $\mathcal{B}$ and, for m sufficiently large,

$$\left| \sum_{k=m}^{n} \varphi_k(B_{kn}C) - \varphi\left(\sum_{k=m}^{n} B_{kn}C\right)\right| \leqq \epsilon \sum_{k=m}^{n} PB_{kn}C \leqq \epsilon \to 0$$

as $n \to \infty$, then $m \to \infty$, and then $\epsilon \to 0$. Theorem **A** applies, and the assertion is proved. Similarly, in the second case, hypothesis (H_0) holds, and theorem $\mathbf{A}_0$ applies.

APPLICATION TO MARTINGALES. 1° Let the r.v.'s X_n form a martingale sequence or a martingale reversed sequence, closed by a r.v. Y whose expectation exists, that is, $X_n = E(Y | X_1, \cdots, X_n)$ a.s. or $X_n = E(Y | X_n, X_{n+1}, \cdots)$ a.s. Then Corollary 2 applies with φ indefinite integral of Y; in fact, in each case $E_B X_n = E_B Y = \varphi(B)/PB$. Thus $X_n \xrightarrow{\text{a.s.}} X = E^{\mathcal{C}}Y$ or $E^{\mathcal{B}}Y$, respectively.

2° Let the r.v.'s X_n with $\sup E| X_n | \leqq c < \infty$ form a martingale sequence. Take for pr. space the "sample pr. space," that is, the range space of the sequence together with its Borel field $\mathcal{A}_\infty$ and the pr. distribution P_∞ of the sequence. Then $\mathcal{B}(X_1, \cdots, X_n)$ is the σ-field $\mathcal{A}_n$ of all Borel cylinders whose bases are Borel sets in the range space of $(X_1, \cdots, X_n)$. Since the X_n form a martingale sequence, we have $\varphi_n(A_n) = \varphi_{n+1}(A_n) = \cdots$ for every $A_n \in \mathcal{A}_n$ and, hence, $\varphi_n \to \varphi$ on $\mathcal{A}_0$—the field of all Borel cylinders in the range space. We apply the extension 4.3**A**, 2°. Since the indefinite expectations $\bar{\varphi}_n$ of $| X_n |$ are bounded by c, and form a nondecreasing sequence, it follows that $\lim \bar{\varphi}_n$ exists and is bounded and σ-additive on $\mathcal{A}_0$. *A fortiori*, φ is bounded and σ-additive on $\mathcal{A}_0$. Thus, Corollary 2 applies and $X_n \xrightarrow{\text{a.s.}} X = \dfrac{d\varphi}{dP_{\mathcal{B}}}$.

We seek now necessary *and* sufficient conditions for a.s. convergence of sequences X_n of r.v.'s.

B. Dominated convergence criterion. *Let* $|X_n| \leq Y$ *integrable. Then* $X_n \xrightarrow{\text{a.s.}} X$ *(necessarily integrable) if, and only if,* (H) *or* (H′) *or* (H_0) *or* (H'_0) *holds.*

Thus, if $|X_n| \leq Y$ integrable and $X_n \xrightarrow{\text{a.s.}} X$, then all the hypotheses H are equivalent.

Proof. The "if" assertion is contained in **A** and $\mathbf{A}_0$. Conversely, let $X_n \xrightarrow{\text{a.s.}} X$ with indefinite expectation φ so that, for every $\epsilon > 0$,

$$PB_\epsilon = P[\sup_{m \leq k \leq n} |X_k - X| \geq \epsilon] \to 0 \quad \text{as} \quad m \to \infty.$$

Whatever be the disjoint events $B_k \in \mathcal{B}$ varying or not with m and n and whatever be $B \in \mathcal{B}$ such that $\sum_{k=m}^{n} B_k \to B$, upon summing over $k = m, \cdots, n$ the relations

$$\varphi_k(B_k) - \varphi(B_k) = \int_{B_k} (X_k - X) = \int_{B_k B_\epsilon^c} (X_k - X) + \int_{B_k B_\epsilon} (X_k - X),$$

it follows from $|X_k| \leq Y$ integrable that

$$\left|\sum_{k=m}^{n} \varphi_k(B_k) - \varphi(B)\right| \leq \epsilon PB_\epsilon^c + 2\int_{B_\epsilon} Y + \left|\varphi(\sum_{k=m}^{n} B_k) - \varphi(B)\right|.$$

Letting $n \to \infty$, then $m \to \infty$, and then $\epsilon \to 0$, the "only if" assertion follows.

C. Convergence criterion. *A sequence X_n of r.v.'s converges a.s. to a r.v. if, and only if, for every $\epsilon > 0$ there exist events B_ϵ with $PB_\epsilon > 1 - \epsilon$ on which $|X_n| \leq Y_\epsilon$ integrable and* (H) *or* (H′) *or* (H_0) *or* (H'_0) *holds.*

Proof. If for $\epsilon_m \downarrow 0$ as $m \to \infty$, we have $|X_n| \leq Y_{\epsilon_m}$ integrable and, say, (H) holds on $B_m = B_{\epsilon_m}$; then by **B**, the sequence X_n converges to a r.v. on $B_m - N_m$ where N_m are null events; hence it converges to a r.v. on $\bigcup B_m - \bigcup N_m$. Since, for every integer m',

$$P \bigcup B_m \geqq PB_{m'} \geqq 1 - \epsilon_{m'},$$

it follows that $P \bigcup B_m = 1$ and the "if" assertion is proved.

Conversely, let $X_n \xrightarrow{\text{a.s.}} X$ r.v. and apply Egorov's theorem which asserts that for every $\epsilon > 0$ there exists an event B with $PB < \frac{\epsilon}{2}$ such that $X_n \to X$ uniformly on B^c. Let n and $c > 0$ be sufficiently large so that, on the one hand, $|X_n| < |X| + 1$ on B^c and, on the other hand,

$$PC = P[|X| > c] < \frac{\epsilon}{2}.$$

Then $|X_n| < c + 1$ on B^cC^c and, by **B**, (H) holds. Since

$$1 - PB^cC^c = P(B \cup C) \leqq PB + PC < \epsilon,$$

the "only if" assertion is proved.

COMPLEMENTS AND DETAILS

1. Let $S_n = \sum_{k=1}^{n} X_k$ where $E(X_n \mid X_1, \cdots, X_{n-1}) = 0$ a.s. $(X_0 = 0)$ and $E(X_n^2 \mid X_1, \cdots, X_{n-1}) = \sigma^2 X_n$ a.s. Find conditions for degenerate and for normal convergence of suitably normed sums.

2. Take one by one the results relative to the degenerate, the normal, and the Poisson convergence obtained in the case of independent summands, and transpose them to the case of dependent summands by using successively the various approaches given and illustrated in the text.

3. Let $S_{\nu_n} = \sum_{k=1}^{\nu_n} X_k$, $X_0 = 0$. The summands are independent with common ch.f. f and finite $a = EX_n$, $\sigma^2 = \sigma^2 X_n$. The ν_n are integer-valued r.v.'s independent of all the summands with $p_{nk} = P[\nu_n = k]$, $k = 0, 1, \cdots$, and with finite $\alpha_n = E\nu_n$, $\beta_n^2 = \sigma^2 \nu_n$. Set $\sigma_n^2 = \sigma^2 S_{\nu_n}$ and let g_n be the ch.f. of $(S_{\nu_n} - E_{\nu_n})/\sigma_n$. Then

$$g_n(u) = \sum_{k=0}^{\infty} p_{nk} e^{-i\alpha_n u/\sigma_n} f_k(u/\sigma_n), \quad \sigma_n^2 = a_n \sigma^2 + a^2 \beta_n^2.$$

Let $\sigma_n^2 \to \infty$, $\beta_n^2 = O(\sigma_n^2)$. Then $g_n(u)$ is of the form

$$g_n(u) = h(c_n u) e^{-\frac{u^2}{2}(1-c_n 2)} + o(1), \quad c_n = a_n \beta_n / \sigma_n.$$

If also $a^2 \beta_n^2 = o(\alpha_n)$, then $g_n(u) \to e^{-u^2/2}$.
If also $\mathfrak{L}(\nu_n) \sim \mathfrak{N}(\alpha_n, \beta_n^2)$, then $\mathfrak{L}(S_{\nu_n}) \sim \mathfrak{N}(a\alpha_n, \sigma_n^2)$.

What happens when $\mathcal{L}(\nu_n)$ is a Poisson law $\mathcal{P}(n)$, or when $\mathcal{L}(\nu_n)$ is a binomial law with

$$p_{nk} = \frac{n!}{k!(n-k)!} p^k q^{n-k}?$$

4. Search for conditions under which the various limit laws obtained in the case of independent summands remain the same when the numbers of summands are r.v.'s ν_n independent of the summands. What happens when $\nu_n \xrightarrow{\text{a.s.}} \infty$?

5. By following the indications given in the text transpose to the case of dependent summands as many as possible of the a.s. convergence and a.s. stability theorems obtained in the case of independence.

6. Let Y be an integrable r.v. defined on a sequence X_n of r.v.'s. Take for pr. space the sample space of the sequence: $\Omega = \prod R_n$, $\mathcal{A}$-Borel field in Ω, P—pr. distribution of the sequence.

If the X_n are independent, with pr. distributions P_n, then

$$E(Y \mid X_1, \cdots, X_n) = \int Y\, dP_{n+1}\, dP_{n+2} \cdots \xrightarrow{\text{a.s.}} Y$$

$$E(Y \mid X_n, X_{n+1}, \cdots) = \int Y\, dP_1 \cdots dP_{n-1} \xrightarrow{\text{a.s.}} EY.$$

What becomes of the integrals when the X_n are dependent?

7. A net is a sequence of countable partitions $\Omega = \sum_j A_{nj}$ into events such that every partition is finer than the preceding one. Every partition determines a σ-field $\mathcal{B}_n$, and $\mathcal{B}_n \uparrow$. Let φ on $\bigcup \mathcal{B}_n$ be bounded and let it be σ-additive on every $\mathcal{B}_n$. Set $X_n = \sum_j x_{nj} I_{A_{nj}}$ with $x_{nj} = \varphi(A_{nj})/P(A_{nj})$ and throw out the null union of all null events A_{nj}.

The sequence X_n is a martingale and its a.s. limit, if it exists, is called the derivative of φ with respect to P given the net.

If φ is σ-additive on $\bigcup \mathcal{B}_n$, then it extends to φ on $\mathcal{B}$ bounded and σ-additive, and $X_n \xrightarrow{\text{a.s.}} X = \dfrac{d\varphi}{dP_{\mathcal{B}}}$. The X_n are uniformly integrable if, and only if, φ is P-continuous and then $X_n \underset{1}{\xrightarrow{\text{a.s.}}} X$.

Particular case. Let $\Omega = [0, 1]$, P is the Lebesgue measure, φ is determined by a function H on Ω of bounded variation. Consider a net of partitions into intervals such that the length of the largest converges to 0. Then $\mathcal{B}$ is the Borel field in Ω and $X_n \xrightarrow{\text{a.s.}} H'$-derivative of H known to exist a.s.

8. If g on R is a Lebesgue integrable function of period 1 and $g_n(x) = \dfrac{1}{2^n} \sum_{k=0}^{2^n-1} g\left(x + \dfrac{k}{2^n}\right)$, then $g \xrightarrow{\text{a.e.}} \int_0^1 g(x)\, dx$. (Let $\Omega = R$, $\mathcal{A} = \sigma$-field of Lebesgue sets A of period 1, PA = Lebesgue measure of a period of A. Let $\mathcal{B}_n$ be a similar σ-field of sets of period $1/2^n$. The σ-field $\bigcap \mathcal{B}_n$ consists only of sets of pr. 0 or 1. Use martingale reversed sequences convergence theorem.)

9. Let φ on $\mathcal{A}$ be bounded and σ-additive. Then $X_n = \dfrac{d\varphi}{dP_{\mathcal{B}_n}} \xrightarrow{\text{a.s.}} X = \dfrac{d\varphi}{dP_{\mathcal{B}}}$ or $\dfrac{d\varphi}{dP_{\mathcal{C}}}$ according as $\mathcal{B}_n \uparrow$ or $\mathcal{B}_n \downarrow$. (Either start with $\varphi \geqq 0$ and observe that the sequence $-X_n$ is a submartingale; or first extend the basic convergence theorem 29.4**A** to $X_n = \dfrac{d\varphi_n}{dP}$ where $\varphi_n = \varphi_n^c + \varphi_n^s$ and X_n a.s. finite with $(\varphi_n^s)^{\pm}(A) = \varphi_n(A[X = \pm\infty])$.)

10. Let $m, n \to \infty$ and let $r \geqq 1$. Let $\mathcal{B}_m$, $\mathcal{B}$ be sub-σ-fields of events with $\mathcal{B}_m \uparrow \mathcal{B}$ or $\mathcal{B}_m \downarrow \mathcal{B}$

(a) $X_n \xrightarrow{r} X \in L_r \Rightarrow E(X_n \mid \mathcal{B}_m) \xrightarrow{r} E(X \mid \mathcal{B})$.

For, by the c_r-inequality and martingales convergence, $E| E(X_n \mid \mathcal{B}_m) - E(X \mid \mathcal{B}) |^r \leqq c_r E| E(X_n - X \mid \mathcal{B}_m) |^r + c_r E| E(X \mid \mathcal{B}_m) - E(X \mid \mathcal{B}) |^r \leqq c_r E| X_n - X |^r + c_r E| E(X \mid \mathcal{B}_m) - E(X \mid \mathcal{B}) |^r \to 0$.

(b) $0 \leqq X_n \uparrow X \in L_r \Rightarrow E(X_n \mid \mathcal{B}_m) \underset{r}{\xrightarrow{\text{a.s.}}} E(X \mid \mathcal{B})$.

Use (a) and, by martingale convergence and conditional monotone convergence,

$$\inf_{j \geqq m, k \geqq n} E(X_k \mid \mathcal{B}_j) \geqq \inf_{j \geqq m} E(X_n \mid \mathcal{B}_j) \Rightarrow \liminf_{m,n} E(X_n \mid \mathcal{B}_m)$$

$$\geqq \lim_n \lim_m E(X_n \mid \mathcal{B}_m) = E(X \mid \mathcal{B}) \text{a.s.}$$

$$\sup_{j \geqq m, k \geqq n} E(X_k \mid \mathcal{B}_j) \leqq \sup_{j \geqq m} E(X \mid \mathcal{B}_j) \Rightarrow \limsup_{m,n} E(X_n \mid \mathcal{B}_m)$$

$$\leqq E(X \mid \mathcal{B}_m) = E(X \mid \mathcal{B}) \text{ a.s.}$$

(c) $\inf_n X_n \in L_r, \liminf_n X_n \in L_r \Rightarrow E(\liminf_n X_n \mid \mathcal{B}) \leqq \liminf_{m,n} E(X_n \mid \mathcal{B}_m)$ a.s.

$\sup_n X_n \in L_r, \limsup_n X_n \in L_r \Rightarrow \limsup_{m,n} E(X_n \mid \mathcal{B}_m) \leqq E(\limsup_n X_n \mid \mathcal{B})$ a.s.

Use (b) as in the Fatou-Lebesgue Theorem.

(d) $\sup | X_n | \in L_r, X_n \xrightarrow{\text{a.s.}} X \Rightarrow E(X_n \mid \mathcal{B}_m) \underset{r}{\xrightarrow{\text{a.s.}}} E(X \mid \mathcal{B})$. Use (a) and (c). What if $X_n \xrightarrow{P} X$?

Chapter X

ERGODIC THEOREMS

Ergodic theory has a phenomenological origin which, on account of the Liouville theorem, leads to the study of measure-preserving transformations. The "classical period" (1930–1944) is concerned with one-to-one measure-preserving transformations T_1 of a measure space onto itself (in what follows we limit ourselves to a pr. space $(\Omega, \mathcal{A}, P)$ and leave out results with which we are not concerned). Let X, Y be r.v.'s and define X_n, Y_n, by $X_n(\omega) = X(T_1{}^{n-1}\omega)$, $Y_n(\omega) = Y(T_1{}^{n-1}\omega)$.

The first two basic results are

von Neumann's: if $X \in L_2$, then $\frac{1}{n}\sum_{k=1}^{n} X_k$ converges in q.m.

Birkhoff's: if $X \in L_1$, then $\frac{1}{n}\sum_{k=1}^{n} X_k$ converges a.s.

Then Khintchine gets rid of the supplementary but unnecessary assumptions made by Birkhoff and, at the same time, simplifies very considerably his proof; Hopf extends this theory to ratios of sequences $\sum_{k=1}^{n} X_k/\sum_{k=1}^{n} Y_k$ and proceeds to a systematic investigation of ergodic properties; Yosida and Kakutani extend von Neumann's theorem to transformations on Banach spaces and apply them to Markov chains.

The "modern period" (1944–) is characterized by several weakenings of the ergodic setup. Hurewicz and Halmos abandon, at least partly, the initial setup and start with set functions from which r.v.'s are derived, whereas Dunford and Miller abandon definitely the measure-preserving property and obtain necessary and sufficient conditions for convergence in the first mean. Then, F. Riesz gives very simple proofs of the Birkhoff theorem and of the sufficiency part of the Dunford-Miller theorem, and, at the same time, he abandons the one-to-one assumption about transformations. Doob applies Birkhoff's theorem to stationary chains. Finally, Hartman and Ryll-Nardzewski, employing methods developed by Y. Dowker in investigating "potentially invariant" measures, obtain necessary and sufficient conditions for a.s. convergence on L_1 to L_1.

In pr. theory we are interested not in the behavior of averages of sequences $\{X_n\}$ due to *any* r.v. $X \in L_1$ but in that of individual sequences. Stripped of all considerations due to their phenomenological origin, *ergodic theorems* assert conditions, in terms of point or, more generally, set transformations, under which averages $\frac{1}{n}\sum_{k=1}^{n} X_k$ of r.v.'s converge in some sense and, then, assert conditions under which the limits are degenerate. In the case of independence, such theorems reduce to limit theorems as expounded at length in Part III. Thus, Bernoulli's law of large numbers is to be thought of as the first ergodic theorem with convergence in pr. and Tchebichev's proof as an improvement of the conclusion—with convergence in q.m. Above all, Borel's strong law of large numbers is to be thought of as the first "best" ergodic theorem—with a.s. convergence.

What characterizes the *ergodic method*—as compared with those of Part III—is that the conditions for convergence are to be in terms of iterated translations, in a sense to be made precise in this chapter. We shall establish a basic ergodic inequality and deduce from it ergodic theorems which contain the mentioned results (sometimes improved or completed). The reader will recognize the proofs (and hence the statements) which remain valid when P is replaced by a σ-finite or an arbitrary measure μ.

§ 33. TRANSLATION OF SEQUENCES; BASIC ERGODIC THEOREM AND STATIONARITY

***33.1. Phenomenological origin.** Let $q = (q_1, \cdots, q_N)$, $p = (p_1, \cdots, p_N)$ be the generalized coordinates and momenta of a "conservative" mechanical system with N degrees of freedom. The equations of motion of the system are

$$\frac{dq_i}{dt} = \frac{\partial H}{\partial p_i}, \quad \frac{dp_i}{dt} = -\frac{\partial H}{\partial q_i}, \quad i = 1, \cdots, N,$$

where $H = H(q, p)$ is the Hamiltonian of the system, independent of time t. The "states" of the system are represented by points $\omega = (q, p)$ of the $2N$-dimensional real space Ω—the "phase space" of the system.

Under the equations of motion, every point $\omega \in \Omega$ which represents the "initial state"—at time $t = 0$—moves along a trajectory and this motion describes the evolution of the system; the state at time t is the point ω_t. Thus the phase space can be envisioned as a fluid in motion within itself. The celebrated Liouville's theorem asserts that in this motion the Lebesgue measure λ of Lebesgue sets A remains invariant. More precisely, in a unit of time the phase space undergoes a one-to-one transformation $\omega \leftrightarrow T_1\omega$ such that $\lambda T_1 A = \lambda A$. To simplify, we consider only discrete values of time $0, 1, 2, \cdots$. In fact, the conservative systems have constant energy E and the possible trajectories lie on the surface $H = E$ whence the term "ergodic" from *ergos* meaning energy; then the invariant measure μ is defined by $d\mu = d\sigma/\text{grad}\, H$ where $d\sigma$ is the differential area of an element of the surface and grad H is taken at a point of the element.

The comparison between theory and reality is made by measuring "observables" or "phase-functions"—λ-measurable functions of the state. The ideal would be to observe the consecutive states (their components are phase functions). Yet, in statistical mechanics, insurmountable difficulties of experimental as well as computational nature arise. In the systems under investigation the number of constituents or "particles" is *extremely large* and so is the number of degrees of freedom, as well as the number of microscopic phenomena in a unit of time of the observer (such as collisions of particles between themselves and with the walls of the container). Thus from the microscopic point of view, the time required by the observer to measure an observable X is extremely large, and its observed values are to be compared not to its instantaneous theoretical values but to the theoretical time-averages

$$\overline{X^n}(\omega) = \frac{1}{n}\{X(\omega) + X(T_1\omega) + \cdots + X(T_1^{n-1}\omega)\}$$

for n extremely large. However, computation of time-averages requires knowledge of consecutive states $T_1^n\omega$ given the initial state ω. Yet the *exact* knowledge of the initial state is experimentally unattainable. Even if it were attainable, then theoretical knowledge of succeeding states would require integration of an extremely large number of equations of motion—which is practically impossible. Thus, some other way of evaluating theoretical time-averages is to be found or postulated. Physicists were led to replace the exact initial state by the set of all possible states compatible with the precision of the experimental data and to postulate equality of time-averages with phase-averages over the multiplicity described by the trajectories of the initial states compatible with the data; this is the "ergodic hypothesis."

From the physicist's point of view, the justification of the ergodic hypothesis lies in its practical success. For the mathematician, ergodic theory was at first an attempt to *justify* theoretically the ergodic hypothesis. But bitter experience imposed supplementary hypotheses. Ergodic theorems assert conditions under which sequences of time-averages converge in some sense and *then* conditions under which the limits are independent of the initial state and reduce to phase averages. From the experimental point of view, comparison between theory and observation will be best when measurements of the observables yield approximate values of the time-averages, that is, when the convergence will be an everywhere or at least an a.e. convergence.

33.2. Basic ergodic inequality. Let the pr. space $(\Omega, \mathcal{A}, P)$ be fixed. On a family of one or several sequences of r.v.'s, say $\{X_n\}$, $\{Y_n\}$, we define Borel functions of the family, say ξ. These functions are measurable and, more precisely, are $\mathcal{B}$-measurable where $\mathcal{B}$ is the sub σ-field of events induced by the family. The *translate by* $k - 1$ of ξ is the function ξ_k (so that $\xi_1 = \xi$) obtained by adding $k - 1 (k = 1 \text{ or } 2 \text{ or } \cdots)$ to the subscripts of all those r.v.'s of the family which figure in the definition of ξ. Thus ξ_k is defined on the family $\{X_{n+k-1}\}$, $\{Y_{n+k-1}\}$—the translate by $k - 1$ of the given family—exactly as ξ is defined on the original one. We say that ξ is *invariant* (under translations) if it coincides with all its translates. Translations of indicators of events of $\mathcal{B}$ define translations of the events; in other words, the above definitions and notation apply to events $B \in \mathcal{B}$—replace ξ by B and ξ_k by B_k. Because of the definition, translations preserve countable set operations on events belonging to $\mathcal{B}$. It follows that the class of all invariant events is closed under countable set operations, hence is a sub σ-field $\mathcal{C}$ of events—the *invariant σ-field* defined on the family, and invariance of measurable functions defined on the family means $\mathcal{C}$-measurability.

The concept of translation can be interpreted by means of the range space of the family: the Borel space of points $(x_1, y_1, x_2, y_2, \cdots)$. For example, the translate $[X_k < Y_k]$ of $[X_1 < Y_1]$ is obtained as follows. The event $[X_1 < Y_1]$ is the inverse image (under the family) of the Borel set $[x_1 < y_1]$ in the range space. Then the translate $[X_k < Y_k]$ is, by definition, the inverse image of the Borel set $[x_k < y_k]$. In fact, *to avoid any ambiguity, in what follows it suffices to think of the pr. space as being the sample pr. space of the double sequence.*

CONVENTION AND NOTATION. In this chapter we shall reserve subscripts to indicate translations and otherwise use superscripts—not to be confused with power indices. Denote by $\mathcal{B}$ the σ-field of events induced by the family $\{X_n\}$, $\{Y_n\}$, by $\mathcal{C}$ the σ-field of invariant events, and set

$$X^n = \sum_{k=1}^{n} X_k, \quad Y^n = \sum_{k=1}^{n} Y_k.$$

To avoid undefined ratios $\frac{X^n}{Y^n}$ and to make the limits independent of an arbitrary but finite number of the summands, we assume once and for all that

$$Y_n > 0, \quad Y^n \uparrow \infty.$$

Then, from

$$\frac{X_2 + \cdots + X_{n+1}}{Y_2 + \cdots + Y_{n+1}} = \frac{X^{n+1} - X_1}{Y^{n+1} - Y_1} = \frac{X^{n+1}}{Y^{n+1}} \cdot \frac{Y^{n+1}}{Y^{n+1} - Y_1} - \frac{X_1}{Y^{n+1} - Y_1}$$

it follows that $\liminf \frac{X^n}{Y^n}$ and $\limsup \frac{X^n}{Y^n}$ are invariant and, hence, so are the sets of convergence and of divergence of the sequence $\frac{X^n}{Y^n}$:

$$C = \left[\liminf \frac{X^n}{Y^n} = \limsup \frac{X^n}{Y^n}\right], \quad D = \left[\liminf \frac{X^n}{Y^n} \neq \limsup \frac{X^n}{Y^n}\right]$$

as well as the events

$$\underline{C}_a = \left[\liminf \frac{X^n}{Y^n} < a\right], \quad \overline{C}_b = \left[\limsup \frac{X^n}{Y^n} > b\right].$$

So far, the defined concepts do not contain the pr. P; they are expressed only in terms of the measurable space $(\Omega, \mathcal{A})$ and of measurable functions on this space. However, when the r.v.'s represent points of L_r-spaces, the transformations have to be equivalence-preserving. It suffices for the translates of null events to be null, *to replace the assumptions on the Y_n's by $Y_n > 0$ a.s. and $Y_n \uparrow \infty$ a.s., and at the same time to replace invariance by a.s. invariance—invariance when a null event is neglected.*

We establish now the basic inequality from which we deduce the basic ergodic theorem. But first we require an elementary lemma. Let $a_1, a_2, \cdots, a_{n+m}$ be finite numbers. We say that a_k is *m-positive* if at least one of the sums $a_k + a_{k+1}, \cdots$ containing no more than m summands is positive. In symbols, a_k is m-positive if $\sup (a_k + \cdots + a_l) > 0$ for $k \leq l \leq \min (k + m - 1, n + m)$.

a. F. RIESZ'S LEMMA. *If there exist m-positive terms, then their sum is positive.*

Proof. Let a_k be the first m-positive term and let $a_k + \cdots + a_l$ be the *shortest* positive sum starting with a_k. If one of the terms a_h of this sum is not m-positive, then $a_h + \cdots + a_l \leqq 0$ so that $a_k + \cdots + a_{h-1} > 0$ and the sum is not the shortest positive one. Thus, all its terms are m-positive. Hence, the successive m-positive terms form disjoint stretches of positive sums. The assertion follows.

A. BASIC ERGODIC INEQUALITY. *If* $B^m = \left[\sup_{j \leqq m} \frac{X^j}{Y^j} > b\right]$ *and* $Z^n > 0$, *then for every integer n and every invariant event C*

$$\sum_{k=1}^{n} \int_{B_k{}^m C} \left(\frac{X_k}{Z^n} - b\frac{Y_k}{Z^n}\right) + \sum_{k=n+1}^{n+m} \int_C \left(\frac{X_k}{Z^n} - b\frac{Y_k}{Z^n}\right)^+ \geqq 0,$$

provided the sum exists.

Proof. Let $k = 1, 2, \cdots, n + m$, $k \leqq l \leqq \min(k + m - 1, n + m)$, and let

$$B^{mk} = [X_k - bY_k \text{ is } m\text{-positive}] = \left[\sup_l \frac{X_k + \cdots + X_l}{Y_k + \cdots + Y_l} > b\right]$$

If $k \leqq n$, then l varies from k to $k + m - 1$ and, hence, $B^{mk} = B_k{}^m$ where $B_k{}^m$ is the translate by $k - 1$ of B^m. Since by Riesz's lemma

$$\sum_{k=1}^{n+m} (X_k - bY_k)I_{B^{mk}} \geqq 0$$

a fortiori

$$\sum_{k=1}^{n} (X_k - bY_k)I_{B_k{}^m C} + \sum_{k=n+1}^{n+m} (X_k - bY_k)^+ I_C \geqq 0,$$

the asserted inequality follows upon dividing by Z^n and integrating.

*Let A with or without affixes denote events defined on the X_n's and Y_n's.

B. BASIC ERGODIC THEOREM. *If*

(i) $$\sum_{k=1}^{n} \int_{A_k} \frac{X_k}{Y^n} \to 0 \quad \textit{and}$$

$$\sum_{k=1}^{n} \int_{A_k} \frac{Y_k}{Y^n} \to 0, \quad \textit{as} \quad n \to \infty \quad \textit{and then} \quad A \downarrow \emptyset,$$

(ii) $$\int \frac{|X_{n+k}|}{Y^n} \rightarrow 0 \quad \textit{and}$$

$$\int \frac{Y_{n+k}}{Y^n} \rightarrow 0, \quad \textit{for every fixed } k \textit{ as} \quad n \rightarrow \infty,$$

then the sequence $\frac{X^n}{Y^n}$ *converges a.s. Assumption* (i) *implies that, from some* n *on, the sum therein exists.*

Proof. We have to prove that the invariant event

$$D = \left[\liminf \frac{X^n}{Y^n} \neq \limsup \frac{X^n}{Y^n}\right]$$

is null. Since $D = \bigcup_{a,b} C_{ab}$ is the denumerable union of the invariant events

$$C_{ab} = \underline{C}_a\overline{C}_b = \left[\liminf \frac{X^n}{Y^n} < a < b < \limsup \frac{X^n}{Y^n}\right]$$

where $a < b$ vary over the set of all rationals, it suffices to prove that every event C_{ab} is null.

We set in the basic inequality $Z^n = Y^n$ and

$$C = C_{ab} = B^m C_{ab} + A^m.$$

Because of the invariance of C_{ab}, translation by $k - 1$ yields

$$C_{ab} = B_k{}^m C_{ab} + A_k{}^m,$$

and the basic inequality becomes

$$\int_{C_{ab}} \left(\frac{X^n}{Y^n} - b\right) - \sum_{k=1}^{n} \int_{A_k{}^m} \left(\frac{X_k}{Y^n} - b\frac{Y_k}{Y^n}\right) + \sum_{k=n+1}^{n+m} \int_{C_{ab}} \left(\frac{X_k}{Y^n} - b\frac{Y_k}{Y^n}\right)^+ \geqq 0.$$

Since by definition of B^m and C_{ab} we have $B^m C_{ab} \uparrow C_{ab}$ and hence $A^m \downarrow \emptyset$ as $m \rightarrow \infty$, it follows because of (i) and (ii) that, upon letting $n \rightarrow \infty$ and then $m \rightarrow \infty$, the foregoing inequality becomes

$$\liminf \int_{C_{ab}} \left(\frac{X^n}{Y^n} - b\right) \geqq 0.$$

By changing the X_k into $-X_k$, b into $-a$, a into $-b$, this inequality becomes

$$\liminf \int_{C_{ab}} \left(a - \frac{X^n}{Y^n}\right) \geqq 0.$$

Therefore, by adding up the two last inequalities, we have

$$(a - b)PC_{ab} \geqq 0, \quad a < b,$$

so that $PC_{ab} = 0$ and the proof is concluded.

Let Z^n be positive r.v.'s such that for every invariant event C defined on the X_n and Y_n

$$\text{(C)} \qquad \liminf \int_C \frac{Y^n}{Z^n} = 0 \Rightarrow PC = 0;$$

this is certainly true if $Z^n = Y^n$. It follows that, if throughout the preceding proof we divide by Z^n instead of by Y^n, this proof remains valid. In other words,

B′. *The basic ergodic theorem remains valid if in the assumptions therein Y^n are replaced by Z^n obeying condition* (C).

33.3. Stationarity. To the concept of invariance correspond weaker ones of invariance in terms of integrals and in terms of pr.'s.

Let $\{X_n\}$, $\{Y_n\}$, be the family of r.v.'s on which the translations are defined. Let the events A, B, C, with or without affixes, be defined on this family, that is, belong to the σ-field $\mathcal{B}$ induced by the family. Let ξ, with or without affixes, be a r.v. defined on the family. We say that ξ is *integral invariant* or that the sequence of translates $\xi_1, \xi_2, \cdots$ is *integral stationary* if the integrals of the ξ_k exist and if, for every A and every k,

$$\int_{A_k} \xi_k = \int_{A_1} \xi_1.$$

We say that ξ is *P-invariant* or that the sequence $\xi_1, \xi_2, \cdots$ is *stationary* if, for every A defined on it,

$$PA_k = PA_1.$$

In particular, the family itself is (*integral*) *stationary*, if the sequence $X_1, X_2, \cdots Y_1, Y_2, \cdots$ is (*integral*) *stationary*. It follows from the definitions that the sequences of translates of all r.v.'s ξ defined on a stationary family are stationary. Furthermore

a. STATIONARITY LEMMA. *The sequences of translates of r.v.'s, defined on a stationary family, and whose expectations exist, are integral stationary.*

Proof. We have to prove that, for every A and every k,

$$\int_{A_k} \xi_k = \int_{A_1} \xi_1,$$

provided the right-hand side exists. As usual, it suffices to prove it for indicators $\xi_1 = I_{B_1}$. Since set operations are preserved under translations, it follows, by stationarity, that

$$\int_{A_k} I_{B_k} = PA_kB_k = P(A_1B_1)_k = PAB = \int_{A_1} I_{B_1},$$

and the lemma follows.

In the integral stationarity case, the basic ergodic inequality takes very simple forms.

b. STATIONARITY INEQUALITIES. *Let the family* $\{X_n\}$, $\{Y_n\}$ *be integral stationary and let* X_1 *or* Y_1 *be integrable. Then for every invariant* C

$$\int_{C\underline{C}_a} (aY_1 - X_1) \geqq 0 \quad \text{and} \quad \int_{C\overline{C}_b} (X_1 - bY_1) \geqq 0$$

where, for X_1 *and* Y_1 *integrable on* C, $\underline{C}_a$ *and* $\overline{C}_b$ ($< a$ *and* $> b$) *can be replaced by* $\underline{B}_a = [\inf X^n/Y^n < a]$ *and* $\overline{B}_b = [\sup X^n/Y^n > b]$($\leqq a$ *and* $\geqq b$).

Proof. On account of integral stationarity and of integrability of X_1 or Y_1 hence of all the X_k or Y_k, the integrals below exist and

$$\int_{A_k} (X_k - bY_k) = \int_{A_1} (X_1 - bY_1).$$

It follows from the hypotheses that the basic ergodic inequality, with $Z^n = n$, can be written

$$\int_{B^mC} (X_1 - bY_1) + \frac{m}{n} \int_C (X_1 - bY_1)^+ \geqq 0$$

where, as $m \to \infty$,

$$B^m = \left[\sup_{j \leqq m} \frac{X^j}{Y^j} > b\right] \uparrow \overline{B}_b.$$

Therefore

$$\int_{C\bar{C}_b} (X_1 - bY_1) \geqq 0$$

since, either $\int_{C\bar{C}_b} (X_1 - bY_1)^+ = \infty$ and this inequality is trivially true, or $\int_{C\bar{C}_b} (X_1 - bY_1)^+ < \infty$ and this inequality is obtained upon replacing C by $C\bar{C}_b$ and letting $n \to \infty$ then $m \to \infty$. By changing b into $-a$ and $X_1, X_2, \cdots$ into $-X_1, -X_2, \cdots$, the other asserted inequality follows; similarly for the remaining assertions.

When the family is integral stationary and X_1 and Y_1 are integrable, then 30.2B$'$, with $Z^n = n$, yields $\frac{X^n}{Y^n} \xrightarrow{\text{a.s.}} U$ invariant. However, by using directly the basic ergodic inequality in its foregoing forms, the assumptions can be weakened and the result be made more precise, as follows.

A. Integral stationarity theorem. *Let the family* $\{X_n\}$, $\{Y_n\}$ *be integral stationary, and let* X_1 *or* Y_1 *be integrable. Then*

$$\frac{X_1 + \cdots + X_n}{Y_1 + \cdots Y_n} \xrightarrow{\text{a.s.}} U \quad \textit{invariant}.$$

If X_1 *is integrable, then* U *is a.s. finite.*
If Y_1 *is integrable, then* $U = E^{\mathfrak{C}}X_1/E^{\mathfrak{C}}Y_1$ *a.s.*

Proof. Upon replacing C by

$$C_{ab} = \left[\liminf \frac{X^n}{Y^n} < a < b < \limsup \frac{X^n}{Y^n}\right],$$

the stationarity inequalities become

$$\int_{C_{ab}} (X_1 - bY_1) \geqq 0, \quad \int_{C_{ab}} (aY_1 - X_1) \geqq 0,$$

so that

$$\int_{C_{ab}} (a - b)Y_1 \geqq 0.$$

Since $a < b$ and $Y_1 > 0$ a.s., it follows that $PC_{ab} = 0$. Since

$$D = \left[\liminf \frac{X^n}{Y^n} \neq \limsup \frac{X^n}{Y^n}\right] = \bigcup_{a,b} C_{ab},$$

where a and $b > a$ vary over all rationals, is a countable union of null sets, $PD = 0$ and the first assertion follows. From now on, we throw out of Ω the null set D; this does not modify the values of the integrals below.

According to the stationarity inequalities, if $\int X_1^+ < \infty$, then, as $b \to \infty$,

$$\int_{[U=+\infty]} Y_1 \leqq \int_{\overline{C}_b} Y_1 \leqq \frac{1}{b}\int_{\overline{C}_b} X_1^+ \to 0,$$

and, since $Y_1 > 0$ a.s., $P[U = +\infty] = 0$; similarly, if $\int X_1^- < \infty$ then $P[U = -\infty] = 0$. The second assertion follows.

Let Y_1 be integrable and set

$$C^m = [(m-1)\epsilon \leqq U < m\epsilon], \quad \epsilon > 0,$$

so that

$$\sum_{m=-\infty}^{+\infty} C^m = [|U| < \infty].$$

The stationarity inequalities with a, b, C replaced by $m\epsilon$, $(m-1)\epsilon$, CC^m, respectively, yield

$$(m-1)\epsilon\int_{CC^m} Y_1 \leqq \int_{CC^m} X_1 \leqq m\epsilon\int_{CC^m} Y_1,$$

while, by definition of C^m,

$$(m-1)\epsilon\int_{CC^m} Y_1 \leqq \int_{CC^m} UY_1 \leqq m\epsilon\int_{CC^m} Y_1.$$

If U is finite, then, by summing over $m = 0, \pm 1, \pm 2, \cdots$ and taking into account that $\int_C X_1$ exists, and $\int_C Y_1 < \infty$, we find that

$$\int_C X_1 - \epsilon\int_C Y_1 \leqq \int_C UY_1 \leqq \int_C X_1 + \epsilon\int_C Y_1$$

and, by letting $\epsilon \to 0$, it follows that

$$\int_C UY_1 = \int_C X_1, \quad C \in \mathcal{C}.$$

Since U is $\mathcal{C}$-measurable, we can write

$$UE^{\mathcal{C}}Y_1 = E^{\mathcal{C}}(UY_1) = E^{\mathcal{C}}X_1 \text{ a.s.},$$

and, since $Y_1 > 0$ a.s. implies that $E^{\mathfrak{C}}Y_1 > 0$ a.s., we have $U = E^{\mathfrak{C}}X_1/E^{\mathfrak{C}}Y_1$ a.s.

If U is not finite, the above equality of integrals continues to hold provided C is replaced by $C[|U| < \infty]$, so that $U = E^{\mathfrak{C}}X_1/E^{\mathfrak{C}}Y_1$ on $[|U| < \infty]$ outside a null subset. But, by the stationarity inequalities with C replaced by $C[U = +\infty]$, we have

$$\int_{C[U=+\infty]} X_1 \geqq b \int_{C[U=+\infty]} Y_1,$$

so that, by letting $b \to \infty$, if $PC[U = +\infty] > 0$, then

$$\int_{C[U=+\infty]} E^{\mathfrak{C}}X_1 = +\infty,$$

and hence $E^{\mathfrak{C}}X_1 = +\infty$ on $[U = +\infty]$ outside a null subset. Since $E^{\mathfrak{C}}Y_1 < \infty$ a.s., it follows that

$$U = +\infty = E^{\mathfrak{C}}X_1/E^{\mathfrak{C}}Y_1 \quad \text{on} \quad [U = +\infty]$$

outside a null subset; similarly, for $[U = -\infty]$. Thus, $U = E^{\mathfrak{C}}X_1/E^{\mathfrak{C}}Y_1$ a.s. whether U is finite or not, and the last assertion is proved.

From now on, we consider only families consisting of one sequence $\{X_n\}$ so that the events and translations are defined in terms of the $\{X_n\}$. We use the fact that if EX_1 exists, then, by the stationarity lemma, stationarity of $\{X_n\}$ is equivalent to integral stationarity of the family $\{X_1, X_2, \cdots\}$ and $\{1, 1, \cdots\}$.

B. STATIONARITY THEOREM. *Let the family $\{X_n\}$ be stationary. If EX_1 exists, then*

$$\frac{X_1 + \cdots + X_n}{n} \xrightarrow{\text{a.s.}} E^{\mathfrak{C}}X_1.$$

If $E|X_1|^r < \infty$ for an $r \geqq 1$, then

$$\frac{X_1 + \cdots + X_n}{n} \underset{\text{a.s.}}{\xrightarrow{r}} E^{\mathfrak{C}}X_1.$$

Proof. The first assertion follows from the integral stationarity theorem with $Y_n = 1$. The second assertion will follow from the first one on account of the L_r-convergence theorem, if we prove that the rth powers of $|\overline{X^n}| = \left|\frac{1}{n}(X_1 + \cdots + X_n)\right|$ are uniformly integrable. But, by the stationarity lemma and integrability of $|X_1|^r$, for every k,

$$\int_{[|X_k|\geqq c]}|X_k|^r = \int_{[|X_1|\geqq c]}|X_1|^r < \epsilon,$$

where $\epsilon > 0$ is arbitrarily small and $c \geqq c_\epsilon$ sufficiently large. Therefore, for PA sufficiently small and for every k,

$$\int_A |X_k|^r = \int_{A[|X_k|\geqq c]}|X_k|^r + \int_{A[|X_k|<c]}|X_k|^r \leqq \epsilon + c^r PA < 2\epsilon$$

and hence, by Minkowski's inequality, $\int_A |\overline{X^n}|^r < 2\epsilon$ whatever be n. By the same lemma and inequality $E|\overline{X^n}|^r \leq E|X_1|^r$. The assertion is proved.

COROLLARY. *If $\{X_n\}$ is stationary, then*

$$\frac{X_1 + \cdots + X_n}{n} \to E^{\mathfrak{C}}X_1^+ - E^{\mathfrak{C}}X_1^-$$

on the set on which the right-hand side generalized c.exp of X_1 exists, outside a null subset.

Apply the stationarity theorem to $\frac{1}{n}(X_1^+ + \cdots + X_n^+)$ and to $\frac{1}{n}(X_1^- + \cdots + X_n^-)$.

REMARK 1. In the case of integrable X_1, the equality $\overline{X} =$ a.s. $\lim \overline{X^n} = E^{\mathfrak{C}}X_1$ a.s., results also from convergence in the first mean, implied, according to the above theorem, by $E|X_1| < \infty$. For then we can pass to the limit under the integration sign, so that

$$\int_C X_1 = \int_C \overline{X^n} \to \int_C \overline{X}, \quad C \in \mathfrak{C}.$$

REMARK 2. It is useful to observe that convergence in the rth mean follows from (1) bounded sequences which converge a.s. (or even only in pr.) converge in the rth mean, and (2) bounded functions are dense in L_r. The first property is immediate. The second property is exploited by setting

$$X_1 = \xi_1 + \eta_1, \quad \xi_1 = X_1 I_{[|X_1|<c]}, \quad \eta_1 = X_1 I_{[|X_1|\geqq c]},$$

so that $E|X_1|^r < \infty$ implies that as $m, n \to \infty$ then $c \to \infty$,

$$\begin{aligned}\|\overline{X^m} - \overline{X^n}\| &\leqq \|\overline{\xi^m} - \overline{\xi^n}\| + \|\overline{\eta^m}\| + \|\overline{\eta^n}\| \\ &\leqq \|\overline{\xi^m} - \overline{\xi^n}\| + 2\|\eta_1\| \to 0.\end{aligned}$$

33.4. Applications; ergodic hypothesis and independence. We say that a sequence X_n of r.v.'s is *indecomposable* if all invariant functions defined on it degenerate into constants or, equivalently, if its invariant σ-field consists of $\emptyset$ and Ω only, up to an equivalence. We say that the *ergodic hypothesis* is true for the sequence X_n if $\frac{1}{n}\sum_{k=1}^{n} \xi_k \xrightarrow{\text{a.s.}} E\xi_1$ whatever be the r.v. $\xi \in L$ defined on the sequence. Since for an indecomposable sequence $E^{\mathfrak{C}}\xi = E\xi$ a.s., we have, on account of the stationarity theorem,

If the sequence X_n is stationary, then the ergodic hypothesis is true if, and only if, the sequence X_n is indecomposable.

Let the r.v.'s X_n be independent. Then the sequence X_n is stationary if, and only if, the X_n are identically distributed. On the other hand, by the zero-one law, its tail σ-field reduces to $\emptyset$ and Ω up to an equivalence and, the invariant events being tail-events, the sequence is indecomposable. Thus

The ergodic hypothesis is true for sequences of independent and identically distributed r.v.'s.

In particular, if $E|X_1| < \infty$, then $\frac{1}{n}\sum_{k=1}^{n} X_k \xrightarrow{\text{a.s.}} EX_1$ finite (and, moreover, converges in the first mean). Conversely, if $\frac{1}{n}\sum_{k=1}^{n} X_k \xrightarrow{\text{a.s.}} c$ finite, then $X_n/n \xrightarrow{\text{a.s.}} 0$ and, by Borel's zero-one law,

$$E|X_1| \leqq 1 + \sum P[|X_n| \geqq n] < \infty.$$

Thus, we have Kolmogorov's strong law of large numbers. In fact, it can be made more precise, as follows:

Let the r.v.'s X_n be independent and identically distributed. If EX_1 exists, then $\frac{1}{n}\sum_{k=1}^{n} X_k \xrightarrow{\text{a.s.}} EX_1$. If $\frac{1}{n}\sum_{k=1}^{n} X_k \xrightarrow{\text{a.s.}} \overline{X}$ necessarily degenerate at some constant c, then c finite implies that EX_1 exists and is finite, while $c = +\infty\ (-\infty)$ implies that $EX_1^+\ (EX_1^-) = +\infty$.

First observe that degeneracy follows by the zero-one law.

The convergence assertion follows from the stationarity theorem.

If $\frac{1}{n}\sum_{k=1}^{n} X_k \xrightarrow{\text{a.s.}} c$ finite, then, according to the converse above, c finite implies that EX_1 exists and is finite. If $\frac{1}{n}\sum_{k=1}^{n} X_k \xrightarrow{\text{a.s.}} c = +\infty$, then

$$EX_1^+ \leftarrow \frac{1}{n}\sum_{k=1}^{n} X_k^+ = \frac{1}{n}\sum_{k=1}^{n} X_k + \frac{1}{n}\sum_{k=1}^{n} X_k^- \xrightarrow{\text{a.s.}} +\infty + EX_1^-$$

and, hence, $EX_1^+ = +\infty$. Similarly, if $c = -\infty$, then $EX_1^- = \infty$. The proposition is proved.

Because the ergodic hypothesis cannot be true for decomposable stationary sequences, the brutal answer to the ergodic problem is in the negative. However, its wreck can be salvaged in various ways. One way is to assert that it is true "in general" with a suitable definition of this term—the best is a category definition. Another approach can be stated as follows: Observe that when the sequence X_n is stationary, then, for all events B defined on it, $\frac{1}{n}\sum_{k=1}^{n} I_{B_k} \xrightarrow{\text{a.s.}} P^{\mathfrak{C}}B$. If the decomposition theorem 26.2**A** applies, then

$$P^{\mathfrak{C}} = \sum_{t \in T} P_{B_t} \cdot I_{B_t}, \quad P_{B_t}B_t = 1,$$

provided a null event N is thrown out of Ω. Then the ergodic hypothesis is true, provided P is replaced by any of the P_{B_t} into which P is decomposed, or Ω is replaced by any of the B_t. This is the *ergodic decomposition.*

***33.5. Applications; stationary chains.** Let X_n be a sequence of r.v.'s (or more generally random vectors) with same range space R and let P_n on the Borel field $\mathfrak{B}$ in R be the distribution of X_n. We recall that when the sequence X_n is a constant chain its law is described by means of the initial distribution P_1 of X_1 and the (one-step) transition probability (tr.pr.)

$$P(x, S) = P(X_{m+1} \in S \mid X_m = x), \quad x \in R, \quad S \in \mathfrak{B}, \quad m = 1, 2, \cdots.$$

We can and do select the tr.pr. to be regular; in other words, $P(x, S)$ is a pr. in S for every fixed x and is a Borel function in x for every fixed S. Then the same is true of its iterates

$$P^n(x, S) = P[X_{m+n} \in S \mid X_m = x]$$

and

$$P^{m+n}(x, S) = \int P^m(x, dy)P^n(y, S)$$

$$P_nS = \int P_1(dx)P^{n-1}(x, S).$$

The chain is stationary if, and only if, the initial distribution is invariant under the tr.pr., that is, under translations: $P_n = \bar{P}$, $n = 1, 2, \cdots$.

From now on, we assume that the X_n form a stationary chain ($P_n = \bar{P}$, $n = 1, 2, \cdots$) with regular tr.pr. $P(x, S)$ and invariant initial distribution $\bar{P}$.

Since the stationary chain is described in terms of the common distribution $\bar{P}$ of single r.v.'s X_n and the common conditional pr.'s $P^n(x, S)$ of events defined on single r.v.'s X_{m+n} given X_m, the limit properties of the chain are essentially related to those of single r.v.'s. All the more so since the invariant events defined on the chain are a.s. defined on any single r.v., say X_1. For, every invariant event C is defined on X_n, $X_{n+1}, \cdots$ whatever be n and, because of stationarity, chain dependence, and martingale convergence theorem,

$$P^{X_1}C = P^{X_n}C = P^{X_1 \cdots X_n}C \to I_C \text{ a.s.}$$

Thus, the inverse image of the sub σ-field $\mathcal{C}$ of Borel sets S such that $P(x, S) = I_S(x) = 1$ or 0 according as $x \in S$ or $x \notin S$ is equivalent to the σ-field of invariant events defined on the stationary chain X_n; by abuse of language, we shall call $\mathcal{C}$ the σ-field of *invariant Borel sets*. We are ready to investigate the asymptotic behavior of the tr.pr.'s of our stationary chain and follow Doob.

A. Invariant tr.pr. theorem. *There exists a tr.pr. $\bar{P}(x, S)$ such that*
(i) *For every S and every $x \notin N_S$ with $\bar{P}N_S = 0$*

$$\frac{1}{n}\sum_{k=1}^{n} P^k(x, S) \to \bar{P}(x, S)$$

and

$$\bar{P}(x, S) = \int \bar{P}(x, dy)P(y, S) = \int P(x, dy)\bar{P}(y, S) = \int \bar{P}(x, dy)\bar{P}(y, S).$$

(ii) *For every S and every invariant S'*

$$\bar{P}SS' = \int_{S'} \bar{P}(dx)\bar{P}(x, S).$$

The equalities in (i) say that, except for $x \in N_S$, the tr.pr.'s $\bar{P}(x, S)$ and $P(x, S)$ are invariant under one another and that $\bar{P}(x, S)$ is idempotent. Property (ii) says that $\bar{P}(x, S)$ is a Radon-Nikodym derivative of $\bar{P}$ given $\mathcal{C}$.

Proof. Since the sequence $I_{[X_n \in S]}$ of translations of the indicator $I_{[X_1 \in S]}$ is defined on the stationary sequence X_n, it follows that

$$\frac{1}{n}\sum_{k=1}^{n} I_{[X_k \in S]} \xrightarrow{\text{a.s.}} \bar{I}(S) \quad \text{invariant.}$$

Therefore, upon taking the conditional exp.'s given X_1 and applying the conditional dominated convergence theorem, we have

$$\frac{1}{n}\sum_{k=1}^{n} P^{X_1}[X_k \in S] \xrightarrow{\text{a.s.}} E^{X_1}\bar{I}(S).$$

Thus, denoting the limit by $\bar{P}(x, S)$, we have

$$\frac{1}{n}\sum_{k=1}^{n} P^k(x, S) \to \bar{P}(x, S)$$

except for $x \in N_S$ such that $\bar{P}_1 N_S = P[X_1 \in N_S] = 0$.

On account of stationarity, if S' is an invariant Borel set, then

$$\begin{aligned}\int_{S'} \bar{P}(dx)\left\{\frac{1}{n}\sum_{k=1}^{n} P^k(x, S)\right\} &= \frac{1}{n}\sum_{k=1}^{n} P[X_k \in S, X_1 \in S'] \\ &= \frac{1}{n}\sum_{k=1}^{n} P[X_k \in S, X_k \in S'] \\ &= P[X_1 \in SS'] = \bar{P}SS'.\end{aligned}$$

Therefore, upon letting $n \to \infty$ and using the dominated convergence theorem, we obtain

$$\int_{S'} \bar{P}(dx)\bar{P}(x, S) = \bar{P}SS'.$$

Since $\bar{P}(x, S)$ is a conditional pr. of the distribution $\bar{P}$ given the sub σ-field $\mathcal{C}$ of Borel sets in R, we can and do regularize it. Then $\bar{P}(x, S)$ is $\mathcal{C}$-measurable in x for every fixed S and the equalities in (i) follow from the fact that the indefinite integrals on $\mathcal{C}$ of their terms coincide. This concludes the proof.

The exceptional $\bar{P}$-null sets N_S of starting points x vary in general with the entrance sets S. The question arises how to recognize points which do not belong to the exceptional sets and to find conditions under which these sets do not vary with the entrance sets. We denote by $P_s^n(x, S)$ the $\bar{P}$-singular part of $P^n(x, S)$ and by S_n a $\bar{P}$-null set such that $P^n(x, S_n) = P_s^n(x, R)$; S_n depends upon n and x.

a. Singular tr.pr. lemma. *For every fixed* $x \in R$, *the sequence* $P_s^n(x, R)$ *is nonincreasing and hence converges.*

For, if $m < n$ and S_0 is the $\bar{P}$-null set of points y for which $P(X_m = y; X_n \in S_n) > 0$, then

$$\begin{aligned} P_s^n(x, R) &= P^n(x, S_n) \\ &= \int P^m(x, dy) P(X_m = y; X_n \in S_n) \leqq P^m(x, S_0) \\ &\leqq P_s^m(x, R). \end{aligned}$$

We set $\bar{P}_s(x, R) = \lim P_s^n(x, R) = \lim P^n(x, S_n)$ and call it *singular tr.pr.*

B. Vanishing singular tr.pr. theorem. *For every* x *such that* $\bar{P}_s(x, R) = 0$ *and for every* S,

$$\frac{1}{n} \sum_{k=1}^{n} P^k(x, S) \to \bar{P}(x, S)$$

where $\bar{P}(x, S)$ *is* $\bar{P}$*-continuous and*

$$\bar{P}(x, S) = \int \bar{P}(x, dy) P(y, S).$$

If $\bar{P}_s(x, R) = 0$ *except for* $x \in N_s$ *with* $\bar{P}N_s = 0$, *then moreover for every* $x \notin N_s$ *and every* S, $\bar{P}(x, S)$ *is idempotent:*

$$\bar{P}(x, S) = \int \bar{P}(x, dy) \bar{P}(y, S).$$

Proof. Fix $x \in [\bar{P}_s(x, R) = 0]$. For $m < n$ and arbitrary S, we have

$$\frac{1}{n} \sum_{k=1}^{n} P^k(x, S) = \frac{1}{n} \sum_{k=1}^{m} P^k(x, S) + \int P^m(x, dy) \left\{ \frac{1}{n} \sum_{k=1}^{n-m} P^k(y, S) \right\}.$$

According to theorem **A**, as $n \to \infty$ the integrand converges to $\bar{P}(y, S)$ except on a $\bar{P}$-null set N_S.

Since the integration measure is $\bar{P}$-continuous for events in S_m^c, we can apply the dominated convergence theorem for the integral taken over S_m^c. As for the integral taken over S_m, it is bounded by $P^m(x, S_m) = P_s^m(x, R) \to 0$ as $m \to \infty$. Therefore, by letting $n \to \infty$ and then

$m \to \infty$, the limit assertion follows. If S_0 is a $\bar{P}$-null set, then

$$\bar{P}(x, S_0) = \lim \frac{1}{n} \sum_{k=1}^{n} P^k(x, S_0) \leqq \lim \frac{1}{n} \sum_{k=1}^{n} P_s{}^k(x, R) = 0$$

so that $\bar{P}(x, S)$ is $\bar{P}$-continuous. Moreover, by letting $n \to \infty$ in

$$\frac{1}{n} \sum_{k=m+1}^{m+n} P^k(x, S) = \int \left\{ \frac{1}{n} \sum_{k=1}^{n} P^k(x, dy) P^m(y, S) \right\},$$

we obtain

$$\bar{P}(x, S) = \int \bar{P}(x, dy) P^m(y, S);$$

hence

$$\bar{P}(x, S) = \int \bar{P}(x, dy) \left\{ \frac{1}{n} \sum_{k=1}^{n} P^k(y, S) \right\}.$$

Now, assume that the set of points x such that what precedes does not hold is a $\bar{P}$-null set. Because of the $\bar{P}$-continuity of $\bar{P}(x, S)$, this exceptional set is also null in the integration measure and we can apply the dominated convergence theorem. This yields idempotency of $\bar{P}(x, S)$ and concludes the proof.

COROLLARY. *If $\bar{P}_s(x, R) = 0$ for every $x \in R$, then*

$$\bar{P}(x, S) = \int P^m(x, dy) \bar{P}(y, S)$$

This follows from the limit assertion in the foregoing theorem upon letting $n \to \infty$ in

$$\frac{1}{n} \sum_{k=1+m}^{n+m} P^k(x, S) = \int P^m(x, dy) \left\{ \frac{1}{n} \sum_{k=1}^{n} P^k(y, S) \right\}.$$

C. DECOMPOSITION THEOREM. *There exists a partition*

$$R = \sum_{t \in T} S_t + N, \quad T \subset R, \quad \bar{P}N = 0$$

such that $P(x, S_t) = 1$ for every $x \in S_t$, and pr.'s $\bar{P}_t$ such that

$$\bar{P}(x, S) = \sum_{t \in T} I_{S_t}(x) \bar{P}_t S, \quad x \notin N, \quad \bar{P}_t B_t = 1$$

and the $\bar{P}_t$ are invariant:

$$\bar{P}_t S = \int \bar{P}_t(dx) P(x, S).$$

In fact, every pr. $\bar{P}'$ of the form

$$\bar{P}'S = \int \bar{P}_t S \mu(dt)$$

where μ is a pr. on a σ-field in T is invariant, and if an invariant pr. is $\bar{P}$-continuous, then the converse is true.

Proof. Since the Borel field $\mathcal{B}$ is generated by a denumerable field $\{S(n)\}$ of Borel sets (say, the field of finite sums of intervals with rational extremities) and the conditional pr. $\bar{P}(x, S)$ given the σ-field of invariant Borel sets is regularized, the decomposition theorem 26.2**A** applies. This yields the asserted decomposition with invariant atoms S_t, and the asserted properties of $P(x, S_t)$ and $\bar{P}(x, S)$. As for the invariance of the $\bar{P}_t$, since, by theorem **A**, for every fixed S,

$$\bar{P}(x, S) = \int \bar{P}(x, dy)P(y, S), \quad x \notin N_S, \quad \bar{P}N_S = 0,$$

it follows that

$$\bar{P}_t S = \int \bar{P}_t(dx)P(x, S)$$

except for indices $t \in T_S \subset T$ corresponding to sets S_t of total $\bar{P}$-pr. zero. We add such sets for $S = S(1), S(2), \cdots$ to the P-null set N of the partition so that the last equality holds for all remaining t's and every $S(n)$. Since the $S(n)$ form a field, it follows as usual that the equality holds for all Borel sets S. This proves the invariance of the pr.'s $\bar{P}_t$, and by integrating the invariance relation with respect to the pr. μ in t we find that the pr. $\bar{P}'$ is invariant:

$$\bar{P}'S = \int \bar{P}'(dx)P(x, S).$$

Conversely, if $\bar{P}'$ is an invariant pr., then for every n

$$\bar{P}'S = \int \bar{P}'(dx) \left\{ \frac{1}{n} \sum_{k=1}^{n} P^k(x, S) \right\}$$

By theorem **A**, the integrand converges to $\bar{P}(x, S)$ except for x belonging to a $\bar{P}$-null set. Thus, if $\bar{P}'$ is $\bar{P}$-null, then the exceptional set is $\bar{P}'$-null, the dominated convergence theorem applies and

$$\bar{P}'S = \int \bar{P}'(dx)\bar{P}(x, S).$$

Upon using the decomposition of $\bar{P}(x, S)$, the converse assertion is proved, and the proof is concluded.

COROLLARY. *Every component chain* $\{\bar{P}_t, P(x, S)\}$ *is stationary and indecomposable.*

*§34. ERGODIC THEOREMS AND L_r-SPACES

Let $(\Omega, \mathcal{A}, P)$ be our pr. space. In the usual ergodic theory the primary datum is a one-to-one transformation T_1 on Ω to Ω. In fact, the basic underlying concept is that of the inverse transformation T_1^{-1} operating on sets and not on points: $T_1^{-1}A = [T_1^{-1}\omega;\ \omega \in A]$. T_1^{-1} preserves all sets operations and is said to be *measurable* if it transforms measurable sets into measurable sets. But this is precisely what the translations along, say, a sequence $X_1, X_2, \cdots$ of r.v.'s do; they transform events defined on the sequence into events defined on the same sequence, that is, they transform the sub σ-field $\mathcal{B}$ of events induced by the sequence into itself. Once the translates of these events, hence of their indicators, are determined, they determine the translates of simple and then measurable functions defined on the sequence and, in particular, determine the sequence itself, given its first term. Thus the primary datum becomes that of translations of events; it is more general than that of point transformations. In the sequel, the σ-field of events to be translated is the whole σ-field $\mathcal{A}$ of events, but it may as well be any fixed sub σ-field $\mathcal{B}$, whether induced by a random function or not.

34.1. Translations and their extensions. We say that a single-valued transformation T on the σ-field $\mathcal{A}$ of events into itself is a *translation* (by 1) if it preserves (commutes with) all countable operations on events and preserves Ω and $\emptyset$: for any $A, A_j \in \mathcal{A}$

$$T(A^c) = (TA)^c,\quad T\bigcap A_j = \bigcap TA_j,\quad T\bigcup A_j = \bigcup TA_j,$$

$$T\sum A_j = \sum TA_j,\quad T\Omega = \Omega,\quad T\emptyset = \emptyset.$$

In fact, it suffices that T preserve complementations and countable intersections (unions); for preservation of countable unions (intersections) follows by the de Morgan rules, while that of Ω, hence of $\emptyset$, and consequently of disjunctions and countable sums, follows by

$$T\Omega = T(A \cup A^c) = TA \cup TA^c = TA \cup (TA)^c = \Omega.$$

Thus, T translates the σ-field $\mathcal{A}$ into a σ-field $T\mathcal{A}$, then $T\mathcal{A}$ into $T(T\mathcal{A})$

$= T^2\mathfrak{A}$, and so on. The translation T^k (by $k = 1, 2, \cdots$) is the kth iterate of T and $T^0 = I$ is the identity transformation: $IA = A, A \in \mathfrak{A}$.

Let ξ on Ω be a measurable function. The *translate* by $k - 1$ of ξ is the measurable function $\xi_k = T^{k-1}\xi$ (so that $\xi_1 = \xi$) determined by

$$(1) \qquad [\xi_k \in S] = T^{k-1}[\xi \in S]$$

for all Borel sets $S \subset \bar{R}$; for, then, ξ_k assigns to any $\omega \in \Omega$ a value $x \in \bar{R}$ by the correspondence

$$\xi_k(\omega) = x \Leftrightarrow \omega \in T^{k-1}[\xi = x];$$

in other words, any atom $[\xi_k = x]$ of the sub σ-field of events induced by ξ_k is the translate by $k - 1$ of the atom $[\xi = x]$ of the sub σ-field of events induced by ξ. Conversely, letting r vary over the rationals,

$T^{k-1}[\xi = y] = \bigcap_{r>y>s} T^{k-1}[s \leq \xi < r]$ implies that

$$[\xi_k \leq x] = \bigcap_{r>s} T^{k-1}[\xi < r] = T^{k-1}[\xi \leq x].$$

Let $\mathfrak{M}$ be the family of measurable functions on $(\Omega, \mathfrak{A})$, to be denoted by ξ with or without affixes. Let $I_{\mathfrak{A}}$ be the subfamily of indicators of events. Translation T on $\mathfrak{A}$ can also be considered as defined on $I_{\mathfrak{A}}$ (to $I_{\mathfrak{A}}$), and relation (1) extends it to a transformation T on $\mathfrak{M}$ (to $\mathfrak{M}$) which we continue to call a translation. We intend to show that this extension is linear: $T(a\xi + a'\xi') = aT\xi + a'T\xi'$ and continuous: $T(\lim \xi^{(n)}) = \lim T\xi^{(n)}$. More precisely

A. Extension theorem. *Relation* (1) *extends the translation T on $I_{\mathfrak{A}}$ to a linear and continuous transformation T on $\mathfrak{M}$, with $T1 = 1$ and $TI_{AB} = TI_A \cdot TI_B$. Conversely, the restriction of such a transformation to $I_{\mathfrak{A}}$ is a translation on $I_{\mathfrak{A}}$.*

Proof. The converse assertion is immediate. As for the direct assertion, it is obvious that $T1 = 1$; linearity follows from the relations below where $a > 0$ and r varies over the set of all rationals:

$$[T(-\xi) < x] = T[-\xi < x] = T[\xi > -x] = [T\xi > -x] = [-T\xi < x],$$

$$[T(a\xi) < x] = T[a\xi < x] = T[\xi < x/a] = [T\xi < x/a] = [aT\xi < x],$$

$$[T(\xi + \xi') < x] = T[\xi + \xi' < x] = T\bigcup_r [\xi < r][\xi' < x - r]$$

$$= \bigcup_r [T\xi < r][T\xi' < x - r] = [T\xi + T\xi' < x];$$

continuity follows from the relations

$$[T \sup \xi^{(n)} > x] = T \bigcup [\xi^{(n)} > x] = \bigcup [T\xi^{(n)} > x] = [\sup T\xi^{(n)} > x]$$

which mean that T commutes with $\sup_n$, hence with $\inf_n \xi^{(n)} = -\sup_n -(\xi^{(n)})$ and, consequently, with $\lim \sup_n = \inf \sup_n$ and $\lim \inf_n = \sup \inf_n$.

COROLLARY 1. *Given a translation T on $I_{\mathcal{A}}$, the translate (by 1) of a simple function $\sum_{k=1}^{n} x_k I_{A_k}$ is $\sum_{k=1}^{n} x_k I_{TA_k}$, and the translate of any ξ is the limit of the translates of any sequence of simple functions which converges to ξ.*

COROLLARY 2. *A translation T on $I_{\mathcal{A}}$ has a unique extension to translation T on $\mathfrak{M}$: a linear continuous transformation with $T1 = 1$ and $TI_{AB} = TI_A \cdot TI_B$.*

This follows from Corollary 1.

34.2. A.s. ergodic theorem. From now on, T denotes a fixed translation, and we set

$$\overline{T^n} = \frac{1}{n} \sum_{k=1}^{n} T^{k-1},$$

so that

$$\overline{T^n}\xi = \overline{\xi^n} = (\xi_1 + \cdots + \xi_n)/n.$$

We denote by $\mathcal{C}$ the sub σ-field of events invariant under T, to be called *invariant events:* $TC = C$, $C \in \mathcal{C}$, set

$$\overline{P^n}A = E\overline{T^n}I_A = \frac{1}{n} \sum_{k=1}^{n} PT^{k-1}A, \quad A \in \mathcal{A},$$

and observe that $\overline{P^n}$ so defined on $\mathcal{A}$ is a pr. coinciding with P on $\mathcal{C}$.

According to the definitions, the translates of events defined on the sequence of translates $X_n = T^{n-1}X$ of a r.v. X coincide with the translates along this sequence as defined in the preceding section. Thus, all propositions therein apply to such sequences. Yet, the primary datum being now the translation and not the sequence, the outlook changes and new problems arise:

Find conditions to be imposed upon T under which the sequences $\overline{T^n}X$ converge in some sense for (as large as possible) families of r.v.'s. Furthermore, find families for which these conditions on T are not only sufficient but also necessary for various types of convergence.

We begin by observing that, upon setting $Y_n = 1$ in the basic ergodic theorem, it yields

a. ERGODIC LEMMA. *Let*

$$\limsup \overline{P^n} A \to 0 \quad as \quad A \downarrow \emptyset.$$

Then $\overline{T^n}X \xrightarrow{\text{a.s.}} \overline{T}X$ *invariant, for every r.v.* X *such that*

$$\frac{1}{n}\int | T^{n-1}X | \to 0 \quad and \quad \frac{1}{n}\sum_{k=1}^{n}\int_{T^{k-1}A} T^{k-1}X \to 0,$$

as $n \to \infty$ *and then* $A \downarrow \emptyset$.

In particular, $\overline{T^n}X \xrightarrow{\text{a.s.}} \overline{T}X$ *bounded, for every bounded r.v.* X.

The particular case follows from $|X| \leqq c < \infty$ and hence $|T^k X| \leqq c$, so that the foregoing integral is bounded by c/n while the sum of integrals is bounded by $c\overline{P^n}A$.

b. INVARIANT PR. LEMMA. *The three properties below are equivalent:*

(i) $\limsup \overline{P^n}A \to 0$ *as* $A \downarrow \emptyset$.

(ii) $\lim \overline{P^n} = \overline{P}$ *exists and is a pr. on* $\mathcal{A}$.

(iii) *There exists on* $\mathcal{A}$ *a pr.* $\overline{P}$ *invariant under* T *and coinciding with* P *on the σ-field* $\mathcal{C}$ *of invariant events* (*and then* $\lim \overline{P^n} = \overline{P}$).

We shall denote by $\overline{E}^{\mathcal{C}}X$ the c.pr. of X given $\mathcal{C}$ with respect to $\overline{P}$, defined by

$$\int_C \overline{E}^{\mathcal{C}} X\, d\overline{P}_{\mathcal{C}} = \int_C X\, d\overline{P}, \quad C \in \mathcal{C},$$

which exists when $\int X\, d\overline{P}$ exists.

Proof. 1° Since a finite measure is continuous at $\emptyset$, (ii) implies (i).

Conversely, if (i) holds, then, by the particular case of the ergodic lemma and the dominated convergence theorem,

$$\overline{P^n}A = \int \overline{T^n} I_A\, dP \to \int \overline{T} I_A\, dP = \overline{P}A, \quad A \in \mathcal{A}.$$

Clearly, $\overline{P}$ so defined on $\mathcal{A}$ is nonnegative and finitely additive, with $\overline{P}\Omega = 1$. Moreover, (i) becomes $\overline{P}A \to 0$ as $A \downarrow \emptyset$, so that $\overline{P}$ is also continuous at $\emptyset$. Thus, $\overline{P}$ is a pr. on $\mathcal{A}$, and (i) implies (ii).

2° If (ii) holds, then, as $n \to \infty$,

$$|\bar{P}TA - \bar{P}A| \leftarrow |\overline{P^n}TA - \overline{P^n}A| = \frac{1}{n}|PT^nA - PA| \leqq \frac{1}{n} \to 0$$

so that $\bar{P}$ is invariant under T. Since for an invariant event C, $PC = \overline{P^n}C = \bar{P}C$, it follows that $\bar{P} = P$ on $\mathcal{C}$. Thus (ii) implies (iii).

Conversely, if there exists on $\mathcal{A}$ an invariant pr. $\bar{P}$ which coincides with P on $\mathcal{C}$, then, by the stationarity theorem, $\overline{T^n}I_A \to \bar{E}^{\mathcal{C}}I_A$ outside an invariant $\bar{P}$-null and hence P-null, event and, by the dominated convergence theorem and $P_{\mathcal{C}} = \bar{P}_{\mathcal{C}}$,

$$\overline{P^n}A = \int \overline{T^n}I_A \, dP \to \int \bar{E}^{\mathcal{C}}I_A \, dP_{\mathcal{C}} = \int \bar{E}^{\mathcal{C}}I_A \, d\bar{P}_{\mathcal{C}} = \bar{P}A.$$

Thus (iii) implies (ii), and the proof is terminated.

A. A.S. ERGODIC THEOREM. *Let*

$$\limsup \overline{P^n}A \to 0 \quad \text{as} \quad A \downarrow \emptyset.$$

Then $\overline{P^n} \to \bar{P}$ *invariant pr. on* $\mathcal{A}$ *with* $\bar{P} = P$ *on* $\mathcal{C}$ *and*

(i) *For every nonnegative r.v.* X

$$\overline{T^n}X \xrightarrow{\text{a.s.}} \bar{E}^{\mathcal{C}}X$$

while for every r.v. X

$$\overline{T^n}X \to \bar{E}^{\mathcal{C}}X^+ - \bar{E}^{\mathcal{C}}X^-$$

on the invariant set on which the right-hand side generalized c.exp. exists, outside an invariant null subset.

(ii) *If* $\int X\,d\bar{P}$ *exists or if the sequences* $\overline{T^n}X^{\pm}$ *converge in pr. (a fortiori, if they converge in the* rth *mean), then* $\bar{E}^{\mathcal{C}}X$ *exists and in the second case is finite, and*

$$\overline{T^n}X \xrightarrow{\text{a.s.}} \bar{E}^{\mathcal{C}}X.$$

Proof. The first assertion follows by the invariant pr. lemma. Since all sequences $\overline{T^n}X$ are $\bar{P}$-stationary, assertion (i) follows by the stationarity theorem and the fact that an invariant $\bar{P}$-null set is P-null. The first case of assertion (ii) is immediate and the second case follows from (i) by the fact that the limits of sequences of r.v.'s which converge in pr. are r.v.'s; hence $\bar{E}^{\mathcal{C}}X = \bar{E}^{\mathcal{C}}X^+ - \bar{E}^{\mathcal{C}}X^-$ a.s. exists, outside an invariant null set.

So far, whenever $\bar{P} = \lim \overline{P^n}$ appeared, we either proved or assumed that $\bar{P}$ is a pr. Since, by Complements and Details, 19, of Chapter I, the limit of the sequence of pr.'s $\overline{P^n}$ is a pr., the fact that $\bar{P} = \lim \overline{P^n}$ *implies* that $\bar{P}$ is a pr. Thus, in this subsection, we can *drop the assumption* that $\lim \overline{P^n}$ is a pr. Then, the a.s. ergodic theorem yields at once the following

B. A.S. ERGODIC CRITERION. *The sequences $\overline{T^n}X$ converge a.s. for every nonnegative r.v. X if, and only if, the sequences $\overline{P^n}A$ converge for every event A.*

The ergodic hypothesis corresponds to T being P-indecomposable, that is, the σ-field $\mathcal{C}$ of invariant sets reducing a.s. to $\emptyset$ and Ω or, equivalently, the invariant functions degenerating into constants (finite or not).

C. INDECOMPOSABILITY THEOREM. *The following properties are equivalent.*

(i) $\overline{T^n}X \xrightarrow{\text{a.s.}} \bar{T}X$ *degenerate for every* $X \geqq 0$.

(ii) T *is P-indecomposable and* $\overline{P^n} \to \bar{P}$.

(iii) $\frac{1}{n}\sum_{k=1}^{n} I_{T^{k-1}A} \xrightarrow{\text{a.s.}} \bar{P}A$ *for every* $A \in \mathcal{A}$.

(iv) $\frac{1}{n}\sum_{k=1}^{n} P(T^{k-1}A)B \longrightarrow \bar{P}A \cdot PB$ *for every pair* $A, B \in \mathcal{A}$.

Proof. (i) $\Leftrightarrow$ (ii) by the a.s. ergodic criterion.

(ii) $\Rightarrow$ (iii) by the a.s. ergodic theorem and the fact that $\bar{P}^{\mathcal{C}}A$ degenerates into $\bar{P}A$.

(iii) $\Rightarrow$ (iv) by integrating over B with respect to P and using the dominated convergence theorem.

(iv) $\Rightarrow$ (ii) by setting $B = \Omega$ so that $\overline{P^n} \to \bar{P}$ and hence $\bar{P} = P$ on $\mathcal{C}$; then, setting $A = B = C \in \mathcal{C}$ so that $PC = \bar{P}C \cdot PC = (PC)^2$, we have $PC = 0$ or 1.

The proof is complete.

34.3. Ergodic theorems on spaces L_r. We can now attack the problem of convergence a.s. or in pr. or in the rth mean of sequences $\overline{T^n}X$ to a limit $\bar{T}X \in L_r$ *for every* $X \in L_r$, $r \geqq 1$. Since a point $X \in L_r$ is an arbitrary element of a class of equivalence and not a specific r.v., the transformation $X \to \bar{T}X$ must not split classes of equivalence, that is, is to be a mapping on L_r to L_r. This is accomplished if we assume, and *we shall do so in the sequel*, that the translation T is *null-preserving*, that is, if $PA = 0$, then $PTA = 0$. For, then $X = X'$ a.s.

implies that $\overline{T^n}X = \overline{T^n}X'$ a.s.; hence $\overline{T}X = \overline{T}X'$ a.s.

We recall that $\| X \|_r = \left(\int | X |^r\right)^{\frac{1}{r}}$ is the *norm* of $X \in L_r$ $(r \geqq 1)$, and a mapping M on L_r to L_r is

linear if $M(aX + a'X') = aMX + a'MX'$ a.s., $a, a' \in R$,
nonnegative if $X \geqq 0$ a.s. $\Rightarrow MX \geqq 0$ a.s.,
bounded (or normed) if $\| MX \|_r \leqq c\| X \|_r$, where c is a finite constant independent of $X \in L_r$; the smallest of these constants is the *norm* $\| M \|_r$ of M, defined by $\| M \|_r = \sup \| MX \|_r / \| X \|_r$ for all $X \neq 0$ a.s.

We drop the subscript r when $r = 1$. Also, denoting by "c" some type of convergence, we write $M_n \xrightarrow{c} M$ on L' when $M_nX \xrightarrow{c} MX$ for every $X \in L' \subset L_r$, and drop "on L'" when $L' = L_r$.

We require three properties of linear mappings, of which the second and the third extend at once to arbitrary Banach spaces (Banach-Steinhaus) and the first extends to partially ordered Banach spaces under a supplementary assumption on the norms.

a. LINEAR MAPPINGS LEMMA. *Let M, M_n be linear mappings on L_r to L_r, $r \geqq 1$.*

(i) *If M is nonnegative, then it is bounded.*

(ii) *If the M_n are bounded and* $\limsup \| M_nX \|_r < \infty$ *for every $X \in L_r$, then they are uniformly bounded.*

(iii) *If the M_n are uniformly bounded and $M_n \xrightarrow{r} M$ on the subspace L_∞ of all bounded r.v.'s, then $M_n \xrightarrow{r} M$ on L_r.*

Proof. 1° Because of the linearity of M, to prove (i) it suffices to show that a nonnegative M is bounded on the subspace of all a.s. nonnegative $X \in L_r$. If M is not bounded, then there exists a sequence $X_n \geqq 0$ a.s. such that $\| X_n \|_r = 1$, while $\| MX_n \|_r > n^2$. Thus

$$\sum \| X_n \|_r / n^2 < \infty; \quad \text{hence} \quad X = \sum X_n/n^2 \in L_r;$$

while, by the elementary inequality $(a + b)^r \geqq a^r + b^r$, $a, b \geqq 0$, $r \geqq 1$, as $n \to \infty$

$$\int (MX)^r \geqq \int \left\{M\left(\sum_{k=1}^{n} X_k/k^2\right)\right\}^r \geqq \sum_{k=1}^{n} \int (MX_k/k^2)^r > n \to \infty.$$

Thus $\| MX \|_r = \infty$, the assertion follows *ab contrario*, and (i) is proved.

2° Let the linear mappings M_n be bounded and let $\limsup \| M_n X \|_r < \infty$ for every $X \in L_r$. To prove that all $\| M_n \| \leqq c < \infty$, it suffices to show that all $\| M_n X' \|_r \leqq c' < \infty$ for all points X' belonging to some sphere $S = [X' : | X' - X_0 | < s]$. For then, if $\| X \| < s$, we have $\| M_n X \|_r = \| M_n(X + X_0) - M_n X_0 \|_r \leqq 2c'$; hence for any $X \in L_r$

$$\| M_n X \|_r = \left\| \frac{\| X \|_r}{s} M_n \left(\frac{sX}{\| X \|_r} \right) \right\|_r \leqq \frac{2c'}{s} \| X \|_r.$$

But $L_r = \bigcup_{m=1}^{\infty} L_r^m$, where L_r^m is the closed set of those points X for which all $\| M_n X \|_r \leqq m$. Since the space L_r is complete, it follows, by Baire's category theorem, that at least one of the L_r^m is of second category and, consequently, contains a sphere S we were looking for. Assertion (ii) is proved.

3° Let all $\| M_n \|_r \leqq c < \infty$ and let $M_n \xrightarrow{r} M$ on L_∞. For every $X \in L_r$ set $X = X' + X''$, where $X' = XI_{[|X| < k]}$ and $X'' = XI_{[|X| \geqq k]}$. Then, by linearity of the M_n and completeness of L_r, as $m, n \to \infty$, then $k \to \infty$,

$$\| M_m X - M_n X \|_r \leqq \| M_m X' - M_n X' \|_r + \| M_m X'' \|_r + \| M_n X'' \|_r$$

$$\leqq \| M_m X' - M_n X' \|_r + 2c \| X'' \|_r \to 0.$$

Assertion (iii) follows.

In what follows, c denotes some finite positive constant *independent of n*.

A. L_r-ERGODIC THEOREM. *The following implications hold on L_r to L_r with $r \geqq 1$:*

$$\begin{array}{ccc} \overline{T^n} \xrightarrow{r} \overline{T} & \Rightarrow & \overline{T^n} \xrightarrow{P} \overline{T} \Leftrightarrow \overline{T^n} \xrightarrow{\text{a.s.}} \overline{T} \Leftrightarrow \limsup \| \overline{T^n} \|_r < \infty \text{ on } L_\infty \\ \Updownarrow & & \Downarrow \end{array}$$

$$\| \overline{T^n} \|_r \leqq c \Rightarrow \overline{P^n} \leqq cP^{\frac{1}{r}} \Rightarrow \overline{P^n} \to \overline{P} \leqq cP^{\frac{1}{r}} \Rightarrow \overline{T^n} \underset{r}{\xrightarrow{\text{a.s.}}} \overline{T} \text{ on } L_\infty.$$

Proof. 1° The implications

$$\overline{T^n} \xrightarrow{r} \overline{T} \Rightarrow \overline{T^n} \xrightarrow{P} \overline{T} \Leftarrow \overline{T^n} \underset{r}{\xrightarrow{\text{a.s.}}} \overline{T}$$

require no proof.

$\overline{T^n} \xrightarrow{P} \overline{T} \Rightarrow \overline{P^n} \to \overline{P} \leqq cP^{\frac{1}{r}} \Rightarrow \overline{T^n} \underset{r}{\xrightarrow{\text{a.s.}}} \overline{T}$ on L_∞. For, $\overline{T}$ being obviously a linear and nonnegative mapping on L_r to L_r, the linear mappings lemma (i) applies, so that $\| \overline{T} \|_r \leqq c$ and, by the dominated convergence theorem,

$$\overline{P^n}A = \int \overline{T^n} I_A \, dP \rightarrow \int \overline{T} I_A \, dP$$

$$\leqq \left(\int (\overline{T} I_A)^r \, dP \right)^{\frac{1}{r}} \leqq c \left(\int (I_A)^r \, dP \right)^{\frac{1}{r}} = c P^{\frac{1}{r}} A.$$

Therefore, $\limsup \overline{P^n}A \rightarrow 0$ as $A \downarrow \emptyset$ and, by the ergodic lemma and the same theorem, $\overline{T^n} \xrightarrow[r]{\text{a.s.}} \overline{T}$ on L_∞.

$\overline{T^n} \xrightarrow{\text{P}} \overline{T} \Rightarrow \overline{T^n} \xrightarrow{\text{a.s.}} \overline{T}$ with $\overline{T} = \overline{E}^{\mathfrak{C}}$. For, then $\overline{P^n} \rightarrow \overline{P} \leqq c P^{\frac{1}{r}}$ while $\overline{T^n} X^{\pm} \xrightarrow{\text{P}} \overline{T} X^{\pm}$ for every $X \in L_r$ and hence, by the a.s. ergodic theorem, $\overline{T^n} \xrightarrow{\text{a.s.}} \overline{T}$ with $\overline{T} = \overline{E}^{\mathfrak{C}}$.

2° $\overline{T^n} \xrightarrow{\text{a.s.}} \overline{T} \Rightarrow \lim \| \overline{T^n} \|_r \leqq c$ on L_∞. For then $\overline{T^n} \xrightarrow{\text{r}} \overline{T}$ on L_∞ and hence $\| \overline{T^n} \|_r \rightarrow \| \overline{T} \|_r$ on L_∞; and we can take $c = \| \overline{T} \|_r$.

$\limsup \| \overline{T^n} \|_r \leqq c$ on $L_\infty \Rightarrow \overline{T^n} \xrightarrow{\text{a.s.}} \overline{T}$. For then, on the one hand, $\overline{P} = \limsup \overline{P^n} \leqq c P^{\frac{1}{r}}$ and hence, by 1°, $\overline{T^n} \xrightarrow{\text{r}} \overline{T}$ on L_∞, lim sup become lim, and $\| \overline{T^n} \|_r \rightarrow \| \overline{T} \|_r$ on L_∞; on the other hand, by the a.s. ergodic theorem, $\overline{T^n} X \xrightarrow{\text{a.s.}} \overline{E}^{\mathfrak{C}} X$ for every nonnegative r.v. X. Thus, to prove the assertion, it suffices to show that, for $0 \leqq X \in L_r$, we have $\overline{E}^{\mathfrak{C}} X \in L_r$. But, setting $X_m = X I_{[|X| < m]}$ so that $0 \leqq X_m \uparrow X$ as $m \rightarrow \infty$ and $X_m \in L_\infty$, we have, by what precedes, $\int (\overline{E}^{\mathfrak{C}} X_m)^r \, dP \leqq c^r \int (X_m)^r \, dP$. Letting $m \rightarrow \infty$, it follows, by the monotone and conditional monotone convergence theorems, that

$$\int (\overline{E}^{\mathfrak{C}} X)^r \, dP \leqq c^r \int (X)^r \, dP < \infty.$$

Thus, $\overline{E}^{\mathfrak{C}} X \in L_r$, and the proof is complete.

3° $\overline{T^n} \xrightarrow{\text{r}} \overline{T} \Rightarrow \| \overline{T^n} \|_r \leqq c$. For, then, clearly for every $X \in L_r$, $TX, T^2X, \cdots \in L_r$ and hence $\overline{T^n} X \in L_r$. Thus every $\overline{T^n}$ is a mapping on L_r to L_r and, these mappings being obviously linear and nonnegative, the linear mappings lemma (i) applies. Therefore, every $\overline{T^n}$ is bounded, while $\limsup \| \overline{T^n} X \|_r = \| \overline{T} X \|_r < \infty$ and hence, by the linear mappings lemma (ii), all $\| \overline{T^n} \|_r \leqq c$.

$\| \overline{T^n} \|_r \leqq c \Rightarrow \overline{P^n} \leqq cP^{\frac{1}{r}} \Rightarrow \overline{P^n} \to \overline{P} \leqq cP^{\frac{1}{r}}$. For, then

$$\overline{P^n} I_A = \int \overline{T^n} I_A \, dP \leqq \left(\int (\overline{T^n} I_A)^r \, dP \right)^{\frac{1}{r}} \leqq c \left(\int (I_A)^r \, dP \right)^{\frac{1}{r}} = cP^{\frac{1}{r}}A,$$

and hence

$$\limsup \overline{P^n} A \leqq cP^{\frac{1}{r}} A \to 0 \quad \text{as} \quad A \downarrow \emptyset.$$

Thus, by the invariant pr. lemma, $\overline{P^n} \to \overline{P} \leqq cP^{\frac{1}{r}}$.

$\| \overline{T^n} \|_r \leqq c \Rightarrow \overline{T^n} \xrightarrow{r} \overline{T}$. For, then $\overline{T^n} \xrightarrow{\text{a.s.}} \overline{T}$ on L_∞ and, hence, by the dominated convergence theorem, $\overline{T^n} \xrightarrow{r} \overline{T}$ on L_∞, the linear mappings lemma (iii) applies, and $\overline{T^n} \xrightarrow{r} \overline{T}$. The theorem is proved.

B. L_r-ERGODIC CRITERIA. *The following equivalences hold on L_r, $r \geqq 1$.*

A.s. ergodic criterion:

$$\overline{T^n} \xrightarrow{P} \overline{T} \Leftrightarrow \overline{T^n} \xrightarrow{\text{a.s.}} \overline{T} \Leftrightarrow \limsup \| \overline{T^n} \|_r \leqq c \text{ on } L_\infty.$$

Mean ergodic criterion:

$$\overline{T^n} \xrightarrow{r} \overline{T} \Leftrightarrow \sup \| \overline{T^n} \|_r \leqq c.$$

And for $r = 1$:

$$\overline{T^n} \xrightarrow{\text{a.s.}} \overline{T} \Leftrightarrow \overline{P^n} \to \overline{P} \leqq cP$$

$$\overline{T^n} \xrightarrow{1} \overline{T} \Leftrightarrow \sup \overline{P^n} \leqq cP$$

Proof. According to the L_r-ergodic theorem, it suffices to prove that

$$\overline{P^n} \to \overline{P} \leqq cP \Rightarrow \overline{T^n} \xrightarrow{\text{a.s.}} \overline{T} \quad \text{and} \quad \overline{P^n} \leqq cP \Rightarrow \| \overline{T^n} \| \leqq c.$$

The first implication follows by the a.s. ergodic theorem from the inequality

$$\int |X| \, d\overline{P} \leqq c \int |X| \, dP < \infty$$

which holds by hypothesis for all indicators $X = I_A$; hence, as usual, it holds for simple r.v. and then for any r.v. $X \in L$. Similarly, the second assertion follows by the L_r-ergodic theorem from the inequality

$$\int | \overline{T^n} X | \, dP \leqq c \int |X| \, dP$$

which holds by hypothesis for all indicators and hence for all r.v.'s $X \in L$.

REMARK. In this subsection, the translation T was assumed to be null-preserving, that is, $PA = 0$ implies that $PTA = PT^2A = \cdots = 0$. Thus, every $\overline{P^n}$ is P-continuous and, if $\overline{P^n} \to \overline{P}$ pr. on $\mathcal{A}$, then $\overline{P}$ is P-continuous. In other words, we can select nonnegative integrable r.v.'s $\overline{p^n}$ and $\overline{p}$ (with $E\overline{p^n} = E\overline{p} = 1$) such that

$$\overline{P^n}A = \int_A \overline{p^n}\, dP, \quad \overline{P}A = \int_A \overline{p}\, dP, \quad A \in \mathcal{A}.$$

The reader is invited to play around with these r.v.'s and the results of this subsection. For example, the following relations hold:

$$E^{\mathcal{C}}\overline{p^n} = E^{\mathcal{C}}\overline{p} = 1 \text{ a.s.}, \quad \overline{T}X = \overline{E}^{\mathcal{C}}X = E^{\mathcal{C}}(\overline{p}X) \text{ a.s.};$$

$$\overline{P^n} \to \overline{P} \text{ pr.} \Leftarrow \overline{p^n} \xrightarrow{1} \overline{p} \Leftarrow \overline{p^n} \xrightarrow{\text{P}} \overline{p};$$

$$\overline{P^n} \leqq cP \Leftrightarrow \overline{p^n} \leqq c \text{ a.s.}, \quad \overline{P} \leqq cP \Leftrightarrow \overline{p} \leqq c \text{ a.s.}$$

*§35. ERGODIC THEOREMS ON BANACH SPACES

The implication $\|\overline{T^n}\|_r \leqq c \Rightarrow \|\overline{T^n}X - \overline{T}X\|_r \to 0$ for every $X \in L_r$ remains meaningful for transformations T on a Banach space B of points X to itself, provided the norms are interpreted as norms in B. The question arises whether a similar ergodic implication can be obtained for Banach spaces and this *without reference to an underlying pr. space*. It is to be expected that some supplementary condition will be required; for sequences of bounded or uniformly bounded transformations have compactness properties in L_r that they do not have in general Banach spaces. We shall follow Yosida and Kakutani.

35.1. Norms ergodic theorem. Let T be a linear transformation on a Banach space B to itself, and set $\overline{T^n} = \frac{1}{n}\sum_{k=1}^{n} T^{k-1}$. Let X, with or without affixes, denote a point in B and let c be some finite positive constant independent of n.

We say that a sequence X_n converges *strongly* to X and write $X_n \xrightarrow{\text{s}} X$, if the convergence is in norm, that is, if $\|X_n - X\| \to 0$. If there exists a transformation $\overline{T}$ on B to B such that $\overline{T^n}X \xrightarrow{\text{s}} \overline{T}X$ for every $X \in B$, then we write $\overline{T^n} \xrightarrow{\text{s}} \overline{T}$. A sequence X_n converges *weakly* to X if $f(X_n) \to f(X)$ for all bounded linear functionals f on B. The term "weakly" as opposed to "strongly" is justified by the fact that $X_n \xrightarrow{s} X$ implies $X_n \xrightarrow{w} X$, for

$$|f(X_n) - f(X)| = |f(X_n - X)| \leqq ||f|| \cdot || X_n - X ||.$$

The *null-transformation*, to be denoted by 0, maps every $X \in B$ onto the null element $\theta \in B$.

A subset B' is *range* of a transformation T' on B to B if T' maps B onto B', and we write $B' = T'B$.

B' is *linear* if it is closed under all (finite) linear combinations of its elements; observe that if T' is linear, then $T'B$ is a linear subspace.

B' is *weakly (strongly) closed* if it is closed under weak (strong) passages to the limit; observe that a strongly closed linear subspace is a Banach space. We denote the strong closure of B' by $\overline{B'}$. In fact, the strong closure of B' is also its weak closure. For, if $X_n \in B'$ and $X_n \xrightarrow{w} X$, then $X \notin \overline{B'}$ implies that the distance d of X to $\overline{B'}$ is positive. But, by Corollary 2 of the Hahn-Banach theorem, there exists a functional f such that $0 = f(X_n) \to f(X) = d > 0$, and we reach a contradiction.

B' is *weakly (strongly) compact* if every sequence in B' contains a weakly (strongly) convergent subsequence; observe that strong compactness implies weak compactness.

a. CONVERGENCE LEMMA. *Let all* $|| T^n || \leqq c$. *Then*

$$(T - I)\overline{T^n} = \overline{T^n}(T - I) \xrightarrow{s} 0$$

and

$$\overline{T^n} \xrightarrow{s} 0 \quad on \quad \overline{(T - I)B}.$$

Proof. Since for every $X \in B$

$$|| (T - I)\overline{T^n}X || = || \overline{T^n}(T - I)X || = \frac{1}{n} || T^nX - X || \to 0,$$

the first assertion is true. Since, given $\epsilon > 0$, for every $X \in \overline{(T - I)B}$ there exists an $X' \in (T - I)B$ such that $|| X - X' || < \epsilon$ and there exists an $X'' \in B$ such that $X' = (T - I)X''$, it follows that, as $n \to \infty$ and then $\epsilon \to 0$,

$$\begin{aligned} || \overline{T^n}X || &\leqq || \overline{T^n}X' || + || \overline{T^n}(X - X') || \\ &\leqq || \overline{T^n}(T - I)X'' || + c\epsilon \to 0, \end{aligned}$$

and the second assertion is proved.

b. NORMS ERGODIC LEMMA. *Let all* $|| T^n || \leqq c$. *Then every weakly compact sequence* $\overline{T^n}X$ *converges strongly to a point invariant under* T.

Proof. If a subsequence $\overline{T^{n'}}X \xrightarrow{w} \overline{X}$ as $n' \to \infty$, then $(T - I)\overline{T^{n'}}X \xrightarrow{w} (T - I)\overline{X}$; hence, by the first part of the convergence lemma, $(T - I)\overline{X} = \theta$, that is, $T\overline{X} = \overline{X}$. Therefore, $\overline{T^n}\overline{X} = \overline{X}$ whatever be n and, setting $X = \overline{X} + (X - \overline{X})$, it remains to be proved that $\overline{T^n}(X - \overline{X}) \xrightarrow{s} 0$. Because of the second part of the convergence lemma, it suffices to prove that $X - \overline{X} \in \overline{(T - I)B}$. But

$$(\overline{T^n} - I)X = (T - I)\left(\frac{n-1}{n}I + \frac{n-2}{n}T + \cdots + \frac{1}{n}T^{n-2}\right)X$$

so that all $(\overline{T^n} - I)X \in (T - I)B$, while a subsequence $(\overline{T^{n'}} - I)X \xrightarrow{w} \overline{X} - X$ as $n' \to \infty$. It follows that $\overline{X} - X \in \overline{(T - I)B}$, and the proof is concluded.

A. Norms ergodic theorem. *Let all* $\| T^n \| \leqq c$. *If all sequences* $\overline{T^n}X$ *are weakly compact, then* $\overline{T^n} \xrightarrow{s} \overline{T}$ *linear and*

$$\| \overline{T} \| \leqq c, \quad \overline{T}T = T\overline{T} = \overline{T}\overline{T} = \overline{T}.$$

Proof. According to the norms ergodic lemma, every sequence $\overline{T^n}X$ converges strongly so that the passage to the limit is a transformation $\overline{T}$ on B to B, obviously linear and of norm bounded by c. Since, on account of the convergence lemma,

$$0 \xleftarrow{s} \overline{T^n}(T - I) \xrightarrow{s} \overline{T}(T - I) = (T - I)\overline{T},$$

it follows that

$$\overline{T} = \overline{T}T = T\overline{T} = \overline{T^n}\overline{T} \xrightarrow{s} \overline{T}\overline{T},$$

and the proof is concluded.

The set B_λ of all points X such that $TX = \lambda X$ is said to be the *proper subspace* of T corresponding to the *proper value* λ of T (λ real or complex), provided this set does not consist of the null-point θ only. Thus, $T - \lambda I = 0$ on $B_\lambda \neq \{\theta\}$, and λ is not a proper value of T if, and only if, $TX = \lambda X$ implies that $X = \theta$. Since, in this section, T is bounded and linear, every B_λ is, clearly, a strongly closed linear subspace of B.

Corollary. *In the norms ergodic theorem,* $\overline{T} \neq 0$ *if and only if* $\lambda = 1$ *is a proper value of* T, *and then* $\overline{T}B = B_1$.

This follows by the implications

$$TX = X \Rightarrow \overline{T^n}X = X \Rightarrow \overline{T}X = X$$

and

$$\overline{T}X = X \Rightarrow TX = T\overline{T}X = \overline{T}X = X.$$

B. Extended norms ergodic theorem. *Let all* $\| T^n \| \leqq c$ *and let all sequences* $\overline{T^n}_\lambda X$ *be weakly compact with* $T_\lambda = T/\lambda$, *where* λ *with or without affixes have modulus* 1. *Then*

(i) $\overline{T_\lambda^n} \xrightarrow{s} \overline{T_\lambda}$ *linear,*

$$\| \overline{T}_\lambda \| \leqq c, \quad \overline{T}_\lambda T_\lambda = T_\lambda \overline{T}_\lambda = \overline{T}_\lambda \overline{T}_\lambda = \overline{T}_\lambda, \quad \overline{T}_\lambda \overline{T}_{\lambda'} = 0 \quad \text{for} \quad \lambda \neq \lambda',$$

$\overline{T}_\lambda \neq 0 \Leftrightarrow \lambda$ *is a proper value of* T, *and then* $B_\lambda = \overline{T}_\lambda B$.

(ii) *If* $T' = T - \sum_{j=1}^{m} \lambda_j \overline{T}_{\lambda_j}$ *with all* $\overline{T}_{\lambda_j} \neq 0$ *and all* λ_j *distinct, then*

$$T^n = T'^n + \sum_{j=1}^{m} \lambda_j^n \overline{T}_{\lambda_j}, \quad \overline{T}_{\lambda_j} T' = T\overline{T}_{\lambda_j} = 0, \quad TT' = T'T = T'T',$$

λ *is a proper value of* $T' \Leftrightarrow \lambda$ *is a proper value of* T *and all* $\lambda_j \neq \lambda$.

Proof. The norms ergodic theorem and its corollary, with T replaced by $T_\lambda = T/\lambda$, yield directly properties (i) except for $\overline{T}_\lambda \overline{T}_{\lambda'} = 0$, $\lambda \neq \lambda'$. The latter follows from

$$\overline{T}_\lambda \overline{T}_{\lambda'} \xleftarrow{s} \overline{T_\lambda^n} \overline{T}_{\lambda'} = \frac{1}{n} \left\{ \sum_{k=1}^{n} (\lambda'/\lambda)^{k-1} \right\} \overline{T}_\lambda \xrightarrow{s} 0.$$

Properties (ii) follow from (i) by elementary computations; only the last one deserves a proof. Let all $\overline{T}_{\lambda_j} \neq 0$, the λ_j be distinct, and $X \neq \theta$. If $TX = \lambda X$ and all $\lambda_j \neq \lambda$, then, by (i), $\overline{T}_{\lambda_j} X = T_{\lambda_j} \overline{T}_\lambda X = \theta$, and forming $T'X$ it follows that $T'X = TX = \lambda X$. Conversely, if $T'X = \lambda X$, then

$$TX = TT'X/\lambda = T'TX/\lambda = T'T'X/\lambda = \lambda X;$$

moreover, if $\lambda_j = \lambda$, then by (i) and the second relation in (ii),

$$\theta \neq X = \overline{T^n_{\lambda_j}} X \xrightarrow{s} \overline{T}_{\lambda_j} X = \overline{T}_{\lambda_j} T'X/\lambda = 0,$$

and the converse follows *ab contrario*.

Remark. The condition (i) $\sup \| T^n \| < \infty$ implies that (i′) $\sup \| \overline{T^n} \| < \infty$ and (i″) $\| T^n \|/n \to 0$. In fact, the foregoing proofs and hence the propositions remain valid if (i) is replaced by (i′) and (i″). Furthermore, if all sequences $\overline{T^n} X$ are weakly compact, then it is easily proved that $\sup \| \overline{T^n} \| < \infty$ (use the Banach-Steinhaus parts of the linear mappings lemma extended to Banach spaces). Thus, in the norms ergodic theorem, $\sup \| T^n \| < \infty$ can be replaced by $\| T^n \|/n \to 0$. This weakening is due to Dunford.

As for the converse, let us only mention that, if $\overline{T^n} \to \overline{T}$, then, by the same lemma, $\sup \| \overline{T^n} \| < \infty$. Also, if the Banach space is reflexive, that is, can be identified with its second adjoint (for example, if it is a space L_r with $r > 1$), then the condition $\sup \| \overline{T^n} \| < \infty$ implies that every sequence $\overline{T^n}X$ is weakly compact.

35.2. Uniform norms ergodic theorems. We recall that in this section all mappings are bounded linear ones on B to B.

The weak compactness assumption in the norms ergodic theorem is fulfilled, under the condition that all $\| T^n \| \leqq c$, if T is *weakly* (*strongly*) *compact*, that is, maps the sphere of points X with $\| X \| \leqq 1$ onto a weakly (strongly) compact set. In fact, as we shall see, it suffices that T be *quasi-weakly* (*strongly*) *compact*, that is, that there be an integer h such that $T^h = V + W$, where V is weakly (strongly) compact and $\| W \| \leqq w < 1$. And it is to be expected that in the quasi-strongly compact case the norms ergodic theorem can be made more precise; it will become the uniform norms ergodic theorem toward which this subsection is directed. This theorem was first given by Krylov and Bogoliubov for Markov chains.

Observe that, if M_1 and M_2 are weakly (strongly) compact and M is bounded, then $M_1 + M_2$ as well as MM_1 and M_1M are weakly (strongly) compact. Therefore, when T is quasi-weakly (strongly) compact, then for every integer k

$$\text{(Q)} \qquad T^{hk} = V_k + W^k, \quad \| W^k \| \leqq w^k, \quad w < 1,$$

where $V_k = (V + W)^k - W^k$, being a finite sum of terms with at least one weakly (strongly) compact factor V, is itself weakly (strongly) compact.

A. *Let all $\| T^n \| \leqq c$ and let T be quasi-weakly compact. Then the conclusions of the extended norms ergodic theorem are valid.*

Proof. It suffices to prove that all sequences $\overline{T^n}X$ are weakly compact. We use repeatedly the relation (Q), denote by f an arbitrary bounded linear functional on B, and set $Y_{m,p} = p\overline{T^p}X/m$. Thus

$$\overline{T^m}X = \frac{p}{m}\overline{T^p}X + \frac{m-p}{m} T^p\overline{T^{m-p}}X, \quad m > p,$$

will become

$$\overline{T^m}X = Y_{m,hk} + V_k Y_{m,m-hk} + W^k Y_{m,m-hk}, \quad m > hk.$$

Since V_k is weakly compact, there exists a subsequence $Y_n = V_k Y_{m_n, m_n - hk}$

such that $f(Y_n) \to f(\bar{Y}_k)$, where $\bar{Y}_k$ is some point which may depend upon k. Since

$$|f(Y_{m_n, hk})| \leqq \frac{chk}{m_n} ||f|| \cdot ||X|| \to 0$$

and

$$|f(W^k Y_{m_n, m_n - hk})| \leqq cw^k ||f|| \cdot ||X||,$$

it follows that

$$\limsup |f(Y_n) - f(\bar{Y}_k)| \leqq cw^k ||f|| \cdot ||X||.$$

By selecting successive subsequences for $k = 1, 2, \cdots$ and applying the diagonal procedure, we may assume that this inequality holds for all k whatever be f. Since, on account of a corollary of the Hahn-Banach theorem, there exists an f such that $|f(\bar{Y}_k) - f(\bar{Y}_l)| = ||\bar{Y}_k - \bar{Y}_l||$, it follows that, as $k, l \to \infty$,

$$||\bar{Y}_k - \bar{Y}_l|| \leqq c(w^k + w^l) ||f|| \cdot ||X|| \to 0$$

so that $\bar{Y}_k \xrightarrow{s} \bar{Y} \in B$ and, whatever be f,

$$\limsup |f(Y_n) - f(Y)| \leqq cw^k ||f|| \cdot ||X|| + |f(Y_k - Y)| \to 0.$$

Thus $f(Y_n) \to f(Y)$, and the assertion is proved.

To pass to the quasi-strongly compact case, we require the following properties.

a. *If B' is a strict Banach subspace of B, then for every $\epsilon \in (0, 1)$ there exists an $X \in B$ such that*

$$||X|| = 1, \quad ||X - X'|| \geqq 1 - \epsilon \quad \textit{for all} \quad X' \in B'.$$

Proof. There exists a point $Y \notin B'$ such that $d = \inf_{Y' \in B'} ||Y - Y'|| > 0$ and hence, given $\epsilon' = d\epsilon/1 - \epsilon$, there exists a point $Y_0 \in B'$ such that $d \leqq ||Y - Y_0|| < d + \epsilon'$. Therefore, if $X = (Y - Y_0)/||Y - Y_0||$ and $X' \in B'$, then $Y' = Y_0 + ||Y - Y_0|| X' \in B'$ and

$$||X|| = 1,$$

$$||X - X'|| = ||Y - Y'|| / ||Y - Y_0|| \geqq d/(d + \epsilon') \doteq 1 - \epsilon.$$

A set S of linearly independent points of B *generates* (*or spans*) a Banach subspace, to be denoted by $B(S)$, if all points of the subspace are linear combinations, or strong limits thereof, of the points of S; when S is a finite set, $B(S)$ is said to be *finite-dimensional.*

b. *If $X_1, X_2, \cdots$ are linearly independent, then there exist points $Y_n \in B(X_1, \cdots, X_n)$ such that $\| Y_n \| = 1$ and $\| X_m - Y_n \| \geqq \frac{1}{2}$ for every $X \in B(X_1, \cdots, X_m)$ and $m < n$.*

It suffices to apply the foregoing lemma with $\epsilon = \frac{1}{2}$ to the strictly increasing sequence of Banach spaces $B(X_1, \cdots, X_n) \subset B$, starting with $Y_1 = X_1/\| X_1 \|$.

c. *Let T be quasi-strongly compact. Then:*

(i) *No sequence of distinct proper values of T may converge to a limit λ with $| \lambda | \geqq 1$, so that the number of distinct proper values of T of modulus 1 is finite and they are isolated proper values.*

(ii) *The proper subspaces B_λ of T with $| \lambda | \geqq 1$ are finite-dimensional.*

Proof. Because of (Q) we can assume, without loss of generality, that $T^h = V + W$, where V is strongly compact and $\| W \| < \frac{1}{4}$.

1° Suppose there is a sequence $\lambda_n \to \lambda$ of proper values λ_n, and $| \lambda | \geqq 1$. Then there is a sequence $X_n \neq \theta$ such that $TX_n = \lambda_n X_n$. If the X_n are not linearly independent, then there exists a smallest integer n such that $X_1, \cdots, X_n$ are linearly independent and $X_{n+1} = \sum_{k=1}^{n} c_k X_k$ with at least one $c_k \neq 0$. But then

$$\sum_{k=1}^{n} c_k(\lambda_{n+1} - \lambda_k) X_k = T(X_{n+1} - \sum_{k=1}^{n} c_k X_k) = 0,$$

so that $X_1, \cdots, X_n$ are linearly dependent and we reach a contradiction. Thus, the X_n are linearly independent and, by **b,** there exist $Y_n = \sum_{k=1}^{n} c_k X_k$ such that $\| Y_n \| = 1$ and $\| X - Y_n \| \geqq \frac{1}{2}$ for every $X \in B(X_1, \cdots, X_m)$ and $m < n$. Let $Z_n = Y_n/\lambda_n{}^h$ and observe that $Y_n - T^h Z_n = \sum_{k=1}^{n-1} d_k X_k$ so that $\| T^h(Z_m - Z_n) \| \geqq \frac{1}{2}$. Since the sequence Z_n is uniformly bounded, there exists a subsequence $Z'_n = Z_{k_n}$ such that $V(Z'_m - Z'_n) \to 0$ as $m \to \infty$, and hence

$$\| T^h(Z'_m - Z'_n) \| \leqq \| V(Z'_m - Z'_n) \| + \| W \| (| \lambda_{k_m}{}^{-h} | + | \lambda_{k_n}{}^{-h} |) \to 2 \| W \| \, | \lambda^{-h} | \geqq 2 \| W \|.$$

It follows that $2 \| W \| \geqq \frac{1}{2}$. Since $\| W \| < \frac{1}{4}$, we reach a contradiction and (i) follows.

2° If the proper space B_λ with $|\lambda| \geqq 1$ is not finite-dimensional, then it contains a sequence X_n of linearly independent points, the argument in 1° applies with $\lambda_n = \lambda$ for every n, and (ii) follows.

d. *Let all* $\| T\|'^n \leqq c' < \infty$ *and let* T' *be quasi-strongly compact. If* T' *has no proper value of modulus* 1, *then there exists two constants* c_1 *and* $c_2 \in (0, 1)$ *such that for all* n

$$\| T'^n \| \leqq c_1(1 - c_2)^n.$$

We give only an indication of the proof. To begin with, T' has no proper values λ with $|\lambda| > 1$. Otherwise, $T'X = \lambda X$ for an $X \neq \theta$ hence $\| T'^n X\| = |\lambda^n| \cdot \| X\| \to \infty$ as $n \to \infty$, and this contradicts $\| T'^n X\| \leqq c' \| X\|$, $c' < \infty$. Since T' has no proper value of modulus 1, it follows, by the preceding lemma, that it has no proper value λ with $|\lambda| \geqq 1 - \gamma$, for some $\gamma > 0$. Now, it is easily seen that it suffices to consider the case $T' = V + W$ where V is strongly compact and $\| W\| \leqq w < 1$. It can then be proved that $T' - \lambda I$ has an inverse for $|\lambda| \geqq \max (1 - \gamma, \frac{1}{2}w) = 1 - c_2$, so that the series $I + \sum T'_\lambda{}^n$ converges ($T'_\lambda = T'/\lambda$). Therefore there exists a c_1 such that $\| T'^n \| \leqq c_1(1 - c_2)^n$.

B. UNIFORM ERGODIC THEOREM. *Let all* $\| T^n \| \leqq c$ *and let* T *be quasi-strongly compact. Then:*

(i) *The conclusions of the extended norms ergodic theorem hold.*

(ii) *T can have only a finite number m of proper values λ_j of modulus* 1 (*and has no proper values of modulus greater than* 1).

For every λ of modulus 1 *there exists a finite constant c' such that for all n*

$$\| \overline{T_\lambda^n} - \overline{T}_\lambda \| \leqq c'/n,$$

$\overline{T}_\lambda \neq 0$ *if, and only if, some* $\lambda_j = \lambda$, *and every* $\overline{T}_{\lambda_j}$ *maps B onto the finite-dimensional proper subspace B_{λ_j} of T.*

(iii) *If* $T' = T - \sum_{j=1}^{m} \lambda_j \overline{T}_{\lambda_j}$, *then T' is quasi-strongly compact and the* $\| T'^n \|$ *are uniformly bounded, in fact there exist constants c_1 and $c_2 \in$ (0, 1) such that for all n*

$$\| T'^n \| \leqq c_1(1 - c_2)^n.$$

Proof. Since a quasi-strongly compact T is quasi-weakly compact, (i) follows by theorem **A**. Together with lemmas **c** and **d**, (i) implies (ii) and (iii) by elementary computations. For (iii), observe that

$$\| T'^n \| \leqq \| T^n \| + \sum_{j=1}^{m} \| \overline{T}_{\lambda_j} \| \leqq (m+1)c = c'$$

while, if $T^h = V + W$ and $V' = V - \sum_{j=1}^{m} \lambda_j^h \overline{T}_{\lambda_j}$, then $\| T'^h - V' \| = \| T^h - V \|$. For (ii), observe that

$$\overline{T^n_\lambda} - \overline{T}_\lambda = \overline{T'^n}_\lambda + \sum_{j=1}^{m} \left(\frac{1}{n} \sum_{k=1}^{n} (\lambda_j/\lambda)^k \overline{T}_{\lambda_j} \right) - \overline{T}_\lambda,$$

where $\| \overline{T'^n} \| \to 0$ faster than any fixed power of $1/n$, $\overline{T}_\lambda = 0$ or $\overline{T}_{\lambda_j}$ according as all $\lambda_j \neq \lambda$ or a $\lambda_j = \lambda$, and $\frac{1}{n} \sum_{k=1}^{n} (\lambda_j/\lambda)^k = O(1/n)$ or 1 according as $\lambda_j \neq \lambda$ or $\lambda_j = \lambda$. One may also (take $\lambda = 1$ to simplify the writing) observe that

$$\overline{T^n} - \overline{T} = (I - \overline{T})\overline{T^n}, \quad (T^n - I)/n = (T - I)(I - \overline{T})\overline{T^n},$$

so that

$$\| (T - I)(\overline{T^n} - \overline{T}) \| = \| (T^n - I)/n \| \leqq (c+1)/n;$$

then examine boundedness of $(T - I)^{-1}$ on $B_1{}^c$.

35.3. Application to constant chains. Let $P^n(x, S)$ be an n-step transition pr.f. from a point $x \in R$ into a Borel set $S \subset R$. It is a Borel function in x for every fixed S, a pr. in S for every fixed x, and

$$P^{m+n}(x, S) = \int P^m(x, dy) P^n(y, S), \quad m, n = 1, 2, \cdots.$$

It is easily checked that:

1. The space of all complex valued σ-additive functions φ of bounded variation on the Borel field in R is a Banach space B when the norm $\| \varphi \|$ of φ is defined by $\| \varphi \| = \text{Var}\ \varphi$.

2. The transformation T with *kernel* $P(x, S)$, defined by

$$\varphi \to T\varphi = \int \varphi(dx) P(x, S)$$

is a linear transformation on B to B of norm 1; its nth iterate T^n is given by

$$\varphi \to T^n\varphi = \int \varphi(dx) P^n(x, S)$$

and has norm 1 and kernel $P^n(x, S)$.

We say that the kernel $P(x, S)$ and the chain are *quasi-strongly compact* if T is quasi-strongly compact. The uniform ergodic theorem, where, obviously, we can replace $\overline{T^n} = \frac{1}{n}\sum_{k=1}^{n} T^{k-1}$ by $\frac{1}{n}\sum_{k=1}^{n} T^k$ without changing the conclusions, yields at once, on account of uniformity in all σ-additive functions of bounded variation, the following result. We set

$$\overline{P^n}_\lambda(x, S) = \frac{1}{n}\sum_{k=1}^{n} P^k(x, S)/\lambda^k, \quad \bar{P}_j(x, S) = \bar{P}_{\lambda_j}(x, S),$$

and omit the subscript if it equals 1.

Let the chain be quasi-strongly compact. Then:

(i) *There exists only a finite number m of proper values λ_j of modulus* 1 *with corresponding finite-dimensional proper subspaces B_{λ_j} of the transformation.*

(ii) *For every λ of modulus* 1, *there exists a constant c' such that for all n*

$$\sup_{x,S} \left| \overline{P^n}_\lambda(x, S) - \bar{P}_\lambda(x, S) \right| \leq c'/n$$

and

$$\int P^n(x, dy)\bar{P}_\lambda(x, S) = \int \bar{P}_\lambda(x, dy)P^n(y, S) = \lambda^n \bar{P}_\lambda(x, S)$$

$$\int \bar{P}_\lambda(x, dy)\bar{P}_{\lambda'}(y, S) = \delta_{\lambda\lambda'}\bar{P}_\lambda(x, S).$$

(iii) $\bar{P}_\lambda(x, S) \neq 0$ *if, and only if, some* $\lambda_j = \lambda$,

$$P^n(x, S) = \sum_{j=1}^{m} \lambda_j^n \bar{P}_j(x, S) + P'^n(x, S)$$

with

$$\int \bar{P}_j(x, dy)P'^n(y, S) = \int P'^n(x, dy)\bar{P}_j(y, S)$$

and constants c_1 and $c_2 \in (0, 1)$ such that for all n

$$\sup_{x,S} \left| P'^n(x, S) \right| \leq c_1(1 - c_2)^n.$$

Let us examine the case $\lambda = 1$. On account of (ii) $\lambda = 1$ *is a proper value of the transformation, and $\bar{P}(x, S)$ is a transition pr.f. which, for every fixed x, is an invariant element of the Banach space, with*

$$\sup_{x,S} \left| \overline{P^n}(x, S) - \bar{P}(x, S) \right| < c'/n.$$

It suffices to observe that from the last relation it follows that $\bar{P}(x, S)$ is a Borel function in x for every fixed S and a pr. in S for every fixed x. Thus, it is a nonvanishing solution ($\bar{P}(x, R) = 1$) of the proper value 1 equation $\varphi(S) = \int \varphi(dy)P(y, S)$.

Observe that $\varphi^+ = (-\varphi)^-$ whatever be $\varphi \in B$, let $\varphi, \varphi' \in B$, and set $\varphi \wedge \varphi' = \varphi - (\varphi - \varphi')^+ = \varphi' - (\varphi' - \varphi)^+$. Nonnegative φ and φ' are *orthogonal* if $\varphi \wedge \varphi' = 0$ or, equivalently, if there exist two disjoint sets S, S' such that $\varphi(S) = \varphi(\Omega)$ and $\varphi'(S') = \varphi'(\Omega)$. Clearly, φ^+ and φ^- are orthogonal and so are $\varphi - \varphi \wedge \varphi'$ and $\varphi' - \varphi \wedge \varphi'$.

There exists a finite number l of mutually orthogonal invariant pr.'s $\bar{P}_i$ such that a pr. $\bar{P}$ is invariant if, and only if,

$$\bar{P} = \sum_{i=1}^{l} p_i \bar{P}_i, \quad p_i \geqq 0, \quad \sum_{i=1}^{l} p_i = 1.$$

Consider the family of all mutually orthogonal pr.'s $\bar{P}_i$ which are invariant, that is, are solutions of the proper value 1 equation. Since mutual orthogonality implies linear independence and the proper space B_1 is finite dimensional, the number of such solutions is finite. The "if" assertion follows and it suffices to prove that if a pr. $\bar{P}$ is invariant, then it can be written in the asserted form; to simplify the writing, we drop the bars over the P's.

First, we show that $Q_i = P \wedge P_i = p_i P_i$. We exclude the trivial case $Q_i = 0$ and set $P'_i = Q_i/\| Q_i \|$. If the pr. P'_i does not coincide with P_i, then $Q'_i = (P_i - P'_i)^+$ and $Q''_i = (P_i - P'_i)$ do not vanish. Since $Q'_i/\| Q'_i \|$, $Q''_i/\| Q''_i \|$, and the P_j with $j \neq i$, form a family of $l + 1$ mutually orthogonal invariant pr.'s, we reach a contradiction. Hence, $Q_i = p_i P_i$ where, necessarily, $0 \leqq p_i \leqq 1$.

It remains to show that $Q = P - \sum p_i P_i = 0$. Either $p_i = 1$ and then $P \wedge P_i = P_i$, hence $P = P_i$; or $p_i < 1$ and then

$$\begin{aligned}(1 - p_i)((P - Q_i) \wedge P_i) &\leqq (P - Q_i) \wedge (1 - p_i)P_i \\ &= (P - Q_i) \wedge (P_i - Q_i) = 0,\end{aligned}$$

hence $(P - Q_i) \wedge P_i = 0$. It follows that every $Q \wedge P_i = 0$, so that, if $Q \neq 0$, then $Q/\| Q \|$ and the P_i form a family of $m + 1$ mutually orthogonal invariant pr.'s. We reach a contradiction and the assertion follows.

There exists a finite measurable partition $\sum_{i=1}^{l} S_i = R$ *such that*

$$\bar{P}_i S_i = 1, \quad \bar{P}(x, S) = \sum_{i=1}^{l} I_{S_i}(x)\bar{P}_i S, \quad P(x, S_i) = 1 \quad \textit{for} \quad x \in S_i;$$

and for all n

$$\sup_{x \in S_i, S} |\overline{P^n}(x, S) - \bar{P}_i S| < c'/n.$$

These relations imply that, if the system starts at any point of S_i, then it remains a.s. in S_i and the uniform limit $\bar{P}_i$ is independent of the starting point.

The proof goes as follows: Since $\bar{P}(x, S)$ is an invariant pr. for every fixed x, we have

$$\bar{P}(x, S) = \sum p_i(x)\bar{P}_i S, \quad p_i(x) \geqq 0, \quad \sum p_i(x) = 1.$$

Since the $\bar{P}_i$ are linearly independent and $\bar{P}(x, S)$ is invariant under $P(x, S)$ and under itself, it follows that

$$p_i(x) = \int P(x, dy)p_i(y), \quad \int \bar{P}_i(dx)p_j(x) = \delta_{ij}.$$

By setting $S_i = [x; p_i(x) = 1]$ and $S'_n = \left[x; \frac{1}{n+1} \leqq 1 - p_i(x) < \frac{1}{n}\right]$, the last relation yields

$$1 = \int \bar{P}_i(dx)p_j(x) \leqq \bar{P}_i S_i + \sum \bar{P}_i S'_n = \bar{P}_i R = 1.$$

But the equality holds only if every $\bar{P}_i S'_n = 0$. Hence $\bar{P}_i S_i = 1$, and the other assertions follow.

Let us mention a few classical cases in which $P(x, S)$ is quasi-strongly compact and invite the reader to check it.

1. Finite constant chains with transition pr. matrix P_{jk}, $j, k = 1, \cdots, l$.

2. Take $P(x, S) = \int_S p(x, y)\mu(dy)$, where μ is a pr. on Borel sets S and $p(x, y)$ is a bounded nonnegative Borel function on $R \times R$; the second iterate is strongly compact.

3. Take $P^h(x, S) \leqq cP^h(y, S)$ for some integer h and all x, y, S.

4. Take $P^h(x, S) \leqq c\mu S$, where μS is a pr., for some integer h and all x, S.

5. Take $P^h(x, S) \leqq 1 - \epsilon$ for $\mu S \leqq \epsilon$, where μS is a pr. and $\epsilon > 0$, for some integer h and all x, S. This is the Doblin condition and encompasses the preceding ones. Form

$$P^h(x, S) = \int_S p_0(x, y)\mu(dy) + P_s^h(x, N_x S)$$

where $p_0(x, y)$ can be selected to be nonnegative and Borel measurable on $R \times R$, and N_x is the μ-null set on which the μ-singular part of $P^h(x, S)$ does not vanish. Then set $p(x, y) = \min(p_0(x, y), 1/\epsilon)$ and observe that

$$P^h(x, S) = \int_S p(x, y)\mu(dy) + Q(x, S)$$

with $0 \leqq Q(x, S) \leqq 1 - \epsilon$ for all x and S. Thus $T^h = T' + Q$ where T' has for kernel $\int_S p(x, y)\mu(dy)$ and Q has for kernel $Q(x, S)$, hence $\| Q \| \leqq 1 - \epsilon$. In the expansion of $T^{hk} = (T' + Q)^k$, all terms containing T' at least twice are strongly compact and there are $k + 1$ other terms of norm $\leqq (1 - \epsilon)^{k-1}$. Thus, for k sufficiently large

$$\| T^{hk} - V_k \| \leqq (k + 1)(1 - \epsilon)^{k-1} < 1$$

and every V_k is strongly compact. Thus, Doblin's condition implies quasi-strong compactness and the foregoing results apply.

COMPLEMENTS AND DETAILS

In what follows T operates on $\mathfrak{A}$ to $\mathfrak{A}$ and t operates on Ω to Ω.

1. Let T be null-preserving. To every P-invariant event A there corresponds an invariant event A' such that $AA'^c + A^cA'$ is null. Example: $\limsup T^nA$.

2. Let $\Omega = [0, 1)$, P = Lebesgue measure, g integrable of period 1.

$$g_n(x) = \frac{1}{n}\sum_{k=0}^{n-1} g(x + kc) \xrightarrow{\text{a.e.}} \int_0^1 g\,dx, \quad c \text{ irrational fixed;}$$

$$g_n(x) = \frac{1}{n}\sum_{k=0}^{n-1} g(2^kx) \xrightarrow{\text{a.e.}} \int_0^1 g\,dx.$$

(In the first case introduce $t(x) = x + c$ modulo 1; in the second case introduce $t(x) = 2x$ modulo 1. Show that the transformations preserve the measure and are indecomposable.)

3. Construct examples of non P-preserving T such that for every $X \in L$, $\overline{T^nX} \xrightarrow[1]{\text{a.s.}} \overline{T}X \in L$. For instance, take suitable T such that $T^2 = I$ on $\Omega = [0, 1)$ with P = Lebesgue measure; in particular, take suitable linear transformations of $[0, c)$ and $[c, 1)$ into $[0, 1)$.

4. Let $Y_1, Y_2, \cdots$ be a stationary sequence of r.v.'s. There exists a stationary sequence $\cdots X_{-1}, X_0, X_1, \cdots$ such that the laws of $(X_1, X_2, \cdots)$ and of $(Y_1, Y_2, \cdots)$ are the same. (Take, for every finite subfamily of X's,

$$\mathfrak{L}(X_{k_1}, \cdots, X_{k_m}) = \mathfrak{L}(Y_{k_1+h}, \cdots, Y_{k_m+h})$$

where h is so large that the subscripts of Y's are positive, and apply the consistency theorem.) Since all pr. properties of the sequence $Y_1, Y_2, \cdots$ are the same as those of $X_1, X_2, \cdots$, we can replace in what follows every X_k with $k > 0$ by Y_k.

If $E|X_1|^r < \infty$ for an $r \geqq 1$, then

$$E(X_1 \mid X_0, \cdots, X_{-n+1}) \xrightarrow[\text{a.s.}]{r} E(X_1 \mid X_0, X_{-1}, \cdots) = U.$$

(Observe that the left-hand side r.v.'s form a martingale sequence.)

If $E|X_1|^r < \infty$ for an $r \geqq 1$, then stationarity of $\cdots X_{-1}, X_0, X_1, \cdots$ entails

$$\mathcal{L}\{E(X_{n+1} \mid X_1, \cdots, X_n)\} \to \mathcal{L}(U)$$

$$U^{(n)} = \frac{1}{n}\sum_{k=1}^{n} E(X_{k+1} \mid X_1, \cdots, X_k) \xrightarrow{r} U.$$

(By stationarity $\mathcal{L}\{E(X_{n+1} \mid X_1, \cdots, X_n)\} = \mathcal{L}\{E(X_1 \mid X_0, \cdots, X_{-n+1})\}$. For every $Z \in L_r$, set $\| Z \| = \left(\int |Z|^r\right)^{1/r}$ and observe that

$$\| U^{(n)} - U \| \leqq \frac{1}{n}\sum_{k=1}^{n} \| E(X_{k+1} \mid X_1, \cdots, X_k) - U_{k+1} \| + \left\| \frac{1}{n}\sum_{k=1}^{n} (U_{k+1} - U) \right\|$$

where the U_{k+1} are the translates of U by k. By the stationarity assumption, the sequence $U_1, U_2, \cdots$ is stationary and, since $E|U|^r < E|X_1|^r < \infty$, the second right-hand side term converges to 0. The first one reduces to

$$\frac{1}{n}\sum_{k=1}^{n} \| E(X_1 \mid X_{-1}, \cdots, X_{-k+1}) - U \|,$$

every term of the sum converges to 0 as $k \to \infty$, and so does their arithmetic mean.)

5. Let Ω be a compact metric space, $\mathcal{A}$ the minimal σ-field over the class of open sets, $PA > 0$ for every open set A. Let $T = t^{-1}$ be P-preserving, indecomposable, and the t^n be equicontinuous.

If g on Ω is continuous and integrable, then $\overline{g^n} \longrightarrow \int g$. The same holds if, for every $\epsilon > 0$, there exist continuous bounds $g' \leqq g \leqq g''$ with $\int |g' - g''| < \epsilon$.

Application. If g on R of period 1 is Riemann integrable, then

$$g_n(x) = \frac{1}{n}\sum_{k=0}^{n-1} g(x + kc) \longrightarrow \int_0^1 g\,dx, \quad c \text{ irrational fixed.}$$

(Replace, in *2*, Ω by a circumference of length 1.)

6. lim sup $\overline{P^n} \leqq cP$ does not imply $\overline{P^n} \leqq cP$ for every n. What are the consequences of this statement? A counter example: Let $\Omega = \{\omega_n\}$, $\Omega = \sum A_n$ where A_n consist of 2^{n+1} points. Denoting by $1, 2, \cdots, 2^{n+1}$ the points of a fixed A_n, set for every $A_n\mu\{k\} = 2^{k-1}$ or 2^{2n-k} or 1 according as $1 \leqq k \leqq n$ or $n < k \leqq 2n$ or $2n < k \leqq 2^{n+1}$, and set $t(k) = 2^{n+1}$ or $k - 1$ according as $k = 1$ or $1 < k \leqq 2^{n+1}$. Finally, set $PA = c\sum \frac{1}{3^n}\mu(AA_n)$.

7. Let Ω be the set of irrationals in (0, 1) and let P = Lebesgue measure. Let $t(x) = 1/x$ modulo 1. Show that with the usual notation for continued fraction

$$t(x) = t\left(\frac{1|}{|c_1(x)} + \frac{1|}{|c_2(x)} + \frac{1|}{|c_3(x)} + \cdots\right) = \frac{1|}{|c_2(x)} + \frac{1|}{|c_3(x)} + \cdots$$

(a) The transformation $T = t^{-1}$ is P-indecomposable.

(Let A be invariant with $p = PA < 1$. To prove that $p = 0$, take a fixed x_0 set $c_n(x_0) = c_n$, and denote by a/b and a'/b' the $(2n - 1)$th and $2n$th approximations of x_0, so that

$$y = \frac{1|}{|c_1} + \cdots + \frac{1|}{|c_{2n-1}} + \frac{1|}{|x + c_{2n}} = \frac{ax + a'}{bx + b'}.$$

Then $t^{2n+1}(y) = x$; hence $I_A(x) = I_A(y)$. Set $\alpha = \dfrac{a}{b}$ and $\beta = \dfrac{a + a'}{b + b'}$. Then

$$\frac{P(A(0, 1))}{\beta - \alpha} = b'(b + b')\int_0^1 I_A(x)\frac{dx}{(bx + b')^2} < 1.$$

As x_0 varies over Ω and n over all the integers, the intervals (α, β) form a Vitali covering of Ω and, by Lebesgue's density theorem, $p = 0$.)

(b) Let $\mu A = \dfrac{1}{\log 2}\displaystyle\int_A \frac{dx}{1 + x}$. Then $T = t^{-1}$ is μ-preserving. (This follows from $\mu(t^{-1}A) = \mu A$ for all $A = (0, x)$ by $\displaystyle\int_0^x \frac{dx'}{1 + x'} = \sum \int_{1/(n+x)}^{1/n} \frac{dx'}{1 + x'}$.)

(c) The null sets and the integrable functions g are the same for μ and P, the a.s.-ergodic theorem holds for $T = t^{-1}$ and both measures, and

$$\frac{1}{n}\sum_{k=0}^{n-1} g(t^k x) \xrightarrow{\text{a.s.}} \frac{1}{\log 2}\int_0^1 \frac{g(x)}{1 + x}\,dx.$$

Applications. 1° For almost all $x \in \Omega$ and every integer p, the frequency of p in $\{c_n(x)\}$ is $\dfrac{1}{\log 2}\log\dfrac{(p + 1)^2}{p(p + 2)}$.

(Take $g = I_A$ where A is the set of all x such that $c_1(x) = p$.)

2° $$\sqrt[n]{c_1(x)c_2(x)\cdots c_n(x)} \xrightarrow{\text{a.s.}} \prod\left(1 + \frac{1}{n^2 + 2n}\right)^{\log n/\log 2}$$

(Take $g(x) = \log c_1(x)$.)

3° $$\frac{1}{n}\sum_{k=1}^{n} c_k(x) \xrightarrow{\text{a.s.}} +\infty.$$

(Take $g(x) = c_1(x)$.)

Chapter XI

SECOND ORDER PROPERTIES

We consider sequences, and more generally random functions, formed by r.v.'s whose second moments and hence mixed second moments are finite.

Their second order properties are those which can be expressed in terms of these moments. Up to equivalences, the r.v.'s in question can be interpreted as points in a Hilbert space, and such spaces are a "natural" generalization of euclidean spaces for which all the classical tools were developed. Thus, it is to be expected that the study of second order properties—to which this chapter is devoted—will only require analytical tools similar to the familiar ones. In fact, except for a few concepts and properties, this chapter is practically independent of the preceding ones.

§36. ORTHOGONALITY

We examine in this section the elementary properties of orthogonal r.v.'s. The concepts become almost obvious if geometric intuition is used; we expect the reader to do it. The only difficulties consist in the justification of the intuitive conclusions in the nonfinite case, and in the use of pr. concepts such as a.s. convergence which have no direct geometric equivalent.

Let $(\Omega, \mathcal{A}, P)$ be a fixed pr. space. Let $X, Y, \cdots$, with or without affixes, be *second order r.v.'s*, in general complex-valued:

$$E|X|^2 < \infty, \quad E|Y|^2 < \infty, \cdots,$$

so that by Schwarz's inequality their mixed second moments $EX\overline{Y}$ exist and are finite. The "bar" means "complex-conjugate."

The space L_2 of equivalence classes of such r.v.'s is a Hilbert space: equivalent r.v.'s represent the same *point*, and $EX\overline{Y}$ defines the *scalar product* of the points represented by X and Y. Scalar products determine the *norms* $\|X\| = (E|X|^2)^{1/2} = (EX\overline{X})^{1/2}$, and norms determine *distances* $\|X - Y\|$. Convergence "in norm" is convergence "in quadratic mean" $X_n \xrightarrow{\text{q.m.}} X$: $\|X_n - X\| \to 0 \Leftrightarrow E|X_n - X|^2 \to 0$. The space L_2 is *complete* in the sense that $X_n \xrightarrow{\text{q.m.}} X \Leftrightarrow X_m - X_n \xrightarrow{\text{q.m.}} 0$, $m, n, \to \infty$.

36.1. Orthogonal r.v.'s; convergence and stability. X and Y are *orthogonal*, and we write $X \perp Y$, if $EX\overline{Y} = 0$. In particular, $X \perp X$ if, and only if, $E|X|^2 = 0$, that is, $X = 0$ a.s.; in fact, $X = 0$ a.s. is orthogonal to every Y. Since independent r.v.'s in L_2 are orthogonal when centered at expectations, we assume, for sake of analogy, that *all our r.v.'s are centered at expectations, unless otherwise stated.* Since our r.v.'s have finite second hence first moments, they can always be so centered and the assumption made does not restrict the generality. Then $E|X|^2$ is the *variance* of X and $EX\overline{Y}$ is the *covariance* of X and Y.

From $X \perp Y$ it follows, upon expanding, that $E|X + Y|^2 = E|X|^2 + E|Y|^2$. More generally, if $X_1, X_2, \cdots$ are orthogonal r.v.'s, then (Pythagorean relation)

$$E\left|\sum_{k=1}^{n} X_k\right|^2 = \sum_{k=1}^{n} E|X_k|^2, \quad n = 1, 2, \cdots,$$

and, as $n \to \infty$,

$$E\left|\sum_{k=1}^{n} X_k\right|^2 \to \sum_{k=1}^{\infty} E|X_k|^2.$$

Since, by the mutual convergence in q.m. criterion, the sequence of sums $\sum_{k=1}^{n} X_k$ converges in q.m. if, and only if, $E\left|\sum_{k=m}^{n} X_k\right|^2 = \sum_{k=m}^{n} E|X_k|^2 \to 0$ as $m, n \to \infty$, we have

A. CONVERGENCE AND STABILITY IN Q.M. *Let the r.v.'s X_n be orthogonal.*

(i) *The series $\sum X_n$ converges in q.m. if, and only if, $\sum E|X_n|^2 < \infty$; and then $E|\sum X_n|^2 = \sum E|X_n|^2$ (Pythagorean relation).*

(ii) *If* $\sum \dfrac{E|X_n|^2}{b_n^2} < \infty$, $b_n \uparrow \infty$, *then* $\dfrac{1}{b_n} \sum_{k=1}^{n} X_k \xrightarrow{\text{q.m.}} 0$.

The second assertion follows from the first by Kronecker's lemma.

The foregoing properties correspond to those of the case of independence if independence and convergence a.s. are replaced by orthogonality and convergence in q.m., respectively. In fact, not much more is needed in the case of orthogonality to obtain an inequality (Rademacher) similar to that of Kolmogorov and, consequently, a.s. convergence and stability theorems well-known in theory of orthogonal functions.

a. *If* $S_n = \sum_{k=1}^{n} X_k$ *are consecutive sums of orthogonal r.v.'s, then*

$$E(\max_{h \leqq n} | S_h |)^2 \leqq (\log 4n/\log 2)^2 \sum_{k=1}^{n} E| X_k |^2.$$

Proof. For $n = 1$ the inequality is trivial. For $n > 1$, let m be the integer such that $2^{m-1} < n \leqq 2^m$, set $X'_k = X_k$ or 0 according as $k \leqq n$ or $n < k \leqq 2^m$, and assign X'_k to the point of abscissa k. Divide the interval $(0, 2^m]$ into intervals $(0, 2^{m-1}]$ and $(2^{m-1}, 2^m]$, each of these two intervals into two halves, and so on; the elements of the $(m - j)$th partition are of length 2^j and $j = 0, 1, \cdots, m$. Every interval $(0, h]$ is the sum of at most m disjoint intervals each of which belongs to a different partition; in other words, we have the dyadic representation of h in geometric terms. We can write $S_h = \sum_{j=0}^{m} Y_{jh}$ where any Y_{jh} is sum of the r.v.'s belonging to the interval of length 2^j which may or may not figure in the representation of h, so that some Y_{jh} may vanish. It follows, by the elementary Schwarz inequality

$$| \sum_{j=0}^{m} a_j |^2 \leqq (m + 1) \sum_{j=0}^{m} | a_j |^2,$$

that, whatever be $h \leqq n$,

$$| S_h |^2 \leqq (m + 1) \sum_{j=0}^{m} | Y_{jh} |^2 \leqq (m + 1)T,$$

where T is the sum of all r.v.'s $| Y_{jh} |^2$ as j and h vary. But the expectation of the sum of all those r.v.'s $| Y_{jh} |^2$ which belong to the jth partition is $\sum_{k=1}^{n} E| X_k |^2$, so that $ET = (m + 1) \sum_{k=1}^{n} E| X_k |^2$. Therefore

$$E(\max_{h \leqq n} | S_h |^2) \leqq (m + 1)^2 \sum_{k=1}^{n} E| X_k |^2$$

and, since

$$(m + 1)^2 \leqq \left(\frac{\log n}{\log 2} + 2\right)^2 = \left(\frac{\log 4n}{\log 2}\right)^2,$$

the asserted inequality is proved.

b. *If* $S_n = \sum_{k=1}^{n} X_k$ *are consecutive sums of orthogonal r.v.'s and* $\sum b_n E|X_n|^2 < \infty$, $b_n \uparrow \infty$, *then* $S_n \xrightarrow{\text{q.m.}} S$ *and there exists a subsequence* $S_{n_k} \xrightarrow{\text{a.s.}} S$ *such that, for every integer* k, b_{n_k} *be the first* $b_n \geq k$.

Proof. The hypothesis implies that $\sum E|X_n|^2 < \infty$ so that **A** applies and $S_n \xrightarrow{\text{q.m.}} S$.

Let

$$r_n = E|S - S_n|^2 = \sum_{j=n+1}^{\infty} E|X_j|^2$$

so that

$$\sum_{k=1}^{\infty} E|S - S_{n_k}|^2 = \sum_{k=1}^{\infty} r_{n_k} = \sum_{k=1}^{\infty} k(r_{n_k} - r_{n_{k+1}}) + \lim_{k \to \infty} k r_{n_{k+1}}$$

$$\leqq \sum_{n=1}^{\infty} b_n E|X_n|^2 < \infty$$

and, hence, $S_{n_k} \xrightarrow{\text{a.s.}} S$ on account of the Borel-Cantelli lemma and the Tchebichev inequality.

B. A.S. CONVERGENCE AND STABILITY. *Let the r.v.'s* X_n *be orthogonal.*

(i) *If* $\sum \log^2 n E|X_n|^2 < \infty$, *then the series* $\sum X_n$ *converges in q.m. and a.s.*

(ii) *If* $\sum \left(\frac{\log n}{b_n}\right)^2 E|X_n|^2 < \infty$, $b_n \uparrow \infty$, *then* $\frac{1}{b_n} \sum_{k=1}^{n} X_k \xrightarrow{\text{a.s.}} 0$.

Proof. Let $S_n = \sum_{k=1}^{n} X_k$. Under hypothesis (i), $S_n \xrightarrow{\text{q.m.}} S$ according to **A**, and lemma **b** with $b_n = \log n/\log 2$ yields $S_{2^k} \xrightarrow{\text{a.s.}} S$; thus (i) will follow if we prove that $T_k = \max_{2^k \leqq n < 2^{k+1}} |S_n - S_{2^k}| \xrightarrow{\text{a.s.}} 0$. But lemma **a**, with $n = 2^{k+1} - 2^k = 2^k$, yields, by elementary computations,

$$\sum_{k=1}^{\infty} E|T_k|^2 \leqq (3/\log 2)^2 \sum_{n=1}^{\infty} \log^2 n E|X_n|^2 < \infty,$$

and the assertion follows by the Borel-Cantelli lemma and the Tchebichev inequality. This proves (i), and (ii) follows by Kronecker's lemma.

COROLLARY. *If the series* $\sum_n X_n$ *of orthogonal r.v.'s converges in q.m., then* $\sum_{k=1}^{n} X_k = o(\log n)$ *a.s.*

This follows by **A**(i) from **B**(ii) with $b_n = \log n$.

36.2. **Elementary orthogonal decomposition.** Let $\{X_j\}$ be a countable family of (second order) r.v.'s. We intend to show that the X_j's are linear combinations of orthogonal r.v.'s.

Without restricting the generality, we can exclude those X_j's which are degenerate at 0 or are linear combinations of other r.v.'s of the family. Let S be an arbitrary finite subset of indices j and let c_j be complex numbers. If $\sum_{j \in S} c_j X_j = 0$ a.s., then

$$E(\sum_{j \in S} c_j X_j)\overline{X}_h = \sum_{j \in S} c_j E X_j \overline{X}_h = 0, \quad h \in S.$$

Conversely, this set of relations implies readily that $E|\sum_{j \in S} c_j X_j|^2 = 0$ and hence $\sum_{j \in S} c_j X_j = 0$ a.s. On the other hand, if there exist some nonvanishing c_j such that the foregoing set of relations holds, then the determinant $\| EX_j\overline{X}_h \| = 0$, $j, h \in S$. Thus, what we have excluded is the possibility for such determinants to vanish. It follows, by elementary computations, that the r.v.'s $Y_1 = X_1$ and, for $n > 1$,

$$Y_n = \begin{vmatrix} X_1 & X_2 & \cdots & X_n \\ EX_1\overline{X}_1 & EX_2\overline{X}_1 & \cdots & EX_n\overline{X}_1 \\ \cdot & \cdot & \cdot & \cdot \\ EX_1\overline{X}_{n-1} & EX_2\overline{X}_{n-1} & \cdots & EX_n\overline{X}_{n-1} \end{vmatrix}$$

are nondegenerate. Furthermore, they are linear combinations of the X_n and are easily verified to be mutually orthogonal. Upon setting $\xi_j = Y_j/(E|Y_j|^2)^{1/2}$, we have $E\xi_j\overline{\xi}_h = \delta_{jh}(= 1$ or 0 according as $j = h$ or $j \neq h)$ and

$$X_j = \sum_{h=1}^{j} c_{jh}\xi_h, \quad c_{jh} = EX_j\overline{\xi}_h.$$

In general, r.v.'s ξ_j, ξ_h such that $E\xi_j\overline{\xi}_h = \delta_{jh}$ are said to be *orthonormal.* Given a finite or denumerable set of orthonormal r.v.'s $\xi_1, \xi_2, \cdots$, we can set

$$X = X'_n + \sum_{k=1}^{n} c_k\xi_k, \quad c_k = EX\overline{\xi}_k.$$

Then $EX'_n\bar{\xi}_k = 0$ for $k \leqq n$ and, by the Pythagorean relation,

$$E|X|^2 = E|X'_n|^2 + \sum_{k=1}^{n} |c_k|^2 \geqq \sum_{k=1}^{n} |c_k|^2.$$

It follows, by letting $n \to \infty$, that, given an orthonormal sequence ξ_n,

$$\sum |c_n|^2 \leqq E|X|^2 < \infty$$

so that, by **A**, the series $\sum c_n\xi_n$ converges in q.m. and hence $X'_n \xrightarrow{\text{q.m.}} X'$ orthogonal to every ξ_n. Thus, we obtain the orthogonal decomposition

$$X = X' + \sum c_n\xi_n \text{ a.s.,} \quad c_n = EX\bar{\xi}_n,$$

with

$$E|X|^2 = E|X'|^2 + \sum |c_n|^2.$$

The r.v. X', being orthogonal to all ξ_n, is orthogonal to all linear combinations of the ξ_n and, hence, to their limits in q.m. This leads to the introduction of linear subspaces; we recall that here all r.v.'s under consideration are of second order.

A *linear subspace* $\mathfrak{L}$ is a family of r.v.'s closed under formation of all a.s. linear combinations of its elements. If, also, $\mathfrak{L}$ is closed under passages to the limit in q.m., then it is a *closed linear subspace.* Given a family $\{X_t\}$ of r.v.'s, the linear space $\mathfrak{L}_0\{X_t\}$ of all their linear combinations and the closed linear subspace $\mathfrak{L}\{X_t\}$, the closure of $\mathfrak{L}_0\{X_t\}$ under passages to the limit in q.m., are said to be *generated* by the r.v.'s X_t. If we keep only those X_t which are linearly independent, we obtain a *base* of $\mathfrak{L}_0\{X_t\}$ and of $\mathfrak{L}\{X_t\}$.

A r.v. X is *orthogonal* to $\mathfrak{L}$, and we write $X \perp \mathfrak{L}$, if $X \perp Y$ whatever be $Y \in \mathfrak{L}$. For example, in what precedes, an arbitrary r.v. X was decomposed into $X' \perp \mathfrak{L}\{\xi_n\}$ and $\sum c_n\xi_n \in \mathfrak{L}\{\xi_n\}$. This orthogonal decomposition with respect to $\mathfrak{L}\{\xi_n\}$ can be generalized for arbitrary closed linear subspaces, as follows:

a. *Let $\mathfrak{L} \subset L_2$ be a closed linear subspace. To every $X \in L_2$ there corresponds an $X_0 \in \mathfrak{L}$ such that $E|X - X_0|^2 = \inf_{Y \in \mathfrak{L}} E|X - Y|^2 = \alpha$, and then $X - X_0 \perp \mathfrak{L}$.*

Proof. Let $\{X_n\} \subset \mathfrak{L}$ be such that $E|X - X_n|^2 \to \alpha$. Since

$$E|X_m - X_n|^2 = 2E|X_m - X|^2 + 2E|X - X_n|^2 - 4E\left|\frac{X_m + X_n}{2} - X\right|^2$$

and $(X_m + X_n)/2 \in \mathfrak{L}$, it follows by letting $m, n \to \infty$ that

$$0 \leqq E|X_m - X_n|^2 \leqq 2E|X_m - X|^2 + 2E|X - X_n|^2 - 4\alpha \to 4\alpha - 4\alpha = 0.$$

Therefore, by the mutual convergence criterion and the closure of $\mathfrak{L}$ under passages to the limit in q.m., $X_n \xrightarrow{\text{q.m.}} X_0 \in \mathfrak{L}$, and

$$E|X - X_0|^2 = \lim E|X - X_n|^2 = \alpha.$$

Since $X_0 + cY \in \mathfrak{L}$ whatever be $Y \in \mathfrak{L}$ and the complex number c, we have $E|X - (X_0 + cY)|^2 \geqq \alpha$. Thus, by taking $c = bE(X - X_0)\bar{Y}$ with $0 < b < 2/E|Y|^2$, we have

$$0 \leqq E|X - (X_0 + cY)|^2 - E|X - X_0|^2 = b(bE|Y|^2 - 2)|E(X - X_0)\bar{Y}|^2 \leqq 0.$$

Hence, $E(X - X_0)\bar{Y} = 0$ and the proof is terminated.

A. PROJECTION THEOREM. *Let $\mathfrak{L}$ be a closed linear subspace. For every X there exists an a.s. unique orthogonal decomposition*

$$X = X' + X'', \quad X' \perp \mathfrak{L}, \quad X'' \in \mathfrak{L}.$$

Proof. There exists such a decomposition: according to **a** it suffices to set $X'' = X_0$. The decomposition is a.s. unique, since if $X = X'_1 + X''_1$ is another such a decomposition, then $X' - X'_1 = X''_1 - X''$ and $X' - X'_1 \perp X''_1 - X''$, so that $X' = X'_1$ a.s. and $X'' = X''_1$ a.s.

COROLLARY. *If $\{\xi_t\}$, $E\xi_t\bar{\xi}_{t'} = \delta_{t,t'}$, is an orthonormal base of the closed linear subspace $\mathfrak{L}$, then, for every X, there exists an a.s. unique orthogonal decomposition*

$$X = X' + \sum c_{t_j}\xi_{t_j}, \quad X' \perp \mathfrak{L}.$$

Proof. It suffices to prove that if $X'' \in \mathfrak{L}$, then there exists a countable subset of indices j such that $X'' = \sum c_{t_j}\xi_{t_j}$. Set $c_t = EX''\bar{\xi}_t$ so that, whatever be the summation set of t's,

$$0 \leqq E|X'' - \sum c_t\xi_t|^2 = E|X''|^2 - \sum |c_t|^2,$$

and hence

$$\sum |c_t|^2 \leqq E|X''|^2 < \infty.$$

Thus, there can be only a countable number of nonvanishing c_t, and the assertion is proved.

36.3. Projection, conditioning, and normality. If $X = X' + X''$ where $X' \perp \mathcal{L}$ and $X'' \in \mathcal{L}$, we say that X' is the *perpendicular* from X to $\mathcal{L}$ and X'' is the *projection* of X on $\mathcal{L}$. According to the projection theorem, the perpendicular X' and the projection X'' exist and are a.s. unique whatever be X and the closed linear subspace $\mathcal{L}$. We denote the projection by $E_2(X \mid \mathcal{L})$ and if $\mathcal{L} = \mathcal{L}\{X_t\}$ we also write it $E_2(X \mid \{X_t\})$. The reasons for this notation similar to that of conditional expectations are many. The operation of projection plays the role of a "second order" conditioning, for, as is readily verified, it is an a.s. linear operation and has the smoothing property of the operation of conditioning if σ-fields are interpreted as closed linear subspaces.

Furthermore, if $\mathcal{L}(\mathcal{B})$ is the space of all $\mathcal{B}$-measurable square integrable r.v.'s then $E(X \mid \mathcal{B}) = E_2(X \mid \mathcal{L}(\mathcal{B}))$ a.s., since, for $B \in \mathcal{B}$, $E\{E(X \mid \mathcal{B})I_B\} = EXI_B = E\{E_2(X \mid \mathcal{L}(\mathcal{B}))I_B\}$. Also, this similarity is related to normality. To begin with, observe that if $X = Y + iZ$, where $Y = \mathcal{R}X$ and $Z = \mathcal{I}X$ with same affixes as X if any, then

$$EX\overline{X}' = EYY' + EZZ' - i(EYZ' - EZY')$$

$$EXX' = EYY' - EZZ' + i(EYZ' + EZY').$$

It follows that $Y \perp X'$ and $Z \perp X'$ if, and only if, $EX\overline{X}' = 0$ *and* $EXX' = 0$ (if X or X' is real-valued, then $EX\overline{X}' = 0$ implies that $EXX' = 0$).

Let the r.v.'s X, X' be jointly normal, that is, let the r.v.'s Y, Z, Y', Z' be jointly normal. If X and X' are orthogonal and real-valued, hence $EXX' = EX\overline{X}' = 0$, then they are independent, since

$$\log Ee^{i(uX+u'X')} = -\frac{u^2}{2}EX^2 - \frac{u'^2}{2}EX'^2 = \log Ee^{iuX} + \log Ee^{iu'X'}.$$

Similarly, if X and X' are orthogonal and complex-valued and $EXX' = 0$, then they are independent, that is, the pairs $\{Y, Z\}$ and $\{Y', Z'\}$ are independent.

We shall say that a family of r.v.'s $X_t = Y_t + iZ_t$, where t with or without affixes varies on some set T, is *strongly normal*, if it is normal (that is, all finite subfamilies of the r.v.'s Y_t, Z_t, are normal) and if either all the X_t are real-valued or all the $EX_tX_{t'} = 0$.

A. *Within a strongly normal family, orthogonality is equivalent to independence and projection is equivalent to conditioning.*

Proof. The first assertion follows from the foregoing discussion. As for the second assertion, let X, X_t, $t \in T$, form a strongly normal family. Since $X' = X - E_2(X \mid \{X_t\}) = X - \sum_j c_{t_j} X_{t_j}$ is orthogonal to and strongly jointly normal with every X_t, it follows, by the first assertion, that X' is independent of every X_t, hence $E(X' \mid \{X_t\}) = EX' = 0$ a.s. Therefore

$$E(X \mid \{X_t\}) = E(\sum_j c_{t_j} X_{t_j} \mid \{X_t\}) = \sum c_{t_j} X_{t_j} = E_2(X \mid \{X_t\}) \text{ a.s.,}$$

and the proof is concluded.

We shall see in the next section that, to any family of second order r.v.'s, we can make correspond a strongly normal family with same second order moments. Thus, a projection can always be considered as a conditioning within suitably selected normal families. Furthermore, to every concept in terms of conditional expectations corresponds a second order concept in terms of projections which, according to what precedes, coincides with the specialization within suitable normality. Let us give two examples.

The concept of a martingale $\{X_n\}$ becomes

$$E_2(X_n \mid X_1, \cdots, X_{n-1}) = X_{n-1} \text{ a.s.}$$

or, setting $Y_n = X_n - X_{n-1}$,

$$E_2(Y_n \mid Y_1, \cdots, Y_{n-1}) = 0 \text{ a.s.,}$$

that is, $Y_n \perp Y_k$ for $k < n$. Thus, a "second order martingale" is simply a sequence of consecutive sums of orthogonal r.v.'s and the results of 33.1 apply. As is to be expected, the a.s. properties of martingales become properties in q.m. of second order martingales. For example,

a. *If the sequence X_n is a second order martingale, then $E|X_n|^2 \uparrow$; if, moreover,* $\lim E|X_n|^2 < \infty$, *then $X_n \xrightarrow{\text{q.m.}} X$ and X closes this martingale.*

Proof. It suffices to write $X_n = \sum_{k=1}^{n} Y_k$ where the Y_k are orthogonal and hence $E|X_n|^2 = \sum_{k=1}^{n} E|Y_k|^2 \uparrow$. If, moreover, $\lim E|X_n|^2 = \sum_{k=1}^{\infty} E|Y_k|^2 < \infty$, then by **A**, $X_n \xrightarrow{\text{q.m.}} X$. Furthermore, $X_n - X_m \perp X_k$

for $k \leqq m \leqq n$ and letting $n \to \infty$, it follows that $X - X_m \perp X_k$ for $k \leqq m$, hence $E_2(X \mid X_1, \cdots, X_m) = X_m$ a.s. This proves the last assertion.

The concept of a chain $\{X_n\}$ yields

$$E(X_n \mid X_1, \cdots, X_m) = E(X_n \mid X_m) \text{ a.s.}$$

for $m < n$, $n = 2, 3, \cdots$. A "second order chain" is defined by replacing E by E_2. (In the strongly normal case, both concepts coincide.)

Set $r_{mn} = E(X_n \bar{X}_m)/E|X_m|^2$ or 0 according as $E|X_m|^2 > 0$ or $= 0$; then $E_2(X_n \mid X_m) = r_{mn} X_m$.

b. *A sequence $\{X_n\}$ is a second order chain if, and only if,*

$$r_{mp} = r_{mn} r_{np}, \quad m < n < p.$$

Proof. The relation is trivially true if $E|X_m|^2 = 0$. Otherwise, if

$$E_2(X_p \mid X_1, \cdots, X_n) = E_2(X_p \mid X_n) = r_{np} X_n \text{ a.s.},$$

then $X_p - r_{np} X_n \perp X_m$ for $m < n$ hence

$$r_{mp} E|X_m|^2 = E(X_p \bar{X}_m) = r_{np} E(X_n \bar{X}_m) = r_{mn} r_{np} E|X_m|^2$$

and the asserted relation holds. Conversely, if this relation is true, then, for every $m < n$,

$$r_{np} E(X_n \bar{X}_m) = E(X_p \bar{X}_m);$$

hence $X_p - r_{np} X_n \perp X_m$, that is,

$$r_{np} X_n = E_2(X_p \mid X_n) = E_2(X_p \mid X_1, \cdots, X_n) \text{ a.s.}$$

The proposition is proved.

§37. SECOND ORDER RANDOM FUNCTIONS

Second order stationary random functions were introduced by Khintchine (1934) who gave the harmonic decomposition of their covariances. Slutsky (1937) obtained a first harmonic decomposition of such random functions. Kolmogorov (1941) proceeded to a detailed study of second order stationary random sequences by means of Hilbert space methods. Cramér (1941) extended Khintchine's results to the vector case and (1942) obtained a decomposition theorem in functional spaces, essentially equivalent to the harmonic decomposition of second order stationary random functions; this decomposition is also an immediate consequence of Stone's theorem (1930) on groups of unitary operators in Hilbert spaces. All this research is limited to second order stationarity.

The author formulated a calculus of general second order random functions and gave (1945–46) the results of this section; they contain as special cases the second order stationarity properties (the foregoing decomposition was stated there explicitly for the first time).

37.1. Covariances. We indulge in the usual abuse of notation by using the same symbol for a function and for its value. The argument t, with or without affixes, will vary over a fixed set T. The only requirement is that the operations performed on the elements of T be meaningful; this will always be so if $T = R = (-\infty, +\infty)$ and, to make his life easier, the reader may assume that $T = R$, unless otherwise stated.

A random function $X(t)$ on T is the family of r.v.'s $\{X(t), t \in T\}$; in general, the r.v.'s will be complex-valued. According to the essential feature of pr. theory, pr. properties are those which are described by the consistent set of laws of all finite subfamilies of the r.v.'s $X(t)$. Conversely, according to the consistency theorem, every such consistent set of laws is the law of some random function.

A *second order random function* $X(t)$ on T is a family of second order r.v.'s: $E|X(t)|^2 < \infty$ whatever be $t \in T$. Without restricting the generality, we can and do assume that the second order random functions under consideration are centered at expectations, unless otherwise stated. Then the second moments $E|X(t)|^2$ are variances and the function defined on $T \times T$ by

$$\Gamma_X(t, t') = EX(t)\overline{X}(t')$$

is, by definition, the *covariance* of the random function $X(t)$ on T. According to the Schwarz inequality, this covariance exists and is finite. Conversely, if $\Gamma_X(t, t')$ on $T \times T$ exists and is finite, then $E|X(t)|^2 = \Gamma_X(t, t) < \infty$, $t \in T$. Thus, second order random functions can be defined as those having covariances. Their *second order properties* are those which can be defined or determined by means of covariances. It is to be expected that to a covariance corresponds more than one random function. For example, the covariances of the random functions $X(t)$ and $Y(t) = \eta X(t)$ on T, where η is a r.v. independent of all $X(t)$, $t \in T$, with $E|\eta|^2 = 1$, coincide. In fact, this example shows that our convention which consists in centering second order random functions at their expectations is immaterial. If we take $E\eta = 0$, then the function $EX(t)\overline{X}(t')$ where the random function $X(t)$ is not centered at its expectation is still a covariance of the random function $Y(t)$ centered at its expectation:

$$EY(t) = E\eta X(t) = E\eta EX(t) = 0,$$

$$EY(t)\bar{Y}(t') = E|\eta|^2 X(t)\bar{X}(t') = E|\eta|^2 EX(t)\bar{X}(t') = EX(t)\bar{X}(t').$$

Since we consider only first and second (mixed or not) moments, it is natural to try to make correspond to a covariance a random function whose law is determined by these moments only. This will be done in the next theorem. We require the following definition.

A function $\Gamma(t, t')$ on $T \times T$ is of *nonnegative-definite type* if, for every finite subset $T_n \subset T$ and every function $h(t)$ on T_n

$$\sum_{t,t' \in T_n} \Gamma(t, t')h(t)\bar{h}(t') \geqq 0.$$

Then it is *hermitian*, that is, $\Gamma(t, t') = \bar{\Gamma}(t', t)$. For, with $T_1 = \{t\}$ and $h(t) = 1$, we have $\Gamma(t, t) \geqq 0$; then, with $T_2 = \{t, t'\}$, the expression $\Gamma(t, t')h(t)\bar{h}(t') + \Gamma(t', t)h(t')\bar{h}(t)$ is real, and hermiticity follows by taking $h(t) = 1$, $h(t') = 1$, i. The reason for the terminology is that a nonnegative-definite type function $\Gamma(t, t') = f(t - t')$ which depends only upon the difference $u = t - t'$ of its arguments reduces to a nonnegative-definite function $f(u)$. We require the following lemma.

a. *Let j, k vary over $1, \cdots, n$ and let the u_k vary over $(-\infty, +\infty)$. If $Q(u) = \sum m_{jk}u_ju_k \geqq 0$, $m_{jk} \in (-\infty, +\infty)$, then there exist jointly normal real-valued r.v.'s X_k (some of which may be degenerate at* 0) *such that $m_{jk} = EX_jX_k$.*

Proof. According to the classical properties of quadratic forms, the assumption means that

$$Q(u) = R(v) = \sum \sigma_k^2 v_k^2, \quad \sigma_k^2 \geqq 0,$$

where the v's are linear combinations of the u's. But $e^{-R(v)/2} = \prod e^{-\sigma_k^2 v_k^2/2}$ is the joint ch.f. of independent normal r.v.'s Y_k (centered at expectations) with $\sigma_k^2 = EY_k^2$. Therefore, by going back to the u's,

$$R(v) = E(\sum Y_k v_k)^2 = E(\sum X_k u_k)^2 = \sum m_{jk}u_ju_k,$$

where the X's are linear combinations of the Y's hence are jointly normal, and $EX_jX_k = m_{jk}$. The assertion is proved.

A. COVARIANCE CRITERION AND NORMALITY. *A function $\Gamma(t, t')$ on $T \times T$ is a covariance if, and only if, it is of nonnegative-definite type.*

And every covariance is also the covariance of a strongly normal random function which can be selected to be real-valued when the covariance is real-valued.

Proof. Let $\Gamma(t, t') = EX(t)\bar{X}(t')$, $t, t' \in T$. Then

$$\sum_{t,t' \in T_n} \Gamma(t, t')h(t)\bar{h}(t') = E \sum_{t,t' \in T_n} X(t)\bar{X}(t')h(t)\bar{h}(t')$$
$$= E\left| \sum_{t \in T_n} X(t)h(t) \right|^2 \geqq 0$$

and the "only if" assertion is proved.

Conversely, let $\Gamma(t, t')$ on $T \times T$ be of nonnegative-definite type. The "if" assertion means that there exists a random function $X(t)$ on T such that $EX(t)\bar{X}(t') = \Gamma(t, t')$ on $T \times T$. By hypothesis,

$$Q(u, v) = \sum_{t,t' \in T_n} \tfrac{1}{2}\Gamma(t, t')h(t)\bar{h}(t') \geqq 0$$

or, setting $h(t) = u_t - iv_t$,

$$Q(u, v) = \sum_{t,t' \in T_n} \tfrac{1}{2}\{\mathcal{R}\Gamma(t, t')(u_t u_{t'} + v_t v_{t'}) - \mathcal{I}\Gamma(t, t')(u_t v_{t'} - u_{t'} v_t)\} \geqq 0$$

whatever be u_t, v_t, $t \in T_n$. Therefore, by **a**, the function $f(u, v) = e^{-\frac{1}{2}Q(u,v)}$ is a normal ch.f. of $2n$ r.v.'s (centered at expectations) $Y(t)$ and $Z(t)$ corresponding to u_t and v_t, respectively, with

$$EY(t)Y(t') = EZ(t)Z(t') = \tfrac{1}{2}\mathcal{R}\Gamma(t, t'), \quad EY(t)Z(t') = -\tfrac{1}{2}\mathcal{I}\Gamma(t, t').$$

It follows by setting $X(t) = Y(t) + iZ(t)$ that $EX(t)X(t') = 0$ and $EX(t)\bar{X}(t') = \Gamma(t, t')$, $t, t' \in T_n$.

The normal laws of finite subfamilies of r.v.'s $X(t)$ so defined for every $T_n \subset T$ are consistent, since the law for $T_m \subset T_n$ coincides with the marginal law on T_m obtained by setting $u_t = v_t = 0$ for $t \in T_n - T_m$. Thus is defined the law of a normal random function $X(t)$ on T with covariance $\Gamma(t, t')$ on $T \times T$. If the covariance is real-valued, we can simply set $EX(t)X(t') = \Gamma(t, t')$ and the ch.f.'s $e^{-Q(u,0)}$ determine the law of a real-valued normal random function $X(t)$. The proof is concluded.

COROLLARY. *The real part of a covariance $\Gamma(t, t')$ is a covariance, while the imaginary part is not a covariance except when it vanishes.*

Proof. The first assertion follows from the above proof by $\mathcal{R}\Gamma(t, t') = E\{\sqrt{2}\,Y(t)\cdot\sqrt{2}\,Y(t')\}$. The second assertion follows from the fact that $\mathcal{I}\Gamma(t, t) = 0$ so that, if $\mathcal{I}\Gamma(t, t') = EX(t)\bar{X}(t')$ on $T \times T$, then $E|X(t)|^2 = 0$ on T, hence $X(t) = 0$ a.s. and $\mathcal{I}\Gamma(t, t') = 0$.

We examine a few operations which preserve covariances (see also the following subsections), and leave to the reader the specialization of what follows to stationary covariances, that is, to nonnegative-definite functions.

B. CLOSURE THEOREM. *The class of covariances is closed under additions, multiplications, and passages to the limit.*

Proof. Let $\Gamma_1(t, t')$ and $\Gamma_2(t, t')$ be two covariances and let $X_1(t)$ and $X_2(t)$ be random functions whose covariances are $\Gamma_1(t, t')$ and $\Gamma_2(t, t')$, respectively. If we select these random functions to be orthogonal, that is, $EX_1(t)\overline{X}_2(t') = 0$ for $t, t' \in T$, then

$$E\{X_1(t) + X_2(t)\}\{\overline{X}_1(t') + \overline{X}_2(t')\} = EX_1(t)\overline{X}_1(t') + EX_2(t)\overline{X}_2(t')$$
$$= \Gamma_1(t, t') + \Gamma_2(t, t').$$

If we select them to be independent, then

$$E\{X_1(t)X_2(t)\}\{\overline{X}_1(t')\overline{X}_2(t')\} = EX_1(t)\overline{X}_1(t')\cdot EX_2(t)\overline{X}_2(t')$$
$$= \Gamma_1(t, t')\Gamma_2(t, t').$$

Such selections are possible, for it suffices to take the random functions $X_1(t)$ and $X_2(t)$ to be normal on pr. spaces $(\Omega_1, \mathfrak{A}_1, P_1)$ and $(\Omega_2, \mathfrak{A}_2, P_2)$, respectively, and then form the product pr. space. Thus, the first two assertions are proved.

Finally, let $\Gamma_s(t, t')$ be covariances for $s \in S$ arbitrary with s_0 a limit point of S (not necessarily in S) and let $\Gamma_s(t, t') \to \Gamma(t, t')$ on $T \times T$ as $s \to s_0$.

Since passage to the limit and finite summation on $T_n \times T_n$ can be interchanged, it follows by **A** that

$$\sum_{T_n}\sum_{T_n} \Gamma(t, t')h(t)\overline{h}(t') = \lim \sum_{T_n}\sum_{T_n} \Gamma_s(t, t')h(t)\overline{h}(t') \geqq 0$$

and by the same theorem $\Gamma(t, t')$ is a covariance.

Applications. 1° If $\Gamma(t, t')$ is a covariance, so is its real part and hence so is the real part $(\mathcal{R}\Gamma(t, t'))^2 - (\mathfrak{J}\Gamma(t, t'))^2$ of $\Gamma^2(t, t')$; similarly for higher powers.

2° Every nonnegative number being a covariance, so is every polynomial in covariances with positive coefficients, and so is every limit of such polynomials. For example, $1/(1 - tt')$ is covariance of a random function $\sum_0^\infty t^n\xi_n$ analytic in $(-1, +1)$, where the ξ_n are orthonormal. This random function was encountered in studying new classes of limit laws and is at the origin of the investigations of this section.

3° If $\Gamma_s(t, t')$ is continuous in $s \in R$ and $F(s)$ on R is nondecreasing, then $\int \Gamma_s(t, t')\, dF(s)$ is a covariance, provided it exists and is finite.

4° Let Δ_h and Δ'_h be difference operators of step h operating on t and $t'(\in R)$, respectively. If $\Gamma(t, t')$ is the covariance of the random function $X(t)$, then, by computing the covariance of $\Delta_h{}^n X(t)$ where $X(t)$ has for covariance $\Gamma(t, t')$, we find that $\Delta_h{}^n \Delta'_h{}^n \Gamma(t, t')$ is a covariance and the variance $\Delta_h{}^n \Delta'_h{}^n \Gamma(t, t) \geqq 0$. It follows that if $\dfrac{\partial^{2n}}{\partial t^n \partial t'^n} \Gamma(t, t')$ exists and is finite, then it is a covariance; we shall see that it is the covariance of the nth derivative in q.m. of $X(t)$.

37.2. **Calculus in q.m.; continuity and differentiation.** Let s, s' vary over some set S and let s_0, s'_0 be limit points of S; they do not necessarily belong to S (for example, $S = (-\infty, +\infty)$ and $s_0 = +\infty$).

a. *If* $X_s \xrightarrow{\text{q.m.}} X$ *as* $s \to s_0$ *and* $X'_{s'} \xrightarrow{\text{q.m.}} X'$ *as* $s' \to s'_0$, *then* $EX_s\bar{X}_{s'} \to EX\bar{X}'$.

Proof. This follows from

$$E(X_s\bar{X}'_{s'} - X\bar{X}') = E(X_s - X)(\bar{X}'_{s'} - \bar{X}') + E(X_s - X)\bar{X}' + EX(\bar{X}'_{s'} - \bar{X}'),$$

since, as $s \to s_0$ and $s' \to s'_0$,

$$| E(X_s - X)(\bar{X}'_{s'} - \bar{X}') |^2 \leqq E| X_s - X|^2 \cdot E|X'_{s'} - X'|^2 \to 0,$$

and similarly for the two other r.h.s. terms.

A. CONVERGENCE IN Q.M. CRITERION. *Second order random functions* $X_s(t)$ *on* T *converge in q.m. as* $s \to s_0$ *to some random function* $X(t)$ *on* T *(necessarily of second order) if, and only if, the functions* $EX_s(t)\bar{X}_{s'}(t)$ *converge to a finite function on* T, *as* $s, s' \to s_0$ *in whatever way* s *and* s' *converge to* s_0. *Then* $\Gamma_{X_s}(t, t') \to \Gamma_X(t, t')$ *on* $T \times T$.

Proof. The "if" assertion follows, by the mutual convergence in q.m. criterion, from

$$\begin{aligned} E| X_s(t) - X_{s'}(t) |^2 &= E| X_s(t) |^2 - EX_s(t)\bar{X}_{s'}(t) - EX_{s'}(t)\bar{X}_s(t) \\ &\quad + E| X_{s'}(t) |^2 \to \Gamma(t, t) - 2\Gamma(t, t) + \Gamma(t, t) \\ &= 0, \quad s, s' \to s_0. \end{aligned}$$

The "only if" and the last assertions follow from the foregoing lemma upon replacing X_s by $X_s(t)$ and $X'_{s'}$ by $X_{s'}(t')$.

A second order random function $X(t)$ on T is *continuous in q.m.* at $t \in T$ if

$$X(t+h) \xrightarrow{\text{q.m.}} X(t) \quad \text{as} \quad h \to 0, \quad t+h \in T.$$

B. CONTINUITY IN Q.M. CRITERION. *$X(t)$ is continuous in q.m. at $t \in T$ if, and only if, $\Gamma_X(t, t')$ is continuous at (t, t).*

Proof. This follows by the convergence in q.m. criterion from

$$\lim_{h,h' \to 0} EX(t+h)\overline{X}(t+h') = \lim_{h,h' \to 0} \Gamma_X(t+h, t+h'), \quad t+h, t+h' \in T.$$

COROLLARY. *If a covariance $\Gamma(t, t')$ on $T \times T$ is continuous at every diagonal point $(t, t) \in T \times T$, then it is continuous on $T \times T$.*

It suffices to observe that if $\Gamma(t, t')$ is the covariance of $X(t)$, then

$$X(t+h) \xrightarrow{\text{q.m.}} X(t), \quad X(t'+h') \xrightarrow{\text{q.m.}} X(t'), \quad h, h' \to 0,$$

imply by a that

$$EX(t+h)\overline{X}(t'+h') \to EX(t)\overline{X}(t').$$

A second order random function $X(t)$ on T has a *derivative in q.m.* $\frac{dX(t)}{dt}$ (or $X'(t)$) at $t \in T$ if

$$\frac{X(t+h) - X(t)}{h} \xrightarrow{\text{q.m.}} X'(t), \quad h \to 0, \quad t+h \in T.$$

C. DIFFERENTIATION IN Q.M. CRITERION. *$X(t)$ has a derivative in q.m. at $t \in T$ if, and only if, the second generalized derivative of $\Gamma_X(t, t')$ exists and is finite at (t, t).*

This follows by the convergence in q.m. criterion from

$$\lim_{h,h' \to 0} E\left\{\frac{X(t+h) - X(t)}{h} \cdot \frac{\overline{X}(t+h') - \overline{X}(t)}{h'}\right\} = \lim_{h,h' \to 0} \frac{1}{hh'} \Delta_h \Delta'_{h'} \Gamma(t, t).$$

COROLLARY 1. *If the second generalized derivative of a covariance $\Gamma(t, t')$ on $T \times T$ exists and is finite at every diagonal point $(t, t) \in T \times T$,* *then the derivatives* $\frac{\partial}{\partial t}\Gamma(t, t')$, $\frac{\partial}{\partial t'}\Gamma(t, t')$, $\frac{\partial^2}{\partial t \partial t'}\Gamma(t, t')$ *exist and are finite on* $T \times T$.

It suffices to observe that, if $\Gamma(t, t')$ is the covariance of a random function $X(t)$, then $X'(t)$ exists, and since "E" and "lim q.m." can be interchanged, it follows by **a** that

$$EX'(t)\overline{X}(t') = \lim_{h \to 0} \frac{\Gamma(t+h, t') - \Gamma(t, t')}{h} = \frac{\partial}{\partial t}\Gamma(t, t').$$

Similarly for $\frac{\partial}{\partial t'}\Gamma(t, t')$, and also for

$$EX'(t)\overline{X}'(t') = \lim_{h' \to 0} \frac{1}{h'}\left\{\frac{\partial}{\partial t}\Gamma(t, t'+h') - \frac{\partial}{\partial t}\Gamma(t, t')\right\} = \frac{\partial^2}{\partial t \partial t'}\Gamma(t, t').$$

COROLLARY 2. *If* $X'(t)$ *on* T *exists, then* $\Gamma_{X'}(t, t') = \frac{\partial^2}{\partial t \partial t'}\Gamma_X(t, t')$ *on* $T \times T$.

This property extends at once to

$$EX^{(n)}(t)\overline{X}^{(n')}(t') = \frac{\partial^{n+n'}}{\partial t^n \, \partial t'^{n'}}\Gamma(t, t').$$

Assume that $\Gamma(t, t')$ is indefinitely differentiable on $T \times T$, select for origin a fixed value of the argument, set

$$X_n(t) = X(0) + \frac{t}{1}X'(0) + \cdots + \frac{t^n}{n!}X^{(n)}(0),$$

and form $E|X(t) - X_n(t)|^2$. Elementary computations yield

COROLLARY 3. *A second order random function* $X(t)$ *on* T *is analytic in q.m. if, and only if,* $\Gamma_X(t, t')$ *is analytic at every diagonal point* $(t, t) \in T \times T$; *and then* $\Gamma_X(t, t')$ *is analytic on* $T \times T$.

37.3. **Calculus in q.m.; integration.** The investigation of integrals in q.m. follows the foregoing pattern but is somewhat more involved.

Let $X(t)$ and $Y(t)$ on T be second order random functions with covariances $\Gamma_X(t, t')$ and $\Gamma_Y(t, t')$. Contrary to our convention, we do not assume that they are centered at expectations. The reason is that we shall have cases in which either one or the other of these random functions degenerates into a nonvanishing sure function, while if they were centered at expectations the sure function would have to vanish.

The *Riemann-Stieltjes integrals in q.m.* are defined as follows: Let

$$D_I: a = t_1 < t_2 < \cdots < t_{n+1} = b$$

be a finite set of consecutive points defining a partition of the finite interval $I = [a, b)$ and set $|D_I| = \max_{k \leqq n} (t_{k+1} - t_k)$. If $X_{D_I}(t) = \sum_{k=1}^{n} X_k I_{[t_k, t_{k+1})}(t)$ is a random step-function, we set

$$\int_I X_{D_I}(t)\, dY(t) = \sum_{k=1}^{n} X_k\{Y(t_{k+1}) - Y(t_k)\}.$$

In the general case of a second order random function $X(t)$, we set $X_k = X(t'_k), t_k \leq t'_k \leq t_{k+1}$

$$\int_I X(t)\, dY(t) = \lim_{|D_I| \to 0} \text{q.m.} \int_I X_{D_I}(t)\, dY(t),$$

$$\int X(t)\, dY(t) = \lim_{\substack{a \to -\infty \\ b \to +\infty}} \text{q.m.} \int_I X(t)\, dY(t),$$

(and similarly if only $a \to -\infty$ or $b \to +\infty$), provided the second limit exists and the first limit exists for the sequence of partitions D_I and is independent of the choices of the corresponding t_k; they are necessarily defined up to an equivalence.

It is important to bear in mind that the preceding definition depends not upon the random function $Y(t)$ but upon its increments $\Delta Y(t)$; in other words, the random functions $Y(t)$ are to be considered as defined up to an additive r.v. Similarly, it is not the covariance $\Gamma_Y(t, t')$ of $Y(t)$ which matters below but its increments $\Delta\Delta'\Gamma_Y(t, t') = E\Delta Y(t)\Delta\overline{Y}(t')$.

A. INTEGRATION IN Q.M. CRITERION. *Let the second order random functions $X(t)$, with or without affixes, be independent of the second order increment function $\Delta Y(t')$ on an interval $I \times I$ finite or not. Then*

$$\int_I X(t)\, dY(t) \text{ exists if, and only if, } \int_I \int_I \Gamma_X(t, t')\, dd'\Gamma_Y(t, t') \text{ exists}$$

and, if the integrals in q.m. which figure below exist, then

$$E\left\{\int_I X^s(t)\, dY(t) \int_{I'} \overline{X}^{s'}(t')\, d\overline{Y}(t')\right\} = \int_I \int_{I'} E\{X^s(t)\overline{X}^{s'}(t')\}\, dd'\Gamma_Y(t, t').$$

The double integrals are usual Riemann-Stieltjes integrals.

Proof. The first assertion follows, upon starting with finite intervals, from the convergence in q.m. criterion applied to

$$E\left\{\int X_{D_I}(t)\,dY(t)\int \overline{X}_{D'_I}(t')\,d\overline{Y}(t')\right\};$$

similarly for the second assertion upon replacing X_{D_I} by $X_{D_I}{}^{s}$ and $\overline{X}_{D'_I}$ by $\overline{X}_{D'_{I'}}{}^{s'}$.

REMARK. The independence condition is certainly fulfilled when the random functions $X(t)$ and $Y(t)$ are independent or when $X(t)$ or $\Delta Y(t)$ degenerate into sure functions. On the other hand, it is used only to assert that for $t, t' \in I$

$$EX(t)\overline{X}(t')\Delta Y(t)\Delta'\overline{Y}(t') = EX(t)\overline{X}(t')\cdot E\Delta Y(t)\Delta'\overline{Y}(t'),$$

and, hence, it can be replaced by this less restrictive condition. Finally, it can be suppressed altogether, provided the elements of double integrals are replaced by, say, $dd'E\{X(t)\overline{X}(t')Y(t)\overline{Y}(t')\}$.

COROLLARY 1. *Formal properties of Riemann-Stieltjes integrals such as finite additivity hold a.s. for corresponding integrals in q.m.*

The corollary follows by elementary computations.

Let $D: a = t_1 < t_2 < \cdots < t_{n+1} = b$ and $D': a = t'_1 < t'_2 < \cdots < t'_{n'+1} = b$ be finite sets of points defining partitions of the finite interval $I = [a, b)$. Let $\Delta Y(t) = Y(t_{k+1}) - Y(t_k)$ and $\Delta' Y(t') = Y(t'_{k'+1}) - Y(t'_{k'})$ denote the corresponding increments of $Y(t)$ for $t = t_k \in D$ and $t' = t'_{k'} \in D'$. We say that the function $\Gamma_Y(t, t')$ is of *bounded variation* on $I \times I$ if there exists a constant c_I such that

$$\sum_{t\in D}\sum_{t'\in D'} |E\,\Delta Y(t)\Delta'\overline{Y}(t')| = \sum_{t\in D}\sum_{t'\in D'} |\Delta\Delta'\Gamma_Y(t, t')| \leqq c_I < \infty$$

whatever be D and D'. We say that $\Gamma_Y(t, t')$ is of *bounded variation* on the infinite interval $I \times I$ if there exists a constant c such that $c_{I'} \leqq c < \infty$ whatever be $I' \subset I$ and then we write, for short, $\int |dd'\Gamma_Y(t, t')| < \infty$. Clearly

COROLLARY 2. *If on $I \times I$ the random function $X(t)$ is continuous in q.m. and independent of the increment random function $\Delta Y(t')$, and if the covariance $\Gamma_X(t, t')$ is bounded while the covariance $\Gamma_Y(t, t')$ is of bounded variation, then $\int_I X(t)\,dY(t)$ exists.*

Let us observe that, under the bounded variation condition, the covariance $\Gamma_Y(t, t')$ can be assumed to have the property $\lim_{t,t' \to -\infty} \Gamma_Y(t, t') = 0$; it can also be normalized, that is, replaced by

$$\hat{\Gamma}_Y(t, t') = \tfrac{1}{4}\{\Gamma_Y(t+0, t'+0) + \Gamma_Y(t-0, t'+0) + \Gamma_Y(t+0, t'-0) + \Gamma_Y(t-0, t'-0)\}$$

where the limits exist and are finite. Furthermore, because of the boundedness and continuity condition on $\Gamma_X(t, t')$, the integral $\int_I \int_I \Gamma_X(t, t')\, dd'\Gamma_Y(t, t')$ remains the same if Γ_Y is replaced by $\hat{\Gamma}_Y$.

37.4. Fourier-Stieltjes transforms in q.m. A covariance $\Gamma(t, t')$ is said to be *harmonizable* if there exists a covariance $\gamma(s, s')$ of bounded variation on $R \times R$ such that

$$(\mathrm{H}_\Gamma) \qquad \Gamma(t, t') = \iint e^{i(ts - t's')}\, dd'\gamma(s, s').$$

A second order random function $X(t)$ is said to be *harmonizable* if there exists a second order random function $\xi(s)$ with a covariance $\gamma(s, s')$ of bounded variation on $R \times R$ such that

$$(\mathrm{H}_X) \qquad X(t) = \int e^{its}\, d\xi(s) \text{ a.s.}$$

We intend to show that harmonizability of a random function implies that of its covariance, and conversely. The direct assertion follows at once from the integration in q.m. criterion, and the problem lies in the proof of the converse assertion. We shall use repeatedly the convergence and integration in q.m. criteria, without further comment.

To begin with, let us observe that the bounded variation condition on $\gamma(s, s')$ implies that harmonizable random functions are continuous in q.m. and harmonizable covariances are continuous and bounded. Moreover, $\gamma(s \pm 0, s' \pm 0)$ and $\xi(s \pm 0)$, $\xi(\mp\infty)$ exist.

We denote by $\hat{\xi}(s)$ and $\Delta_0\xi(s)$ the "normalized" $\xi(s)$ and the "jump" of $\xi(s)$ at s, defined by

$$\hat{\xi}(s) = \tfrac{1}{2}\{\xi(s+0) + \xi(s-0)\}, \quad \Delta_0\xi(s) = \xi(s+0) - \xi(s-0).$$

Similarly, we denote by $\hat{\gamma}(s, s')$ and $\Delta_0\Delta'_0\gamma(s, s')$ the "normalized" $\gamma(s, s')$ and the "jump" of $\gamma(s, s')$ at (s, s'), defined by

$$\hat{\gamma}(s, s') = \tfrac{1}{4}\{\gamma(s+0, s'+0) + \gamma(s+0, s'-0) \\ + \gamma(s-0, s'+0) + \gamma(s-0, s'-0)\}$$

$$\Delta_0\Delta'_0\gamma(s, s') = \gamma(s+0, s'+0) - \gamma(s+0, s'-0) \\ - \gamma(s-0, s'+0) + \gamma(s-0, s'-0).$$

It follows that

$$E\hat{\xi}(s)\bar{\hat{\xi}}(s') = \hat{\gamma}(s, s'), \quad E\Delta_0\hat{\xi}(s)\Delta'_0\hat{\xi}(s') = \Delta_0\Delta'_0\hat{\gamma}(s, s').$$

Let Δ_h and $\Delta'_{h'}$ be difference operators of step h and h' acting on s and s', respectively. Let $a_\tau(u) = \sin \tau u/\tau u$ and let

$$b_\tau(v, h) = \frac{1}{\pi}\int_{\tau(v-h)}^{\tau v} \frac{\sin u}{u}\, du.$$

We use repeatedly the fact that $b_\tau(v, h) \to 0$, 1, or $\frac{1}{2}$, according as $v(v-h) > 0$, <0, or $=0$.

a. INVERSION OF COVARIANCES. *If a covariance $\Gamma(t, t')$ is harmonizable, then, as $\tau, \tau' \to \infty$,*

$$\frac{1}{4\tau\tau'}\int_{-\tau}^{+\tau}\int_{-\tau'}^{+\tau'} e^{-i(st-s't')}\Gamma(t, t')\, dt\, dt' \to \Delta_0\Delta'_0\gamma(s, s')$$

and

$$\frac{1}{4\pi^2}\int_{-\tau}^{+\tau}\int_{-\tau'}^{+\tau'} \frac{1}{tt'}\Delta_h\Delta'_{h'}e^{-i(st-s't')}\Gamma(t, t')\, dt\, dt' \to \Delta_h\Delta'_{h'}\hat{\gamma}(s, s').$$

Proof. (H_Γ) entails, by elementary transformations, that the integrals are, respectively,

$$\iint a_\tau(u-s)a_{\tau'}(u'-s')\, dd'\gamma(u, u'),$$

$$\iint b_\tau(u-s, h)b_{\tau'}(u'-s', h')\, dd'\gamma(u, u'),$$

and the assertions follow by letting $\tau, \tau' \to \infty$.

b. INVERSIONS OF RANDOM FUNCTIONS. *If $X(t)$ is a random function with a harmonizable covariance $\Gamma(t, t')$, then there exist random functions*

$\Delta_0\xi(s)$ with covariance $\Delta_0\Delta'_0\gamma(s, s')$ and $\hat{\xi}(s)$ with covariance $\hat{\gamma}(s, s')$ such that, as $\tau \to \infty$,

$$\frac{1}{2\tau}\int_{-\tau}^{+\tau} e^{-ist}X(t)\,dt \overset{\text{q.m.}}{\longrightarrow} \Delta_0\xi(s)$$

$$\frac{1}{2\pi}\int_{-\tau}^{+\tau} -\frac{1}{it}\Delta_h e^{-ist}X(t)\,dt \overset{\text{q.m.}}{\longrightarrow} \Delta_h\hat{\xi}(s).$$

If the random function $X(t)$ is harmonizable with respect to $\xi(s)$, then $\Delta_0\xi(s)$ is the jump function of $\xi(s)$ and $\hat{\xi}(s)$ is the normalized $\xi(s)$.

Proof. The first assertions follow readily from the foregoing lemma. The remaining ones can be deduced from the first ones or follow directly from (H_X), upon observing that, by elementary transformations, the foregoing integrals become, respectively,

$$\int a_\tau(u - s)\,d\xi(u), \quad \int b_\tau(u - s, h)\,d\xi(u).$$

A. HARMONIZABILITY THEOREM. *A random function is harmonizable if, and only if, its covariance is harmonizable.*

Proof. From

$$X(t) = \int e^{its}\,d\xi(s), \quad \iint |\,dd'\gamma(s, s')\,| < \infty$$

where $\gamma(s, s')$ is the covariance of $\xi(s)$, it follows at once that

$$EX(t)\bar{X}(t') = \iint e^{i(ts - t's')}\,dd'\gamma(s, s').$$

Conversely, let $X(t)$ have for covariance

$$\Gamma(t, t') = \iint e^{i(ts - t's')}\,dd'\gamma(s, s') \quad \textit{with} \quad \iint |\,dd'\gamma(s, s')\,| < \infty.$$

Since the integrand is continuous and bounded and $\gamma(s, s')$ is of bounded variation, we can assume without restricting the generality that $\gamma(s, s')$ is normalized. According to the foregoing lemma, there exists a random function $\xi(s)$ whose covariance is $\gamma(s, s')$ such that

$$\frac{1}{2\pi}\int_{-\tau}^{+\tau} -\frac{1}{it}\Delta_h e^{-ist}X(t)\,dt \overset{\text{q.m.}}{\longrightarrow} \Delta_h\xi(s), \quad \tau \to \infty.$$

Upon applying the second parts of the foregoing lemmas, it follows by elementary computations that $Y(t) = \int e^{its}\,d\xi(s)$ exists and $E|\,X(t)\,|^2$

$= EX(t)\overline{Y}(t) = E|\,Y(t)\,|^2$, so that $E|\,X(t) - Y(t)\,|^2 = 0$, and the proof is concluded.

Particular cases

1° If s varies over a countable set only, then the integrals with respect to $\xi(s)$ and $\gamma(s, s')$ reduce to countable sums, and what precedes continues to hold. In fact, the proofs reduce to those of the first parts of **a** and **b**.

2° If t varies over a countable set only, then the integrals with respect to dt, $dt\,dt'$ reduce to countable sums and what precedes continues to hold. In fact, the same proofs apply, provided $\Gamma(t, t')$ and $X(t)$ are first extended by letting t, t' vary over R in (H_Γ) and (H_X).

REMARK. The following analytical problem is of interest: characterize harmonizable covariances $\Gamma(t, t')$, that is, harmonizable functions of nonnegative-definite type. The answer ought to reduce to Bochner's theorem in the particular case of a continuous covariance $\Gamma(t, t') = f(t - t')$. The necessary condition is that $\Gamma(t, t')$ be continuous and bounded. Is this condition sufficient? If not, what supplementary conditions—which ought to disappear in the preceding particular case—are required?

37.5. Orthogonal decompositions. Among various decompositions of second order random functions, the orthogonal ones play a prominent role. The physical reason is that orthogonal components can be isolated experimentally by means of suitable "filters." The mathematical reason is that orthogonal decompositions correspond to the introduction of a general form of cartesian frames of reference which allow the use of a general form of Pythagorean relation. We saw in the preceding subsection that in the case of a random function defined on a countable set of values of the argument such a decomposition is always possible and the frame of reference can be obtained by linear combinations of the random values of the function. We intend to proceed to more general orthogonal decompositions of the same character. First let us give two countable decompositions. The one below follows from 34.2**B**, Corollary 3 by elementary computations.

A. ORTHOGONAL EXPANSION THEOREM. *The expansion*

$$X(t) = X(0) + \frac{t}{1!}X'(0) + \frac{t^2}{2!}X''(0) + \cdots, \quad t \in I,$$

of a second order analytic random function $X(t)$ is an orthogonal decom-

position if, and only if, its (analytic) covariance $\Gamma(t, t')$ *is a function of the product* tt' *of its arguments in* $I \times I$.

In fact, every random function $X(t)$ continuous in q.m. on a closed interval I has a countable orthogonal decomposition. We shall use Mercer's theorem which states that if a nonnegative-definite type function $\Gamma(t, t')$ is continuous on $I \times I$, then

$$\Gamma(t, t') = \sum |\lambda_n|^2 \psi_n(t) \bar{\psi}_n(t'),$$

where the series converges absolutely and uniformly on $I \times I$, and the continuous functions $\psi_n(t)$ are "proper functions" of $\Gamma(t, t')$ corresponding to "proper values" $|\lambda_n|^2$:

$$\int \Gamma(t, t') \psi_n(t') \, dt' = |\lambda_n|^2 \psi_n(t).$$

Proper functions which correspond to (necessarily finitely) multiple proper values are written with distinct indices, and all proper functions are orthonormalized on I:

$$\int \psi_m(t) \bar{\psi}_n(t) \, dt = \delta_{mn}.$$

B. PROPER ORTHOGONAL DECOMPOSITION THEOREM. *A random function* $X(t)$ *continuous in q.m. on a closed interval* I *has on* I *an orthogonal decomposition*

$$X(t) = \sum \lambda_n \xi_n \psi_n(t)$$

with

$$E\xi_m \bar{\xi}_n = \delta_{mn}, \quad \int \psi_m(t) \bar{\psi}_n(t) \, dt = \delta_{mn},$$

if, and only if, the $|\lambda_n|^2$ *are the proper values and the* $\psi_n(t)$ *are the orthonormalized proper functions of its covariance. Then the series converges in q.m. uniformly on* I.

Proof. Let t, t' vary over I, and

$$X_n(t) = \sum_{k=1}^{n} \lambda_n \xi_n \psi_n(t).$$

If $X(t)$ has the asserted decomposition, then

$$EX(t)\bar{X}(t') = \lim EX_n(t)\bar{X}_n(t') = \sum |\lambda_n|^2 \psi_n(t) \bar{\psi}_n(t'),$$

and the "only if" assertion follows. Conversely, let the $\psi_n(t)$ be the orthonormalized proper functions of the covariance $\Gamma(t, t')$ of $X(t)$, and form the integrals

$$\lambda_n \xi_n = \int X(t) \bar{\psi}_n(t)\, dt.$$

These integrals exist, since $X(t)$ (in q.m.) and $\psi_n(t)$ are continuous on the closed interval I, and

$$E\xi_m \bar{\xi}_n = \delta_{mn}, \quad EX(t)\bar{\xi}_n = \lambda_n \psi_n(t).$$

It follows from Mercer's theorem that $E| X(t) - X_n(t) |^2 \to 0$ uniformly on I, and the "if" assertion is proved.

In physics, the most important orthogonal decomposition is the harmonic one, for, loosely speaking, it yields "amplitudes," and hence "energies," corresponding to the various parts of the "spectrum" of the random function, and we seek it now. But, first, we have to introduce random functions which correspond to sums of orthogonal r.v.'s. It will be convenient to denote the increment of a function, say $\xi(t)$, on an interval $[a, b)$ by $\xi[a, b) = \xi(b) - \xi(a)$. Increment functions are characterized by their additivity:

$$\xi[a, b) + \xi[b, c) = \xi[a, c)$$

and determine the point functions $\xi(t)$ up to additive quantities. While what follows is valid for more general ordered sets T, we shall assume, to simplify the language, that $T \subset R$.

A second order random function $\xi(t)$ has *orthogonal increments* if, for disjoint intervals $[a, b)$, $[a', b')$,

$$E\xi[a, b)\bar{\xi}[a', b') = 0.$$

Then

$$E| \xi[a, b) \pm \xi[a', b') |^2 = E| \xi[a, b) |^2 + E| \xi[a', b') |^2,$$

and it follows, by setting, for some fixed a,

$$E| \xi[a, t) |^2 = F(t), \quad E| \xi[t, a) |^2 = -F(t),$$

that

$$E| \xi[t, t') |^2 = F[t, t');$$

for short,

$$E| d\xi(t) |^2 = dF(t).$$

If $\gamma(t, t')$ is the covariance of $\xi(s)$, this relation becomes

$$\gamma(t', t') - \gamma(t', t) - \gamma(t, t') + \gamma(t, t) = F(t') - F(t), \quad t' \geqq t;$$

for short,

$$dd'\gamma(t, t') = dF(t), \quad t' \geqq t.$$

In words, the increment of the foregoing covariance over a two-dimensional rectangle with sides parallel to the axes of t and of t' reduces to its increment over the square whose diagonal is the part of the line $t = t'$ belonging to the rectangle (draw the figure).

Observe that the second moments of the increments of $\xi(t)$ are bounded (by some fixed finite number) if, and only if, the finite nondecreasing function $F(t)$ is bounded on T. Whether bounded or not, this nondecreasing function extends to a nondecreasing function (finite or not) on $R = (-\infty, +\infty)$ and, in fact, on $\overline{R} = [-\infty, +\infty]$. This leads to

C. EXTENSION THEOREM. *If the random function $\xi(t)$ on T has orthogonal increments with bounded second moments, then, preserving this property, $\xi(t)$ can be extended by continuity to the closure of T and can be extended on $\overline{R}$.*

Proof. Let τ be a limit point of T from the left (it may or may not belong to T and may be $+\infty$). Since $F(t)$ is nondecreasing and bounded on T, $F(\tau - 0) = \lim\limits_{t \uparrow \tau} F(t)$ exists and is finite, and as $t, t' \uparrow \tau (t > t')$

$$E|\,\xi(t) - \xi(t')\,|^2 = F(t) - F(t') \to 0. \tag{1}$$

It follows that the r.v. $\xi(\tau - 0) = \lim\limits_{t \uparrow \tau}$ q.m. $\xi(t)$ exists, and for $t \geqq \tau$

$$E|\,\xi(t) - \xi(\tau - 0)\,|^2 = \lim_{t' \uparrow \tau} E|\,\xi(t) - \xi(t')\,|^2 = F(t) - F(\tau - 0).$$

Similarly if τ is a limit point of T from the right. If τ is a limit point from both sides, then, by letting $t \downarrow \tau$ and $t' \uparrow \tau$ in (1), we also have

$$E|\,\xi(\tau + 0) - \xi(\tau - 0)\,|^2 = F(\tau + 0) - F(\tau - 0).$$

Now, if $\tau \not\in T$ is a limit point from the left set $F(\tau) = F(\tau - 0)$, $\xi(\tau) = \xi(\tau - 0)$, and if it is a limit point from the right but not from the left set $F(\tau) = F(\tau + 0)$, $\xi(\tau) = \xi(\tau + 0)$. This provides the asserted extension on the closure $\overline{T}$ of T. For it suffices to write that the increments are orthogonal on T and let one or more end points approach points $\tau \in \overline{T} - T$; in particular $\xi(\tau + 0) - \xi(\tau - 0)$ is orthogonal to increments on intervals disjoint from $\{\tau\}$.

Finally, $\overline{R} - \overline{T}$ is a countable sum of intervals I with at least one end point τ of I belonging to $\overline{T}$. By setting $F(t) = F(\tau)$, $\xi(t) = \xi(\tau)$ on the I's, we obtain the asserted extension on $\overline{R}$.

When the boundedness condition is not satisfied, then the same proof shows that the asserted extension exists on the smallest interval containing T, provided the end points which do not belong to T are excluded.

COROLLARY 1. *Under the hypotheses of the extension theorem, the random function $\xi(t)$ is decomposable into two parts with mutually orthogonal increments:*

$$\xi(t) = \xi_d(t) + \xi_c(t),$$

where $\xi_d(t)$ is a sum of orthogonal r.v.'s (converging in q.m. when denumerable) and $\xi_c(t)$ is continuous in q.m., and

$$E|\, d\xi_d(t)\,|^2 = dF_d(t), \quad E|\, d\xi_c(t)\,|^2 = dF_c(t),$$

where $F_d(t)$ and $F_c(t)$ are the purely discontinuous and the continuous parts of $F(t)$, respectively.

Extend on $\overline{R}$, let $\{t_j\}$ be the (countable) discontinuity set of F, and set

$$\xi_d(t) = \sum_{t_j < t} (\xi(t_j + 0) - \xi(t_j - 0)), \quad \xi_c(t) = \xi(t) - \xi_d(t).$$

The assertions follow by elementary computations.

We say that a random function $\xi(t)$ on T with covariance $\gamma(t, t')$ is *orthogonal to its increments*, if for $t < t'$

$$E\xi(t)(\overline{\xi}(t') - \overline{\xi}(t)) = 0, \quad \text{that is,} \quad E\xi(t)\overline{\xi}(t') = E|\, \xi(t)\,|^2.$$

Clearly, every such random function $\xi(t)$ has orthogonal increments and the definition is equivalent to

$$\gamma(t, t') = F(t) \text{ nondecreasing}, \quad t \leqq t';$$

if, moreover, $F(t)$ is bounded on T, then the foregoing extension on $\overline{R}$ preserves this relation. Conversely

COROLLARY 2. *If a random function $\xi(t)$ on T has orthogonal increments with bounded second moments, then it can be made orthogonal to its increments by a suitable change of origin of its random values.*

Extend $\xi(t)$ on $\overline{R}$ and subtract $\xi(-\infty)$.

Thus, in the integrals below with respect to random functions $\xi(t)$ which have orthogonal increments with bounded second moments, we can assume that the $\xi(t)$ are orthogonal to their increments.

Consider now the integration in q.m. criterion and assume that the random function $Y(t)$ therein is a function $\xi(t)$ with orthogonal increments. Then, clearly, the double integrals therein reduce to simple integrals with respect to the nondecreasing function $F(t)$ defined above. In particular, the harmonizability theorem becomes

a. *Let* $E|d\xi(s)|^2 = dF(s)$. *A second order random function* $X(t)$ *is of the form*

$$X(t) = \int e^{its}\, d\xi(s)$$

if, and only if, its covariance $\Gamma(t, t')$ *is of the form*

$$\Gamma(t, t') = \int e^{i(t-t')s}\, dF(s).$$

For such covariances, the question raised at the end of the preceding subsection has a very simple answer (Khintchine, Kolmogorov).

b. *A covariance* $\Gamma(t, t')$ *is of the form*

$$\Gamma(t, t') = \int e^{i(t-t')s}\, dF(s), \quad \operatorname{Var} F < \infty$$

where t, t' *vary over* R *or over* $\{\cdots -1, 0, +1, \cdots\}$ *if, and only if, it depends only upon the difference* $t - t'$ *of its arguments, and when* t, t' *vary over* R *it is, moreover, continuous.*

Proof. The "only if" assertion is obvious. The "if" assertion follows from the fact that a covariance is of nonnegative-definite type. For, the definition of this type reduces to that of nonnegative-definite functions when $\Gamma(t, t') = f(t - t')$, and then Herglotz's and Bochner's theorems apply.

A covariance which depends only upon the difference of its arguments is said to be *stationary*. A random function with a stationary covariance is said to be *second order stationary*. This concept is closely related to the usual concept of stationarity (of distributions); for, if a random function is stationary in the usual sense and is of second order, then it is *a fortiori* second order stationary, and in the strongly normal case the converse is true. A stationary random function need not be second order stationary since it may not be of second order.

Together, the preceding two lemmas and the inversion lemma for random functions become

D. HARMONIC ORTHOGONAL DECOMPOSITION THEOREM. *A second order random function $X(t)$ on $T = R$ or $\{\cdots -1, 0, +1, \cdots\}$ has a harmonic orthogonal decomposition*

$$X(t) = \int e^{its}\, d\xi(s), \quad t \in T,$$

where $E|\, d\xi(s)\,|^2 = dF(s)$, and $F(s)$ is of bounded variation on T

if, and only if,

$X(t)$ is second order stationary and in the case $T = R$ also continuous in q.m. at one point t.

Then, as $\tau \to \infty$,

$$\frac{1}{2\tau}\int_{-\tau}^{+\tau} e^{-ist}X(t)\, dt \xrightarrow{\text{q.m.}} \Delta_0\xi(s),$$

$$\frac{1}{2\pi}\int_{-\tau}^{+\tau} -\frac{1}{it}\Delta_h e^{-ist}X(t)\, dt \xrightarrow{\text{q.m.}} \Delta_h\hat{\xi}(s),$$

where the integrals reduce to sums when $T = \{\cdots -1, 0, 1, \cdots\}$.

COROLLARY 1. *A second order stationary random function on $\{\cdots -1, 0, +1, \cdots\}$ extends on R to a second order stationary random function continuous in q.m.*

COROLLARY 2. *A second order stationary random function $X(t)$ continuous in q.m. is decomposable into two orthogonal and second order stationary parts $X(t) = X_d(t) + X_c(t)$, with*

$$X_d(t) = \int e^{its}\, d\xi_d(s), \quad X_c(t) = \int e^{its}\, d\xi_c(t)$$

where the first integral reduces to a countable sum converging in q.m.

What precedes applies to second order r.f.'s of the form $X(t) = \{X_u(t),\ u \in U\}$, $t \in T$, that is, whose random values $X(t)$ are second order random functions of t: $E|\, X_u(t)\,|^2 < \infty$ whatever be t and u. It suffices to consider the argument $(u, t) \in U \times T$. However, the arguments u and t may play a nonsymmetric role. For example, the above random function is *second order stationary in t* if

$$EX_u(t)\overline{X}_{u'}(t') = f_{u,u'}(t - t'), \quad t, t' \in R, \quad u, u' \in U.$$

Clearly, this property is equivalent to the following one: the random functions $Y_{U_n}(t) = \sum_{u \in U_n} c_uX_u(t)$, $t \in T$, are second order stationary whatever be the complex numbers c_u and whatever be the finite subset $U_n \subset U$. Upon applying the harmonic orthogonal decomposition theorem to the random functions $Y_{U_n}(t)$, we obtain by elementary computations

Corollary 3. *A random function* $X(t) = \{X_u(t), u \in U\}$ *continuous in q.m. on R is second order stationary if, and only if,*

$$EX_u(t)\overline{X}_{u'}(t') = \int e^{i(t-t')s}\, dF_{u,u'}(s), \quad u, u' \in U$$

where the functions $F_{u,u'}(s)$ *are of bounded variation on R and the functions* $\Delta_h F_{u,u'}(s)$ *are of nonnegative-definite type in* u, u' *for any s and* $h > 0$

or if, and only if,

$$X_u(t) = \int e^{its}\, d\xi_u(s), \quad u \in U$$

where the second order random functions $\xi_u(s)$ *have mutually orthogonal increments with*

$$E\xi_u(s)\overline{\xi}_{u'}(s') = F_{u,u'}(s) \quad \text{for} \quad s \leqq s'.$$

If U contains only two elements, the first assertion reduces to a result of Khintchine. If U contains only a finite number of elements, then it reduces to a result of Cramér.

Another nonsymmetric role of the two arguments u and t appears in considering them as the real and imaginary parts of a complex argument $z = u + it$. The definition of second order stationarity becomes: the random function $X(z)$ is *second order stationary* if its covariance

$$\Gamma_X(z, z') = f(z + \bar{z}') = f(u + u' + i(t - t'))$$

depends only upon $z + \bar{z}'$. Then, for every fixed $u + u'$, the covariance is stationary in t, t', the harmonic orthogonal decomposition theorem applies, and it is not difficult to obtain

Corollary 4. *A random function* $X(z)$, $z = u + it$, *continuous in q.m. in the complex-plane strip S:* $a < u < b$, *is second order stationary if, and only if, for* $z, z' \in S$,

$$\Gamma_X(z, z') = \int e^{(z+\bar{z}')s}\, dF(s), \quad \operatorname{Var} F < \infty,$$

or, if and only if, for $z \in S$,

$$X(z) = \int e^{izs}\, d\xi(s),$$

where the random function $\xi(s)$ *is orthogonal to its increments with* $\Gamma_\xi(s, s') = F(s)$ *for* $s < s'$.

REMARK. By replacing the argument t by a complex argument z in the preceding extension, Corollary 3 extends to random functions of z. It suffices to replace therein t by z and t' by $\bar{z}'$.

37.6. **Normality and almost-sure properties.** The class of normal r.v.'s is closed under linear combinations and passages to the limit in q.m. Since differentiation and integration in q.m. (with one of the two functions being a sure function) are obtained by such operations, it follows that the stability of normal laws (so far considered only for sums of independent r.v.'s) extends to the calculus in q.m.:

A. NORMAL STABILITY THEOREM. *Normality is preserved under differentiations and integrations in q.m.*

In fact, the foregoing remarks apply word for word to second order random functions which obey infinitely decomposable laws—necessarily with finite variances (that is, all the linear combinations of the random values obey such laws). But more is true when the second order random functions are normal: many of the properties in q.m. established in this section become then a.s. properties. We saw that to every covariance there corresponds a strongly normal random function and, for the random values of such functions, orthogonality becomes independence. If a series of these orthogonal values converges in q.m., then, the summands being independent, the convergence is almost sure. Therefore, upon applying the foregoing theorem to the three orthogonal decomposition theorems of the preceding subsection, we obtain

COROLLARY 1. *If the covariance $\Gamma_X(t, t')$ is an analytic function of tt', then the random function $X(t)$ is normal if, and only if, its derivatives in q.m. are normal. If, moreover, $X(t)$ is real-valued, then its derivatives are independent and the Taylor expansion of $X(t)$ converges a.s.*

COROLLARY 2. *If the covariance $\Gamma_X(t, t')$ is continuous on a closed interval I, then the random function $X(t)$ is normal if, and only if, the r.v.'s $\lambda_n\xi_n = \int_I X(t)\bar{\psi}_n(t)\,dt$ are normal.*

If, moreover, $X(t)$ is real-valued, then the ξ_n are independent and the proper decomposition of $X(t)$ converges a.s.

COROLLARY 3. *If the continuous covariance $\Gamma_X(t, t')$ is stationary, then*

(i) *The random function $X(t)$ is normal if, and only if, the random function $\xi(s)$ which figures in its harmonic decomposition $X(t) = \int e^{its}\,d\xi(s)$*

is normal. If $\xi(t)$ is strongly normal, then its increments are independent and $\int_a^b \xrightarrow{\text{a.s.}} \int$ *as* $a \to -\infty, b \to +\infty$.

(ii) *If* $\xi(t) = \xi_d(t) + \xi_c(t)$ *is the decomposition of* $\xi(t)$ *with independent increments into its purely discontinuous and its continuous in q.m. parts, then* $X(t) = X_d(t) + X_c(t)$ *is decomposed into two independent parts:* $X_d(t) = \int e^{its}\, d\xi_d(s)$ *which is an a.s. convergent series of independent r.v.'s, and* $X_c(t) = \int e^{its}\, d\xi_c(s)$ *which obeys an infinitely decomposable law with finite variances.*

Only the very last assertion deserves proof. It is due to the fact that $E|\,d\xi(s)\,|^2 = dF_c(s)$, where the function $F_c(s)$ on R is nondecreasing, bounded, and continuous. Thus, R can be subdivided into intervals I on which the increments of $F_c(s)$ are bounded by $\epsilon > 0$ arbitrarily small. It follows that $X_c(t)$ is a sum of an arbitrarily large number of independent r.v.'s $\int_I e^{its}\, d\xi_c(s)$ whose variances are uniformly bounded by an arbitrarily small number, and this implies the assertion for every r.v. $X_c(t)$. Since the same is true of every linear combination of these r.v.'s, the random function $X_c(t)$ obeys an infinitely decomposable law.

Observe that the "if and only if" assertions remain valid when "normal" is replaced by "infinitely decomposable."

37.7. **A.s. stability.** If the second order random function $X(t)$ with covariance $\Gamma(t, t')$ is not normal, we have to impose supplementary restrictions upon the covariance in order to transform properties in q.m. into a.s. properties. We shall content ourselves with conditions for a.s. stability corresponding to the strong law of large numbers. If the random function is defined on the set of all integers, we look for $Y(n) = \frac{1}{n}\sum_{k=1}^{n} X(k) \xrightarrow{\text{a.s.}} 0$, and if it is defined on $T = [0, +\infty)$, we look for $Y(\tau) = \frac{1}{\tau}\int_0^\tau X(t)\, dt \xrightarrow{\text{a.s.}} 0$ where we assume $X(t)$ continuous in q.m. on T.

In both cases, the conditions we shall find and the proof are the same, except that in the second case the integrals are to be replaced by sums. Therefore, we shall give the proof in the slightly more involved case of $T = [0, +\infty)$.

First, we reduce the random function $Y(\tau)$ to a sequence of r.v.'s. We consider a sequence m^a, $m = 1, 2, \cdots$, and the symbols a, c, c'

are finite positive constants while b is a nonnegative constant. For $m^a \leqq \tau < (m+1)^a$, we can set

$$\frac{\tau}{m^a} Y(\tau) = Y(m^a) + Z(m^a, \tau), \quad Z(m^a, \tau) = \frac{1}{m^a} \int_{m^a}^{\tau} X(t)\, dt.$$

Since

$$U(m^a) = \sup_{m^a \leqq \tau < (m+1)^a} | Z(m^a, \tau) | \leqq \frac{1}{m^a} \int_{m^a}^{(m+1)^a} | X(t) |\, dt,$$

we have

$$E| U(m^a) |^2 \leqq \frac{1}{m^{2a}} \int_{m^a}^{(m+1)^a} \int_{m^a}^{(m+1)^a} E| X(t)\bar{X}(t') |\, dt\, dt'$$

$$\leqq \left(\frac{1}{m^a} \int_{m^a}^{(m+1)^a} \sqrt{\Gamma(t, t)}\, dt \right)^2.$$

We are led to assume that $\Gamma(t, t) \leqq ct^{2b}$ so that

$$\sum_{m=1}^{\infty} E| U(m^a) |^2 \leqq \sum_{m=1}^{\infty} \frac{c'}{m^{2a}} | (m+1)^{a(b+1)} - m^{a(b+1)} |^2 \sim \sum_{m=1}^{\infty} \frac{1}{m^{2(1-ab)}}$$

and the series converges when $ab < 1/2$. Then by Tchebichev's inequality and the Borel-Cantelli lemma $U(m^a) \xrightarrow{\text{a.s.}} 0$.

Similarly, by taking the sequence $q^m = (1+\epsilon)^m$ with $\epsilon > 0$, we find that

$$U(q^m) \leqq \frac{1}{q^m} \int_{q^m}^{q^{m+1}} | X(t) |\, dt \text{ a.s.}$$

and are led to assume that $| X(t) | \leqq c$, so that

$$U(q^m) \leqq c \frac{q^{m+1} - q^m}{q^m} = c\epsilon \text{ a.s.}$$

and, hence, $U(q^m)$ is arbitrarily small for $\epsilon > 0$ sufficiently small. Thus

a. *If, for t sufficiently large,*

(i) $\Gamma(t, t) \leqq ct^b$ *and* $ab < 1/2$, *then* $Y(\tau) - Y(m^a) \xrightarrow{\text{a.s.}} 0$ *as* $\tau \to \infty$.

(ii) $| X(t) | \leqq c$, *then* $Y(\tau) - Y(q^m)$ *becomes a.s. arbitrarily small for* $q - 1 > 0$ *sufficiently small.*

It remains to insure the a.s. convergence of the sequences $Y(m^a)$ or $Y(q^m)$ to zero. This is obtained as follows:

A. A.S. STABILITY THEOREM. *Let the random function $X(t)$ on $T = [0, +\infty)$ with covariance $\Gamma(t, t')$ be continuous in q.m., and let c, c', γ be finite positive constants.*

If, for large $\tau > 0$

(i) $$\Gamma(t, t) \leqq c \quad \text{and} \quad \frac{1}{\tau^2}\int_0^\tau\int_0^\tau \Gamma(t, t')\, dt\, dt' \leqq \frac{c'}{\tau^\gamma}$$

or

(ii) $$|X(t)| \leqq c \quad \text{and} \quad \int_0^\infty \frac{d\tau}{\tau^3}\int_0^\tau\int_0^\tau |\Gamma(t, t')|\, dt\, dt' \leqq c,$$

then, as $\tau \to \infty$

(iii) $$\frac{1}{\tau}\int_0^\tau X(t)\, dt \xrightarrow{\text{a.s.}} 0.$$

The same is true if $X(t)$ is defined on the set of all integers, the integrals being replaced by the corresponding sums.

Proof. The first condition in (i) permits to apply the first part of the preceding lemma with $b = 0$ and arbitrary $a > 0$. The second condition in (i) yields

(1) $$\sum_p E|Y(p)|^2 = \sum_p \frac{1}{p^2}\int_0^p\int_0^p \Gamma(t, t')\, dt\, dt' < \infty$$

for $p = m^a$ with $a > 1/\gamma$. The first assertion follows. Similarly, the first condition in (ii) permits to apply the second part of the preceding lemma, and the second condition in (ii) yields by elementary computations (1) with $p = q^m$ however small be $q - 1 > 0$. The second assertion follows.

If the random function $X(t)$ is second order stationary either on $T = \{\cdots -1, 0, +1, \cdots\}$ or continuous in q.m. on $T = R$, then the first condition of (i) holds for the random function $e^{-ist}X(t)$ whose covariance is $\int e^{-i(t-t')(s-s')}\, dF(s')$ with the function $F(s')$ of bounded variation. Upon replacing $\Gamma(t, t')$ by this expression in the second condition of (i), we obtain

COROLLARY. *Let the random function $X(t)$ be second order stationary and continuous in q.m. on $T = R$. If there exist two finite positive constants c and γ such that, for large $\tau > 0$*

$$\int \frac{\sin^2 \frac{\tau}{2}(s - s')}{\frac{\tau^2}{4}(s - s')^2}\, dF(s') \leqq \frac{c}{\tau^\gamma},$$

then as $\tau \to \infty$,

$$\frac{1}{\tau}\int_0^\tau e^{-ist}X(t)\, dt \xrightarrow{\text{a.s.}} 0.$$

The same is true if $X(t)$ *is defined on* $T = \{\cdots -1, 0, +1, \cdots\}$, τ *being replaced by* n *and the last integral being replaced by the corresponding sum.*

The reader is invited to compare the last assertion with the second part of the stationarity theorem (with $r = 2$).

It is easily seen that the conclusion holds a.e. (in Lebesgue measure) in s. In fact, if the symmetric derivative

$$F'(s_0) = \lim_{h \to 0} \frac{F(s_0 + h) - F(s_0 - h)}{2h}$$

exists and is finite (and this is true a.e., since the d.f. F has a.e. a finite derivative), then the integral which figures in the hypothesis is $O(F'(s_0)/\tau)$, and hence the conclusion holds with $s = s_0$. It suffices to take s_0 for origin of values of s and use the relation

$$\lim_{T \to \infty} \frac{1}{\pi}\int \frac{\sin^2 Ts}{Ts^2}\, dF(s) = F'(0),$$

which can be proved as follows: Write the integral in the form

$$\int_0^\infty \frac{\sin^2 Ts}{Ts^2}\, d\{F(s) - F(-s)\},$$

split it into $\int_0^a + \int_a^\infty$, select $a > 0$ for a given $\epsilon > 0$ so as to have $\left|\frac{F(s) - F(-s)}{2s} - F'(0)\right| < \epsilon$ on $(0, a)$, integrate by parts the integral $\int_0^a$, and let $T \to \infty$ and then $\epsilon \to 0$.

COMPLEMENTS AND DETAILS

In what follows $\Gamma(t, t')$ denotes the covariance of a second order random function $X(t)$ with $EX(t) = 0$ on $T \subset R$, unless otherwise stated.

1. $\Gamma(t,t')$ on $R \times R$ is said to be a triangular covariance if $\Gamma(t,t') = F_1(t)\overline{F}_2(t')$, $t \leqq t'$. Such a product, with $F_2 \neq 0$, is a covariance if, and only if, F_1/F_2 is nonnegative and nondecreasing on R. Construct a few random functions with triangular covariances. What about random functions orthogonal to their increments?

2. Let $\tilde{M}(t, t')$ denote the complex-conjugate of the function $M(t, t')$ on $T \times T$.

If $\int_T\int_T M(t, \tau)\Gamma(\tau, \tau')\tilde{M}(\tau', t')\, d\tau\, d\tau'$ exists and is finite, then it is covariance of the random function $\int_T M(t, \tau)X(\tau)\, d\tau$. What about $\sum M(t, t_j)X(t_j)$?

Application. The iterated $\Gamma^{(n)}(t, t')$ is defined by

$$\Gamma^{(m+n)}(t, t') = \int_T \Gamma^{(m)}(t, \tau)\Gamma^{(n)}(\tau, t')\, d\tau$$

assumed to exist and be finite. Every iterate of a covariance is a covariance. (This follows for $\Gamma^{(2n+1)}(t, t')$ from what precedes. For $\Gamma^{(2n+2)}(t, t')$ begin by verifying directly that $\Gamma^{(2)}(t, t')$ is a covariance.)

3. Let H be a family of functions h on T forming an euclidean or, more generally, a Hilbert space; denote—by overabusing the notations—the scalar product by $(h(t), h'(t))$. A function $\Gamma(t, t')$ on $T \times T$ is said to be a *reproducing kernel* of H, if $\Gamma(t, t') \in H$ for every fixed t' and $\Gamma(t, t')$ reproduces every $h \in H$, that is, $h(t') = (h(t), \Gamma(t, t'))$.

$\Gamma(t, t')$ is a reproducing kernel of some family H if, and only if, it is covariance of a random function $X(t)$ on T. If the Y's are limits in q.m. of all possible linear combinations of random values of $X(t)$, then there exist Y's such that $h(t) = EY\overline{X}(t)$, $(h_1(t), h_2(t)) = EY_1Y_2$.

Examples

1° Let $T = \{t_1, t_2, \cdots\}$. Consider the space of all sequences $\{h(t_1), h(t_2), \cdots\}$ with $\sum |h(t_n)|^2 < \infty$, $(h_1(t), h_2(t)) = \sum h_1(t_n)h_2(t_n)$. Then $\Gamma(t_m, t_n) = \delta_{m,n}$, $X(t)$ on T are second order random functions with orthonormal random values.

2° Let $T = R$ and F on R be a nondecreasing function of bounded variation. Consider all functions of the form $h(t) = \int e^{-itx}g(x)\, dF(x)$ with $\int |g(x)|^2\, dF(x) < \infty$. Then $\Gamma(t, t') = \int e^{i(t-t')x}\, dF(x)$, $X(t)$ on T are second order stationary and continuous in q.m.

4. Let $\{t_1, t_2, \cdots,\} \subset T$, set

$$D\begin{pmatrix} t, t_1, \cdots, t_n \\ t', t_1, \cdots, t_n \end{pmatrix} = \begin{vmatrix} \Gamma(t, t') & \Gamma(t, t_1) & \cdots & \Gamma(t, t_n) \\ \Gamma(t_1, t') & \Gamma(t_1, t_1) & \cdots & \Gamma(t_1, t_n) \\ \cdot & \cdot & \cdots & \cdot \\ \Gamma(t_n, t') & \Gamma(t_n, t_1) & \cdots & \Gamma(t_n, t_n) \end{vmatrix}$$

and denote by $D(t, t_1, \cdots, t_n)$ the square root of this determinant when $t' = t$.

Assume that the $X(t_n)$ are nondegenerate and are linearly independent. Then $D^2(t_1, \cdots, t_n) > 0$.

(a) The foregoing determinant is a covariance, say, of the random function $D(t_1, \cdots, t_n)Y_n(t)$ where

$$Y_n(t) = \frac{1}{D^2(t_1, \cdots, t_n)} \begin{vmatrix} X(t) & \Gamma(t, t_1) & \cdots & \Gamma(t, t_n) \\ X(t_1) & \Gamma(t_1, t_1) & \cdots & \Gamma(t_1, t_n) \\ \cdot & \cdot & \cdot & \cdot \\ X(t_n) & \Gamma(t_n, t_1) & \cdots & \Gamma(t_n, t_n) \end{vmatrix}.$$

Set $Y_0(t) = X(t)$, $Y_n(t) = X(t) - \sum_{k=1}^{n} c_{nk}(t)X(t_k)$, $n = 1, 2, \cdots$. Then $Y_n(t)$ are determined either by the condition that they be orthogonal to the $X(t_k)$ or by the condition that its variance be the smallest possible. We have

$$D^2(t) \geqq D^2(t, t_1)/D^2(t_1) \geqq \cdots \geqq D^2(t, t_1, \cdots, t_n)/D^2(t_1, \cdots, t_n) \geqq \cdots$$

and

$$D^2(t_1)D^2(t_2) \cdots D^2(t_n) \geqq D^2(t_1, \cdots, t_n).$$

(b) Set $\xi_k = Y_{k-1}(t_k)/\sqrt{E|Y_{k-1}(t_k)|^2}$. Then $E\xi_k\bar{\xi}_l = \delta_{k,l}$ and

$$Y_n(t) = X(t) - \sum_{k=1}^{n} a_k(t)\xi_k, \quad a_k(t) = EX(t)\bar{\xi}_k.$$

The sequence $Y_n(t) \xrightarrow{\text{q.m.}} Y(t)$ such that

$$Y(t_n) = 0,\ EY(t)\bar{X}(t_n) = 0,\ EY(t)\bar{Y}(t') = \lim D\begin{pmatrix} t, t_1, \cdots, t_n \\ t', t_1, \cdots, t_n \end{pmatrix}/D^2(t_1, \cdots, t_n)$$

and

$$X(t) = X_\infty(t) + Y(t)$$

with

$$EX_\infty(t)\bar{Y}(t') = 0, \quad X_\infty(t) = \sum_{n=0}^{\infty} a_n(t)\xi_n.$$

Let $t_0 = t$, $t_n = U^n t$ where U is a transformation on T to T, so that $T_\infty = \{t_n\}$ moves with t. Let $\eta(t_n) = Y(t_n)/\sqrt{E|Y(t_n)|^2}$, $n = 0, 1, \cdots$, so that the $\eta(t_n)$ form an orthonormal system moving with t.

The random function $X(t)$ is decomposable into $X(t) = X_1(t) + X_2(t)$ where the random functions $X_1(t)$ and $X_2(t)$ are orthogonal on $\{U^n t\}$,

$$X_1(t) = \sum_{n=0}^{\infty} a_n(t)X_1(t_n), \quad X_2(t) = \sum_{n=0}^{\infty} b_n(t)\eta(t_n)$$

with

$$\frac{D(t, t_1, \cdots)}{D(t_1, t_2, \cdots)}\eta(t) = X(t) - \sum_{k=1}^{\infty} a_k(t)\xi_k, \quad E\xi_k\bar{\xi}_l = \delta_{k,l}.$$

(c) What are the properties of ch.f.'s and second order stationary random functions which follow from (a) and (b).

5. To every continuous stationary covariance there corresponds a random function of the form $X(t) = \alpha e^{\beta t}$ where α and β are independent r.v.'s. Find its harmonic decomposition.

6. Let $\xi(t)$ on $[0, +\infty)$ be a random function with orthogonal increments and $E|\xi(t)|^2 = t$. Investigate the random functions $X(t) = e^{-t}\xi(e^{2t})$, $X(t) = \xi(t+1) - \xi(t)$, $X(t) = \int g(t+\tau)\,d\xi(\tau)$ with $\int |g(t)|^2\,dt = 1$,

$$X(t) = \int_{-\infty}^{t} e^{\tau - t}(t-\tau)^n\,d\xi(t), \quad X(t) = \int_0^{\infty} e^{-\tau - \frac{1}{\tau}}\,d\xi(t-\tau).$$

7. Let the random function $X(t)$ be defined for $t = n = 0, 1, 2, \cdots$. $X(t)$ is second order stationary and a second order chain if, and only if, $\Gamma(m, n) = f(m-n) = f(0)e^{(a+ib)(m-n)}$ with $a \leqq 0$. What can be said about the harmonic decomposition of such random functions and of their covariances?

Extend to $X(t)$ on $[0, +\infty)$.

8. It is assumed in what follows that the random functions under consideration are of second order and that the integrals and derivatives are in q.m. and exist. In every case the reader shall write the conditions for existence in terms of covariances and assume or prove that they are fulfilled.

Let the random functions $\xi(s)$ and $\eta(s)$ on R be independent. Let $X(t) = \int e^{its}\,d\xi(s)$ and $Z(t) = \int e^{its}\,\eta(s)\,d\xi(s)$. We say that $Z(t)$ is obtained from $X(t)$ by a linear operation with gain $\eta(s)$. Usually, it is assumed that $\eta(s)$ is degenerate (into a sure function) and that $X(t)$ is second order stationary and continuous in q.m., that is, $E|d\xi(s)|^2 = dF(s)$, $\operatorname{Var} F < \infty$.

(a) Express Γ_Z in terms of Γ_ξ and Γ_η; find its form in the usual case.

Interpret $\dfrac{dX(t)}{dt}$ as a linear operation; what is the corresponding gain?

$\{X(t), Z(t)\}$ is second order stationary and continuous in q.m., if, and only if, so is $X(t)$.

(b) Express $\Delta_\theta(t) = X(t+\theta) - Z(t)$ in terms of $\xi(s)$ and $\eta(s)$. We say that $\Delta_\theta(t)$ is the error function when $X(t+\theta)$ is replaced by $Z(t)$. Express Γ_θ in terms of Γ_ξ and Γ_η. In the usual case $\Gamma_\theta(t, t) = \int |e^{i\theta s} - \eta(s)|^2\,dF(s)$ and is minimized for $\eta(s) = \eta_0(s)$ such that for all t, $\int e^{its}\eta_0(s)\,dF(s) = \int e^{its}e^{i\theta s}\,dF(s)$, provided such an $\eta_0(s)$ exists; the difference for $\eta_0(s)$ and any $\eta(s)$ is given by $\int |\eta_0(s) - \eta(s)|^2\,dF(s)$. The linear prediction problem is that of existence and determination of $\eta_0(s)$.

(c) If $\eta(s) = \int e^{-ist}\,dY(t)$, with same affixes for $\eta(s)$ and $Y(t)$ if any, then $Z(t) = \int X(t-\tau)\,dY(\tau)$ is equivalent to $Z(t) = \int e^{its}\eta(s)\,d\xi(s)$. We say that the convolution is a filtering of $X(t)$ by $Y(t)$ with gain $\eta(s)$. What are $X(t)$ and $Y(t)$ in the usual case?

Filtering by $Y_1(t) + Y_2(t)$ is a filtering with gain $\eta_1(s) + \eta_2(s)$. Filtering by $Y_1(t)$ and then by $Y_2(t)$ is a filtering with gain $\eta_1(s)\eta_2(s)$; find the corresponding $Y(t)$.

(d) A convolution defined by $Z(t) = \int X(t-\tau)\,dY(\tau)$ (without reference to $\eta(s)$ which may not exist) will be called averaging of $Y(t)$ by $X(t)$. Usually, this terminology is used when $X(t)$ is degenerate and $E|dY(t)|^2 = dt$.

Let $X(t)$ be degenerate, let $E|\,dY(t)\,|^2 = dG(t)$ and denote by μ the σ-finite measure on the Borel field in R determined by the finite function G on R. Then, up to a constant factor Γ_Z is a ch.f. with a μ-continuous d.f. and we can write $Z(t) = \int e^{its}g(s)\,dY(s)$; find $g(s)$ in the usual case. Conversely, if Γ_Z has the foregoing properties, then $Z(t)$ is an averaging of a $Y(t)$ with the foregoing property by a degenerate $X(t)$.

Part Five

ELEMENTS OF RANDOM ANALYSIS

As soon as random functions on more general sets than sets of integers appear, random analysis comes into its own. It is concerned with analytical properties and, in particular, with local ones such as continuity. The most important types of random functions isolated so far are the decomposable, martingale, and Markov ones. Foundations of random analysis and analysis of decomposable and martingale types are due primarily to P. Lévy and to Doob. Analysis of the Markov type was founded primarily by Kolmogorov and by Feller.

Investigations of random functions rely very heavily upon the particular case of random sequences. But, by their very nature, they are on a higher level of mathematical sophistication. The less involved portions may be covered first: 38.1, 38.4, 39, 41, 43.1, 44.1.

Chapter XII

FOUNDATIONS; MARTINGALES AND DECOMPOSABILITY

§38. FOUNDATIONS

Random functions were defined as families $X_T = (X_t, t \in T)$ of r.v.'s on some pr. space $(\Omega, \mathcal{A}, P)$. According to convenience, the random values at t are denoted by X_t or $X(t)$, the values of X_t at ω are denoted by $X_t(\omega)$ or $X(\omega, t)$ and, unless otherwise stated, ω and s, t, u with or without affixes, are elements of Ω and of T, respectively.

Since a function is a mapping of a space—its domain, to a space—its range space, the above definition is to be completed by specifying the domain and the range space of the random function. We are at liberty to select them according to the argument ω, t or (ω, t). Then, in order to proceed to the analytical study of random functions or *random analysis*, concepts such as extrema, continuity, measurability, are to be introduced. Thus σ-fields and/or topologies (that is, concepts of limit) are to be selected in the domains and/or range spaces. There was no such problem for random sequences: The domain was the set of natural integers $n = 1, 2, \cdots$ with limit as $n \to \infty$. The range space was the space of r.v.'s on some pr. space with limits in pr., a.s., in the rth mean. Analytical questions, such as continuity in the argument or measurability of the limits, did not arise except for the existence of limits as $n \to \infty$. However, for more general families of r.v.'s these questions are to be given a precise meaning before proceeding to the study of analytical properties of various types of r.f.'s.

In the literature, the terms "random function," "random process" and "stochastic process" are treated as synonymous. In fact, "process" means sometimes a family of r.v.'s and sometimes a class of such families, and it is deemed preferable to separate these meanings. A *random function* (*r.f.*) will be a family $X_T = (X_t, t \in T)$ of r.v.'s. A (*random* or *stochastic*) *process* will be a class of r.f.'s with a common conditional law (given the "initial" or "boundary" or "lateral" conditions). In intuitive terms, consider the argument t belonging to, say, $T = [0, \infty)$ as "time" and the conditional law of a r.f. X_T given, say, X_{T_0}, $T_0 \subset T$, as its "law of evolution." Then a process $(X_T \mid X_{T_0})$ is the class of all r.f.'s on our pr. space with the same law of evolution $\mathcal{L}(X_T \mid X_{T_0})$, and to every choice of X_{T_0} there corresponds a r.f. of the process. "Markov" processes, whose analytical properties are investigated in the next chapter, are of this type with $T_0 = \{0\}$ and those which lend themselves to a detailed analysis are regular:

We say that $(X_T \mid X_0)$ is a *regular process* if there exists a regular c.pr. P^{X_0} of events defined on the process. There are two ways of viewing a regular process. The first one corresponds to underlying laws only: the process is a family of r.f.'s X_T with laws determined by the family of laws $\mathcal{L}(X_T \mid X_0 = x)$, $x \in \mathfrak{X}$, and the choice of the initial law $\mathcal{L}(X_0)$ on $\mathcal{S}$. The second one corresponds to underlying functions—the process is a family X_T of measurable functions X_t, $t \in T$, on a measurable space $(\Omega, \mathcal{A})$ to a measurable space $(\mathfrak{X}, \mathcal{S})$ and a family $(P^x, x \in \mathfrak{X})$ of pr.'s on $\mathcal{B}(X_T)$; to every choice of the initial distribution P_0 there corresponds a r.f. X_T on the pr. space $(\Omega, \mathcal{B}, P)$ with $\mathcal{B} = \mathcal{B}(X_T)$ and P on $\mathcal{B}$ defined by $PB = \int P_0(dx)P^xB$, $B \in \mathcal{B}$. We shall choose the point of view according to convenience. But first we have to install the apparatus of random analysis.

38.1. Generalities. We examine possible complete definitions of r.f.'s considered as mappings. We take them on some unspecified but fixed pr. space $(\Omega, \mathcal{A}, P)$.

Analogy with random sequences yields the following

$\mathcal{R}$-DEFINITION. *A r.f. X_T is a function on a set T to the space $\mathcal{R}$ of r.v.'s on the pr. space.*

In general, T is some set in the Euclidean line $R = (-\infty, +\infty)$ or the compactified Euclidean line $\bar{R} = [-\infty, +\infty]$. Unless otherwise stated, it will be so, and we shall denote by $\bar{T}$ the closure (or adherence) of T in the corresponding topology.

The values X_t at t are r.v.'s, that is, finite measurable functions on the pr. space to the Borel line. We have encountered more general random elements such as complex-valued r.v.'s or random vectors. Their common feature is that they take their values in (linear complete) separable metric spaces $\mathfrak{X}$ and the Borel sets in these spaces are topological Borel sets—the elements of the σ-fields $\mathcal{S}$ generated by the topology (the class of open sets). What follows, either is directly applicable to such random elements or can be transposed without difficulty. Therefore, we shall denote the *state space*—the measurable space of values of each r.v. by $(\mathfrak{X}, \mathcal{S})$ but, to fix the ideas, $\mathfrak{X}$ *will be a Borel set in* R *or* $\bar{R}$, unless otherwise stated.

In the range space $\mathcal{R}$ we have at our disposal various types of limit based upon probability. However, in order that such limits, when they exist, be unique (that is, the corresponding topology be separated) we are led, as in the case of random sequences, to identify equivalent r.v.'s and, more generally, equivalent measurable functions: X and $\tilde{X}$ are equivalent when $X = \tilde{X}$ on N^c; N, *with or without affixes, will denote a null event.* In turn, this identification already led us to a slight extension of the concept of r.v., to be considered either as a representative element only of a class of equivalence or as an a.s. defined, a.s. finite, real-valued measurable function on the pr. space.

In the case of limits in pr. and in the rth mean, we know that the spaces of equivalence classes are linear complete metric spaces with the corresponding distances defined by $d(X, Y) = E\{|X - Y|/(1 + |X - Y|)\}$ and $d(X, Y) = E|X - Y|^r$ or $E^{1/r}|X - Y|^r$ according as $r < 1$ or $r \geqq 1$; for $r \geqq 1$ the spaces L_r are Banach spaces with norm $\|X\|_r = E^{1/r}|X|^r$. Classes of equivalence are partially ordered by the relation $X \leqq Y$ a.s. or, equivalently, $g(X) \leqq g(Y)$ a.s. where g is some real-valued strictly increasing finite Borel function, otherwise arbitrary; in particular, we can select a bounded and continuous function g, say, $g =$ Arctan. Thus, whenever we are concerned with order relations or existence and uniqueness of limits along T, we may assume without loss of generality that our r.f. X_T is bounded, that is, all its values X_t are uniformly bounded by some finite constant.

At first sight, identification of equivalent r.v.'s leads to identification of equivalent r.f.'s: We say that X_T and $\tilde{X}_T$ are *equivalent r.f.'s*, and write, $X_T \sim \tilde{X}_T$, if $X_t = \tilde{X}_t$ a.s., that is, $X_t = \tilde{X}_t$ on N_t^c, for every $t \in T$; thus, the equivalence class of a r.f. X_T is characterized by the common law of all the elements of the class. When, moreover, the set $N = \bigcup_{t \in T} N_t$ is null, that is, $X_t = \tilde{X}_t$ on N^c for all $t \in T$, we say that X_T and $\tilde{X}_T$ are *a.s. equal r.f.'s.* When the set T is not countable, the foregoing union set may be nonnull or even nonmeasurable, so that equivalence of r.f.'s does not imply their a.s. equality and a.s. equality classes are subclasses of equivalence classes of r.f.'s. Identification of equivalent r.f.'s looks natural since the pr. space is to be but a frame of reference, that is, the pr. properties of r.f.'s are to be describable in terms of their laws. Yet, while no difficulties arise in the case of random sequences, in the general case we are faced with analytical properties of r.f.'s which vary within the equivalence classes and thus are not describable in terms of the common laws alone. For example, let $T = [0, 1]$, take for pr. space the Lebesgue interval $[0, 1]$, and consider r.f.'s X_T defined as follows: $X_t(\omega) = 0$ except at ω_t where $X_t(\omega_t) = 1$; the ω_t can be varied in any way from one r.f. to another without altering their equivalence. Consider $Y = \sup_{t \in T} X_t$, which takes at most two values 0 and 1, and set $A = [Y = 1] = (\omega_t, t \in T)$. We can select the ω_t so that A be any set in Ω. For instance, $A = \emptyset$ and $Y \equiv 0$ for the r.f. with every $\omega_t \in \emptyset$, $A = \Omega$ and $Y = 1$ for the r.f. with every $\omega_t = t$, and Y is not measurable for a r.f. with ω_t ranging over a nonmeasurable set. Thus we are forced to back down and to limit ourselves to subclasses of equivalence classes of r.f.'s, their choice to be based upon the requirement that limits along T be measurable, hence the possibility of expressing their analytical properties in terms of the common laws. Luckily such a choice is always possible within any equivalence class and the next subsection is devoted to a specific choice—that of "separable" r.f.'s, due to Doob.

Instead of emphasizing the argument "t" we may emphasize the argument "ω" and thus consider r.f.'s X_T as functions on Ω to the space $\mathfrak{X}_T = \prod_{t \in T} \mathfrak{X}_t$ where the $\mathfrak{X}_t$ are replicas of the range space $\mathfrak{X}$ of the X_t. The values $X_T(\omega)$ at ω of a r.f. X_T will be called *sample functions* or *trajectories* or *paths* of the r.f.; they are elements of the space $\mathfrak{X}_T$, that is, functions on T with values $X_t(\omega) \in \mathfrak{X}_t$. However, we have to insure that the sections $X_t = (X_t(\omega), \omega \in \Omega)$ at every t be r.v.'s. This necessitates the introduction of the Borel field $\mathcal{S}_T$ in $\mathfrak{X}_T$ generated by the class of Borel cylinders $C(S_{T_n})$ whose bases S_{T_n} are Borel sets in finite-dimensional subspaces $\mathfrak{X}_{T_n} = \prod_{t \in T_n} \mathfrak{X}_t$; it suffices to take finite product bases $S_{T_n} = \prod_{t \in T_n} S_t$ of Borel sets S_t in $\mathfrak{X}_t$. But the class of cylinders with countably-dimensional Borel bases is contained in $\mathcal{S}_T$, contains the generating class of cylinders, and is itself a σ-field, hence

The Borel field $\mathcal{S}_T$ in $\mathfrak{X}_T$ coincides with the class of cylinders with countably-dimensional Borel bases.

The measurable space $(\mathfrak{X}_T, \mathcal{S}_T)$ will be called the *sample space* of X_T.

SAMPLE DEFINITION. *A r.f. X_T is a measurable function on a pr. space $(\Omega, \mathcal{A}, P)$ to a sample space $(\mathfrak{X}_T, \mathcal{S}_T)$; in symbols $X_T^{-1}(\mathcal{S}_T) \subset \mathcal{A}$.*

If a r.f. X_T is so defined then, for every Borel set $S_t \subset \mathfrak{X}_t$, $[X_t \in S_t] = X_T^{-1}(C(S_t)) \in \mathcal{A}$ so that the X_t are r.v.'s. Conversely, if according to the $\mathcal{R}$-definition the X_t are r.v.'s then, for every finite product $S_{T_n} = \prod_{t \in T_n} S_t$ of Borel sets S_t in $\mathfrak{X}_t$, $X_T^{-1}(C(S_{T_n})) = \bigcap_{t \in T_n} X_T^{-1}(C(S_t)) \in \mathcal{A}$ so that X_T is a r.f. according to the sample definition. Thus these definitions are equivalent.

The $\mathcal{R}$-definition leads to analytical properties of r.f.'s X_T in terms of limits in $\mathcal{R}$, say, X_T is *continuous in pr. or a.s. or in the rth mean at t* according as $X_{t'} \xrightarrow{P} X_t$ or $X_{t'} \xrightarrow{\text{a.s.}} X_t$ or $X_{t'} \xrightarrow{r} X_t$ as $t' \to t$; we drop "at t" when the property holds for all t. The sample definition leads to analytical properties of r.f.'s X_T in terms of those of their sample functions $X_T(\omega)$, say, X_T is *sample continuous or sample measurable or sample integrable at* ω if the property is true for $X_T(\omega)$; we replace "at ω" by "on A" when the property holds for all $\omega \in A$, drop it when $A = \Omega$, and say that the sample property is *almost sure* (*a.s.*), or holds for *almost all* sample functions, when $A = N^c$.

In general, analytical properties of a r.f. relative to the sample space are "finer" than the corresponding properties relative to the space $\mathcal{R}$ of r.v.'s. Let us consider a.s. continuity properties: *a.s. sample continuity of X_T implies a.s. continuity of X_T* but the converse is not necessarily true. For, as $t' \to t$, in the first case $X_{t'} \to X_t$ on N^c where the null event N is independent of t, while in the second case $X_{t'} \to X_t$ on N_t^c where the null events N_t may depend upon t and $\bigcup_{t \in T} N_t$ may not be a null event or may even not be an event.

It may be convenient to describe a.s. continuity and (a.s.) sample continuity in negative terms. When $X_{t'}$ does not converge a.s. to X_t as $t' \to t$, we say that t is a *fixed* discontinuity point of X_T. A discontinuity point $t = t(\omega)$ of $X_T(\omega)$ which is not a fixed discontinuity point of X_T will be called a *moving* (with ω) discontinuity point of X_T. Thus a.s. continuity of X_T means no fixed discontinuity points while (a.s.) sample continuity of X_T means neither fixed nor (outside a null event) moving discontinuity points.

An advantage of the sample definition is that it introduces directly the σ-field $\mathcal{B}(X_T) = X_T^{-1}(\mathcal{S}_T)$ of events induced by the r.f. X_T—the union σ-field of the σ-fields $\mathcal{B}(X_t)$ induced by the r.v.'s X_t. As long as we are concerned with the properties of X_T alone, the only events and (not necessarily finite) r.v.'s we have to consider are those *defined on* X_T, that is, events belonging to $\mathcal{B}(X_T)$ and $\mathcal{B}(X_T)$-measurable r.v.'s. Since the Borel sets in the sample space are cylinders with countably-dimensional Borel bases S_{T_c} and $X_T^{-1}(C(S_{T_c})) = X_{T_c}^{-1}(S_{T_c})$, events defined on X_T are defined on countable sections X_{T_c} of X_T (those sections varying in general with the events). More generally

a. COUNTABILITY LEMMA. *A r.v. ξ is defined on X_T if and only if it is a Borel function g of X_T, in fact, of a countable section X_{T_c} of X_T. In particular, if $\xi = E(Y \mid X_T)$ a.s. then $\xi = E(Y \mid X_{T_c})$ a.s.*

Proof. If $\xi = g(X_T)$ then $\xi^{-1} = X_T^{-1}g^{-1}$ implies that $\mathcal{B}(\xi) \subset \mathcal{B}(X_T)$ and the "if" assertion is proved.

Conversely, let $\mathcal{B}(\xi) \subset \mathcal{B}(X_T)$ so that the events $A_{nk} = \left[\frac{k}{2^n} \leq \xi < \frac{k+1}{2^n}\right]$ for k finite and $A_{nk} = [\xi = -\infty]$ or $[\xi = +\infty]$ for $k = -\infty$ or $+\infty$, belong to $\mathcal{B}(X_T)$. By the sample definition there exist Borel sets $S_{nk} \in \mathcal{S}_T$ such that $A_{nk} = X_T^{-1}(S_{nk})$ and, since the events A_{nk} are disjoint in k, we also have $A_{nk} = X_T^{-1}(S'_{nk})$ where the Borel sets $S'_{nk} = S_{nk}(\bigcup_{j \neq k} S_{nj})^c$ are disjoint in k. The functions $g_n = \sum_k \frac{k}{2^n} I_{S'_{nk}}$ are Borel functions on $\mathfrak{X}_T$ and $g_n(X_T) \to \xi$ on the range $S' = X_T(\Omega)$ of X_T. Thus, $g = \lim g_n$ exists on some Borel set $S \supset S'$ and setting, say, $g = 0$ on S^c, we obtain a Borel function g on $\mathfrak{X}_T$ with $\xi = g(X_T)$. This proves the "only if" assertion.

The "countable section" assertion follows from what precedes since the events A_{nk} are defined on countable sections of X_T; or it suffices to observe that the events $[\xi < r]$ where r varies over the rationals generate $\mathcal{B}(\xi)$ and, every $[\xi < r]$ being defined on a countable section X_{T_r}, the r.v. ξ is defined on the countable section X_{T_c} where $T_c = \bigcup_r T_r$. In particular, if $\xi = E(Y \mid X_T)$ a.s. then $\xi = E(Y \mid X_{T_c})$ a.s. upon conditioning by X_{T_c}. The proof is terminated.

We may emphasize the argument (ω, t) and consider r.f.'s as functions on $\Omega \times T$ to $\mathfrak{X}$ with values $X(\omega, t)$. But then we have to insure that sections $X_t = (X(\omega, t), \omega \in \Omega)$ at every t be r.v.'s, and this brings us back to the previous definitions. Yet, the present interpretation leads to an important type of r.f. as follows. Let $\mathfrak{I}$ be some σ-field in T.

MEASURABLE R.F. DEFINITION. *An $(\mathcal{A} \times \mathfrak{I})$-measurable r.f. is a measurable function on $(\Omega \times T, \mathcal{A} \times \mathfrak{I})$ to the state space $(\mathfrak{X}, \mathcal{S})$.*

Since every section at t of an $(\mathfrak{A} \times \mathfrak{J})$-measurable function X_T is $\mathfrak{A}$-measurable, it follows that the sections X_t are $\mathfrak{A}$-measurable; similarly for $\mathfrak{J}$-measurability of sections $X_T(\omega)$ at ω:

$(\mathfrak{A} \times \mathfrak{J})$-measurable r.f.'s are r.f.'s and their sample functions are $\mathfrak{J}$-measurable.

A measurable X_T is a *Borel* r.f. if T is a Borel set and $\mathfrak{J}$ is the σ-field of Borel sets in T. Then we introduce on $\mathfrak{J}$ the Lebesgue measure λ (that is, its restriction to Borel measure on $\mathfrak{J}$). If X_T coincides with a Borel r.f. outside a $(P \times \lambda)$-null set, we say that X_T is an *a.e. Borel* r.f. We emphasize that measurability of r.f.'s is relative to the product σ-field $\mathfrak{A} \times \mathfrak{J}$ and not, as usual, to the completion of this σ-field with respect to the product measure $P \times \mu$ where μ is a measure on $\mathfrak{J}$. The importance of measurable r.f.'s is due to the fact that, as we shall see later, under some continuity conditions a r.f. is equivalent to a measurable one, and to the following immediate consequences of measurability and of Fubini's theorem.

b. Measurability lemma. *If X_T is an $(\mathfrak{A} \times \mathfrak{J})$-measurable r.f. then the sample functions are $\mathfrak{J}$-measurable and, for $X_T \geqq 0$ or $\int_T E|X_t|\,d\mu(t) < \infty$ where μ is a σ-finite measure on $\mathfrak{J}$,*

$$E\left(\int_T X_t\,d\mu(t)\right) = \int_T (EX_t)\,d\mu(t).$$

If X_T is a Borel r.f. then, for every r.v. τ with range in T, the function $X_\tau = (X_{\tau(\omega)}(\omega),\ \omega \in \Omega)$ is a r.v.

Application. Let X_T be a Borel or, more generally, an a.e. Borel stationary r.f. with $T = [0, \infty)$; stationarity means as usual that $\mathfrak{L}(X_{t_1}, \cdots, X_{t_n}) = \mathfrak{L}(X_{t_1+h}, \cdots, X_{t_n+h})$ for all finite subsets $(t_1, \cdots, t_n) \subset T$ and all $h > 0$. It follows at once that if a r.v. ξ is defined on X_T and ξ_h is its translate by h, then ξ and ξ_h have the same distribution. Let $E|X_0|^r$ be finite for some $r \geqq 1$ so that the r.v.'s $Y_n = \int_{n-1}^n X_s\,ds$ and $Z_n = \int_{n-1}^n |X_s|\,ds$ exist and $EY_n = EX_0$, $EZ_n = E|X_0|$, $E|Y_n|^r \leqq EZ_n^r \leqq E|X_0|^r$ are finite. Since the sequences are stationary, it follows that $(Y_1 + \cdots + Y_n)/n \xrightarrow[r]{\text{a.s.}} \overline{Y}$ and $(Z_1 + \cdots + Z_n)/n \xrightarrow[r]{\text{a.s.}} \overline{Z}$ hence $Z_n/n \xrightarrow[r]{\text{a.s.}} 0$. Therefore, if $n = n_t$ is the largest integer contained in t then, as $t \to \infty$,

$$U_t = \frac{1}{t}\int_0^t X_s\,ds = \frac{n}{t}\left(\frac{1}{n}\int_0^n X_s\,ds + \frac{1}{n}\int_n^t X_s\,ds\right) \xrightarrow[r]{\text{a.s.}} \bar{Y},$$

since $n/t \to 1$, the first divided integral in parentheses is $(Y_1 + \cdots + Y_n)/n$ and the second one is bounded by Z_{n+1}/n. Finally, if $\mathcal{C}$ is the σ-field of invariant (under translations) events on X_T then, for every $C \in \mathcal{C}$,

$$\int_C E^{\mathcal{C}} X_0\,dP = \int_C X_0\,dP = \int_C X_s\,dP = \int_C U_t\,dP \to \int_C \bar{Y}\,dP$$

so that $\bar{Y} = E^{\mathcal{C}} X_0$ a.s. Thus

A. R.F.'S STATIONARITY THEOREM. *If $X_{[0,\infty)}$ is a stationary a.s. Borel r.f. with $E|X_0|^r < \infty$ for an $r \geqq 1$, then*

$$\frac{1}{t}\int_0^t X_s\,ds \xrightarrow[r]{\text{a.s.}} E^{\mathcal{C}} X_0$$

where $\mathcal{C}$ is the σ-field of invariant events on X_T.

38.2. Separability. Let $X_T = (X_t, t \in T)$ be a r.f. with domain T. Let $\bar{T}$ be the closure of T and set $I_n = \left(t - \frac{1}{n}, t + \frac{1}{n}\right)$. By definition, for every $t \in \bar{T}$,

$$\underline{X}_t = \liminf_{t' \to t} X_{t'} = \sup_n \inf_{t' \in I_n T} X_{t'}$$

$$\bar{X}_t = \limsup_{t' \to t} X_{t'} = \inf_n \sup_{t' \in I_n T} X_{t'}.$$

Thus, to every r.f. X_T on T there correspond two limit functions $\underline{X}_T$ and $\bar{X}_T$ on $\bar{T}$, respectively, lower and upper semi-continuous. Since the limits are taken using nondeleted neighborhoods,

$$\underline{X}_t \leqq X_t \leqq \bar{X}_t, \quad t \in T,$$

and $\underline{X}_t = X_t = \bar{X}_t$ at every isolated point $t \in T$. Since the X_t are a.s. finite, it follows that $\underline{X}_t < +\infty$ a.s., $\bar{X}_t > -\infty$ a.s. Also

If $\lim_{t' \to t} X_{t'}$ exists at $t \in T$, then it coincides with X_t.

The basic difficulty consists in that the functions $\underline{X}_t$ and $\bar{X}_t$ of $\omega \in \Omega$ may not be measurable, for then analytical properties of the r.f. X_T would not be expressible in probability terms. Since these functions of

ω are formed by means of sequences of extrema of the X_t on sets of the form IT where the I are open intervals, we are led to require that all such extrema be measurable. This requirement is automatically fulfilled when T is countable and would be fulfilled were it possible to replace T by some fixed countable subset S of T in the formation of these extrema. If there is such a set S, we say that the r.f. X_T is *separable* and is *separated by* S—*a separating set* of X_T. In fact

A. SEPARABILITY CRITERIA. *The following properties are equivalent and define separability of a r.f.* X_T: *There exists a countable subset* $S = \{s_j\}$ *of* T *such that*

—*for every open interval* I *whose intersection with* T *is not empty*

$$(\mathrm{S}_1)\quad \inf_{s_j \in IS} X_{s_j} = \inf_{t \in IT} X_t, \quad \sup_{t \in IT} X_t = \sup_{s_j \in IS} X_{s_j}$$

$$(\mathrm{S}_2)\quad \inf_{s_j \in IS} X_{s_j} \leqq \inf_{t \in IT} X_t, \quad \sup_{t \in IT} X_t \leqq \sup_{s_j \in IS} X_{s_j}$$

$$(\mathrm{S}_3)\quad \inf_{s_j \in IS} X_{s_j} \leqq X_t \leqq \sup_{s_j \in IS} X_{s_j}, \quad t \in IT$$

—*for every* $t \in \overline{T}$

$$(\mathrm{S}'_1)\quad \liminf_{s_j \to t} X_{s_j} = \liminf_{t' \to t} X_{t'}, \quad \limsup_{t' \to t} X_{t'} = \limsup_{s_j \to t} X_{s_j}$$

$$(\mathrm{S}'_2)\quad \liminf_{s_j \to t} X_{s_j} \leqq \liminf_{t' \to t} X_{t'}, \quad \limsup_{t' \to t} X_{t'} \leqq \limsup_{s_j \to t} X_{s_j}$$

$$(\mathrm{S}'_3)\quad \liminf_{s_j \to t} X_{s_j} \leqq X_t \leqq \limsup_{s_j \to t} X_{s_j}, \quad t \in T.$$

Proof. The three nonprimed properties are equivalent: For, $(\mathrm{S}_1) \Rightarrow (\mathrm{S}_2)$ while $(\mathrm{S}_2) \Rightarrow (\mathrm{S}_1)$ since for $S \subset T$ the reverse inequalities are always true, and $(\mathrm{S}_2) \Rightarrow (\mathrm{S}_3)$ while $(\mathrm{S}_3) \Rightarrow (\mathrm{S}_2)$ upon taking extrema over $t \in IT$.

The three primed properties are equivalent: For, $(\mathrm{S}'_1) \Rightarrow (\mathrm{S}'_2)$ while $(\mathrm{S}'_2) \Rightarrow (\mathrm{S}'_1)$ since for $S \subset T$ the reverse inequalities are always true, and $(\mathrm{S}'_2) \Rightarrow (\mathrm{S}'_3)$ while $(\mathrm{S}'_3) \Rightarrow (\mathrm{S}'_2)$ upon replacing t by t' in (S'_3) letting $t' \to t \in I\overline{T}$ and using the semi-continuity of inferior and superior limits.

It follows that the primed and unprimed properties are equivalent: For, $(\mathrm{S}_2) \Rightarrow (\mathrm{S}'_2)$ upon replacing I by $I_n = \left(t - \frac{1}{n}, t + \frac{1}{n}\right)$ and letting $n \to \infty$, and $(\mathrm{S}'_3) \Rightarrow (\mathrm{S}_3)$ since $I_n \subset I$ from some n on for any fixed

$t \in IT$ so that the extreme terms of (S_3) are then farther away from X_t than those of (S'_3). The proof is terminated.

The use of separable r.f.'s is justified in the sense that every r.f. is equivalent to a separable one. In fact, more is true, as follows.

Separability criterion (S_1) means that for every open interval I and every closed interval C

$$\text{(S)} \qquad [X_t \in C, t \in IT] = [X_{s_j} \in C, s_j \in IS](\in \mathcal{A}),$$

that is, *separability as defined is separability for closed intervals.* This interpretation leads to the general concept of separability for sets of a given class. In particular, if the above equality holds for all closed sets C, the r.f. is *separable for closed sets;* it is then *a fortiori* separable (for closed intervals). Yet, given any r.f. X_T, there exists an equivalent r.f. separable for closed sets, because of the following

a. SEPARABILITY LEMMA. *Given a r.f. X_T and a Borel set C in the state space, there exists a countable subset $S = \{s_j\}$ of T such that for all $t \in T$ the intersections $A_t = [X_{s_j} \in C, s_j \in S][X_t \notin C]$ are null events.*
In fact, given a class $\mathcal{C}$ of countable intersections of a countable class $\{C_k\}$ of Borel sets, there exists a countable subset S of T such that the intersections A_t are subsets of null events N_t which do not vary with $C \in \mathcal{C}$.

Proof. Set $A_{nt} = [X_{s_k} \in C, k \leqq n][X_t \notin C]$ and $B_n = A_{ns_{n+1}}$. The PA_{nt} and $p_n = \sup\limits_{t \in T} PA_{nt}$ form nonincreasing sequences and the B_n being disjoint $\sum PB_n \leqq 1$ hence $PB_n \to 0$. Select $s_1 \in T$ arbitrarily and, for every n such that $p_n > 0$, select s_{n+1} so that $PB_n \geqq \left(1 - \dfrac{1}{n}\right) p_n$. If $p_n = 0$ the asserted set is $(s_1, \cdots, s_n)$, and if no p_n vanishes the asserted set is $(s_1, s_2, \cdots)$ since

$$PA_t \leqq PA_{nt} \leqq p_n \leqq PB_n \Big/ \left(1 - \frac{1}{n}\right) \to 0.$$

If $C \in \mathcal{C}$ is intersection of some sets $C_{k'}$ of a countable class $\{C_k\}$, denote by A_{tk} and S_k the sets A_t and S relative to C_k. Then the set $S = \bigcup\limits_k S_k$ is a countable subset of T and the event $N_t = \bigcup\limits_k A_{tk}$ is null. Since $[X_t \notin C] \subset \bigcup\limits_{k'} [X_t \notin C_{k'}]$ and $[X_{s_j} \in C, \; s_j \in S][X_t \notin C_{k'}] \subset A_{tk'} \subset N_t$, the event $[X_{s_j} \in C, s_j \in S][X_t \notin C]$ is a subset of N_t, and the proof is terminated.

B. SEPARABILITY EXISTENCE THEOREM. *Given a r.f. X_T, there exists a separable for closed sets r.f. $\tilde{X}_T$ equivalent to X_T and defined on it.*

Proof. We seek a r.f. $\tilde{X}_T$ equivalent to X_T with $\mathcal{B}(\tilde{X}_T) \subset \mathcal{B}(X_T)$, and a countable subset S of T such that for every open interval I (with $IT \neq \emptyset$) and every closed set C

$$[\tilde{X}_{s_j} \in C, s_j \in IS] = [\tilde{X}_t \in C, t \in IT].$$

Since the left side event contains the right side one, it suffices that it be contained in the right side one. Since every open interval I is a countable union of open intervals I_r with rational or infinite endpoints, it suffices to realize this property simultaneously for all I_r.

The class of closed sets is contained in the class $\mathcal{C}$ of countable intersections C of finite unions C_k of open or closed intervals with rational or infinite endpoints, so that we can apply the separability lemma with T replaced by I_rT, S by I_rS, and N_t by N_{tr}. Then the set $S = \bigcup_r S_r$ is a countable subset of T, the event $N_t = \bigcup_r N_{tr}$ is null and is the same for all C and all I_r and, for $t \in I_rT$,

$$[X_{s_j} \in C, s_j \in I_rS][X_t \notin C] \subset N_t.$$

Given I_r, let $C_r(\omega)$ be the closure of the set $\{X_{s_j}(\omega), s_j \in I_rS\}$ in the closure $\bar{\mathfrak{X}} \subset \bar{R}$ of the state space. The set $C_t(\omega) = \bigcap_{I_r \ni t} C_r(\omega)$ is nonempty and closed, and $X_t(\omega) \in C_t(\omega)$ for $\omega \notin N_t$. We set $\tilde{X}_t(\omega) = X_t(\omega)$ for $t \in S$, $\omega \in \Omega$ and for $t \notin S$, $\omega \notin N_t$ and we set, say, $\tilde{X}_t(\omega) = \liminf_{s_j \to t} X_{s_j}(\omega)$ for $t \notin S$, $\omega \in N_t$. Thus $\tilde{X}_T$ is equivalent to X_T with $\mathcal{B}(\tilde{X}_T) \subset \mathcal{B}(X_T)$ and $\tilde{X}_t(\omega) \in C_t(\omega)$ for every ω and t. Given $C \in \mathcal{C}$, if ω is such that $\tilde{X}_{s_j}(\omega) \in C$ for all $s_j \in I_rS$, then $C_t(\omega) \subset C$ for all $t \in I_rS$ and, by definition of $\tilde{X}_T$, $\tilde{X}_t(\omega) \in C$ for all $t \in I_rT$. Therefore the set $\{\tilde{X}_{s_j}(\omega) \in C, s_j \in I_rS\}$ is contained in the set $\{\tilde{X}_t(\omega) \in C, t \in I_rT\}$, hence coincides with it. The proof is terminated.

Separability implications. Separability implies properties of separating sets and of one-sided limits, and it will be convenient to weaken slightly this concept, as follows.

1° A.S. SEPARABILITY. The equivalent definitions of separability were assumed to hold on Ω. Yet, the proof of their equivalence remains valid when Ω is replaced by any fixed event. If, in particular, they hold outside some fixed null event, we say that the r.f. X_T is *a.s. separable* and continue to say that S *separates it.* When X_T is a.s. separable so is every r.f. a.s. equal to it, and there always exists a separable r.f. a.s. equal to X_T—it suffices to change X_T to a constant on the exceptional null event. Thus

In the study of a.s. properties of a given r.f., a.s. separability may be replaced by separability without loss of generality.

Note that in **A**

If the separability criteria in terms of open intervals I hold outside null events $N(I)$ which may vary with I, then the r.f. is a.s. separable.

For then, they hold for all I outside the fixed null event $N = \bigcup_r N(I_r)$ where the I_r are all open intervals with rational or infinite endpoints.

2° SEPARATING SETS. Since every $t \in \overline{T}$ has to be a limit along any separating set of a separable r.f. X_T, such sets are to be dense in T and, in particular, contain the set of isolated points of T. On account of (S_2), the union of a separating set with any countable subset of T is also a separating set. Therefore,

A separable r.f. X_T is separated by a sufficiently large countable subset of T dense in it and by any larger countable subset of T.

Thus, whenever convenient, we may include within a separating set a suitable countable set, say, that of all rationals or of all dyadic numbers belonging to T. Note that

If the separability criteria in terms of limits hold outside null events N_t which may depend upon t and the r.f. X_T is known to be a.s. separable, then the set S therein separates X_T.

For, (S'_3) outside N_t implies (S_3) outside N_t, while a.s. separability implies (S_2) outside a fixed null event N' with a separating set $S' = \{s'_k\}$ (in lieu of S). Therefore, outside the fixed null event $N = N' \cup (\bigcup_k N_{s'_k})$,

$$\inf_{s_j \in IS} X_{s_j} \leq \inf_{s'_k \in IS'} X_{s'_k} \leq \inf_{t \in IT} X_t,$$

and similarly for the suprema.

3° ONE-SIDED LIMITS. Separability allows us to replace two-sided limits along the domain of a r.f. by two-sided limits along a suitable countable set. The question arises whether the same is true of one-sided limits which are defined like the two-sided ones but with $I'_n = \left(t - \frac{1}{n}, t\right)$

for left limits and $I''_n = \left(t, t + \frac{1}{n}\right)$ for right limits, in lieu of $I_n = \left(t - \frac{1}{n}, t + \frac{1}{n}\right)$. The answer is as follows.

Let X_T be a r.f. separated by $S = \{s_j\}$ and let t be a left limit point of T. There exists in S a sequence $t_n \uparrow t$ such that, outside a null event N_t,

$$\underline{X}_{t-0} = \liminf_{t' \to t-0} X_{t'} = \liminf_{t_n \uparrow t} X_{t_n}, \quad \overline{X}_{t-0} = \limsup_{t' \to t-0} X_{t'} = \limsup_{t_n \uparrow t} X_{t_n}.$$

In particular, if $\lim_{t_n \uparrow t} X_{t_n}$ exists a.s. for every sequence $t_n \uparrow t$, then $X_{t-0} = \lim_{t' \to t-0} X_{t'}$ exists a.s. even when the exceptional null events vary with the sequence.

Similarly for right limit points of T.

It will suffice to prove the general assertion for, say, a left limit point t. We may assume that the r.f. is bounded (replacing, if necessary, X_T by Arctan X_T), so that the limits are finite. Because of separability there exist finite subsets $S_{nm} = (s_{nk}, k \leqq m)$ of S in $I'_n T$ such that

$$P[Y_{nm} - Y_n > 1/n] < 1/n, \quad P[Z_n - Z_{nm} > 1/n] < 1/n$$

where

$$Y_n = \inf_{t' \in I'_n T} X_{t'} \leqq Y_{nm} = \inf_k X_{s_{nk}},$$

$$Z_{nm} = \sup_k X_{s_{nk}} \leqq Z_n = \sup_{t' \in I'_n T} X_{t'}.$$

The elements of all these subsets can be reordered into a sequence $t_n \uparrow t$, for they are all less than t and for every n only a finite number of them are less than $t - \frac{1}{n}$. Since all $S_{nm} \subset I'_n S$, it follows that

$$Y_n \leqq Y'_n = \inf_{t_k \in I'_n S} X_{t_k} \leqq Y_{nm}$$

so that $P[Y'_n - Y_n > 1/n] < 1/n$ and

$$0 \overset{P}{\leftarrow} Y'_n - Y_n \to \liminf_{t_n \uparrow t} X_{t_n} - \underline{X}_{t-0}.$$

Thus, $\underline{X}_{t-0} = \liminf_{t_n \uparrow t} X_{t_n}$ outside a null event N'_t and similarly $\overline{X}_{t-0} = \limsup_{t_n \uparrow t} X_{t_n}$ outside a null event N''_t, so that both equalities hold outside $N_t = N'_t \cup N''_t$ and the general assertion is proved.

If the one-sided limits $g_{t\pm 0}$ of a numerical function exist and are finite but unequal, the function is said to have a *simple* (or *first kind*) *discontinuity* at t. If, moreover, g_t lies between the one-sided limits (inclusively), the discontinuity is said to be a *jump*.

If a r.f. X_T is separable then the simple discontinuities of almost all its sample functions are jumps except perhaps at fixed discontinuity points, and if X_T is separable for closed sets then at these jumps the sample functions are left or right continuous.

Let $S = \{s_j\}$ separate X_T and use separability relation (S) for closed intervals and for closed sets with all open intervals $I \ni t$ where $t \notin S$ is a simple discontinuity point of $X_T(\omega)$. Upon taking the closed interval with endpoints $X_{t\pm 0}(\omega)$, (S) implies that $X_t(\omega)$ belongs to the interval so that t is a jump point. Upon observing that these endpoints are the common limit points of all sets $(X_{s_j}(\omega), s_j \in IS)$, separability for closed sets implies that $X_t(\omega)$ coincides with one of these common limit points. Thus the assertions are true for all sample functions except perhaps at the s_j. At those s_j which are not fixed discontinuity points the sample functions $X_T(\omega)$ are continuous for ω's outside null events N_{s_j}, hence outside their null union. The assertions are proved.

Random analysis is based upon the fact that limits along T for a r.f. X_T separated by S are limits along S—necessarily measurable. The immediate problem is that of finding separating sets. There is no difficulty whenever the r.f. is continuous in pr.:

C. CONTINUITY SEPARATION THEOREM. *Let X_T be an a.s separable r.f. continuous in pr. or a.s. Then every countable subset dense in T separates X_T. Let $I = [a, b]$ be intervals with endpoints in T and let $S_n = (s_{nk}, k \leq k_n)$ form sequences of finite subsets of IT becoming dense in it: $\sup_{t \in IT} \inf_k |t - s_{nk}| \to 0$. Then*

$$\inf_k X_{s_{nk}} \xrightarrow{P} \inf_{t\in IT} X_t, \quad \sup_k X_{s_{nk}} \xrightarrow{P} \sup_{t\in IT} X_t$$

and if X_T is continuous a.s., then the convergence is a.s., while if $S_1 \subset S_2 \subset \cdots$ then the convergence is a.s. monotone.

Proof. Let $S = \{s_j\}$ be a subset of T dense in it. Convergence in pr. along S is implied by convergence a.s. and implies convergence a.s. of a subsequence. Thus, if $s_j \to t$ then there exists a subsequence $s_{j'} \to t$ such that $X_{s_{j'}} \to X_t$ outside a null event N_t so that on N_t^c

$$\liminf_{s_j \to t} X_{s_j} \leq X_t \leq \limsup_{s_j \to t} X_{s_j}.$$

The separation assertion results then from separability implications 2° for separating sets and implies the pointwise convergence assertion.

It remains to prove the convergence assertions. It will suffice to do so for, say, the infima. We can assume X_{IT} bounded. Let $S = \{s_j\}$ with $a, b \in S$ denote now a separating set of X_{IT} and set

$$Y = \inf_{t \in IT} X_t, \quad Y_n = \inf_k X_{s_{nk}}, \quad Z_m = \inf_{j \leq m} X_{s_j}.$$

Since $Z_m \downarrow Y$ a.s. as $m \to \infty$, it follows that for every $\epsilon > 0$, as $n \to \infty$ then $m \to \infty$,

$$P[Y_n - Y \geqq 2\epsilon] \leqq P[Y_n - Z_m \geqq \epsilon] + P[Z_m - Y \geqq \epsilon] \to 0,$$

so that $Y_n \xrightarrow{P} Y$. If X_T is continuous a.s. hence limsup $Y_n \leqq X_{s_j}$ outside null events N_{s_j} then, outside their null union, limsup $Y_n \leqq Y$ while $Y_n \geqq Y$, so that $Y_n \xrightarrow{\text{a.s.}} Y$. The proof is terminated.

In fact, less than continuity in pr. is required provided a suitable countable subset is excluded, because of

D. Continuity extension theorem. *Let X_T be an a.s. separable r.f. such that limits in pr. (in the rth mean) X_{t-0} or X_{t+0} exist for every $t \in T' \subset \overline{T}$. Then there exists a countable subset T_c of T' such that for every $t \in T' - T_c$, $X_{t-0} = X_{t+0}$ and $= X_t$ for $t \in T$, outside a null event N_t. In particular, if $T' = T$ then X_T is continuous in pr. (in the rth mean) on $T - T_c$, and if $T' = \overline{T}$ then X_T extends on $\overline{T} - T_c$ to a r.f. continuous in pr. (in the rth mean) and determined up to an equivalence.*

Proof. It suffices to prove the general assertion with convergence in pr. The space of equivalence classes of r.v.'s on our pr. space, with distance $d(X, Y) = E(|X - Y|/1 + |X - Y|)$ is a complete metric space in which convergence in distance is equivalent to convergence in pr. The family of r.f.'s equivalent to X_T is a function ξ_T on T to this space, and the general assertion reduces to a classical one about functions ξ_T which take their values in a complete metric space, proved as follows.

At least one of the one-sided limits $\xi_{t\pm0}$ has no meaning at isolated or one-sided limit points of $T(\subset R)$. But their set is countable and thus may be included in T' provided it is also included in the asserted exceptional countable set T_c. Thus it suffices to consider two-sided limit points of T belonging to T'. But the set T'_n of such points and at which the oscillation of ξ_T is at least $1/n$ is countable for, by hypothesis, if $t \in T'_n$ then at least one of the one-sided limits, say, ξ_{t-0} exists, so that t is right endpoint of an interval containing no other points of T'_n. Thus we include the countable set T'_n in T_c. The remaining set $T' - T_c$ is that of two-sided limit points at which the oscillation of ξ_T vanishes, and the general assertion follows. The proof is terminated.

Continuous numerical functions on a Borel set in R are Borel functions. The question arises whether similar properties hold for r.f.'s X_T with some continuity on T based upon pr. The answer is in the affirmative in the following sense.

Let $\mathfrak{T}$ be the σ-field of Borel sets in a Borel set T and λ the Lebesgue measure on $\mathfrak{T}$.

E. MEASURABILITY THEOREM. *Let X_T be a r.f. with a Borel set T. If X_T is a.s. continuous and separable, then it is an a.e. Borel r.f. and the discontinuity sets of almost all its sample functions are λ-null. If X_T is continuous in pr., then there exists an equivalent a.e. Borel r.f. separable for closed sets.*

Proof. 1° Let $T_k^{(n)} = \left[\dfrac{k-1}{2^n}, \dfrac{k}{2^n}\right) T$, $k = 0, \pm 1, \pm 2, \cdots,$ and form the infimum $\underline{X}_k^{(n)}$ and the supremum $\overline{X}_k^{(n)}$ over each nonempty $T_k^{(n)}$. Set $\underline{X}_T^{(n)} = \sum_k \underline{X}_k^{(n)} I_{T_k^{(n)}}$, $\overline{X}_T^{(n)} = \sum_k \overline{X}_k^{(n)} I_{T_k^{(n)}}$ and note that

$$\underline{X}_T \uparrow \underline{X}_T^{(n)} \leqq X_T \leqq \overline{X}_T^{(n)} \downarrow \overline{X}_T.$$

Since $\underline{X}_T^{(n)}$ and $\overline{X}_T^{(n)}$ are Borel r.f.'s so are their limits $\underline{X}_T$ and $\overline{X}_T$, hence the set $L^c = [(\omega, t): \underline{X}_t(\omega) = \overline{X}_t(\omega)]$ is $\mathcal{A} \times \mathfrak{T}$-measurable and, by the preceding inequality $\underline{X}_t(\omega) = X_t(\omega) = \overline{X}_t(\omega)$ for $(\omega, t) \in L^c$.

If X_T is a.s. continuous then $\underline{X}_t(\omega) = X_t(\omega) = \overline{X}_t(\omega)$ for $\omega \notin N_t$, $t \in T$, hence the ($\mathcal{A}$-measurable) sections $L_t \subset N_t$ are null events. It follows that L is $(P \times \lambda)$-null, hence almost all its sections L_ω are λ-null. Thus X_T is an a.e. Borel r.f. and the discontinuity sets L_ω of almost all its sample functions $X_T(\omega)$ are λ-null. The first assertion is proved.

2° It suffices to prove the second assertion for X_T separated for closed sets by $S = \{s_j\}$ (on account of the separability theorem and the fact that equivalence preserves continuity in pr.) with X_T and T bounded (replace, if necessary, X_T by Arctan X_T and transform T into a bounded set, say, by $x' = -2 - \dfrac{1}{x}$ for $x < -1$, $x' = x$ for $-1 \leqq x \leqq 1$, $x' = 2 - \dfrac{1}{x}$ for $x > 1$). Reorder the n first s_j into $s_1^{(n)} < \cdots < s_n^{(n)}$ and set $s_0^{(n)} = -\infty$, $T_k^{(n)} = [s_{k-1}^{(n)}, s_k^{(n)})T$.

The $Y_T^{(n)} = \sum_k X_{s_k(n)} I_{T_k(n)}$ are Borel r.f.'s and, by continuity in pr., $Y_t^{(n)} \xrightarrow{P} X_t$ hence $Y_t^{(m)} - Y_t^{(n)} \xrightarrow{P} 0$ as $m, n \to \infty$. Since T and the sequence $Y_T^{(n)}$ are bounded it follows, by the Fubini theorem, that

$$\int_{\Omega \times T} | Y_t^{(m)}(\omega) - Y_t^{(n)}(\omega) | \, d(P\lambda)(\omega, t) = \int_T E| Y_t^{(m)} - Y_t^{(n)} | \, d\lambda(t) \to 0.$$

A fortiori, $Y_T^{(n)} \xrightarrow{P \times \lambda} Y_T$ where Y_T is a Borel r.f. and there is a subsequence $Y_T^{(n')} \to Y_T$ outside a $(P \times \lambda)$-null set M. It follows that $Y_t^{(n')} \xrightarrow{\text{a.s.}} Y_t$ for all t outside a λ-null set T_0 into which we can and do include the countable separating set S. But, by hypothesis, $Y_t^{(n')} \xrightarrow{P} X_t$ and therefore $X_t = Y_t$ outside a null event N_t for every $t \in T - T_0$. Thus, the a.e. Borel r.f. $\tilde{X}_T$ with values $Y_t(\omega)$ for $(\omega, t) \in M^c \cap (\Omega \times (T - T_0))$ and values $X_t(\omega)$ elsewhere is equivalent to X_T. Since we included S in T_0 so that $\tilde{X}_S = X_S$ and outside the section M_t every Y_t is limit of X_{s_j} while on these sections $\tilde{X}_t = X_t$, it follows that the set S which separates X_T for closed sets does the same for $\tilde{X}_T$. The proof is terminated.

VARIANTS. 1° The theorem remains valid when continuity in pr. or a.s. holds outside a λ-null Borel subset of T: it suffices to increase the exceptional $(P \times \lambda)$-null set.

2° The theorem remains valid when continuity in pr. is one-sided. The continuity extension theorem reduces it to 1°.

3° The theorem and its foregoing variants remain valid when "Borel set T" is replaced by "Lebesgue set T": proceed as for 1°.

4° The Borel r.f.'s of the theorem are a.e. limits of sample continuous, in fact, of sample polygonal r.f.'s: We may replace the approximating simple measurable r.f.'s by polygonal ones with vertices $\underline{X}_k^{(n)}$, $\overline{X}_k^{(n)}$, or $X_{s_k}^{(n)}$.

38.3. Sample continuity. The presence in r.f.'s of two arguments ω and t and the requirements of uniqueness then of measurability in ω of limits in t led us to equivalence classes of r.f.'s then to a partial retreat to their never empty subclasses of separable r.f.'s. The *a priori* weakest type of limit in t of r.v.'s X_t, based upon pr. and yielding r.v.'s, is the limit in pr. Random analysis is concerned primarily with separable r.f.'s X_T such that one-sided limits in pr. exist on T and hence, by the continuity extension lemma, with r.f.'s continuous in pr. outside countable subsets of T. When investigating specific types of r.f.'s, our first problem will thus be that of existence of one-sided limits in pr. or a.s.

But the next problem and, in fact, the essential problem of random analysis is that of the behavior of sample functions and especially that of the a.s. sample behavior, that is, behavior common to almost all sample functions. In particular, it seeks conditions for some kind of continuity: a.s. sample continuity, or a.s. sample continuity except for simple discontinuities (that is, except for jumps—because of separability) or except for nonsimple discontinuities. In this subsection we are concerned with some general conditions for such kinds of behavior when the types of r.f.'s are not specified.

In general, pointwise continuity or continuity based upon pr. is required or reduced to continuity on compact domains—when it becomes uniform. For any type "c" of convergence, X_T is "c"-*uniformly continuous* if $X_{t'} - X_t \xrightarrow{c} 0$ uniformly in $t \in T$, as $t' \to t$. For ordinary pointwise convergence hence for sample functions, the classical fact that continuity is uniform on compact domains is due to the Heine-Borel characterization of compact sets and to the equivalence of convergence and mutual convergence—which is also true of convergences based upon pr. Thus

a. UNIFORM CONTINUITY LEMMA. *For separable r.f.'s on compact domains, continuity of a sample function and continuities based upon pr. are uniform.*

In what follows T will be compact so that there will be no distinction between continuity and uniform continuity. To simplify the writing we shall take $T = [a, b]$ and, whenever convenient, replace it by $[0, 1]$—which does not restrict the generality. The reader is invited to rewrite the results for any compact T and examine their validity for noncompact T.

Let $\alpha(I) = \sup_{t',t'' \in IT} |X_{t'} - X_{t''}|$ be the oscillation of X_T on an interval I and let $\beta(I) = \inf_{t \in IT} \{\alpha(I \cap (-\infty, t)) + \alpha(I \cap (t, +\infty))\}$ be the "left-right" oscillation of X_T on I. Given $t \in T$, let $I_1 \supset I_2 \supset \cdots$ be a nonincreasing sequence of intervals converging to $\{t\}$ and to which t is interior. Then $\alpha(I_n)$ converges to the oscillation α_t of X_T at t and every sample function $X_T(\omega)$ for which $\alpha_t(\omega) = 0$ is continuous at t, and conversely. Similarly $\beta(I_n)$ converges to the "left-right" oscillation of X_T at t and every sample function $X_T(\omega)$ for which $\beta_t(\omega) = 0$ is free of nonsimple discontinuity at t, and conversely. Finally, if $X_{t'_n}(\omega) - X_{t''_n}(\omega) \to 0$ for some $t'_n \uparrow t$ and $t''_n \downarrow t$ then the sample function $X_T(\omega)$ cannot have a simple discontinuity at t.

A. SAMPLE CONTINUITY THEOREM. *Let a r.f. X_T, $T = [a, b]$, be separable for closed sets, let $a, b \in S_n = (s_{n1}, \cdots, s_{nk_n}) \uparrow S$ dense in T, and let $I_{nk} = [s_{nk}, s_{n,k+1}]$.*

If $P[\max_k \alpha(I_{nk}) \geqq \epsilon] \to 0$ for every $\epsilon > 0$, then X_T is a.s. sample continuous.

If $P[\max_k \beta(I_{nk}) \geqq \epsilon] \to 0$ for every $\epsilon > 0$, then X_T is a.s. sample continuous except perhaps for simple discontinuities.

If $P[\max_k | X_{s_{n,k}} - X_{s_{n,k+1}} | \geqq \epsilon] \to 0$ for every $\epsilon > 0$, then X_T is a.s. sample continuous except perhaps for nonsimple discontinuities.

The $\beta(I)$ are assumed measurable. Otherwise, in their definition replace T by S, or in the corresponding condition replace P by its outer extension.

Proof. If $p_n = P[\max_k \alpha(I_{nk}) \geqq \epsilon] \to 0$ then there exists a subsequence n' such that $\sum p_{n'} < \infty$. It follows, by the Borel-Cantelli lemma that there exists a finite integer-valued r.v. ν_ϵ such that $\max_k \alpha(I_{n'k}) \leqq \epsilon$ outside a fixed null event N_ϵ for all $n' \geqq \nu_\epsilon$. But, given $t \in [a, b]$, there exists a subsequence $I_{n'k}$ converging nonincreasingly to $\{t\}$ so that $\alpha_t \leqq \epsilon$ and $\epsilon > 0$ being arbitrary, $\alpha_t = 0$. For, if $t \notin S$ then it is interior to the $I_{n'k}$ and if $t \in S$ then for n' sufficiently large it suffices to replace ϵ by 2ϵ and $I_{n'k}$ by its union with the other interval of which t is the endpoint, unless $t = a$ or b—in which case the asserted continuity is automatically one-sided.

Similarly for the other assertions, and the proof is terminated.

PARTICULAR CASES. 1° Let the suprema be taken over all intervals $[t, t + h]$ in $[a, b]$ and $o_\epsilon(h)/h \to 0$ as $h \to 0$ for every $\epsilon > 0$.

If $\sup P[\alpha[t, t + h] \geqq \epsilon] = o_\epsilon(h)$ or $\sup P[\beta[t, t + h] \geqq \epsilon] = o_\epsilon(h)$, then the r.f. is a.s. sample continuous or a.s. sample continuous except perhaps for simple discontinuities.

If $\sup P[| X_{t+h} - X_t | \geqq \epsilon] = o_\epsilon(h)$ then the r.f. is a.s. sample continuous, except perhaps for nonsimple discontinuities.

Replace $[a, b]$ by $[0, 1]$, take for S the set of dyadic numbers kh_n, $h_n = 2^{-n}$, and note that

$$P[\max \alpha(I_{nk}) \geqq \epsilon] \leqq \sum_k P[\alpha(I_{nk}) \geqq \epsilon] \leqq o_\epsilon(h_n)/h_n,$$

and similarly for the other assertions. The first proposition is due to Dynkin and the second one to Dobrushin.

2° In the case of consecutive sums of r.v.'s, convergence in pr. entailed a.s. convergence under conditions which for $T = [a, b]$ take the form

(C) $p_{\epsilon,t}(h)P[\sup_{t' \in [t,t+h]} |X_{t'} - X_t| \geqq g(\epsilon)] \leqq P[|X_{t+h} - X_t| \geqq \epsilon]$ for every $[t, t+h] \subset T$ and every $\epsilon > 0$, with $g(\epsilon) \to 0$ as $\epsilon \to 0$.

Inequalities of this type were obtained in the case of sums $X_1, X_2, \cdots$ by a procedure which, similarly, yields the following property:

If $P(|X_n - X_k| < \epsilon \mid X_1, \cdots, X_k) \geqq p_\epsilon$ *a.s.,* $k = 1, \cdots, n$, *then* $p_\epsilon P[\max_k |X_k| \geqq 2\epsilon] \leqq P[|X_n| \geqq \epsilon]$.

For, setting $A_k = [|X_k| \geqq 2\epsilon, \max_{j<k} |X_j| < 2\epsilon]$ (with $X_0 = 0$) and $B_k = [|X_n - X_k| < \epsilon]$ so that $\sum_k A_k = [\max |X_k| \geqq 2\epsilon]$ and $|X_n| \geqq \epsilon$ on the $A_k B_k$, we have

$$P[|X_n| \geqq \epsilon] \geqq \sum_k PA_kB_k = \sum_k E(I_{A_k}E(I_{B_k} \mid X_1, \cdots, X_k)) \geqq p_\epsilon \sum_k PA_k.$$

Upon setting $X_k = X_{t_k} - X_t$ it follows, by separability and the usual limiting procedure, that

If
(C′) $P(|X_{t+h} - X_{t_k}| < \epsilon \mid X_{t_1} - X_t, \cdots, X_{t_k} - X_t) \geqq p_{\epsilon,t}(h)$ *a.s.,* $k = 1, \cdots, n$, *for every finite subset* $t_1 < \cdots < t_n$ *in every* $(t, t+h) \subset T$, *then* (C) *holds with* $g(\epsilon) > 2\epsilon$.

Therefore, on account of 1° and

$$[\sup_{t',t'' \in [t,t+h]} |X_{t'} - X_{t''}| \geqq 2g(\epsilon)] \subset [\sup_{t' \in [t,t+h]} |X_{t'} - X_t| \geqq g(\epsilon)],$$

Under (C) *or* (C′) *with* $p_{\epsilon,t}(h) \geqq p_{\epsilon,t} > 0$ *for h sufficiently small, the r.f. is (left, right) a.s. continuous if and only if it is (left, right) continuous in pr., and if* $p_{\epsilon,t} \geqq p_\epsilon > 0$ *then* $\sup_t P\|X_{t+h} - X_t| \geqq \epsilon| = o_\epsilon(h)$ *entails a.s. sample continuity.*

In its turn, (C′) is implied by

(C″) $P(|X_{t+h} - X_{t_k}| < \epsilon \mid X_{t_1}, \cdots, X_{t_k}) \geqq p_{\epsilon,t}(h)$ a.s. for every finite subset $t_1 < \cdots < t_n$ in every $[t, t+h) \subset T$.

Then, taking $n = 1$ and $t_1 = t$, it follows that

$$P[|X_{t+h} - X_t| \geqq \epsilon] = EP(|X_{t+h} - X_t| \geqq \epsilon \mid X_t) \leqq 1 - p_{\epsilon,t}(h)$$

and hence

Under (C″) *with* $1 - p_{\epsilon,t}(h) \leqq o_\epsilon(h)$, *the r.f. is a.s. sample continuous.*

Conditions on moduli of continuity in pr. yield moduli of sample continuity, as follows:

Let $g(h)$ with $|h| \leqq h_0$ be an even, nondecreasing in $h > 0$ function such that $g(h) \to 0$ as $h \to 0$. Let $h_n = q^{-n}$, $q > 1$ integer.

B. SAMPLE CONTINUITY MODULI THEOREM. *Let X_T, $T = [a, b]$, be a separable r.f. such that, for every $t, t + h \in T$,*

$$P[|X_{t+h} - X_t| \geqq g(h)] \leqq q(h) \to 0 \text{ as } h \to 0.$$

(i) *If $\sum q(h_n)/h_n < \infty$ and $\sum g(h_n) < \infty$, then X_T is a.s. sample continuous.*

(ii) *If $\sum q(jh_n)/h_n < \infty$ for every integer j and $\sum_{r=1}^{\infty} g(h_{n+r})/g(h_n) \leqq \alpha < \infty$ for all n, then X_T is a.s. sample continuous and there exists an a.s. positive r.v. H and a finite constant c such that*

$$|X_{t+h} - X_t| < cg(h), \quad |h| < H.$$

If, moreover, $g(h_n)/g(jh_n)$ is arbitrarily small for n, j sufficiently large, then for every $\epsilon > 0$ there exists an a.s. positive r.v. H_ϵ such that

$$|X_{t+h} - X_t| < (1 + \epsilon)g(h), \quad |h| < H_\epsilon.$$

Proof. To simplify the writing we take $T = [0, 1]$. Since the hypotheses about $q(h)$ imply continuity in pr., we can and do separate the r.f. by the dense subset of all q-adic numbers kh_n, $k = 0, \cdots, 1/h_n$, $n = 1, 2, \cdots$, $h_n = q^{-n}$, $q > 1$ integer.

1° By the covering rule and the first hypothesis in (i)

$$\sum_n P[\max_k |X_{(k+1)h_n} - X_{kh_n}| \geqq g(h_n)]$$
$$\leqq \sum_n \sum_k P[|X_{(k+1)h_n} - X_{kh_n}| \geqq g(h_n)]$$
$$\leqq \sum_n q(h_n)/h_n < \infty$$

so that, by the Borel-Cantelli lemma, $\max_k |X_{(k+1)h_n} - X_{kh_n}| \geqq g(h_n)$ finitely often, that is, there exists an a.s. finite and integer-valued r.v. ν

such that for all k and $n \geqq \nu$

$$| X_{(k+1)h_n} - X_{kh_n} | < g(h_n).$$

Since every q-adic number in $I_{nk} = [kh_n, (k+1)h_n]$ is of the form $t = kh_n + \sum_{r=1}^{m} \theta_r h_{n+r}$, where $\theta_r = 0$ or 1 or $\cdots$ or $q-1$ and m is some integer, it follows, by repeated applications of the triangular inequality that for all k and $n \geqq \nu$

$$| X_t - X_{kh_n} | \leqq \sum_{r=1}^{m} \theta_r g(h_{n+r}) \leqq q \sum_{r=1}^{\infty} g(h_{n+r})$$

and, because of separability, the same is true for every $t \in I_{nk}$. By the second hypothesis in (i), $q \sum_{r=1}^{\infty} g(h_{n+r}) < \epsilon/2$ for any given $\epsilon > 0$ and $n \geqq n_\epsilon$ sufficiently large. Therefore, by applying the triangular inequality, for $n \geqq \max(n_\epsilon, \nu)$ hence $|h| < H_\epsilon$ a.s. positive r.v. and all $t \in T$, it follows that $|X_{t+h} - X_t| < \epsilon$ hence X_T is a.s. sample continuous.

2° The first two hypotheses in (ii) imply those in (i). Therefore, by the second of these hypotheses in (ii), there exists an a.s. positive r.v. ν_0 such that, for all kh_n, $t \in I_{nk}$ and $n \geqq \nu_0$,

$$| X_t - X_{kh_n} | < q \sum_{r=1}^{\infty} g(h_{n+r}) \leqq \alpha q g(h_n).$$

On the other hand, by the first hypothesis in (ii), for every fixed j,

$$\sum_n P[\max_k | X_{(k+j)h_n} - X_{kh_n}] \geqq g(jh_n) \,| \leqq \sum_n q(jh_n)/h_n < \infty$$

so that there exists a similar r.v. ν_j such that, for all k and $n \geqq \nu_j$

$$| X_{(k+j)h_n} - X_{kh_n} | < g(jh_n).$$

Let m be an integer to be selected later and let $\nu = \max(\nu_0, \cdots, \nu_{qm})$. Given t and $h > 0$ there exists an n such that $mh_n < h < qmh_n$, so that there exist k and j with $m \leqq j \leqq qm$ such that

$$(k-1)h_n < t \leqq kh_n < (k+j)h_n \leqq t + h < (k+j+1)h_n.$$

Since

$$| X_{t+h} - X_t | \leqq | X_{t+h} - X_{(k+j)h_n} | + | X_{(k+j)h_n} - X_{kh_n} | + | X_{kh_n} - X_t |,$$

it follows that for $n \geqq \nu$ hence $h \leqq H$ a.s. positive r.v.

$$|X_{t+h} - X_t| < g(jh_n) + 2\alpha q g(h_n) \leqq (1 + 2\alpha q)g(h) = cg(h).$$

If, moreover, given $\epsilon > 0$, for $n(\geqq n_\epsilon)$ and m hence j sufficiently large, $g(h_n)/g(jh_n) < \epsilon/2\alpha q$ then, for $n \geqq \max(n_\epsilon, \nu)$, hence $h < H_\epsilon$ a.s. positive r.v.

$$|X_{t+h} - X_t| < (1 + \epsilon)g(jh_n) \leqq (1 + \epsilon)g(h),$$

and the proof is terminated.

COROLLARY. *Let X_T with $T = [a, b]$ be a separable r.f. such that for some $r(> 0)$ and all $t, t + h \in T$ with h sufficiently small*

$$E|X_{t+h} - X_t|^r \leqq \rho(h).$$

If $\rho(h) = c|h|^{1+s}$ with $s > 0$ or $\rho(h) = c|h|/|\log|h||^{1+s}$ with $s > r$, then X_T is a.s. sample continuous and in the first case for every $0 < \alpha < s/r$ and $\epsilon > 0$ there exists an a.s. positive r.v. H_ϵ such that for all $t, t + h \in T$ and $|h| < H_\epsilon$

$$|X_{t+h} - X_t| < (1 + \epsilon)|h|^\alpha.$$

The assertions follow from the theorem and Markov inequality:

$$P[|X_{t+h} - X_t| \geqq g(h)] \leqq q(h) = \rho(h)/g^r(h),$$

upon taking in the first case $g(h) = |h|^\alpha$ so that $q(h) = c|h|^{1+s-\alpha r}$, and in the second case $g(h) = |\log|h||^{-\beta}$ with $1 < \beta < s/r$ so that $q(h) = c|h|/|\log|h||^{1+s-\beta r}$.

The fact that $\rho(h) = c|h|^{1+s}$, $s > 0$, implies a.s. sample continuity is due to Kolmogorov.

Application to second order calculus. Analytical properties of second order r.f.'s which can be described in terms of their covariances constitute the second order calculus. This was done in Section 34 for continuity, differentiability and integrability, all in q.m., and required no concepts introduced in this section. We now proceed to use them and go back to the usual abuse of notation: $X(t)$ represents a second order r.f.—complex-valued with t varying over an interval $[a, b]$, and $\Gamma(t, t') = EX(t)\overline{X}(t')$ represents its covariance; it is not assumed that $EX(t) = 0$. We recall that $\Delta_h X(t) = X(t + h) - X(t)$, $\Delta_h \Delta'_{h'} \Gamma(t, t') = E\Delta_h X(t)\Delta'_{h'}\overline{X}(t')$, $\operatorname{Var}\Gamma = \iint |dd'\Gamma(t, t')|$ and that if $\Gamma(t, t')$ is of bounded variation, then the limits $\Gamma(t \pm 0, t' \pm 0)$ exist and differ from $\Gamma(t, t')$ on a countable set of (t, t') only, so that the one-sided limits in q.m. $X_{t\pm 0}$ exist and $X(t)$ is continuous in q.m. outside a countable set of values of t. This permits us to extend, with obvious modifications, Gavce's theorem which follows to the case of covariances of bounded variation. Since all relations below between r.v.'s will be a.s. relations, we drop "a.s.".

C. SECOND ORDER CALCULUS THEOREM. *Let $X(t), t \in [a, b]$, be a second order r.f. with a continuous covariance $\Gamma(t, t')$. Then, up to equivalences,*

(i) *The indefinite integral in q.m. $Y(t) = \int_a^t X(s)\,ds$ exists, $X(t)$ is its derivative in q.m., and if $Y_0(t)$ is a primitive in q.m. of $X(t)$ then $Y(t) = Y_0(t) - Y_0(a)$.*

The r.f. $X(t)$ continuous in q.m. is Borelian, sample integrable and sample square integrable, and its indefinite sample and q.m. integrals coincide.

(ii) *If $\Delta_h\Delta'_h\Gamma(t, t) \leqq ch^2$ for h sufficiently small, then $X(t)$ is sample continuous, is the sample derivative of its indefinite sample integral, and for every $\epsilon > 0$ and $0 < \alpha < \frac{1}{2}$ there exists an a.s. positive r.v. $H_{\epsilon,\alpha}$ such that $|X(t+h) - X(t)| < (1+\epsilon)|h|^\alpha$ for $|h| < H_{\epsilon,\alpha}$.*

If the derivative $\frac{\partial^2}{\partial t\,\partial t'}\Gamma(t, t')$ exists and is finite then $X(t)$ is differentiable in q.m., and if $\Delta_h\Delta'_h\frac{\partial^2}{\partial t\,\partial t'}\Gamma(t, t') \leqq c'h^2$ for h sufficiently small then $X(t)$ is sample differentiable.

(iii) *If the derivative $\frac{\partial^{2n+2}}{\partial t^{n+1}\partial t'^{n+1}}\Gamma(t, t')$ exists and is finite then $X(t)$ is n times sample differentiable, and if $\Gamma(t, t')$ is infinitely differentiable, then $X(t)$ is infinitely sample differentiable.*

Proof. We use throughout the convergence in q.m. criterion without further comment.

1° Since $\Gamma(t, t')$ is continuous hence bounded on $[a, b] \times [a, b]$, the indefinite integral in q.m. $Y(t)$ exists and, as $h \to 0$,

$$E\left|\frac{1}{h}\int_t^{t+h} X(s)\,ds - X(t)\right|^2$$

$$= \frac{1}{h^2}\int_t^{t+h}\int_t^{t+h}\Gamma(s, s')\,ds\,ds' - \frac{1}{h}\int_t^{t+h}\Gamma(s, t)\,ds - \frac{1}{h}\int_t^{t+h}\Gamma(t, s')\,ds' + \Gamma(t, t) \to 0,$$

so that $X(t)$ is the derivative in q.m. of $Y(t)$. Therefore, if $X(t)$ is also the derivative in q.m. of $Y_0(t)$ then the derivative $\Delta'(t)$ in q.m. of $\Delta(t) =$

$Y(t) - (Y_0(t) - Y_0(a))$ is 0 while $\Delta(a) = 0$. It follows that $\Delta(t) = 0$ since $\frac{d}{dt} E|\Delta(t)|^2 = E\{\Delta(t)\bar{\Delta}'(t) + \Delta'(t)\bar{\Delta}(t)\} = 0$, and the first set of assertions in (i) is proved.

$X(t)$ is continuous in q.m. hence in pr. and consequently, by the measurability theorem, the r.f. $X(t)$ is equivalent to a separable a.e. Borel r.f. Boundedness of $E|X(t)|^2 = \Gamma(t, t)$ implying that of $E|X(t)| \leqq E^{1/2}|X(t)|^2$, the second set of assertions follows from the measurability lemma, except for the assertion of equivalence of $Y(t)$ and the sample integral $Z(t)$.

Since $Y_n(t) \xrightarrow{\text{q.m.}} Y(t)$ where

$$Y_n(t) = \sum_k X(t_{nk})(t_{nk} - t_{n,k-1}),$$

$$a = t_{n0} < \cdots < t_{nk_n} = t, \quad \max(t_{nk} - t_{n,k-1}) \to 0$$

the assertion will be proved if we show that

$$E|Y_n(t) - Z(t)|^2$$

$$= E|Y_n(t)|^2 - EY_n(t)\bar{Z}_n(t) - E\bar{Y}_n(t)Z(t) + E|Z(t)|^2 \to 0.$$

But

$$E|Y_n(t)|^2 \to \int_a^t \int_a^t \Gamma(s, s')\, ds\, ds'$$

while, by the measurability lemma,

$$E|Z(t)|^2 = E\int_a^t \int_a^t X(s)\bar{X}(s')\, ds\, ds' = \int_a^t \int_a^t \Gamma(s, s')\, ds\, ds',$$

and

$$EY_n(t)\bar{Z}(t) = \sum_k \left(\int_a^t \Gamma(t_{nk}, t)\, dt\right)(t_{nk} - t_{n,k-1}) \to \int_a^t \int_a^t \Gamma(t, t')\, dt\, dt'.$$

The assertion is proved and the proof of (i) is terminated.

2° Since, by the first hypothesis in (ii),

$$E|X(t+h) - X(t)|^2 = \Delta_h \Delta'_h \Gamma(t, t) \leqq ch^2,$$

the first and third assertions in (ii) result from the Corollary of the sample continuity moduli theorem with $r = 2$ and $s = 1$ applied to an equivalent separable r.f. The second one results from the fundamental theorem of ordinary calculus, since $Z(\omega, t) = \int_a^t X(\omega, s)\, ds$ where $X(\omega, s)$

is continuous in s. The fourth assertion is immediate. The fifth and last one results from

$$E|\,X'(t+h) - X'(t)\,|^2 = \Delta_h \Delta'_h \left[\frac{\partial^2}{\partial t\,\partial t'}\,\Gamma(t,t')\right]_{t'=t} \leqq c'h^2$$

upon applying what precedes with $X(t)$ replaced by its derivative in q.m. $X'(t)$, and assertions (iii) readily follow.

38.4. Random times. Random times are at the center of random Analysis and, especially, of sample Analysis. We already encountered them in the case of sequences in §26 and in 32.4. Moreover, random times are of the essence in the continuous parameter case. They were introduced and used systematically by P. Lévy.

Statistical Physics has a familiar—the "Maxwell demon" who travels along the individual paths of particles subject to deterministic laws of mechanics; his clock is the same along all paths. In sample Analysis, there is now also a familiar—the "Lévy demon" who travels along individual sample paths of r.f.'s, and his "random time" clock varies with the paths. In fact, the Maxwell demon is but a degenerate form of the Levy demon.

Let X_T be a r.f. on a "time" domain $T \subset \bar{R}$. This r.f. is automatically accompanied by the nondecreasing family of σ-fields $(\mathcal{B}(X_s, s \leqq t), t \in T)$. Intuitively, the observable events up to time t (included), determined by observations of $(X_s, s \leqq t)$, are members of $\mathcal{B}(X_s, s \leqq t)$. Let, say, τ be the "time" X_T first reaches a value $c(\in R)$. This time depends upon the "state of Nature" $\omega \in \Omega$, that is, $\tau(\omega)$ is the time the sample function $X_T(\omega)$ first reaches the value c.

It may and does happen that the observable events belong to larger σ-fields than those determined by the r.f. X_T. Then X_T is accompanied by a σ-field-valued nondecreasing function $\mathcal{B}_T = (\mathcal{B}_t, t \in T)$ on T and we write $(X_T, \mathcal{B}_T)$ in lieu of X_T. *Once and for all, for all* $r \leqq s \leqq t$ *belonging to* T,

$$\mathcal{B}_t \supset \mathcal{B}_s \supset \mathcal{B}(X_r, r \leqq s).$$

These larger σ-fields may happen in various ways:
If $g_T = (g_t, t \in T)$ here the g_t are Borel functions on R to R, then the r.f. $g_T(X_T) = (g_t(X_t), t \in T)$ is accompanied automatically by the σ-fields of its own observable events—for every $t \in T$,

$$\mathcal{B}(g_s(X_s), s \leqq t) \subset \mathcal{B}(X_s, s \leqq t) \subset \mathcal{B}_t.$$

Yet, such a transformation of $(X_T, \mathcal{B}_T)$ with some property defined in terms of X_T *and* $\mathcal{B}_T$ may preserve this property in terms of $g_T(X_T)$ *and the same* $\mathcal{B}_T$. For example, if $(X_T, \mathcal{B}_T)$ is a "submartingale" so is $(X_T^+, \mathcal{B}_T)$—see 39.1. A much more impelling situation arises when a property is first defined in terms of $\mathcal{B}_T$ *alone* and then every pair $(X_T, \mathcal{B}_T)$, or X_T alone, is said to have this property. For example, "independence" as well as "conditional independence" are defined in terms of σ-fields alone—see 16.1 as well as 28.3, and so is "Markov independence" —see 43.1. This is also the case of "random times" that we introduce and study in what follows.

Let $\Omega^\tau \subset \Omega$ be a nonnull event which we assign the trace σ-field $\mathcal{A}^\tau = \mathcal{A} \cap \Omega^\tau$ of events in Ω^τ. To every Borel subset $S \subset R$ we assign the trace σ-field $\mathfrak{I}_S = \mathfrak{I} \cap S$ of Borel sets in S. *Once and for all, T denotes a Borel set in R.*

We say that a measurable function τ *on* Ω^τ *to* T is a $\mathcal{B}_T$*-time* if, for every $t \in T$,

$$[\tau \leqq t] \in \mathcal{B}_t.$$

Outside of Ω^τ with $P\Omega^\tau < 1$, τ may have no meaning or may not exist. It is frequently convenient then to assign an exceptional value "t_e" to τ on $(\Omega^\tau)^c$, and the definition becomes accordingly: A measurable function *on* Ω *to* $T \cup \{t_e\}$ is a $\mathcal{B}_T$*-time* if, for every $t \in T$,

$$[\tau \leqq t] \in \mathcal{B}_t.$$

This modification is particularly useful in the usual cases of $T = [0, \infty)$ or $T = \{1,2, \cdots\}$ where we shall take automatically $t_e = +\infty$. However, *unless otherwise stated, our* $\mathcal{B}_T$*-times will be on* Ω *to* T. Given $\mathcal{B}_T = (\mathcal{B}_t, t \in T)$, *we set once and for all*

$$\mathcal{B}_{t+0} = \bigcap_{u>t} \mathcal{B}_u, \quad \mathcal{B}_{t-0} = \bigvee_{s<t} \mathcal{B}_s, \quad \mathcal{B} = \bigvee_{t \in T} \mathcal{B}_t,$$

where "$\bigvee$" stands for the σ-fields generated by the union fields. In the "discrete case" of $T = (1,2, \cdots)$, as used in §26, the condition $[\tau \leqq t] \in \mathcal{B}_t$ for every $t \in T$ is clearly equivalent to $[\tau = t] \in \mathcal{B}_t$ for every $t \in T$, and the $\mathcal{B}_{t-0}$, $\mathcal{B}_{t+0}$ are of no interest since for $t = n$ they reduce to $\mathcal{B}_{n-1}$ and to $\mathcal{B}_{n+1}$.

To every $\mathcal{B}_T$-time τ we assign

$$\mathcal{B}_\tau = \{B \in \mathcal{B}: B[\tau \leqq t] \in \mathcal{B}_t, t \in T\}.$$

Note that every $t \in T$ is a degenerate $\mathcal{B}_T$-time with corresponding family $\mathcal{B}_t$ of events. Intuitively, $\mathcal{B}_\tau$ consists of observable events up to time τ (included).

Events belonging to any σ-field with or without subscripts will be given the same subscripts, if any.

A. ELEMENTARY PROPERTIES OF $\mathcal{B}_T$-TIMES.

I. *Let τ be a $\mathcal{B}_T$-time.*

(i) *$\mathcal{B}_\tau$ is a σ-field and τ is $\mathcal{B}_\tau$-measurable.*

(ii) *$\tau \wedge t = \min(\tau,t)$ and $\tau \vee t = \max(\tau,t)$ are $\mathcal{B}_t$-measurable.*

(iii) *$[\tau < t]$, $[\tau = t]$, $[\tau > t]$ belong to $\mathcal{B}_t$ for every $t \in T$.*

(iv) *If $\mathcal{B}_\tau$-measurable $\tau' \geqq \tau$, then τ' is a $\mathcal{B}_T$-time.*

(v) *When T is a finite interval closed on the right, say, $T = (0,1]$ or $[0,1]$, then the simple $\mathcal{B}_T$-times*

$$\tau_n = \sum_{k=1}^{2^n} \frac{k}{2^n} I_{\left[\frac{k-1}{2^n} < \tau \leqq \frac{k}{2^n}\right]} \downarrow \tau.$$

When T is an infinite interval, say, $T = [0, \infty)$ or $T = [o, \infty]$, then the elementary $\mathcal{B}_T$-times

$$\tau_n = \sum_{k=1}^{\infty} \frac{k}{2^n} I_{\left[\frac{k-1}{2^n} \leqq \tau < \frac{k}{2^n}\right]} \downarrow \tau.$$

or

$$\tau'_n = \tau_n + (\infty) I_{[\tau=\infty]} \downarrow \tau.$$

II. *Let σ and τ be $\mathcal{B}_T$-times.*

(i) *If $\sigma \leqq \tau$ then $\mathcal{B}_\sigma \subset \mathcal{B}_\tau$.*

(ii) *$\sigma \wedge \tau$ and $\sigma \vee \tau$ are $\mathcal{B}_T$-times.*

(iii) $[\sigma < \tau]$, $[\sigma = \tau]$, $[\sigma > \tau]$ belong to $\mathcal{B}_{\sigma\wedge\tau} = \mathcal{B}_\sigma \cap \mathcal{B}_\tau$.

(iv) *$B_\sigma[\sigma \leqq \tau] \in \mathcal{B}_\tau$ for every $B_\sigma \in \mathcal{B}_\sigma$.*

Proof. When not immediate, the proofs are elementary.

1°. We prove I. Properties (i) and (ii) are immediate and property (iii) obtains by

$$[\tau > t] = [\tau \leqq t]^c \in \mathcal{B}_t,\ [\tau < t] = \bigcup_n \left[\tau \leqq 1 - \frac{1}{n}\right] \in \mathcal{B}_t,$$
$$[\tau = t] = [\tau \leqq t] - [\tau < t] \in \mathcal{B}_t.$$

In fact, (ii) and (iii) are particular cases of (ii) and (iii) in II.

Property (iv) obtains by $[\tau' \leqq t] \in \mathcal{B}_\tau$ since τ' is $\mathcal{B}_\tau$-measurable and $\tau' \geqq \tau$ implies that, for every $t \in T$,

$$[\tau' \leqq t] = [\tau' \leqq t]\,[\tau \leqq t] \in \mathcal{B}_t.$$

Property (v) is immediate, provided one notices that, by (iii), the τ_n are $\mathcal{B}_T$-times.

2°. We prove II. Property (i) is proved as follows. Given $B_\sigma \in \mathcal{B}_\sigma$, for every $t \in T$, $[\sigma \leqq t] \supset [\tau \geqq t]$ since $\sigma \leqq \tau$, hence

$$B_\sigma[\tau \leqq t] = (B_\sigma[\sigma \leqq t])[\tau \leqq t] \in \mathcal{B}_t$$

since

$$B_\sigma[\sigma \leqq t] \in \mathcal{B}_t \text{ and } [\tau \leqq t] \in \mathcal{B}_t.$$

Property (ii) results from

$$[\sigma \wedge \tau \leqq t] = [\sigma \leqq t] \cup [\tau \leqq t] \in \mathcal{B}_t$$

and

$$[\sigma \vee \tau \leqq t] = [\sigma \leqq t] \cap [\tau \leqq t] \in \mathcal{B}_t.$$

We shall return to property (iii) after establishing (iv): $B_\sigma[\sigma \leqq \tau] \in \mathcal{B}_\tau$ means that, for every $t \in T$,

$$(B_\sigma[\sigma \leqq \tau])[\tau \leqq t] = (B_\sigma[\sigma \leqq t])([\tau \leqq t])([\sigma \wedge t \leqq \tau \leqq t])$$

belongs to $\mathcal{B}_t$, which follows from the fact that each of the three right side events belongs to $\mathcal{B}_t$: $B_\sigma[\sigma \leqq t]$ since $B_\sigma \subset \mathcal{B}_\sigma$, $[\tau \leqq t]$ by definition of $\mathcal{B}_T$-time τ, and $[\sigma \wedge t \leqq \tau \leqq t]$ since, by (ii), both sides of the inequality are $\mathcal{B}_t$-measurable.

Finally, we prove (iii): $[\sigma \leqq \tau]$ belongs to $\mathcal{B}_\tau$ by (iv), and so does its complement $[\sigma > \tau]$. Since $\sigma \wedge \tau$ is $\mathcal{B}_{\sigma \wedge \tau}$-measurable hence also, by (i), $\mathcal{B}_\tau$-measurable, the events

$$[\sigma \wedge \tau = \tau] = [\sigma = \tau] \text{ and } [\sigma \wedge \tau < \tau] = [\sigma < \tau]$$

belong to $\mathcal{B}_\tau$. They also belong to $\mathcal{B}_\sigma$ since σ and τ can be interchanged. It remains to show that $\mathcal{B}_{\sigma \wedge \tau} = \mathcal{B}_\sigma \cap \mathcal{B}_\tau$. On the one hand, $\sigma \wedge \tau \leqq \sigma$ and $\sigma \wedge \tau \leqq \tau$ imply that $\mathcal{B}_{\sigma \wedge \tau} \subset \mathcal{B}_\sigma \cap \mathcal{B}_\tau$. On the other hand, for $B \in \mathcal{B}_\sigma \cap \mathcal{B}_\tau$,

$$B[\sigma \wedge \tau \leqq t] = (B[\sigma \leqq t] \cup (B[\tau \leqq t]) \in \mathcal{B}_t,$$

hence $\mathcal{B}_\sigma \cap \mathcal{B}_\tau \subset \mathcal{B}_{\sigma \wedge \tau}$.

The proof is terminated.

The elementary properties in **A** will be used without further comment.

We consider now pairs $(X_T, \mathcal{B}_T)$; recall that every X_t is $\mathcal{B}_t$-measurable. To $\mathcal{B}_T$-time τ—to be also called *time of* $(X_T, \mathcal{B}_T)$ we associate the function X_τ on Ω defined by

$$X_\tau(\omega) = X_{\tau(\omega)}(\omega), \quad \omega \in \Omega.$$

X_τ is a r.v. since it is a measurable function on $(\Omega,\mathcal{A})$ to $(\Omega \times T, \mathcal{A} \times \mathcal{I}_T)$ of a measurable function on $(\Omega \times T, \mathcal{A} \times \mathcal{I}_T)$ to $(T, \mathcal{I}_T)$ or, for short, the composition of two measurable maps: $\omega \to (\omega, \tau(\omega))$ and $(\omega, t) \to X_t(\omega)$. We saw that to τ is associated the σ-field $\mathcal{B}_\tau$ with respect to which it is measurable. The question which arises is when X_τ is $\mathcal{B}_\tau$-measurable, that is, when the observable events in terms of X_τ are observable events up to time τ (included)? In order to answer it in the affirmative, in the most important case of sample rightcontinuous X_T on an interval $T \subset \bar{R}$, it is convenient to consider "progressively Borel" r.f.'s or, more precisely, "progressively Borel" pairs $(X_T, \mathcal{B}_T)$ since they are defined in terms of X_T *and* $\mathcal{B}_T$. To simplify the writing, set $T_t = \{s \in T\colon s \leqq t \in T\}$, $X = X_T$ and $\mathcal{I} = \mathcal{I}_T$, $\mathcal{I}_t = \mathcal{I}_{T_t}$.
$(X_T, \mathcal{B}_T)$ is said to be *progressively Borelian* if, for every $t \in T$, the restriction of X on $\Omega \times T_t$ to R is $(\mathcal{B}_t \times \mathcal{I}_t)$-measurable. Note that when every $\mathcal{B}_t = \mathcal{A}$, this definition becomes that of Borelian X_T. Clearly, if $(X_T, \mathcal{B}_T)$ is progressively Borelian then X_T is Borelian.

a. *A sample rightcontinuous $(X_T, \mathcal{B}_T)$ on an interval T is progressively Borelian.*

Proof. To simplify the writing and also because it is the usual case, take $T = [0, \infty)$. Given $t \in T$, for every $k \leqq 2^n$ and every $s \in \left[\frac{k-1}{2^n}t, \frac{k}{2^n}t\right]$, set $X_s^{(n)} = X_{kt/2^n}$ and set $X_t^{(n)} = X_t$ for $s = t$. The map $(\omega, s) \to X_s^{(n)}(\omega)$ on $(\Omega \times [0, t], \mathcal{B}_t \times \mathcal{I}_t)$ to $(\bar{R}, \mathcal{I})$ is measurable hence, by sample rightcontinuity, so is the limit $X_s^{(n)}$ of X_s for every $s \in T$. Thus $(X_T, \mathcal{B}_T)$ is progressively Borelian.

b. *If τ is a time of progressively Borelian $(X_T, \mathcal{B}_T)$, then X_τ is $\mathcal{B}_\tau$-measurable.*

Proof. To prove that for every Borel set $S \subset \bar{R}$ and every $t \in T$, upon setting $\tau_t = \tau \wedge t$,

$$[X_t \in S][\tau \leqq t] = [X_{\tau_t} \in S][\tau_t < t] \cup [X_t \in S][\tau = t]$$

belongs to $\mathcal{B}_t$, it suffices to show that X_{τ_t} is $\mathcal{B}_t$-measurable. But this obtains by hypothesis since X_{τ_t} is the composition of the measurable maps $\omega \to (\omega, \tau_t(\omega))$ on $(\Omega, \mathcal{B}_t)$ to $(\Omega \times T_t, \mathcal{B}_t \times \mathcal{I}_t)$ and $(\omega,s) \to X_s(\omega)$ on $(\Omega \times T_t, \mathcal{B}_t \times \mathcal{I}_t)$ to $(T, \mathcal{I})$.

B. $\mathcal{B}_\tau$-MEASURABILITY THEOREM. *If τ is a time of $(X_T, \mathcal{B}_T)$ then X_τ is $\mathcal{B}_\tau$-measurable when τ is an elementary time or when X_T is sample right-continuous on an interval T.*

Proof. When τ is elementary, that is, has a countable set of distinct values $t_1, t_2, \cdots$ then, for every Borel set $S \subset \overline{R}$ and every $t \in T$,

$$[X_\tau \in S][\tau \leqq t] = \bigcup_{t_j < t} [X_t \in S][\tau = \tau_j] \in \mathcal{B}_T.$$

while the sample rightcontinuity case obtains by **a** and **b**.

§ 39. MARTINGALES

The main r.f.'s and processes studied in this Part V are, by increasing order of generality, Brownian, decomposable, martingale and semimartingale, and Markovian types. We begin with martingales and semimartingales (submartingales and supermartingales). For, while important in their own right, they are extremely useful in investigating the other types as well as in the applications of pr. methods to various branches of mathematical analysis.

P. Lévy introduced, studied, and used the martingale concept in the discrete case of sequences of r.v.'s. Then Doob deepened his results, introduced semimartingales, proceeded to a systematic investigation of these concepts and, thus, made them into the powerful tool (presented here and farther sharpened by P. A. Meyer) for pr. theory and for classical Analysis.

Section 32 was devoted to a direct study and applications of the discrete case; the results therein will be obtained here and, in fact, completed—as very particular cases.

The approach will be centered on random times and the main theme will be preservation of (semi)martingale property under various transformations of r.f.'s, passages to the limit and randomization of time.

Unless otherwise stated, *our r.f.'s are taken to be separable, and our σ-fields are subσ-fields of σ-field of events of some underlying pr. space* $(\Omega, \mathcal{A}, P)$.

We use the following notation and terminology:

1°. $X_T^+ = (X_t^+, t \in T), X_T \vee c = (X_t \vee c, t \in T), c \in R,$

$$|X_T|^r = (|X_t|^r, t \in T), \quad EX_T = (EX_t, t \in R).$$

2°. $\mathcal{B}_T = (\mathcal{B}_T,\ t \in T)$ where the $\mathcal{B}_T$ are σ-fields nondecreasing with increasing t.

3°. $X_T \geqq 0$ and $EX_T \geqq 0$ means that the $X_t \geqq 0$ and the $EX_t \geqq 0$. X_T is integrable means that the X_t are integrable. X_t is uniformly integrable means that the X_t are uniformly integrable. Recall—to be used without further comment, that uniform integrability of X_T means that

$$\sup_t \int_{[|X_t|>c]} |X_t| \to 0 \text{ as } c \to \infty$$

equivalently,

$$\sup_t \int_B |X_t| \to 0 \text{ as } PB \to 0 \text{ and } \sup_t E|X_t| < \infty;$$

and uniform integrability of X_T is implied by $\sup_t E|X_t|^r < \infty$ for some $r > 1$.

39.1. Closure and Limits. *Subscripts of r.v.'s X will belong to T, unless otherwise stated.*

We say that a r.f. X_T is a *martingale* if, for every $s < t$ and every $B_s \in \mathcal{B}_s = \mathcal{B}(X_r, r \leqq t)$,

$$\text{(M)} \qquad \int_{B_s} X_s = \int_{B_s} X_t, \text{ equivalently, } X_s = E(X_t \mid \mathcal{B}_s) \text{ a.s.}$$

In fact, it suffices to require that for every finite set $r_1 < \cdots r_n < s$ and every $B \in \mathcal{B}(X_{r_1}, \cdots, X_{r_n}, X_s)$

$$\text{(M}'\text{)} \quad \int_B X_s = \int_B X_t, \text{ equivalently, } X_s = E(X_t \mid X_{r_1}, \cdots, X_{r_n}, X_s) \text{ a.s.}$$

For, (M) implies (M′) since $\mathcal{B}(X_{r_1}, \cdots, X_{r_n}, X_s) \subset \mathcal{B}_s$, and (M′) implies (M) on account of the following property, already used in various guises, that we isolate now.

a. EXTENSION OF MEASURES LEMMA. *Let a σ-field $\mathcal{B}$ be generated by a field $\mathcal{C}$ which consists of finite sums of events belonging to a class $\mathfrak{D}$. If ϕ and ϕ' are signed measures on $\mathcal{B}$ σ-finite on $\mathcal{C}$ then $\phi \leqq \phi'$ ($\phi = \phi'$) on $\mathfrak{D}$ implies the same relation on $\mathcal{B}$, and if ϕ and ϕ' are σ-finite measures on $\mathfrak{D}$ then the same implication holds for their unique extensions to measures on $\mathcal{B}$.*

Proof. The assertion about measures on $\mathfrak{D}$ reduces to the one about signed measures. The equality assertion reduces to the inequalities one upon applying that one to $\phi \leqq \phi'$ and to $\phi' \leqq \phi$. Since the class of events on which $\phi \leqq \phi'$ is closed under countable summations and contains $\mathfrak{D}$, it contains the field $\mathcal{C}$ over $\mathfrak{D}$. Furthermore, upon taking a countable partition of Ω, we may assume that ϕ and ϕ' are finite on this class so that it is also closed under monotone passages to the limit, hence contains the monotone field $\mathcal{B}$ over the field $\mathcal{C}$. The inequalities assertion follows, and the proof is terminated.

We replace the term "*martingale*" by "*sub(super)martingale*" whenever in the defining relations the sign "$=$" is replaced by "$\leqq$" ("$\geqq$").

Various transformations of submartingales yield submartingales for which the defining inequalities hold for larger σ-fields of events than those induced by the transformed r.v.'s, in fact, for the σ-fields of the initial submartingales. Thus, they leave invariant the corresponding random times—(see 38.4 and **b** below). This leads to extend the foregoing concept as follows.

Let $(X_T, \mathcal{B}_T)$ be a r.f. X_T with accompanying $\mathcal{B}_T$ such that the X_t are $\mathcal{B}_t$-measurable. We say that $(X_T, \mathcal{B}_T)$ is a *martingale* if, for every $s < t$ and every $B_s \in \mathcal{B}_s$,

$$\int_{B_s} X_s = \int_{B_s} X_t, \text{ equivalently, } X_s = E(X_t \mid \mathcal{B}_s) \text{ a.s.}$$

$(X_T, \mathcal{B}_T)$ is said to be a *sub(super)martingale* when "$=$" is replaced by "$\leqq$" ("$\geqq$"). When $(X_T, \mathcal{B}_T)$ is one of them so is X_T since $\mathcal{B}_s \supset \mathcal{B}(X_r, r \leqq s)$ but, in general this inclusion relation is strict so that $(X_T, \mathcal{B}_T)$ has more properties—which may be lost if only X_T were considered.

ELEMENTARY PROPERTIES. The following properties are mostly immediate and will be used without further comment.

1. *A martingale is a submartingale and a supermartingale.*

2. *A semimartingale $(X_T, \mathcal{B}_T)$ with finite EX_T is a martingale if and only if EX_T is a constant function on T.*

In the semimartingale case, for $s < t$,

$$X_s \leqq E(X_t \mid \mathcal{B}_s) \text{ a.s. or } X_s \geqq E(X_t \mid \mathcal{B}_s) \text{ a.s.}$$

and both sides cannot be equal a.s. unless the finite $EX_s = EX_t$ finite.

3. *If $(X_T, \mathcal{B}_T)$ and $(X'_T, \mathcal{B}_T)$ are submartingales (martingales) then $(cX_T + c'X'_T, \mathcal{B}_T)$ is a submartingale (martingale) for any constants c, c' in* $[0, \infty)$ (in R).

4. *If $(X_T, \mathcal{B}_T)$ is a submartingale (supermartingale), then $(-X_T, \mathcal{B}_T)$ is a supermartingale (submartingale).*

5. *Martingale and submartingale property is preserved when a same constant is added to all its r.v.'s.*

6. *If $(X_T, \mathcal{B}_T)$ is a submartingale (martingale) then, for $s < t$,*

$$EX_s \leqq EX_t \;(EX_s = EX_t)$$

and more generally, for every $\mathcal{B}_s$-measurable $Y \geqq 0$,

$$EX_sY \leqq EX_tY \; (EX_sY = EX_tY).$$

REMARK. According to 4, properties of submartingales $(X_T, \mathcal{B}_T)$ yield those of supermartingales upon changing X_T into $-X_T$, and conversely. Thus, it suffices to study, say, submartingales.

Let g on R be convex with $g(+\infty) = \lim_{x\to\infty} g(x) = \infty$. Recall that if $E^{\mathcal{B}}X > -\infty$ a.s. then $g(E^{\mathcal{B}}X) \leqq E^{\mathcal{B}}\, g(X)$ a.s. and note that if $(X_T, \mathcal{B}_T)$ is a submartingale then, for $s < t$, $E^{\mathcal{B}_s}X_t \geqq X_s > -\infty$.

A. SUBMARTINGALE PRESERVING TRANSFORMATIONS. *Let $(X_T, \mathcal{B}_T)$ be a martingale (submartingale) and g be convex (and nondecreasing) with $g(+\infty) = \lim_{x\to\infty} g(x) = \infty$. Then $(g(X_T), \mathcal{B}_T)$ is a submartingale.*
In particular, $(X_T \vee c, \mathcal{B}_T)$ is a submartingale for every $c \in R$ and if $X_T \geqq 0$ or $(X_T, \mathcal{B}_T)$ is a martingale, then $(|X_T|^r, \mathcal{B}_T)$ is a submartingale for every $r \geqq 1$.

Proof. Let $s < t$. For g convex (*and nondecreasing*) if $(X_T, \mathcal{B}_T)$ is a martingale (submartingale) then

$$g(X_s) = (\leqq)\, g(E^{\mathcal{B}_s}X_t) \leqq E^{\mathcal{B}_s}\, g(X_t) \text{ a.s.}$$

and $(g(X_T), \mathcal{B}_T)$ is a submartingale.
For $g(x) = x^+$ we obtain

$$X_s^+ \leqq E(X_t^+ \mid \mathcal{B}_s) \text{ a.s.}$$

and $(X_T^+, \mathcal{B}_T)$ is a submartingale. Since $X_t \vee c = (X_t - c)^+ + c$ and $(X_T - c, \mathcal{B}_T)$ is a submartingale, so is $((X_T - c)^+, \mathcal{B}_T)$ hence so is $(X_T \vee c, \mathcal{B}_T)$.
When $X_T \geqq 0$ then, for $g(x) = 0$ or x^r according as $x \leqq 0$ or $x > 0$,

$$X_s^r \leqq E^{\mathcal{B}_s} X_t^r \text{ a.s.}$$

and $(X_T^r, \mathcal{B}_T)$ is a submartingale.
When $(X_T, \mathcal{B}_T)$ is a martingale then, for $g(x) = |x|^r$ with $r \geqq 1$,

$$|X_s|^r = |E^{\mathcal{B}_s} X_t|^r \leqq E^{\mathcal{B}_s}|X_t|^r \text{ a.s.}$$

and $(|X_T|^r, \mathcal{B}_T)$ is a submartingale. The proof is terminated.

Closure. Let $(X_T, \mathcal{B}_T)$ be a martingale or a submartingale. We say that it *is closed on the left (right) by X_α (X_β)* if T has a first (last) element $\alpha(\beta)$. Note that for every $t \in T$, $(X_u, \mathcal{B}_u, u \geqq t)$ $((X_s, \mathcal{B}_s, s \leqq t))$ is closed on the left (right) by X_t.

More generally, we say that $(X_T, \mathcal{B}_T)$ *is closed on the left (right) by* Y if there are $\alpha \leqq \inf T (\beta \geqq \sup T)$ and a σ-field $\mathcal{B}_\alpha(\mathcal{B}_\beta)$ with $\mathcal{B}(Y) \subset \mathcal{B}_\alpha \subset \mathcal{B}_t (\mathcal{B}(Y) \subset \mathcal{B}_\beta \supset \mathcal{B}_t)$ such that, for all $t \in T$, setting $X_\alpha = Y (X_\beta = Y)$, $(X_{\{\alpha\} \cup T}, \mathcal{B}_{\{\alpha\} \cup T})$ $((X_{T \cup \{\beta\}}, \mathcal{B}_{T \cup \{\beta\}}))$ is a martingale or a submartingale as is $(X_T, \mathcal{B}_T)$.

ELEMENTARY CLOSURE PROPERTIES. The following properties are almost immediate and will be used without further comment.

1. *A submartingale* $(X_T, \mathcal{B}_T)$ *is closed on the right by* Y *if and only if, for all* $t \in T$,

$$X_t \leqq E(Y \mid \mathcal{B}_t) \text{ a.s.}$$

2. $(X_T, \mathcal{B}_T)$ *is a martingale closed on the right by* Y *if and only if* $X_t = E(Y \mid \mathcal{B}_t)$ *a.s. for all* $t \in T$.

The "only if" assertion is immediate and the "if" assertion obtains from $\mathcal{B}_s \subset \mathcal{B}_t$ when $s < t$ by

$$X_s = E(Y \mid \mathcal{B}_s) = E(E(Y \mid \mathcal{B}_t) \mid \mathcal{B}_s) = E(X_t \mid \mathcal{B}_s) \text{ a.s.}$$

We shall use a *standard limiting procedure* for separable r.f.'s: Let $S_n = \{s_1, \cdots, s_n\} \uparrow S = \{s_1, s_2, \cdots\}$ which separates X_T. To establish a relation for X_T, do it for $(X_k, k = 1, \cdots, n)$, then for X_{S_n} by replacing k by t_k where the t_k are points of S_n reordered increasingly, and pass to the limit.

b. RIGHT CLOSURE LEMMA. *Let* $(X_T, \mathcal{B}_T)$ *be a nonnegative submartingale closed on the right by a nonnegative r.v.* Y, *and let* $0 < c < \infty$. *Then*

$$cP(\sup_t X_t > c) \leqq \int_{[\sup_t X_t > c]} \leqq EY$$

and, for $r > 1$,

$$E(\sup_t X_t)^r \leqq s^r EY^r, \text{ where } \frac{1}{r} + \frac{1}{s} = 1.$$

If $EY < \infty$ *then* $\sup_t X_t < \infty$ *a.s. and* X_T *is uniformly integrable.*

Proof. 1°. To prove the first inequality, it suffices to show that for a submartingale $(X_k, \mathcal{B}_k, k = 1, \cdots, n)$ closed on the right by Y,

$$cPA_n \leqq \int_{A_n} Y \leqq EY, \text{ where } A_n = [\max_k X_k > c]. \tag{1}$$

For,

$$\max_{t \in S_n} X_t \uparrow \sup_{t \in S} X_t = \sup_t X_t \tag{2}$$

and the standard limiting procedure applies.
But, upon setting

$$B_1 = [X_1 > c],\ B_k = [X_k > c,\ \max_{j<k} X_j \leq c] \text{ for } k > 1,$$

$A_n = \sum_k B_k$ hence

$$EY \geqq \int_{A_n} Y = \sum_k \int_{B_k} Y \geqq \sum_k \int_{B_k} X_k \geqq c \sum_k PB_k = cPA,$$

and (i) is proved.
Let $EY < \infty$ and let $c \to \infty$. Then

$$P(\sup_t X_t > c) \leqq EY/c \to 0,$$

hence $\sup_t X_t < \infty$ a.s. Since

$$\sup_t P(X_t > c) \leqq \sup_t EX_t/c \leqq EY/c \to 0$$

and Y is integrable,

$$\sup_t \int_{[X_t>c]} X_t \leqq \sup_t \int_{[X_t>c]} Y \to 0,$$

hence X_T is uniformly integrable.

2°. Upon setting $X = \sup_t X_t$, it remains to prove

(3) $$EX^r \leqq s^r EY^r, \text{ where } EX^r > 0 \text{ and } EY^r < \infty$$

since the inequality is trivially true when $EX^r = 0$ or $EY^r = \infty$.
It suffices to show that (3) holds for nonnegative r.v.'s X and Y such that, for every $c > 0$,

(4) $$cP(X > c) \leqq \int_{[X>c]} Y.$$

Set $q(c) = P(X > c)$ and leg g on $[0, \infty)$ be nondecreasing with $g(0) = 0$. Then

$$Eg(X) = -\int_0^\infty g(c)dq(c) \leqq \int_0^\infty q(c)dg(c)$$

$$= \int_0^\infty \left(\int_{[X>c]} YdP\right)\frac{dg(c)}{c} = \int_\Omega \left(\int_0^X \frac{dg(c)}{c}\right)YdP.$$

Thus, for $g(c) = c^r$ with $r < 1$, by Hölder inequality,

$$EX^r \leqq sE(YX^{r-1}) \leqq s(EY^r)^{1/r}E(X^{(r-1)s})^{1/s} = s(EY^r)^{1/r}(EX^r)^{1/s}.$$

and (3) obtains upon dividing by $(EX^r)^{1/s}$ provided $EX^r < \infty$. To obviate this difficulty, replace X by $X \wedge c$ so that $E(X \wedge c)^r \leqq c^r < \infty$ and $E(X \wedge c)^r > 0$ for c sufficiently large since, as $c \uparrow \infty$, $X \wedge c \uparrow X$ hence $E(X \wedge c)^r \uparrow EX^r \uparrow 0$. Inequality (4) holds a fortiori for $X \wedge c$, the division mentioned above is valid, and then let $c \uparrow \infty$.
The proof is terminated.

B. SUPREMA THEOREM. *Let $(X_T, \mathcal{B}_T)$ be a submartingale and let $c \in R$. Then*

$$cP(\sup_t X_t > c) \leqq \sup_t EX_t^+$$

and, for $r > 1$,

$$E(\sup_t X_t^+ > c)^r \leqq \sup_t E(X_t^+)^r.$$

If $\sup_t EX_t^+ < \infty$ then $\sup_t X_t < \infty$ a.s., $(X_s \vee c, s \leqq t)$ is uniformly integrable for every $t \in T$, and

$$\sup_t EX_t^+ \leqq 2 \sup_t EX_t^+ - \inf_t EX_t.$$

Proof. We use the preceding lemma without further comment. The first inequality is trivially true when $c \leqq 0$. Thus, let $c > 0$. Since the submartingale $(X_s^+, \mathcal{B}_s, s \leqq t)$ with $t \in T$ is closed by X_t^+ and $X_s \leqq X_s^+$,

$$cP(\sup_{s \leqq t} X_s > c) \leqq cP(\sup_{s \leqq t} X_s^+ > c) \leqq EX_t^+ \leqq \sup_t EX_t^+$$

and, for $r > 1$,

$$E(\sup_{s \leqq t} X_s^+)^r \leqq E(X_t^+)^r \leqq \sup_t E(X_t^+)^r.$$

Thus, the first two asserted inequalities obtain upon letting t vary over T. Let $\sup_t EX_t^+ < \infty$. The first inequality yields, as $c \to \infty$,

$$P(\sup_t X_t > c) \leqq \sup_t EX_t^+/c \to 0,$$

hence $\sup_t X_t < +\infty$ a.s. The last inequality obtains by

$$E|X_t| = 2EX_t^+ - EX_t \leqq 2 \sup_t EX_t - \inf_t EX_t$$

and it remains to prove the asserted uniform integrability.
Since the martingale $(X_s^+, \mathcal{B}_s, s \leqq t)$ with $t \in T$ is closed on the right by integrable X_t^+, the assertion is valid for $(X_s^+, s \leqq t)$. Since $|X \vee c| = |(X - c)^+ + c| \leqq X^+ + |c|$, the asserted uniform inte-

grability for $(X_s \vee c, s \leqq t)$ with $t \in T$ follows, and the proof is completed.

Martingale times. When $(X_T, \mathcal{B}_T)$ is a martingale or a semimartingale it is convenient to say that $\mathcal{B}_T$-times (defined on Ω to T) are *its times* or *martingale times.* The following proposition is central to the whole section.

c. Central lemma. *Let $\sigma \leqq \tau \leqq n$ be times of a submartingale (martingale) $(X_k, \mathcal{B}_k)$, $k = 1, \cdots, n$.*

If EX_σ and EX_τ exist, then $(X_\sigma, \mathcal{B}_\sigma)$ and $(X_\tau, \mathcal{B}_\tau)$ form a submartingale (martingale) and

$$EX_1 \leqq EX_\sigma \leqq EX_\tau \leqq EX_n (EX_1 = EX_\sigma = EX_\tau = EX_n).$$

If $EX_1 > -\infty$ or $EX_n^+ < +\infty$ then all EX_τ exist and

$$E|X_\tau| \leqq 2EX_n^+ - EX_1.$$

Proof. If $B_k \in \mathcal{B}_k$ and $j > k$ then $B_k[\tau < j - 1] \in \mathcal{B}_{j-1}$, hence

$$\int_{B_k[\tau > j-1]} X_{j-1} \leqq \int_{B_k[\tau > j-1]} X_j = \int_{B_k[\tau = j]} X_j + \int_{B_k[\tau > j]} X_j$$

and, summing over $j = k + 1, \cdots, n$,

$$\int_{B_k[\tau > k]} X_k \leqq \sum_{j=k+1}^{n} \int_{B_k[\tau = j]} X_j = \int_{B_k[\tau > k]} X_\tau.$$

Since for $B \in \mathcal{B}_\sigma$, we can take $B_k = B_\sigma[\sigma = k]$ and, for $j > k$, $[\sigma = k][\tau < k] = \emptyset$, it follows that

$$\int_{B_\sigma[\sigma = k]} X_\sigma = \int_{B_\sigma[\sigma = k][\tau = k]} X_k + \int_{B_\sigma[\sigma = k][\tau > k]} X_k \leqq \int_{B_\sigma[\sigma = k]} X_\tau$$

and, summing over k,

$$\int_{B_\sigma} X_\sigma \leqq \int_{B_\sigma} X_\tau.$$

In particular, $EX \leqq EX_\tau$ so that, taking $1 = \sigma' \leqq \sigma \leqq \tau \leqq \tau' = n$,

$$EX_1 \leq EX_\sigma \leqq EX_\tau \leqq EX_n.$$

In the martingale case all foregoing inequalities become equalities.

Let $EX_1 > -\infty$ hence the $EX_k > -\infty$, or $EX_n^+ < \infty$ hence the $EX_k^+ < \infty$. Then, for every martingale time $\tau \leqq n$,

$$EX_\tau = \sum_k EX_k I_{[\tau=k]} > -\infty \text{ or } < +\infty \text{ exists.}$$

Finally, since $(X_k^+, \mathcal{B}_k, k = 1, \cdots, n)$ is a submartingale hence $EX_\tau^+ \leqq EX_n^+$, it follows that

$$E\,|\,X_\tau\,| = 2EX_\tau^+ - EX_\tau \leqq 2EX_n^+ - EX_1.$$

Applications. The basic properties of submartingales stem from the above lemma. On the one hand, it yields the inequality (and more) from which **A** stems. On the other hand, it yields the "crossings" inequality from which the limit properties stem.

Let $(X_k, \mathcal{B}_k, k = 1, \cdots, n)$ be a submartingale with $EX_1 > -\infty$ or $EX_n^+ < +\infty$.

1°. *Two inequalities.* Let $\sigma(\omega)$ $(\tau(\omega))$ be the smallest $k \leqq n$, if any, such that $X_k(\omega) > c$ $(X_k(\omega) < c)$ or be n if there is not such k. Set

$$B_n = [\sup_k X_k > c](C_n = [\inf_k X_k < c]).$$

Apply the lemma to $\sigma \leqq n$ and $1 \leqq \tau$ so that

$$cPB_n \leqq \int_{B_n} X_\sigma \leqq \int_{B_n} X_n$$

and

$$EX_1 \leqq \int_{C_n} X_\tau + \int_{C_n^c} X_\tau \leqq cPC_n + \int_{C_n^c} X_n.$$

Let now $(X_T, \mathcal{B}_T)$ be a submartingale with $EX_T > -\infty$ or $EX_T^+ < +\infty$ and use the standard limiting procedure with separating set $S = \{s_1, s_2, \cdots\}$ and $S_n = \{s_1, \cdots, s_n\} \uparrow S$. If the submartingale is closed on the right by Y, then the first inequality yields

$$cP\,(\sup_{t\in S_n} X_t > c) \leqq \int_{[\sup_{t\in S} X_t > c]} Y$$

and, passing to the limit,

$$cP(\sup_t X_t > c) \leqq \int_{[\sup_t X_t > c]}.$$

This is the main inequality in **b**.

Similarly, by the same procedure, the second inequality yields

$$\inf_t EX_t \leqq \int_{[\inf_t X_t > c]} Y + cP(\inf_t X_t < c).$$

This new inequality completes **b.**

2°. *Crossings.* The number $H_n(\omega)$ of *crossings* (from the left) by $(X_1(\omega), \cdots, X_n(\omega))$ of a finite interval $[r, s]$ is the number of times that starting with $X_1(\omega)$ and proceeding to $X_n(\omega)$, we pass from the left to the right of $[r, s]$. Let $\tau_1(\omega) = 1$, $\tau_2(\omega)$ be the smallest $k \geqq \tau_1(\omega)$, if any, for which $X_k(\omega) \leqq r$, $\tau_3(\omega)$ the smallest $k \geqq \tau_2(\omega)$, if any, for which $X_k(\omega) \geqq s$ and so on, proceeding alternately up to $\tau_n(\omega) = n$; from the first undefined $\tau_j(\omega)$, if any, set $\tau_j(\omega) = \cdots = \tau_n(\omega) = n$. H_n is also the (random) number of crossings of $[0, s - r]$ by the submartingale formed by the $((X_k - r)^+, \mathcal{B}_k)$. Set $Y_k = (X_k - r)^+$ so that

$$Y_n = \sum_{j=2}^{n} (Y_{\tau_j} - Y_{\tau_{j-1}}) + Y_1$$

and, by the lemma, the $(Y_{\tau_j}, \mathcal{B}_{\tau_j})$ form a submartingale.
Let $EY_n < \infty$ so that $0 \leqq EY_{\tau_j} < \infty$ and, in

$$EY_n = \sum_{j=2}^{n} E(Y_{\tau_j} - Y_{\tau_{j-1}}) + EY_1,$$

the sum of the first H_n summands with odd j is at least $(s - r)EH_n$ while the remaining ones are nonnegative. Thus

$$(s - r)EH_n \leqq E(X_n - r)^+$$

and the inequality is trivially true when $E(X_n - r)^+ = +\infty$. Note that H_n does not decrease as n increases.

Let now $(X_T, \mathcal{B}_T)$ be a submartingale with $EX_1 > -\infty$ or $EX_T^+ < +\infty$. The standard limiting procedure starting with H_n replaced by $H_{S_n}(r,s)$ yields the basic *crossings inequality*

$$(s - r)EH_T(r,s) \leqq \sup_t E(X_t - r)^+;$$

we shall omit the subscript T, unless confusion is possible.

Limits. We begin with

d. SAMPLE LIMITS LEMMA. *If $(X_T, \mathcal{B}_T)$ is a submartingale with* $\sup_t EX_t^+ < \infty$, *then there is a null event N such that for $\omega \notin N$, the discontinuities of sample functions $X_T(\omega)$ are simple.*

Proof. Since

$$(s - r)EH(r,s) \leqq \sup_t E(X_t - r)^+ \leqq \sup_t EX_t^+ + |r| < \infty,$$

$H(r,s) < \infty$ outside a null event $N(r,s)$ for every finite interval $[r,s]$. Thus $X_T(\omega)$ crosses every such interval a finite number of times only for $\omega \notin N(r,s)$. It follows that its discontinuities are simple for

$$\omega \in N = \bigcup_{\text{rational } r,s} N(r,s).$$

For, if it is a left limit point of T with

$$\liminf_{t' \uparrow t} X_{t'}(\omega) \equiv \underline{X}_{t-0}(\omega) < \overline{X}_{t-0}(\omega) \equiv \limsup_{t' \uparrow t} X_{t'}(\omega)$$

then there are rationals r, s such that

$$\underline{X}_{t-0}(\omega) < r < s < \overline{X}_{t-0}(\omega)$$

so that, as $t' \uparrow t$, $X_t'(\omega)$ is less tnan r and greater than s infinitely often hence $\omega \in N(r,s)$. Thus, $\underline{X}_{t-0}(\omega) = \bar{X}_{t-0}(\omega)$ for $\omega \notin N$. Similarly for rightlimit points, and the lemma is proved.

Discussion. To avoid repeated mention of exceptions, *we include all null* events in every $\mathcal{B}_t$ (replacing the initial $\mathcal{B}_t$, if necessary, by the σ-field generated by it and the class of all null events. Thus, the above exceptional null event N belongs to every $\mathcal{B}_t$ and we either speak of *almost all sample functions or we replace $X_T(\omega)$ for $\omega \in N$ by the constant function zero*, according to convenience. The r.f. so obtained is a.s. equal to the initial one and the submartingale or martingale property is preserved. In what follows, to return to the initial one it suffices to add "for almost all sample functions."

Let T' be the set of (left or right or both) limit points of T so that the closure of $\overline{T}$ is $T = T \cup T' = TT'^c + T'$; set *once and for all* $a = \min T$ and $b = \max T$. To avoid further exceptions, note that the statement of the following theorem is trivial when a (b) belongs to T but not to T', provided we set $X_{a+0} = X_a$ ($X_{b-0} = X_b$). Thus, we can assume therein that $a(b)$ is a limit point.

C. SAMPLE LIMITS THEOREM. *Let $(X_T, \mathcal{B}_T)$ be a submartingale.*

I. *Let X_T^+ be integrable.*

(i) *At leftlimit points $t < b$ or ($t \leqq b$ when $b \in T$ and then X_T^+ is uniformly integrable), sample leftlimits X_{t-0} exist, are $\mathcal{B}_{t-0}$-measurable and, for $s \leqq t < u$ with $s, u \in T$,*

$$X_s \leqq E(X_{t-0}|\mathcal{B}_s) \text{ a.s.}, \; X_{t-0} \leqq E(X_u|\mathcal{B}_{t-0}) \text{ a.s.} \tag{1}$$

X_T^+ is uniformly integrable if and only if X_{b-0} exists and closes on the right the submartingale $(X_t, \mathcal{B}_t, t < b)$.

(ii) *At rightlimit points t, sample rightlimits X_{t-0} exist, are $\mathcal{B}_{t+0}$-measurable and, for $s \leqq t < u$ with $s, u \in T$,*

$$X_s \leqq E(X_{t+0}|\,\mathcal{B}_s) \text{ a.s.}, \quad X_t \leqq E(X_u|\,\mathcal{B}_{t+0}) \text{ a.s.} \tag{2}$$

the submartingale is closed on the left by X_{a+0} or by X_a when $a \in T$.

II. *Let X_T be integrable.*

(i) *Inequalities* (1) *and* (2) *hold*, X_{a+0} *is also a limit in the first mean if and only if* $\lim\limits_{t \downarrow a} EX_t > -\infty$; *then all sample rightlimits are also limits in the first mean.*

In the martingale case, inequalities (1) *and* (2) *become equalities, the if and only if condition is satisfied and its consequences hold.*

Furthermore, integrable X_{b-0} *exists and closes on the right the martingale* $(X_t, \mathcal{B}_t, t < b)$ *if and only if* X_T *is uniformly integrable; then all sample left and right limits are also limits in the first mean and inequalities* (1) *with* $t \leqq b$, *and inequalities* (2) *become equalities.*

(ii) *If* $|X_T|^r$ *is uniformly integrable for some* $r \geqq 1$ *or* $\sup_t E|X_t|^r < \infty$ *for some* $r > 1$ *when* $X_T \geqq 0$ *(when* $(X_T, \mathcal{B}_T)$ *is a martingale), then all sample left and right limits are also limits in the* r-th *mean and inequalities (equalities)* (1) *with* $t \leqq b$ *and inequalities (equalities)* (2) *hold.*

Proof. 1°. *We prove* I; *by hypothesis* X_T^+ *is integrable.*

For every leftlimit point $t < b \in T'$ there is $u \in T$ with $t < u < b$ since b is a leftlimit point. Thus, the submartingale $(X_s^+, \mathcal{B}_s, s \leqq u)$ is closed X_u^+ hence $\sup\limits_{s \leqq u} EX_s^+ = EX_u^+ < \infty$ and, by **d**, X_{t-0} exists; since all X_s are $\mathcal{B}_{s-0}$ measurable for $s < t$, so is X_{s-0}. By **B**, $(X_s \vee c, s \leqq u)$ is uniformly integrable hence, for $s < t' < t$ and $B_s \in \mathcal{B}_s$, as $t' \uparrow t$,

$$\int_{B_s} (X_s \vee c) \leqq \int_{B_s} (X_{t'} \vee c) \rightarrow \int_{B_s} (X_{t-0} \vee c).$$

By the Fatou-Lebesgue theorem, X_{t-0}^+ is integrable since the X_s^+ are integrable. Hence so is $X_{t-0} \vee c$ and, letting $c \downarrow -\infty$,

$$\int_{B_s} X_s \leqq \int_{B_s} X_{t-0}.$$

Similarly, for $t \leqq u \in T$ and $B \in \bigcup\limits_{s<t} \mathcal{B}_s \subset \bigvee\limits_{s<t} \mathcal{B}_s = \mathcal{B}_{t-0}$,

$$\int_B (X_{t-0} \vee c) \leftarrow \int_B (X_{t'} \vee c) \leqq \int_B (X_u \vee c),$$

and, by **a**, this inequality holds for every $B_{t-0} \in \mathcal{B}_{t-0}$ hence, letting $c \downarrow -\infty$,

$$\int_{B_{t-0}} X_{t-0} \leqq \int_{B_{t-0}} X_u.$$

Thus, inequalities (1) and, proceeding similarly, inequalities (2) obtain.

Finally, if X_T^+ is uniformly integrable so that $\sup_t EX^+ < \infty$, **c** applies and X_{b-0} exists. Thus, the above proof for leftlimit points holds for $t = b$ and $(X_t, \mathcal{B}_t, t < b)$ is closed on the right by X_{b-0} with $EX_{b-0}^+ < \infty$. The converse obtains by **b** applied to $(X_t^+, \mathcal{B}_t, t < b)$ closed by X_{b-0}^+.

2°. *We prove* II; *by hypothesis X_T is integrable.* We use I without further comment.

Let $t < u$ with $t, u \in T$ so that $EX_t \leqq EX_u < +\infty$ hence $\alpha = \lim_{t \downarrow a} EX_t < +\infty$. Thus, $\alpha > -\infty$ is equivalent to: α is finite. But then

$$E|X_t| \leqq 2EX_t^+ - EX_t \leqq 2EX_u^+ - \alpha < \infty$$

so that $\beta = \sup_{t \leqq u} E|X_t| < \infty$. Upon setting $B_t = [|X_t| > c]$, it follows that, as $c \to +\infty$,

$$PB_t \leqq E|X_t|/c \leqq \beta/c \to 0 \text{ uniformly in } t \leqq u.$$

But, given $\epsilon < 0$, there is $v \leqq u$ with $v \in T$ such that $EX_t - \alpha < \epsilon$ hence $EX_v - EX_t < \epsilon$ for $t \leqq v$. Therefore,

$$\int_{B_t} |X_t| = 2\int_{B_t} X_t^+ + \int_{B_t^c} X_t - EX_t \leqq 2\int_{B_t} X_v^+ + \int_{B_t^c} X_v - EX_t$$

$$\leqq \int_{B_t} |X_v| + EX_v - EX_t < \int_{B_t} |X_v| + \epsilon$$

so that, letting $c \to +\infty$ then $\epsilon \to 0$,

$$\sup_{t \leqq v} \int_{B_t} |X_t| \leqq \int_{B_t} |X_v| + \epsilon \to 0,$$

Thus, the X_t are uniformly integrable for $a < t \leqq v$ and $X_t \xrightarrow{1} X_{a+0}$ as $t \downarrow a$ hence $EX_t \to EX_{a+0}$ finite. Conversely, if $X_t \xrightarrow{1} X_{a+0}$ as $t \downarrow a$ then $\alpha = EX_{a+0}$ is finite so that the condition $\alpha > -\infty$ is equivalent to: there is a $v \in T$ such that the X_t are uniformly integrable for $a < t \leqq v$.

In fact, the preceding argument applies to every rightlimit point $s > a$ since a is a limit point so that there is $u \in T$ with $s > u > a$, hence

$$\alpha_s = \lim_{t \downarrow s} EX_t \geqq EX_u,$$

and the first part of (i) obtains.

For the martingale part of (i), let $(X_T, \mathcal{B}_T)$ be a martingale. Since all EX_t are equal, $\lim_{t \downarrow a} EX_t$ is finite and the first part applies. Furthermore, $(X_t, t \leqq u \in T)$ being uniformly integrable since X_u integrable closes it on the right, sample leftlimits are also limits in the first mean for all leftlimit points $t < b$ for, b being a limit point for every $t < b$, there is $u \in T$ with $t \leqq u < b$. Thus, inequalities (1) become equalities. In fact, these equalities (1) hold for $t \leqq b$ when X_T is uniformly integrable so that $X_s \xrightarrow{1} X_t$ as $s \uparrow t \leqq b$. Conversely, if integrable X_{b-0} exists and closes the martingale $(X_t, \mathcal{B}_t, t < b)$ then, clearly, X_T is uniformly integrable, and the martingale part obtains.

We prove (ii). The case of $|X_T|^r$ uniformly integrable for some $r \geqq 1$ is immediate. The case of $\sup_t E|X_t|^r < \infty$ for some $r > 1$ when $X_T \geqq 0$ or $(X_T, \mathcal{B}_T)$ is a martingale follows, by **A**, from the fact that $(|X_T|^r, \mathcal{B}_T)$ is a submartingale with, by **B**, $E(\sup_t |X_t|)^r > \infty$.

The proof is terminated.

Consequences. We use **C** without further comment.

1°. EXTENSION. *Let $(X_T, \mathcal{B}_T)$ be a positively integrable submartingale (an integrable submartingale). Then it can be extended to a submartingale (a martingale) on the interval J with endpoints $a < b$ provided b is excluded unless $b \in T$ and a is excluded unless* $\lim_{t \downarrow a} EX_t > -a$.

For every rightlimit point $t \in J - T$ set $X_t = X_{t+0}$ and $\mathcal{B}_t = \mathcal{B}_{t+0}$ so that this extension is a submartingale (or a martingale). Then, for every $[c, d]$ whose endpoints only belong to $\overline{T}$, set $X_t = X_d$, $\mathcal{B}_t = \mathcal{B}_d$ for $c < t < d$, and also for $t = c$ unless X_c is already defined.

The *continuous parameter* case is that of intervals T on $\overline{R}$ with endpoints $a < b$ which may or may not belong to T. According to 1°, this can always be achieved under minimal usual assumptions. Then

2°. RIGHT REGULARIZATION. *Let $(X_T, \mathcal{B}_T)$ with continuous parameter be a positively integrable submartingale (an integrable martingale). Set $X_{b+0} = X_b$ when $b \in T$ and $X_t' = X_{t+0}$, $\mathcal{B}_t' = \mathcal{B}_{t+0}$ for every $t \in T$.*

$(X_T', \mathcal{B}_T')$ is a submartingale (martingale) which is sample rightcontinuous with sample leftlimits.

If X_T is rightcontinuous in pr., or EX_T and $\mathcal{B}_T$ are rightcontinuous, then the r.f.'s X'_T and X_T are equivalent.

The first assertion is immediate. If X_T is sample rightcontinuous then, for every $t \in T$, $X_{t_n} \xrightarrow{P} X_t$ as $t_n \downarrow t$ hence $X_{t_{n'}} \xrightarrow{a.s.} X_t$ as $t_{n'} \downarrow t$ along some subsequence $(t_{n'})$ of (t_n). Since $X'_t = X_{t+0} = \lim_{t' \downarrow t} X_{t'}$, it follows that $X_t' = X_t$ a.s., and X_T' is equivalent to X_T.

If $\mathcal{B}_T$ is rightcontinuous ($\mathcal{B}_{t+0} = \mathcal{B}_t$ for every $t \in T$), then X_{t+0} is $\mathcal{B}_t$-measurable, hence $X_t = E(X_{t+0} \mid \mathcal{B}_t) = X_{t+0}$ a.s. Thus, if moreover EX_T is rightcontinuous ($EX_u \to EX_t$ as $u \downarrow t$ for every $t \in T$) then, for $t < u$,

$$X_t \leqq X_{t+0} \leqq E(X_u \mid \mathcal{B}_t) \text{ a.s.}$$

hence $X_t = X_{t+0}$ a.s. and X'_T is equivalent to X_T.

The *discrete parameter* case will be that of intervals T in the extended line of integers $\{-\infty, \cdots, -1, 0, 1, \cdots, +\infty\}$; leaving out finite sequences which have no limit point, there are two-sided sequences $(\cdots, -n, \cdots, -1, 0, 1, \cdots, +n, \cdots)$ with two (left and right) limit points $-\infty$ and $+\infty$ and one-sided sequences either beginning or ending at some integer and, without loss of generality, we can take the integer to be 0. Thus, we have $(0, 1, \cdots)$ with one (left) limit point $+\infty$ or $(\cdots, -1, 0)$ with one (right) limit point $-\infty$; they may or may not belong to T. If $(X_n, \mathcal{B}_n, n = (\cdots, -1, 0))$ is a submartingale or a martingale then, setting $Y_n = X_{-n}$ and $\mathcal{C}_n = \mathcal{B}_{-n}$, we have a submartingale or martingale *reversed sequence* $(Y_n, \mathcal{C}_n, n = 0, 1, \cdots)$ with $\mathcal{C}_n \downarrow$ as n.

The reader is invited to restate the preceding propositions in the discrete case and compare with 32.2 and 32.3.

The foregoing discussion following **C** leads us to consider *from now on* two types of martingales and submartingales and two minimal kinds of integrability.

The two types are discrete and rightcontinuous $(X_T, \mathcal{B}_T)$; we say that $(X_T, \mathcal{B}_T)$ is *rightcontinuous* if $T \subset \bar{R}$ is an interval with endpoints $a < b$ and X_T is sample rightcontinuous.

The two minimal integrability conditions are positive integrability for submartingales (X_T^+ integrable) and integrability for martingales (X_T is integrable); positive integrability for submartingales implies negative integrability for supermartingales hence integrability for martingales.

39.2 Martingale times and stopping. In the preceding subsection it was shown that submartingale or martingale property of $(X_T, \mathcal{B}_T)$ is preserved by a family of transformations of X_T and, under some integrability conditions, by sample left or right passages to the limit. In this subsection it will be shown that this property is preserved, under similar integrability conditions, by "randomizing" the times—replacing times $t \in T$ by martingale times τ. Recall, to be used without further comment that, by 38.4**B**, X_τ are $\mathfrak{B}_\tau$-measurable in the discrete and in the rightcontinuous cases—which are the cases investigated herein.

The approach consists in extending the central lemma and then using a martingale time τ as *stopping time* $(\tau \wedge t, t \in T)$, so called because

$$X_{\tau\wedge t} = X_t I_{[t<\tau]} = X_\tau I_{[t\geqq\tau]}$$

yields r.f. $(X_t, t \in T)$ *the stopped at time* τ.

Lemma 39.1**c** is still central. It has two characteristics: Its martingale times are simple times and the submartingale is closed on the right. Taking these into account, we reformulate the lemma for the general setup $(X_T, \mathfrak{B}_T)$ with the "usual" integrability assumptions: *Positive integrability*, that is, X_T^+ integrable, for submartingales and *integrability*, that is, X_T integrable, for martingales.

a. EXTENDED CENTRAL LEMMA. *Let* $\sigma \leqq \tau$ *be times of a positively integrable submartingale (an integrable martingale)* $(X_T, \mathfrak{B}_T)$.

I. *Let* $\sigma \leqq \tau$ *be simple times. Then* $(X_\sigma, \mathfrak{B}_\sigma)$, $(X_\tau, \mathfrak{B}_\tau)$ *form a positively integrable submartingale (an integrable martingale).*

II. *Let* $(X_T, \mathfrak{B}_T)$ *be closed on the right by* X_b.

(i) *The r.v.'s* $X_\tau \vee c$ *(the r.v.'s* X_τ*) where* τ *varies over all simple times, in fact, over all* $\tau \leqq b$ *such that* $(X_\tau, \mathfrak{B}_\tau)$, $(X_b, \mathfrak{B}_b)$ *form a submartingale (martingale) are uniformly integrable.*

(ii) *If* $\sigma \leqq \tau \leqq b$ *are times of rightcontinuous* $(X_T, \mathfrak{B}_T)$, *the foregoing conclusions hold.*

Proof. 1°. Assertion I obtains upon replacing $X_1, \cdots, X_n$ in the central lemma by $X_{t_1}, \cdots, X_{t_n}$ with $t_1 < \cdots < t_n$ being all the possible values of σ and τ. Assertion I(i) obtains exactly as for t in lieu of τ in 39.1**B**; we repeat it for completeness: In the submartingale case, it suffices to prove uniform integrability when $c = 0$, that is, for the X_τ^+. This results, upon letting $c \to +\infty$, from

$$cP(X_\tau > c) \leqq \int_{[X_\tau^+ > c]} X_\tau^+ \leqq \int_{[X_\tau^+ > c]} X_b^+ \leqq EX_b^+ < \infty$$

hence

$$\sup_\tau P(X_\tau^+ > c) \leqq EX_b^+/c \to 0,$$

by

$$\sup_\tau \int_{[X_\tau^+ > c]} X_\tau^+ \leqq \sup_\tau \int_{[X_\tau^+ > c]} X_b^+ \to 0.$$

The martingale case follows from the fact that then $(|X_T|, \mathcal{B}_T)$ is a martingale; the simple times case obtains from I by taking $\tau \leqq \tau' = b$.

2°. It remains to prove II(ii). T is an interval with end points $a < b \in T$. Without restricting the generality, we can take $a = 0$, $b = 1$.
(For, we use only the linear order structure of the time set $T \subset \bar{R}$, and $\bar{R}$ is order isomorphic to $[-1, +1]$ by $t \to \frac{2}{\pi}$ Arctan t so that T becomes its subinterval with endpoints $a' < b'$ which is order isomorphic to the interval with endpoints 0 and 1 by $t \to (t - a')/(b' - a')$.
The argument parallels the one for **36.1C**. By **38.4A** there are sequences of simple times $\sigma_n \leqq \tau_n$ with $\sigma_n \downarrow \sigma$, $\tau_n \downarrow \tau$ so that, by rightcontinuity, $X_{\sigma_n} \to X_\sigma$, $X_{\tau_n} \to X_\tau$ and, by II(i), the $X_{\sigma_n} \vee c$ and the $X_{\tau_n} \vee c$ are uniformly integrable. Since $\mathcal{B}_{\sigma_n} \supset \mathcal{B}_\sigma$, by I, for $B_\sigma \in \mathcal{B}_\sigma$,

$$\int_{B_\sigma} (X_{\sigma_n} \vee c) \leqq \int_{B_\sigma} (X_{\tau_n} \vee c)$$

and, upon letting $n \to \infty$ then $c \downarrow -\infty$,

$$\int_{B_\sigma} X_\sigma \leqq \int_{B_\sigma} X_\tau$$

obtains so that, X_σ being $\mathcal{B}_\sigma$-measurable according to **38.4B**, $(X_\sigma, \mathcal{B}_\sigma)$, $(X_\tau, \mathcal{B}_\tau)$ is a submartingale. Upon taking $\tau \leqq \tau' = b$, II(i) applies.

In the martingale case, the above inequalities with $X_{\sigma_n} \vee c$, $X_{\tau_n} \vee c$ become equalities with X_{σ_n}, X_{τ_n} instead and lead to final equality. The proof is terminated.

A. Martingale times theorem. *Let $\sigma \leqq \tau < b$ be times of a positively integrable submartingale (an integrable martingale) $(X_T, \mathcal{B}_T)$, either discrete with $T = (0, 1, \cdots)$ and $b = +\infty$ or rightcontinuous with $T =$* (a, b) or $[a, b)$ in $\bar{R}$.

(i) *If*

$$\liminf_{t \to b} \int_{[\tau > t]} X_t^+ = 0 \quad \left(\liminf_{t \to b} \int_{[\tau > t]} |X_t| = 0\right)$$

then $(X_\sigma, \mathcal{B}_\sigma)$, $(X_\tau, \mathcal{B}_\tau)$ form a positively integrable submartingale (an integrable martingale).

(ii) *For all $\tau \leqq b$ when $X_T^+(X_T)$ is uniformly integrable also for all $\tau \leqq u \in T$, the foregoing conclusion holds and the $X_\tau^+ \vee c$ (the X_τ) are uniformly integrable.*

Note that $(X_T, \mathcal{B}_T)$ is closed on the right by a positively integrable (an integrable) r.v. if and only if $X_T^+(X_T)$ is uniformly integrable.

Proof. 1°. We prove (i). Let $\sigma_t = \sigma \wedge t, \tau_t = \tau \wedge t, t \in T$, so that $\sigma_t \leqq \tau_t \leqq t$ are stopping times of the submartingale (the martingale) $(X_s, \mathcal{B}_s, s \leqq t)$. Since

$$B_\sigma[\sigma \leqq t] \in \mathcal{B}_{\sigma_t} \text{ for } B_\sigma \in \mathcal{B}_\sigma \text{ and } [\sigma \leqq t][\tau \leqq t] = [\tau \leqq t],$$

(1) $$\int_{B_\sigma[\sigma \leqq t]} X_\sigma \leqq \int_{B_\sigma[\tau \leqq t]} X_\tau + \int_{B_\sigma[\sigma \leqq t][\tau > t]} X_t$$

obtains from **a**II. The last integral is bounded from above by $\int_{[\tau > t]} X_n^+$ hence, by hypothesis, upon letting $t \to b$ along a suitable sequence, (1) yields

(2) $$\int_{B_\sigma} X_\sigma \leqq \int_{B_\tau} X_\tau.$$

In the martingale case (1) is an equality and the last integral being bounded by $\int_{[\tau > t]} |X_t|$, by hypothesis, upon letting $t \to b$ along a suitable sequence, equality in (2) obtains.

2°. It remains to prove (ii). If $b \in T$ or all $\tau \leqq u \in T$, (ii) obtains from **a**II. If $b \notin T$ and $X_T^+(X_T)$ is uniformly integrable then, by 39.1**C** the sample limit X_{b-0} at b exists and we can close the submartingale (the martingale) on the right by $X_b = X_{b-0}$ and, thus, have $b \in T$. The proof is terminated.

COROLLARY. *Stopping at $\mathcal{B}_T$-time τ of a positively integrable submartingale or an integrable martingale $(X_T, \mathcal{B}_T)$ preserves these properties.*

REMARK. Since the martingale or submartingale property is in terms of ordered pairs, whenever there is a martingale or submartingale assertion with $\sigma \leq t$ then, under the same hypotheses, they can be replaced by pairs τ_{u_1}, τ_{u_2} with $u_1 < u_2$ from times $(\tau_u, u \in \overline{R})$ with τ_u nondecreasing with u increasing.

We are now ready for P. Lévy's martingale characterization of Brownian motion (see Chapter XIII).
$(X_T, \mathcal{B}_T)$ with $T = [0, \infty)$ and $X_0 = 0$ is a *Brownian motion* if it is sample continuous (right continuous at 0) and decomposable, that is, increments X_{st} on disjoint intervals $[s, t)$ are independent, with $\mathcal{L}(X_{st}) = \mathfrak{N}(0, t - s)$. Thus, for $s < t$, it is square integrable,

$$E(X_t|\,\mathcal{B}_s) = E(X_t - X_s|\,\mathcal{B}_s) + X_s = X_s \quad \text{a.s.}$$

so that $(X_T, \mathcal{B}_T)$ is a martingale, and

$$E(X_{st}^2|\,\mathcal{B}_s) = EX_{st}^2 = t - s \quad \text{a.s.}$$

so that these conditional variances of the increments degenerate into variances. We prove the converse:

B. *If* $(X_T, \mathcal{B}_T)$ *with* $T = [0, \infty)$ *and* $X_0 = 0$ *is a square integrable sample continuous martingale with*

$$E(X_{st}^2|\,\mathcal{B}_s) = t - s \quad \text{a.s.,}$$

then $(X_T, \mathcal{B}_T)$ is a Brownian motion.

Proof. It suffices to prove the assertion for every interval $[a, b] \subset [0, \infty)$. In fact, it suffices to show that $\mathcal{L}(X_b - X_a) = \mathfrak{N}(0, b - a)$. For, the same argument applies, with obvious modifications, to every fixed linear combination and then normal increments on disjoint intervals being orthogonal, will be independent.
Without loss of generality, we can take $a = 0$, $b = 1$ to simplify the notation and, to avoid repetitions in what follows, r, s, t will belong to $[0, 1]$.

Given $\epsilon > 0$, let $\tau_n(\omega)$ be the first t, if any, for which $\max\{|X_r(\omega) - X_s(\omega)| = \epsilon\}$—the maximum being taken over all $0 \leqq r,\ s \leqq t$ *with* $|r - s| \leqq 1/n$; if there is no such t let $\tau_n(\omega) = 1$. By sample continuity τ_n is time of the martingale $(X_t, \mathcal{B}_t, 0 \leqq t \leqq 1)$ since given $0 < t < 1$, for all rationals r, s with $0 \leqq r, s \leqq t$, $|r - s| \leqq 1/n$ and integers $m = 1, 2, \cdots$

$$[\tau_n > t] = \bigcup_m \bigcap_{r,s} [|\,X_r - X_s\,| \leqq \epsilon - 1/m] \in \mathcal{B}_t;$$

also τ_n depends upon the given ϵ, $0 < \tau_n \leqq 1$ and $\tau_n \wedge 1 = \tau_n \to 1$ as $n \to \infty$.
The preceding Corollary applied to the integrable martingale $(X_t, \mathcal{B}_t, 0 \leqq t \leqq 1)$ and submartingale $(X_t^2, \mathcal{B}_t, 0 \leqq t \leqq 1)$ yields the sample continuous martingale $(X_{\tau_n \wedge t}, \mathcal{B}_{\tau_n \wedge t}, 0 \leqq t \leqq 1)$ and $EX_{\tau_n \wedge t} = 0$, $EX_{\tau_n \wedge t}^2 \leqq EX_1^2 = 1$.

Partition $[0, 1]$ into n equal subintervals and denote by $Y_{nk} = X_{\tau_n \wedge k/n} - X_{\tau_n \wedge (k-1)/n}$ the increments of $X_{\tau_n \wedge t}$ on the k-th interval, $k = 1, \ldots, n$. It follows that $X_{\tau_n} = \sum_k Y_{nk}$ and $|\,Y_{nk}\,| \leqq \epsilon$ with

$$EY_{nk} = E(Y_{nk}|\,Y_{nj}, j < k) = 0 \text{ a.s.}$$

$$EY_{nk}^2 = E(Y_{nk}^2|\,Y_{nj}, j < k) \leqq 1/n \text{ a.s.}$$

Therefore, as $\epsilon \to 0$, if $n = n(\epsilon) \to \infty$, then

$$\sum_k E|Y_{nk}|^3 \leqq \epsilon \sum_k EY_{nk}^2 \leqq \epsilon \to 0$$

and, by 31.3**A,** $\mathfrak{L}(X_{\tau_n})$ converges to a normal law. Since, given $\epsilon > 0$, by sample continuity, $X_{\tau_n} \to X_1$, there is a sequence $n = n(\epsilon) \to \infty$ as $\epsilon \to 0$ such that $\mathfrak{L}(X_{\tau_n}) \to \mathfrak{L}(X_1)$ with $EX_1 = 0$, $EX_1^2 = 1$. Therefore $\mathfrak{L}(X_1) = \mathfrak{N}(0, 1)$, and the proof is terminated.

§ 40. DECOMPOSABILITY

Decomposability sprang forth fully armed from the forehead of P. Lévy (1934). His analysis of "integrals with independent elements" or "r.f.'s with independent increments" or "additive r.f.'s" or "differential r.f.'s" or "P. Lévy r.f.'s" or, as we shall call them, "decomposable r.f.'s" was so complete that since then only improvements of detail have been added. Before his work there were only pioneering ones by de Finetti (1929) and Kolmogorov (1932). The only decomposable r.f.'s known were the Poisson and the Brownian processes, both born from physical phenomena. (After a pioneering work by Bachelier (1900), the first rigorous study of the Brownian process was by Wiener (1923) who discovered its a.s. sample continuity.) Furthermore, the basic concepts and problems of random analysis appeared in and were born from the P. Lévy analysis of decomposability. Thus, decomposability is at the root of the concepts and problems of random analysis.

40.1. Generalities. A r.f. X_T is said to be *decomposable* if its increments $X_{st} = X_t - X_s$ for disjoint intervals $[s, t)$ are independent. A *decomposable process* on T is the family of decomposable r.f.'s (on some pr. space) with the same increments. But there is a one-to-one correspondence between the increment function on T—an additive r.f. X_{st} of intervals whose endpoints belong to T—and the family of r.f.'s on T defined by $X_t = X_a + X_{at}$ or $X_a - X_{ta}$ according as $t \geqq a$ or $t < a$, where $a \in T$ and X_a are selected arbitrarily. Therefore, we may consider a decomposable process on T as a decomposable r.f. on T defined up to an arbitrarily selected value X_a at an arbitrarily selected point $a \in T$. Thus, a decomposable process will be *represented* by one of its r.f.'s X_T either an unspecified one or selected according to convenience, but *always separable.*

The law of a decomposable process determined by the joint laws of all its finite sections is, in fact, determined by the individual laws of its increments. For, because of decomposability, setting $f_{st}(u) = E \exp \{iuX_{st}\}$, the joint law of, say, X_{ac} and X_{bd} with $a < b < c < d$ is given by

$$E \exp \{iu_1X_{ac} + iu_2X_{bd}\} = E \exp \{iu_1X_{ab} + i(u_1 + u_2)X_{bc} + iu_2X_{cd}\}$$
$$= f_{ab}(u_1)f_{bc}(u_1 + u_2)f_{cd}(u_2).$$

Also, if X_T is one of the r.f.'s of a decomposable process then, upon setting $f_s(u) = E \exp \{iuX_s\}$ and $v_k = u_k + \cdots + u_n$, we have $f_t = f_sf_{st}$ and, for $t_1 < \cdots t_n$,

$$E \exp \{iu_1X_{t_1} + \cdots + iu_nX_{t_n}\}$$
$$= E \exp \{iv_1X_{t_1} + iv_2X_{t_1t_2} + \cdots + iv_nX_{t_{n-1}t_n}\}$$
$$= f_{t_1}(v_1)f_{t_1t_2}(v_2 \cdots f_{t_{n-1}t_n}(v_n).$$

Therefore, if the individual laws of all increments are of a specific type, say, normal or Poisson or infinitely decomposable, we shall say that the process is of this type: *normal decomposable* or *Poisson decomposable* or *infinitely decomposable*. In fact, we shall find that deleting the fixed discontinuities the remaining part of any decomposable process is infinitely decomposable. It is by a deep sample analysis of this remaining part that P. Lévy discovered the general form of infinitely decomposable laws with ch.f.'s e^{ψ}, $\psi = (\alpha, \Psi)$. Since we have at our disposal this general form (§22), we shall proceed from it to the sample interpretation. In order to do so, it will prove convenient to write ψ in P. Lévy's form $\psi = (\alpha, \beta^2, L)$, explicitly

$$\psi(u) = i\alpha u - \frac{\beta^2}{2}u^2 + \int_{-\infty}^{+\infty} \left(e^{iux} - 1 - \frac{iux}{1 + x^2}\right) dL(x)$$

where the bar which crosses the integral excludes the origin from the domain of integration and where we can and do take $L(-\infty) = L(+\infty) = 0$. The correspondence between the two forms of ψ is given by

$$\beta^2 = \Psi(+0) - \Psi(-0), \quad dL(x) = \frac{1 + x^2}{x^2} d\Psi(x) \text{ for } x \neq 0$$

and

$$\operatorname{Var} \Psi < \infty \Leftrightarrow \int_{-1}^{+1} x^2 \, dL(x) < \infty.$$

In order to reach the infinitely decomposable part of the process, we shall have to delete its degenerate discontinuities by "centering" it, then delete the "fixed discontinuities" part. This will require recalling (Sections 16 and 17) and adding to convergence properties of series of independent summands, as follows.

Let $X_1, X_2, \cdots$ be a sequence of independent r.v.'s with ch.f.'s $f_1, f_2, \cdots$. The series $\sum X_n$ is said to be *convergent* if it converges a.s. to a r.v., equivalently, if $\prod_{k=1}^{n} f_k \to f$ ch.f., or if $\prod_n f_n > 0$ on an argument set of positive Lebesgue measure. The series $\sum X_n$ is said to be *essentially convergent* if there exist centering constants c_n such that the series $\sum (X_n - c_n)$ is convergent, equivalently, if the series $\sum X_n^s$ of symmetrized summands is convergent, or if $\prod |f_n|^2 > 0$ on an argument set of positive Lebesgue measure. Clearly, if c_n are centering constants then c'_n are also centering constants if and only if the series $\sum (c_n - c'_n)$ converges (to a finite limit).

a. *If the series $\sum X_n$ is essentially convergent, then the constants $c_n = s_n - s_{n-1}(s_0 = 0)$, determined by the relation E Arctan $(S_n - s_n) = 0$ where $S_n = X_1 + \cdots + X_n$, are centering constants.*

Note that since Arctan is a bounded increasing and continuous function, the stated relation determines the (finite) constant s_n.

Proof. By hypothesis, there exists a sequence s'_n such that $S_n - s'_n \xrightarrow{\text{a.s.}} S$ r.v. Therefore, for every subsequence n' such that $s_{n'} - s'_{n'} \to s$ finite or not,

$$S_{n'} - s_{n'} = (S_{n'} - s'_{n'}) - (s_{n'} - s'_{n'}) \xrightarrow{\text{a.s.}} S - s$$

and, by the dominated convergence theorem,

$$E \operatorname{Arctan} (S - s) = \lim_{n'} E \operatorname{Arctan} (S_{n'} - s_{n'}) = 0.$$

Thus, the constant s is finite and independent of the subsequence n' so that the whole sequence $s_n - s'_n \to s$, and $S_n - s_n \xrightarrow{\text{a.s.}} S - s$ r.v. The assertions follow.

b. *If there exists a r.v. X such that the r.v.'s Y_n defined by $X_1 + \cdots + X_n + Y_n = X$ are independent of $X_1, \cdots, X_n$, then the series $\sum X_n$ is essentially convergent.*

Proof. Let $g_n = \prod_{k=1}^{n} f_k$ and let h_n and f be the ch.f.'s of Y_n and X.

By hypothesis and because ch.f.'s are continuous and bounded by $f(0) = 1$, $|g_n|^2 \geqq |g_n|^2 |h_n|^2 = |f|^2 > 0$ on some neighborhood of the origin. Since $|g|^2 = \lim |g_n|^2 = \prod |f_n|^2$ exists, it follows that $|g|^2 > 0$ on this neighborhood. The assertion follows.

The series $\sum X_n$ is *unconditionally convergent* if it is convergent under all reorderings. Constants $\bar{c}_n$ such that the series $\sum (X_n - \bar{c}_n)$ is unconditionally convergent are said to be *unconditionally centering*, and constants $\bar{c}'_n$ are so if and only if the series $\sum (\bar{c}_n - \bar{c}'_n)$ is absolutely convergent, since for numerical series unconditional and absolute convergence are the same. For example, if the second moments EX_n^2 are finite and $\sum \sigma^2 X_n < \infty$ then the series $\sum (X_n - EX_n)$ converges in q.m. hence is convergent. In fact, it is unconditionally convergent since so is the series $\sum \sigma^2 X_n$ (and the EX_n are unconditionally centering constants), while the series $\sum X_n$ itself is unconditionally convergent if and only if $\sum |EX_n| < \infty$.

c. *If the series $\sum X_n$ is essentially convergent, then the constants $\bar{c}_n = \mu X_n + E(X_n - \mu X_n)^c$ (where μX_n is a median of X_n and c is some truncating constant) are unconditionally centering.*

Proof. According to the two-series criterion, essential convergence of the series $\sum X_n$ is equivalent to convergence of the two series $\sum P[|X_n - \mu X_n| \geqq c]$ and $\sum \sigma^2 (X_n - \mu X_n)^c$, and then the series $\sum (X_n - \bar{c}_n)$ is convergent. Since the two series are convergent under all reorderings so is the series $\sum (X_n - \bar{c}_n)$, and the assertion is proved.

d. *The series $\sum X_n$ is unconditionally convergent if and only if the series $\sum |f_n - 1|$ converges on some argument set U of positive Lebesgue measure.*

Proof. If $\sum |f_n - 1| < \infty$ on some U hence converges on U under all reorderings, then so does $\prod f_n$ and the "if" assertion follows.

Conversely, if the series $\sum X_n$ is unconditionally convergent, then so is the series $\sum (X_n - \bar{c}_n)$ and $\sum |\bar{c}_n| < \infty$. Let $\bar{f}_n(u) = e^{-i\bar{c}_n u} f_n(u)$ so that, for $|u| \leqq b$,

$$|f_n(u) - 1| = |\bar{f}_n(u) - e^{-i\bar{c}_n u}| \leqq |\bar{f}_n(u) - 1| + b|\bar{c}_n|.$$

According to 22.2$\mathbf{B}_1$ and 22.2$\mathbf{B}_2$ where we take $\tau = c$ and center at a median μ—which does not change $|f|$, for $|u| \leqq b$ sufficiently small so that $\log |f_n(u)|$ exists and is finite, there exists a constant a such that

$$|\bar{f}_n(u) - 1| \leqq a \int_0^b |\log |f_n(v)|| \, dv.$$

But the series $\sum (X_n - \bar{c}_n)$ being convergent, we can take b sufficiently small so that $\prod |f_n(u)| > \frac{1}{2}$ for $|u| \leqq b$, and then

$$\sum |f_n(u) - 1| \leqq (a \log 2 + \sum |\bar{c}_n|)b < \infty.$$

The "only if" assertion is proved.

Let $\sum T_j$ be an arbitrary partition of the set of integers $(1, 2, \cdots)$. The series $\sum_j (\sum_{k \in T_j} X_k)$ is a "partitioned summation" of the series $\sum X_n$.

e. *If the series $\sum X_n$ is unconditionally convergent then it converges a.s. to a same limit r.v. under all reorderings and partitioned summations, and so does any of its subseries.*

Proof. If $\sum |f_n(u) - 1| < \infty$ then $\sum |f_{n'}(u) - 1| < \infty$ for any subsequence n' hence, by **d**, the subseries $\sum X_{n'}$ is unconditionally convergent.

The difference between the series $\sum X_n$ and the reordered series $\sum X'_n$ is defined on the independent r.v.'s X_n and is independent of $X_1, \cdots, X_n$ for any n. Therefore, by the zero-one law, this difference degenerates into a constant c. Since $\sum |f_n - 1|$ is absolutely convergent so that $\prod f_n = \prod f'_n$, it follows that $c = 0$. Similarly for partitioned summations because of the elementary inequality $\sum_j |\prod_{k \in T_j} f_k - 1| \leqq \sum_j \sum_{k \in T_j} |f_k - 1|$ valid for arbitrary complex numbers $f_k(u)$ bounded by one (proceed by induction).

40.2. Three parts decomposition. We separate decomposable processes into three parts: a numerical function, an interpolated series of fixed discontinuities and an a.s. continuous part. More precisely

A. THREE PARTS DECOMPOSITION THEOREM. *Every decomposable process is the sum*

$$X_T = x_T + X_T{}^d + X_T{}^c$$

of three independent parts (not necessarily all present):

(i) *A centering function x_T.*

(ii) *A fixed discontinuities part $X_T{}^d$, centered decomposable, with almost all sample functions continuous except at the countable fixed discontinuities set.*

(iii) *An a.s. continuous part $X_T{}^c$, centered decomposable, with almost all sample functions continuous except for countable sets of jumps.*

We proceed in steps and, first, we have to give a meaning to the term "centered."

Let a, b, be the extreme points of the closure $\overline{T}$ of T and let $\overline{T}_{ab}$ be this closure with a and/or b excluded unless they belong to T. We intend to show that we can delete from X_T its degenerate discontinuities, if any, by including them within a *centering function* x'_T so as to leave a *centered* decomposable process X'_T: a process such that for every $t \in \overline{T}_{ab}$ which is a left (right) limit point of T the a.s. limits $X_{t-0}(X_{t+0})$ exist and there are no degenerate discontinuities, that is, if $X_{t-0,t} = X_t - X_{t-0}(X_{t,t+0} = X_{t+0} - X_t)$ degenerates then it degenerates at zero. Note that

The fixed discontinuities set of a centered decomposable process is countable and $X_{t-0,t+0} = X_{t+0} - X_{t-0}$ degenerates at zero if and only if $X_{t-0,t}$ and $X_{t,t+0}$ degenerate at zero.

The first assertion results from the continuity extension theorem. The second assertion results from the fact that $X_{t-0,t+0} = X_{t-0,t} + X_{t,t+0}$ is the sum of two independent r.v.'s, since $f_{t-0,t+0} = f_{t-0,t}f_{t,t+0} = 1$ if and only if $f_{t-0,t}(u) = e^{-iuc}, f_{t,t+0}(u) = e^{+iuc}$ where $c = 0$ by definition of a centered process.

a. CENTERING LEMMA. *Let X_T be decomposable. There exist centering functions x'_T such that $X'_T = X_T - x'_T$ is centered decomposable. The fixed discontinuities set of X'_T is independent of the choice of the centering function and the set D_s of its points to the right of any $s \in T$ forms the discontinuity set of the function $d_{st} = \int_0^1 |f_{st}(u)|\, du, t > s$.*

Proof. Let x'_T be determined by the relation E Arctan $(X_T - x'_T) = 0$. If $s_n \uparrow t \in \overline{T}_{ab}$ then, from $t \leqq t' \in T$ and the equality $\sum_{k=1}^{n} (X_{s_{k+1}} - X_{s_k}) + (X_{t'} - X_{s_{n+1}}) = X_{t'} - X_{s_1}$ it follows, by 37.1**a** and **b**, that the sequence $X'_{s_{n+1}} = X'_{s_1} + \sum_{k=1}^{n} (X'_{s_{k+1}} - X'_{s_k})$ converges a.s. to a r.v. If $s'_n \to t - 0$ and $s''_n \to t - 0$ then both sequences can be reordered and combined into a sequence $s_n \uparrow t$ so that, because of separability, the one-sided a.s. limit r.v. X_{t-0} exists; similarly for X_{t+0}. Since E Arctan $X'_T = 0$ hence, by the dominated convergence theorem, E Arctan $X'_{t\pm 0} = 0$, it follows that all degenerate discontinuities and, in fact, all degenerate increments degenerate at zero. Thus the decomposable process X'_T is centered. Since any change of centering function changes any given discontinuity by a constant hence cannot reduce nondegenerate discontinuities to zero, it follows that the fixed discontinuities set of X'_T does not vary with the centering function. The first assertion is proved.

The relation $f_{s,t+h} = f_{st}f_{t,t+h}$ implies that $|f_{st}|$ hence d_{st} is nonincreasing in t. But, setting $f'_{st}(u) = E \exp\{iuX'_{st}\}$ we have $|f'_{t-0,t+0}(u)| = 1$ for $t \notin D_s$ and for all u, and $|f'_{t-0,t+0}(u)| < 1$ for $t \in D_s$ and an u-set of positive Lebesgue measure in $[0, 1]$. Therefore, from $|f'_{st}| = |f_{st}|$ it follows that d_{st} is discontinuous at t if and only if $t \in D_s$. The proof is terminated.

So far, T was an arbitrary set in $\bar{R}$. In fact, we can and whenever convenient we shall take as domain an interval in the interval I_{ab} whose endpoints a, b are those of $\bar{T}$ except that they are to be excluded from I_{ab} unless they belong to T. For

b. DECOMPOSABILITY EXTENSION LEMMA. *A decomposable process X_T can be extended on I_{ab} with preservation of decomposability, of centering, and of type provided the type is invariant under additions and passages to the limit.*

Proof. Let X'_T be the process after centering. Set $X_t = X'_{t+0}$ for right limit points $t \in I_{ab} - T$. The remaining points of $I_{ab} - T$ form intervals (c, d) or $[c, d)$ where the d are right limit points belonging or not to T. Set $X_t = X_d$ on such intervals. The process so extended has the asserted properties.

c. CENTERED SAMPLE LEMMA. *Almost all sample functions of a centered decomposable process X'_T are bounded on every set $[c, d]T$, $c, d \in T$, and are continuous except for countable sets of jumps outside the fixed discontinuities set.*

Proof. We can take T to be $[c, d]$, without restricting the generality.

1° Let μ_t be a median of $X'_d - X'_t$. If $s_n \uparrow t$ then every limit value of the sequence μ_{s_n} is a median of $X'_d - X'_{t-0}$, and similarly for $s_n \downarrow t$. Therefore, the function μ_t is bounded by some constant α. Since the decomposable process defined by $Y_t = X'_t - X'_c + \mu_t$ is such that $Y_d - Y_t$ has 0 for a median it follows by 17.1**c** that, for $C = s_0 < \cdots < s_n = d$,

$$P[\sup_{k \leq n} |Y_{s_k}| \geqq \beta] \leqq 2P[|Y_d| \geqq \beta]$$

hence

$$P[\sup_{k \leq n} |X'_{s_k} - X'_c| \geqq \alpha + \beta] \leqq 2P[|X'_d - X'_c| \geqq \beta]$$

and, by separability,

$$P[\sup_t |X'_t - X'_c| \geqq \alpha + \beta] \leqq 2P[|X'_d - X'_c| \geqq \beta],$$

which implies the boundedness assertion.

2° The second assertion will be proved if we show that one-sided limits exist for almost all sample functions. Let u vary over $[-\gamma, +\gamma]$ where γ is sufficiently small so that $f'_{cd}(u) \neq 0$. Since $f'_{ct}(u)f'_{td}(u) = f'_{cd}(u)$, the same is true of all $f'_{ct}(u)$. Then the $Z_t(u) = \exp\{iuX'_{ct}\}/f'_{ct}(u)$ are bounded (complex-valued) r.v.'s. Since a.s.

$$E(Z_t(u) \mid Z_r(u), r \leqq s) = \exp\{iX'_{cs}u\}f'_{st}(u)/f'_{cs}(u)f'_{st}(u) = Z_s(u),$$

the $Z_t(u)$ form a (complex-valued) martingale for every $u \in [-\gamma, +\gamma]$. But, the decomposable process X'_T being centered, the functions $f'_{ct}(u)$ of t have one-sided limits. Thus, to prove that almost all sample functions $X'_T(\omega)$ have one-sided limits, it suffices to show that the same is true of every martingale $Z_t(u)$. If these martingales are separable, this follows from 36.1**C**. However, the martingales may not be separable. But, X'_T being separable, it suffices to prove the assertion for its restriction X'_S on a separating set S and hence for the restrictions $Z_S(u)$ of the martingales on this countable set, and then what precedes applies. The proof is terminated.

For the next proposition, it will be convenient to think of T as an interval. Let $\{t_j\}$ be the fixed discontinuities in $[a, b]T$ of a centered decomposable process X'_T and let $U'_j = X'_{t_j-0,t_j}$, $V'_j = X'_{t_j,t_j+0}$; at most one of these one-sided jumps may degenerate and then it will be at zero. Let $U_j = U'_j - \bar{c}_j$ where $\bar{c}_j = \mu U'_j + E(U'_j - \mu U'_j)^c$ and let V_j be similarly defined in terms of V'_j. Since the one-sided jumps are all independent and, setting $\sum_{k=1}^{n} U'_k + Y_n = X'_b - X'_a$, lemma 37.1**b** applies, and the same is true for the V'_j, it follows, by 37.1**c** and **e**, that the series $\sum_{j\in I} U_j$ and $\sum_{j\in I} V_j$ are unconditionally convergent for every interval $I \subset T$ and converge to the same a.s. limit under all reorderings and partitioned summations. Thus, selecting arbitrarily $t_0 \in T$ and setting

$$X_t^d = \sum_{t_0\leqq t_j\leqq t} U_j + \sum_{t_0\leqq t_j<t} V_j \quad \text{or} \quad -\sum_{t<t_j<t_0} U_j - \sum_{t\leqq t_j<t_0} V_j$$

according as $t \geqq t_0$ or $t < t_0$, the X_t^d are a.s. determined and we can and do select them so that X_T^d be separable.

d. DECOMPOSITION LEMMA. *A centered decomposable process X'_T is the sum*

$$X'_T = x''_T + X_T^d + X_T^c$$

of three independent parts: a centering function x''_T, a centered decomposable process X_T^d corresponding to the fixed discontinuities of X'_T, and a centered decomposable process X_T^c with no fixed discontinuities.

Proof. If $s_n \uparrow t \notin \{t_j\}$, then a.s. $X_{s_{n+1}}{}^d = X_{s_1}{}^d + \sum_{k=1}^{n} (X_{s_{k+1}}{}^d - X_{s_k}{}^d)$, hence $X_{s_n}{}^d \xrightarrow{\text{a.s.}} X_t{}^d$. Similarly, if $s_n \uparrow t_j$, then $X_{s_n}{}^d \xrightarrow{\text{a.s.}} X_{t_j}{}^d - U_j$. In the same manner, if $s_n \downarrow t \notin \{t_j\}$, then $X_{s_n}{}^d \xrightarrow{\text{a.s.}} X_t{}^d$; and if $s_n \downarrow t_j$, then $X_{s_n}{}^d \xrightarrow{\text{a.s.}} X_{t_j}{}^d + V_j$. Thus $U_j = X_{t_j-0,t_j}{}^d$, $V_j = X_{t_j,t_j+0}{}^d$, $X_{t_j-0,t_j+0}{}^d = U_j + V_j$ nondegenerate, and $X_T{}^d$ is centered with fixed discontinuities set $\{t_j\}$. If x''_T is a centering function of $X'_T - X_T{}^d$ it follows that the process $X_T{}^c = X'_T - X_T{}^d - x''_T$ is centered decomposable and has no fixed discontinuity points. The proof is terminated.

e. DISCONTINUOUS PART SAMPLE LEMMA. *Almost all sample functions of the centered decomposable process $X_T{}^d$ are continuous outside the fixed discontinuities set $\{t_j\}$.*

Proof. We already know that for a centered decomposable process almost all sample functions are continuous except for countable sets of jumps outside the fixed discontinuities set $\{t_j\}$. Yet, this does not imply that they are continuous except at the t_j's.

To prove the assertion, we take $T = [a, b]$, denote by D the ω-set of sample functions with discontinuities outside $\{t_j\}$, and note that $\sup_t | X_t{}^d(\omega) | \geq \epsilon$ for those sample functions $X_T{}^d(\omega)$ which have a discontinuity outside the t_j at which their oscillation is at least 2ϵ. Furthermore, if a finite number n of t_j's with the corresponding U_j and V_j are deleted from T and from $X_T{}^d$, the ω-set D_ϵ of these sample functions remains the same. Thus, to prove the assertion, we may assume this deletion made (including a and V_a if necessary).

Since the series $\sum U_j$ and $\sum V_j$ are unconditionally convergent, it follows from the three series criterion that

$$\sum_{j>n} \sigma_j^2 = \sum_{j>n} \sigma^2(U_j^c + V_j^c) \to 0$$

and, for $n \geq n(\epsilon)$ sufficiently large,

$$\sum_{j>n} \{P[U_j \neq U_j^c] + P[V_j \neq V_j^c]\} < \epsilon, \quad \sum_{j>n} \{| EU_j^c) | + | EV_j^c|\} < \epsilon$$

Let Y_T be a process defined by means of the U_j^c, V_j^c as $X_T{}^d$ was defined by means of the U_j, V_j, with same deletion. According to Kolmogorov's inequality, if $a = s_0 < \cdots < s_m$ then

$$P[\sup_{j \leqq m} | Y_{s_j} - EY_{s_j} | > \epsilon] \leqq \sigma^2 Y_{s_m}/\epsilon^2 \leqq \sum_{j>n} \sigma_j{}^2/\epsilon^2$$

hence

$$P[\sup_{j \leqq m} | X_{s_j}{}^d | > 2\epsilon] \leqq \epsilon + \sum_{j>n} \sigma_j{}^2/\epsilon^2$$

and, by separability,

$$P[\sup_t | X_t{}^d | > 2\epsilon] \leqq \epsilon + \sum_{j>n} \sigma_j{}^2/\epsilon^2.$$

Since the bracketed condition defines an event which contains the set of sample functions with $\omega \in D_\epsilon$ and we can let $n \to \infty$, it follows that $D_\epsilon \subset A_\epsilon$ with $PA_\epsilon \leqq \epsilon$. If $\epsilon_n = \epsilon/2^n$, then $D_{\epsilon_n} \uparrow D = \bigcup D_{\epsilon_n} \subset \bigcup A_{\epsilon_n} = B$ with $PB \leqq \sum \epsilon_n = \epsilon$. Thus the set D is contained in events of pr. at most equal to $\epsilon > 0$ arbitrarily small and hence is contained in a null event. The proof is terminated.

Upon gathering the foregoing lemmas and setting $x_T = x'_T + x''_T$, the three-parts decomposition follows.

40.3. Infinite decomposability; normal and Poisson cases. We proceed now to the analysis of the a.s. continuous decomposable parts. To simplify the writing, unless otherwise stated, the domain T will be an interval $[a, b]$ and *the processes X_T will be represented by their r.f.'s determined by the condition $X_a = 0$*, so that $X_{at} = X_t$ and $f_{at} = f_t$. We shall also use the following notation: partitions of $[a, t]$ are given by $a = s_{n0} < \cdots < s_{nn} = t$ with $\max_k (s_{nk} - s_{n,k-1}) \to 0$ and the increments $X_{nk} = X_{s_{nk}} - X_{s_{n,k-1}}$ have for d.f.'s F_{nk} and for ch.f.'s f_{nk}.

a. CONTINUITY EQUIVALENCE LEMMA. *For a decomposable process X_T, $T = [a, b]$, continuity in law, in pr., a.s., are equivalent and are uniform, and imply that the process is centered.*

Proof. Since continuity a.s. implies continuity in pr. which implies continuity in law, the equivalence assertion will be proved by showing that continuity in law, that is, continuity of the ch.f.'s f_t in t implies a.s. continuity. Since the fixed discontinuities set of the centered process $X'_T = X_T - x_T$ is the discontinuities set of the function $d_t = \int_0^1 | f_t(u) | \, du$

and f_t is continuous in t, it is empty. Therefore X'_T is continuous a.s. hence in law, that is, the function $e^{-iux_t}f_t(u)$ as well as the function $f_t(u)$ are continuous in t. Thus the centering function is continuous, hence $X_T = x_T + X'_T$ has no fixed discontinuities and is centered. The uniformity assertion follows from 35.3**a**, and the proof is terminated.

Because of the above lemma, we drop "a.s." in "a.s. continuous decomposable process" and write *c.d. process*, for short.

We say that $\Psi_t(x)$ is continuous in t if $\Psi_{t'}(x) \to \Psi_t(x)$ as $t' \to t$ for every continuity point x of the function Ψ_t, $t \in T$.

A. CONTINUOUS DECOMPOSABILITY THEOREM. *Let X_T, $T = [a, b]$, be a separable c.d. process. Then*

(i) *X_T is infinitely decomposable and* $\log f_t = \psi_t = (\alpha_t, \beta_t^2, L_t) = (\alpha_t, \Psi_t)$ *with α_t continuous and $\Psi_t(x)$ continuous in t and nondecreasing in t and x.*

(ii) *Almost all sample functions of X_T are bounded and are continuous except for countable sets of jumps, and $|L_t(x)| = E\nu_t(x)$ is the expectation of the number $\nu_t(x)$ of jumps of these sample functions in $[a, t)$ of height less than $x < 0$ or at least equal to $x > 0$.*

Proof. 1° Since X_T is uniformly continuous in law, $f_{ss'} \to 1$ uniformly in s, s' as $s - s' \to 0$ so that, taking partitions of $[a, t]$, $X_t = \sum_{k=1}^{n} X_{nk}$ is sum of uan independent r.v.'s. Therefore, by the central limit theorem, X_T is infinitely decomposable and $\log f_t = \psi_t = (\alpha_t, \beta_t^2, L_t) = (\alpha_t, \Psi_t)$. Since $\log f_t$ is continuous in t, it follows from the convergence theorem 22.1**D** that α_t, $\Psi_t(x)$ are continuous in t. Since $\log f_{st} = \psi_{st} = (\alpha_{st}, \Psi_{st})$ with $\Psi_{st}(x)$ nonnegative and nondecreasing in x, the function $\Psi_t(x)$ is nondecreasing in t and x.

2° Let $x < 0$ be a continuity point of $L_t(x)$, take partitions of $[a, t]$ and set $\nu_t^{(n)}(x) = \sum_k I_{[X_{nk} < x]}$. Consider the (almost all) sample functions which are bounded and continuous except for countable sets of jumps, according to the centered sample lemma 37.2**c**.

Then $\nu_t^{(n)}(x) \xrightarrow{\text{a.s.}} \nu_t(x)$ while, by the central limit theorem,

$$E\nu_t^{(n)}(x) = \sum_{k=1}^{n} P[X_{nk} < x] = \sum_{k=1}^{n} F_{nk}(x) \to L_t(x)$$

and

$$\sigma^2(\nu_t^{(n)}(x)) = \sum_{k=1}^{n} F_{nk}(x) - \sum_{k=1}^{n} F_{nk}^2(x) \leq E\nu_t^{(n)}(x),$$

so that the sequence of second moments

$$E(\nu_t^{(n)}(x))^2 \leqq E\nu_t^{(n)}(x) + (E\nu_t^{(n)}(x))^2 \to L_t(x) + L_t^2(x)$$

is bounded. Therefore $E\nu_t(x) = \lim E\nu_t^{(n)}(x) = L_t(x)$, and the nondecreasing and continuous from the left functions $E\nu_t(x)$ and $L_t(x)$ coincide for all continuity points $x < 0$ of $L_t(x)$ hence for all $x < 0$. Similarly for $x > 0$, and the proof is terminated.

Brownian and Poisson cases. Leaving out the trivial degenerate c.d. case which corresponds to vanishing $\Psi_t(x)$, the foregoing theorem yields properties of two extreme cases corresponding to $\Psi_t(x)$ with one fixed point of increase only: The *normal c.d. process* corresponds to the point of increase $x = 0$, that is, to vanishing $L_t(x)$, so that

$$\psi_t(u) = i\alpha_t u - \frac{\beta_t^2}{2} u^2$$

with α_t continuous and β_t^2 continuous and nondecreasing. The *Poisson c.d. process* corresponds to the point of increase $x = c \neq 0$ and, setting $\lambda_t = L_t(c + 0) - L_t(c - 0)$, $\gamma_t = \alpha_t - \dfrac{c}{1 + c^2} \lambda_t$, to

$$\psi_t(u) = i\gamma_t u - \lambda_t(e^{iuc} - 1)$$

with γ_t continuous and λ_t continuous and nondecreasing. The "reduced" forms obtained by centering and changing the instantaneous time-scale are called the *Brownian process*, $\psi_t(u) = -\dfrac{\sigma^2 t}{2} u^2$, $\sigma^2 > 0$, and the *Poisson process*, $\psi_t(u) = \lambda t(e^{iu} - 1)$, $\lambda > 0$. We denote by $\mathfrak{L}(\alpha, \beta)$ a law with two values α and β only.

B. Normal and Poisson c.d. criteria. *Let X_T, $T = [a, b]$ be a c.d. process and take partitions of $[a, b]$.*

($\mathfrak{N}$) *X_T is a normal c.d. process if and only if almost all its sample functions are continuous, or X_{ab} is a normal r.v., or $\mathfrak{L}(\max_k |X_{nk}|) \to \mathfrak{L}(0)$.*

($\mathfrak{P}$) *X_T is a Poisson c.d. process if and only if almost all its sample functions are step-functions with jumps of constant height, or X_{ab} is a Poisson type r.v., or $\mathfrak{L}(\min_k X_{nk}) \to \mathfrak{L}(0)$ and $\mathfrak{L}(\max_k X_{nk}) \to \mathfrak{L}(0, c)$ where $c > 0$ or $\mathfrak{L}(\min_k X_{nk}) \to \mathfrak{L}(c, 0)$ and $\mathfrak{L}(\max_k X_{nk}) \to \mathfrak{L}(0)$ where $c < 0$, with X_{ab} lattice-valued.*

Proof. In ($\mathfrak{N}$) and in ($\mathcal{P}$) the sample functions assertions follow from **A**(ii), the normality and Poisson type assertions follow from **A**(i) or from the composition and decomposition theorem 19.2**A**, and the extrema assertions follow from the extrema criterion 22.4**C**.

Brownian processes are born from and are used in physics to describe motions of particles such as molecules of a gas. However, by their very nature, they are first approximations only, for, while almost all sample functions are continuous, they are extremely irregular in nature. They are investigated in detail in the next chapter.

Poisson processes are also born from and serve to describe various physical phenomena such as radioactive disintegrations. In fact, we can and do consider only the step sample functions (whose jumps correspond to disintegrations) taken to be rightcontinuous. Then

b′. POISSON SAMPLE JUMPS LEMMA. *Let X_T, $T = [0, \infty)$, be a Poisson process.*

(i) *The c. law of jumps in $I = [s, s + t)$ given that n occurred is that of n independent r.v.'s uniformly distributed in I.*

(ii) *The times $\tau_n(\tau_0 = 0)$ between the $(n - 1)$th and nth jumps are independent r.v.'s with $P[\tau_n > t] = e^{-\lambda t}$.*

Proof. The numbers of jumps Y and Z in $I' = [s', s' + t') \subset I$ and in $I - I'$ are independent Poisson r.v.'s with parameters $\lambda t'$ and $\lambda(t - t')$ and $Y + Z$ is also Poisson with parameter λt. Thus, by elementary computations,

$$P(Y = m \mid Y + Z = n) = P[Y = m, Z = n - m]/P[Y + Z = n]$$
$$= C_n^m (t'/t)^m (1 - t'/t)^{n-m},$$

and (i) is proved. Therefore, taking $t > t_1 + t_2$,

$$P(\tau_1 > t_1, \tau_2 > t_2 \mid X_t = n) = n \int_{t_1}^{t-t_2} d\left(\frac{s_1}{t}\right)\left(1 - \frac{s_1 + t_2}{t}\right)^{n-1}$$
$$= \left(1 - \frac{t_1 + t_2}{t}\right)^n = p^n$$

and $P[\tau_1 > t_1, \tau_2 > t_2] = e^{-\lambda t}(1 + \lambda tp + (\lambda tp)^2/2! + \cdots) = e^{-\lambda t_1} e^{-\lambda t_2}$; similar computations yield independence of any number of τ's, and (ii) is proved. Or, note that 41.1**A** case 2° or 38.4**A** corollary applies to X_T, set $\sigma_n = \tau_1 + \cdots + \tau_{n-1}$, and

$$P(\tau_n > t_n \mid \tau_{n-1}, \tau_{n-2}, \cdots) = P(X_{\sigma_n+t_n} - X_{\sigma_n} = 0 \mid \sigma_n)$$
$$= P[X_{t_n} - X_0 = 0] = e^{-\lambda t_n}.$$

Stationarity and law derivatives. The ch.f.'s of the increments $X_{t,t+h}$ of Brownian and Poisson processes are $f_{t,t+h} = f_h = e^{h\psi}$ with $\psi(u) = -\frac{\sigma^2}{2}u^2$ and $\psi(u) = \lambda(e^{iu} - 1)$. Thus, these processes are stationary and (equivalently) have stationary law derivatives, according to what follows.

The general concept of stationarity along an index set T is that of invariance under translations along T or, to use an intuitive terminology when $T = [0, \infty)$, of invariance under translations in time. Thus, *a process X_T on $T = [0, \infty)$ is stationary* if its law of evolution is stationary under translations in time. For a centered decomposable X_T it suffices that the individual laws of its increments $X_{t,t+h}$ be stationary: $f_{t,t+h} = f_h$ for all $h > 0$. For, it follows at once that the joint laws of its increments hence its law of evolution are stationary. But then, by decomposability, $f_{h+k} = f_h f_k$ for all $h, k > 0$ so that $f_h = f_{h/n}{}^n$ for n as large as we wish. Therefore $f_h = e^{\psi_h}$ is infinitely decomposable and $f_h \to 1$ as $h \to 0$. Thus the relation becomes $\psi_{h+k} = \psi_h + \psi_k$ with $\psi_h \to 0$ as $h \to 0$. It follows that a stationary centered decomposable is a c.d. process with $f_{t,t+h} = f_h = e^{h\psi}$, $\psi = (\alpha, \Psi)$.

We say that a ch.f. $\dot{f}_t$ represents the (*right*) (*left*) *law derivative* at t of a centered decomposable X_T if $(f_{t-h,t+k})^{[1/(h+k)]} \to \dot{f}_t$ as $h + k \to 0$ with $h, k \geqq 0, h + k > 0$ $(h \equiv 0)$ $(k \equiv 0)$. According to the central convergence theorem, when the limit ch.f. exists, then it is necessarily an i.d. ch.f. $\dot{f}_t = e^{\dot{\psi}_t}$, $\dot{\psi}_t = (\dot{\alpha}_t, \dot{\Psi}_t)$ and the process is (right) (left) law continuous; in particular, if X_T is stationary then $\dot{f}_t \leftarrow (e^{h\psi})^{1/h} = e^{\psi}$ so that the law derivative exists and is stationary. Thus, in searching for conditions of existence of law derivatives, we may assume that X_T is a c.d. process with $f_{st} = e^{\psi_{st}}$, $\psi_{st} = (\alpha_{st}, \Psi_{st})$. Then it follows from 22.1**D** that the law derivatives exists if and only if $\frac{1}{h+k}\psi_{t-h,t+k} \to \dot{\psi}_t$ or, equivalently, $\frac{1}{h+k}\Psi_{t-h,t+k} \xrightarrow{c} \dot{\Psi}_t$ and $\frac{1}{h+k}\alpha_{t-h,t+k} \to \dot{\alpha}_t$. In particular, if the law derivative exists and is stationary hence for every t and every u the derivative $\dot{\psi}(u)$ of $\psi_t(u)$ exists and is independent of t, it follows that the process is stationary.

We collect what precedes into

C. STATIONARY DECOMPOSABILITY CRITERION. *A centered decomposable process X_T, $T = [0, \infty)$ is stationary if and only if the ch.f.'s of its increments are of the form $f_{t,t+h} = f_h = e^{h\psi}$, $\psi = (\alpha, \Psi)$ or, equivalently, its law derivative exists and is stationary.*

Integral decomposition. A look at the P. Lévy form of $\psi_t = (\alpha_t, \beta_t^2, L_t)$ makes one think of a c.d. process as a "sum" of a normal c.d. process and of the Poisson c.d. processes corresponding to all points of increase of the P. Lévy functions. In fact, P. Lévy showed that in a sense it was true and *then* obtained the general form for infinitely decomposable laws. Itō made this analysis precise. We shall give the P. Lévy-Itō result but proceeding *from* the general form of the infinitely decomposable laws—to be specific, from the continuous decomposability theorem to which it led—to the reconstruction of c.d. processes by means of their normal and Poisson "components," as follows.

Let X_T be a c.d. process, use the notation introduced in this subsection, and take partitions of $[a, t)$.

c. NUMBER OF JUMPS LEMMA. *The numbers $\nu_{st}[x, y)$ of jumps in $[s, t)$ of height in $[x, y)$, $xy > 0$, are Poisson r.v.'s with parameter $L_{st}[x, y)$ independent for disjoint time-intervals $[s, t)$ and independent for disjoint height-intervals $[x, y)$.*

Thus, the processes $(\nu_t(x), t \in T)$, $x \neq 0$, are Poisson c.d. processes and the processes $(\nu_t(x), x \in (-\infty, 0) \cup (0, +\infty))$, $t \in T$, are Poisson decomposable.

Proof. Independence of the $\nu_{st}[x, y)$ for disjoint time-intervals $[s, t)$ results from independence of the corresponding increments X_{st} on which they are defined $x < y < 0$,

For the $x, y \in C(L_t)$ the continuity set of L_t with $x < y < 0$,

$$\varphi_t^{(n)}(u) = E \exp\{iu\nu_t^{(n)}[x, y)\} = \prod_k E \exp\{iuI_{[x \leq X_{nk} < y]}\}$$

$$= \prod_k \{1 + (e^{iu} - 1)F_{nk}[x, y)\}$$

where $\max_k F_{nk}[x, y) \to 0$ and $\sum_k F_{nk}[x, y) \to L_t[x, y)$. Upon taking n sufficiently large and setting $\varphi_t(u) = E \exp\{iu\nu_t[x, y)\}$, it follows from $\nu_t^{(n)}[x, y) \xrightarrow{\text{a.s.}} \nu_t[x, y)$ that

$$\log \varphi_t(u) \leftarrow \log \varphi_t^{(n)}(u) = (1 + o(1))(e^{iu} - 1)\sum_k F_{nk}[x, y)$$
$$\to (e^{iu} - 1)L_t[x, y).$$

Therefore, upon letting $x' \uparrow x$ along $C(L_t)$ when $x \notin C(L_t)$, and similarly when $y \notin C(L_t)$, the result is valid for all $x, y < 0$, by continuity from the left of L_t; similarly for all $x, y > 0$. The Poisson r.v.'s assertion follows. In fact, it is an immediate consequence of the first criterion in $\mathbf{B}(\mathcal{P})$. We gave a direct proof because the one below generalizes it:

If $[x_j, y_j)$, $x_j y_j > 0$, are m disjoint height-intervals and all the x_j, $y_j \in C(L_t)$ then, as above, for n sufficiently large,

$$\begin{aligned}\log E \exp \{\sum_j iu_j \nu_t^{(n)}[x_j, y_j)\} &= \sum_j \log \{1 + (e^{iu_j} - 1) \sum_k F_{nk}[x_j, y_j)\} \\ &= (1 + o(1)) \sum_j (e^{iu_j} - 1) \sum_k F_{nk}[x_j, y_j) \\ &\to \sum_j (e^{iu_j} - 1)\, L_t[x_j, y_j),\end{aligned}$$

and the restriction to continuity endpoints is removed as above. The Poisson r.v.'s and independence assertions follow, and the proof is terminated.

d. INTEGRATION LEMMA. *The a.s. integrals*

$$I_t = \int_{-\infty}^{+\infty} \{x\, d\nu_t(x) - \frac{x}{1 + x^2} dL_t(x)\}$$

exist and are i.d. r.v.'s with

$$\log E \exp \{iuI_t\} = \int_{-\infty}^{+\infty} \left(e^{iux} - 1 - \frac{iux}{1 + x^2}\right) dL_t(x).$$

Proof. We drop the subscript t and consider only the almost all sample functions for which $\nu[a, b)$ has the properties stated in $\mathbf{A}$(ii), so that we drop "a.s."

Let $\alpha < \beta < 0$. If the jumps of the step-function $\nu(\omega, [\alpha, \beta))$ occur at $x_1(\omega), \cdots, x_{n(\omega)}(\omega)$ then $I_\alpha^\beta(\omega) = \int_\alpha^\beta x\, d\nu(\omega, x) = \sum_{k=1}^{n(\omega)} x_k(\omega)$ defines a finite function I_α^β of ω. Set $x_{nk} = \alpha + k(\beta - \alpha)/2^n$, $\nu_{nk} = \nu[x_{nk}, x_{n,k+1})$, and $L_{nk} = E\nu_{nk} = L[x_{nk}, x_{n,k+1})$. Since $S_n = \sum_k x_{nk}\nu_{nk} \uparrow I_\alpha^\beta$, it follows that the integral I_α^β is measurable,

$$\begin{aligned}\log E \exp \{iuS_n\} \sim \sum_k (e^{iux_{nk}} - 1)L_{nk} &\to \int_\alpha^\beta (e^{iux} - 1)\, dL(x) \\ &= \log E \exp \{iuI_\alpha^\beta\},\end{aligned}$$

and this integral is an i.d. r.v. We can define the integral $I_{-\infty}^\beta$ by $I_\alpha^\beta \uparrow I_{-\infty}^\beta$ as $\alpha \downarrow -\infty$ or directly by means of partitions of $[\alpha_n, \beta]$ where

$\alpha_n \downarrow -\infty$. In either case, it follows that $I_{-\infty}{}^{\beta}$ is an i.d. r.v. with $\log E \exp \{iuI_{-\infty}{}^{\beta}\} = \int_{-\infty}^{\beta} (e^{iux} - 1)\, dL(x)$. Similarly for $\alpha, \beta > 0$ then $\beta \uparrow +\infty$.

We cannot do the same with, say, $I_{-\infty}{}^{\beta}$ as $\beta \uparrow 0$, since $\int_{-1}^{-0} (e^{iux} - 1)\, dL(x)$ may not exist. However, as $\beta \uparrow 0$, by the convergence in q.m. criterion, there is a limit in q.m.:

$$(1) \qquad J_{-1}{}^{\beta} = \int_{-1}^{\beta} x\, d\{\nu(x) - L(x)\} \xrightarrow{\text{q.m.}} \int_{-1}^{-0} x\, d\{\nu(x) - L(x)\},$$

since as $\beta, \beta' \uparrow 0$

$$\int_{-1}^{\beta}\int_{-1}^{\beta'} xx'\, dd'E\{\nu(x) - L(x)\}\{\nu(x') - L(x')\} \to \int_{-1}^{-0} x^2\, dL(x) < \infty.$$

But the left side of (1) is a martingale in $\beta \uparrow 0$ or may be considered as a sequence of consecutive sums of independent summands $\xi_n = \int_{\beta_n}^{\beta_{n+1}}$ with $\beta_n \uparrow 0$. Either way, the limit in q.m. is also an a.s. limit and

$$\log E \exp \{iuJ_{-1}{}^{\beta}\} = \int_{-1}^{\beta} (e^{iux} - 1 - iux)\, dL(x) \to$$

$$\int_{-1}^{-0} (e^{iux} - 1 - iux)\, dL(x)$$

$$= \log E \exp \{iuJ_{-1}{}^{-0}\};$$

similarly for $\int_{+0}^{+1} x\, d\{\nu(x) - L(x)\}$. Finally, upon adding to $\int_{-\infty}^{-1} x\, d\nu(x)$ the finite constant $-\int_{-\infty}^{-1} \frac{x}{1+x^2}\, dL(x)$ and to $\int_{-1}^{-0} x\, d\{\nu(x) - L(x)\}$ the finite constant $-\int_{-1}^{-0} \frac{x^3}{1+x^2}\, dL(x)$ so as to obtain under the integral signs the same expression $x\, d\nu(x) - \frac{x}{1+x^2}\, dL(t)$, similarly for integrals over $(0, 1)$ and $[1, \infty)$, and adding the four integrals, the lemma follows.

We are now ready for the "integral decomposition" of c.d. processes. We go back to our partitions of $[a, t)$ and recall that the X_{nk} are uan independent (in k) r.v.'s with ch.f.'s $f_{nk} = e^{\psi_{nk}}$, $\psi_{nk} = (\alpha_{nk}, \Psi_{nk})$ and $\max_k |\psi_{nk}(u)| \to 0$. Let $\alpha_t \equiv 0$. Since $\psi_t = \sum_k \psi_{nk}$ where $\psi_t = (0, \Psi_t)$, it follows that there exist a finite $c(u)$ such that

$$\sum_k |f_{nk}(u) - 1| = \sum_k |e^{\psi_{nk}(u)} - 1| = (1 + o(1)) \sum_k |\psi_{nk}(u)|$$

$$\leqq (1 + o(1)) \left\{ \sum_k \left| \int_{-\infty}^{+\infty} \left(e^{iuy} - 1 - \frac{iuy}{1+y^2}\right) \frac{1+y^2}{y^2} \, d\Psi_{nk}(y) \right| \right\}$$

$$\leqq (1 + o(1)) \{ c(u) \operatorname{Var} \Psi_t \}.$$

Thus,

$$\sum_k |f_{nk}(u) - 1|^2 \leqq \max_k |f_{nk}(u) - 1| \sum_k |f_{nk}(u) - 1| \to 0$$

and therefore

$$\sum_k (f_{nk}(u) - 1) = (1 + o(1)) \sum_k \psi_{nk}(u) \to \psi_t(u).$$

D. INTEGRAL DECOMPOSITION THEOREM. *Every c.d. process X_T, $T = [a, b]$, is sum of two independent processes*

$$X_T = \eta_T + \int_{-\infty}^{+\infty} \left\{ x \, d\nu_T(x) - \frac{x}{1+x^2} \, dL_T(x) \right\}$$

where η_T is a normal c.d. process, and ν_T has the properties stated in the number of jumps lemma; and conversely.

Proof. The converse is immediate and, by **A** and **C**, the direct assertion requires only the proof of the independence and normality parts. Since the integral process values I_t are defined on the process $(\nu_t(x), x \neq 0)$, it suffices to prove that η_t and $\nu_t(x)$ are independent for $x \in C(L_t)$. The proof is the same for $x < 0$ and $x > 0$. Let, say, $x < 0$. We can and do assume that $\alpha_t \equiv 0$, without restricting the generality.

If

$$I_{nt}(\epsilon) = \int_{|x| \geqq \epsilon} \left\{ x \, d\nu_t^{(n)}(x) - \frac{x}{1+x^2} \, dL_t(x) \right\}$$

then it is easily verified that $I_{nt}(\epsilon) \to I_t$ as $n \to \infty$ then $\epsilon \to 0$ so that, setting $\eta_t = X_t - I_t$, it follows that

$$Y_{nt}(\epsilon) = X_t - I_{nt}(\epsilon) \xrightarrow{\text{a.s.}} \eta_t$$

and

$$\begin{aligned}\varphi_{n,\epsilon}(u, v) &= E \exp \{iuY_{nt}(\epsilon) + iv\nu_t^{(n)}(x)\} \to \varphi(u, v)\\ &= E \exp \{iu\eta_t + iv\nu_t(x)\}.\end{aligned}$$

We drop t, write ν_n in lieu of $\nu^{(n)}$, and can and do take $x < -\epsilon$, $\pm\epsilon \in C(L_t)$. Since

$$\int_{-\infty}^{-\epsilon-0} x\, dI_{[X_{nk}<x]} = X_{nk}I_{[X_{nk}<-\epsilon]}, \quad \int_{\epsilon+0}^{+\infty} x\, dI_{[X_{nk}\geqq x]} = -X_{nk}I_{[X_{nk}>\epsilon]},$$

we have

$$Y_n(\epsilon) = \sum_k X_{nk}I_{[|X_{nk}|<\epsilon]} + \int_{|y|\geqq\epsilon} \frac{y}{1+y^2}\, dL(y)$$

and

$$\varphi_{n,\epsilon}(u, v)$$

$$= \exp\left\{iu\int_{|y|\geqq\epsilon} \frac{y}{1+y^2}\, dL(y)\right\} \cdot \prod_k E \exp\{iuX_{nk}I_{[|X_{nk}|<\epsilon]} + ivI_{[X_{nk}<x]}\}.$$

Elementary computations yield for the factors in $\prod_k$ the expression

$$1 + (f_{nk}(u) - 1) - \int_{|y|\geqq\epsilon} (e^{iuy} - 1)\, dF_{nk}(y) + (e^{iv} - 1)F_{nk}(x).$$

Since

$$\max_k |f_{nk}(u) - 1| \to 0, \quad \max_k \int_{|y|\geqq\epsilon} dF_{nk}(y) \to 0, \quad \max_k F_{nk}(x) \to 0,$$

it follows that for n sufficiently large

$$\begin{aligned}\log \varphi_{n\epsilon}(u, v) = &+iu\int_{|y|\geqq\epsilon} \frac{y}{1+y^2}\, dL(x)\\ &+ (1 + o(1))\left\{\sum_k (f_{nk}(u) - 1) - \sum_k \int_{|y|\geqq\epsilon} (e^{iuy} - 1)\, dF_{nk}\right.\\ &\left. + (e^{iv} - 1)\sum_k F_{nk}(x)\right\}.\end{aligned}$$

Therefore, as $n \to \infty$ then $\epsilon \to 0$,

$$\log \varphi(u, v) \leftarrow \log \varphi_{n\epsilon}(u, v) \to -\frac{\beta^2}{2}u^2 + (e^{iv} - 1)L(x).$$

The independence and normality assertions follow, and the proof is terminated.

COMPLEMENTS AND DETAILS

All r.f.'s are selected to be separable on some unspecified but fixed pr. space.

1. Let $X_T = (X_t, t \geqq 0)$ be separated for closed sets by S. Set $\underline{X}_t = \liminf_{t' \to t} X_{t'}$, $\underline{X}_{t\mp 0} = \liminf_{t' \to t\mp 0} X_{t'}$.

a) If $\underline{X}_t = X_t$ outside a null event N_t, then the r.f. $(\underline{X}_t, t \geqq 0)$ is equivalent to X_T, is separated for closed sets by S, and is a Borel r.f. with almost all sample functions lower semi-continuous. What if the limits are defined in terms of deleted neighborhoods?

b) If $\underline{X}_{t-0} = \underline{X}_{t+0} = X_t$ outside N_t, then the r.f. $(X_{t+0}, t \geqq 0)$ has the above asserted properties except that lower semi-continuity is replaced by right lower semi-continuity.

c) What if liminf is replaced by limsup?

2. If there exist constants c, p, $q \geqq 0$, $r > 0$ such that

$$E|X_{t_1} - X_{t_2}|^p |X_{t_2} - X_{t_3}|^q < c|t_1 - t_3|^{1+r}, \quad 0 \leqq t_1 < t_2 < t_3 \leqq 1,$$

then almost all sample functions of $(X_t, 0 \leqq t \leqq 1)$ are continuous except perhaps for simple discontinuities.

3. Let X_T, $T = [0, \infty)$ be a Borelian stationary r.f.

a) If X_0 is integrable then, for every $u \in R$,

$$\frac{1}{t}\int_0^t e^{-ius} X_s\, ds \xrightarrow{a.s.} \xi_u \text{ integrable,}$$

$E\xi_u = 0$ for $u \neq 0$ and ξ_u degenerates at zero at every u except for a countable set of values of u.

b) If X_0^2 is integrable, then the convergence is also in quadratic mean and $E\xi_u\bar{\xi}_v = 0$ for $u \neq v$. What if above e^{-ius} is replaced by $(e^{-ius} - e^{-iu(s+h)})/iu$?

4. $\mathfrak{B}_T$ *and* $\mathfrak{B}_{T+}$ *times.* Let $\mathfrak{B}_T = (\mathfrak{B}_t, t \geqq 0)$ $\mathfrak{B}_{T+} = (\mathfrak{B}_{t+0}, t \geqq 0)$.

a) If σ, τ, τ_n are $\mathfrak{B}_T$-times then so are $\sigma + \tau$ and $\sup \tau_n$; in particular, limits of nondecreasing sequences of $\mathfrak{B}_T$-times are $\mathfrak{B}_T$-times.

b) If τ is a $\mathfrak{B}_T$-time, then $[\tau < t] \in \mathfrak{B}_t$ for every t, but in general, the converse is not true unless $\mathfrak{B}_T$ is rightcontinuous. Example: Let $(X_t, \mathfrak{B}_t, t \geqq 0)$ be a Brownian motion. Then $\tau = \min\{t\colon X_t = 1\}$ is $\mathfrak{B}_T$-time, $\tau' = \inf\{t\colon X_t > 0\}$ is $\mathfrak{B}_{T+}$ (but not $\mathfrak{B}_T$)-time.

c) $\mathfrak{B}_{T+}$ is rightcontinuous; to simplify the writing assume that $\mathfrak{B}_T$ is rightcontinuous. If τ_n are $\mathfrak{B}_T$-times then so are $\liminf \tau_n$ and $\limsup \tau_n$. If, moreover, $\tau_n \downarrow \tau$ then $\mathfrak{B}_\tau = \bigcap_n \mathfrak{B}_{\tau_n}$.

d) Let $(X_t, \mathfrak{B}_t, t \geqq 0)$ with X_T sample rightcontinuous ($T = [0, \infty)$). For Borel $S \subset R$ set

$$\tau_S = \inf\{t\colon X_t \in S\} \quad \text{or} \quad +\infty,$$

according as this set is nonempty or empty.
Let $U \subset R$ be open. Then τ_{U^c} is $\mathfrak{B}_T$-time and, when $\mathfrak{B}_T$ is rightcontinuous so is τ_U.

e) Let the foregoing X_T be rightcontinuous with sample left limits and let $\mathcal{B}_T$ be rightcontinuous. Let every $\mathcal{B}_T$ contain all subsets of all null events. Then τ_S is $\mathcal{B}_T$-time for every Borel S.

5. Let $(X_T, \mathcal{B}_T)$, $T = [0, 1]$, be a martingale.

a) Let X_T^2 be integrable and set $F_t = EX_t^2$. Then X_T has orthogonal increments for $s < t$, $E(X_t - X_s)^2 = F_t - F_s$, and the fixed discontinuity points of X_T are the discontinuity points of F_t. $(X_T^2, \mathcal{B}_T)$ is a submartingale. What about $(X_T^2 - F_T, \mathcal{B}_T)$? Is it a martingale if and only if, for $s < t$,

$$E((X_t - X_s)^2 \mid \mathcal{B}_s) = E(X_t - X_s)^2 \quad \text{a.s.}$$

If X_T has no fixed discontinuity points, what does the change of parameter to t' given by $t = F(t')$ do?

b) Let X_T be sample continuous. By partitioning suitably T, find $\mathfrak{L}(X_T)$ under various hypotheses which permit the use of results and/or methods of §28. When is $\mathfrak{L}(X_T)$ normal? Can it be Poisson?

6. Second order integral decomposition. Given a second order r.f. X_T with $EX_t\bar{X}'_{t'} = \iint g_t(s)\bar{g}_{t'}(s')\, dd'\gamma(s, s')$ where $\gamma(s, s')$ is of bounded variation on every finite square, there exists a second order r.f. $\xi(s)$ with $E\Delta_h\xi(s)\Delta'_{h'}\bar{\xi}(s') = \Delta_h\Delta'_{h'}\gamma(s, s')$ such that $X_t = \int g_t\, d\xi$.

Set $(X_t, X_{t'}) = (g_t, g_{t'})$, let c_k's be constants, $L(X)$ be the family of all linear combinations $\sum c_k X_{t_k}$, $L(g)$ be the family of all linear combinations $\sum c_k g_{t_k}$, and let $L_2(X)$ be the closure in q.m. of $L(X)$ and $L_\gamma(g)$ be the γ-closure of $L(g)$.

a) Extend the correspondence $X_t \leftrightarrow g_t$, $t \in T$, to $L(X)$ and $L_\gamma(g)$ preserving linearity and then extend it to $L_2(X)$ and $L_\gamma(g)$ preserving inner products. The correspondence between $L_2(X)$ and $L_\gamma(g)$ preserves linearity and inner products.

b) If all indicators $I_{[a,b)} \in L_\gamma(g)$, denote by $\xi[a, b)$ the corresponding elements of $L_2(X)$. Then $X_t = \int g_t\, d\xi$.

c) If some indicators $I_{[a,b)} \notin L_\gamma(g)$ introduce a set T' whose elements t are the missing indicators, extend g on $T + T'$ by taking g_t, $t \in T'$, such that $(g_t, g_{t'})$ exist and are finite for $t, t' \in T \cup T'$, extend X_T on $T \cup T'$ by taking X_t, $t \in T'$ so that $(X_t, X_{t'}) = (g_t, g_{t'})$, $t, t' \in T \cup T'$ and $X_{T'}$ independent of X_T, and apply *b*).

d) Give as particular cases the decompositions theorems in Section 34.

7. Let $E|\, d\xi(s)\,|^2 = dF(s)$ and let $g_t(s)$ be measurable with respect to the $dt\, dF(s)$-product measure with $|\, g_t(s)\,|^2\, dF(s) < \infty$ for almost all t.

a) The r.f. X_T with $X_t = \int g_t(s)\, d\xi(s)$ can be selected so as to be measurable: Start with $g_t(s) = \sum_k f_{k,t} h_k(s)$.

b) If $\int dF(s) \left(\int |\, g_t(s)\,|\, dt\right)^2 < \infty$, $\int dt \left(\int |\, g_t(s)\,|^2\, dF(s)\right)^{1/2} < \infty$, then the iterated integrals $\int d\xi(s) \left(\int g_t(s)\, dt\right)$ (where the bracketed integral is selected so as to be measurable), $\int dt \left(\int g_t(s)\, d\xi(s)\right)$ exist, are a.s. equal, and denoted by

$\iint g_t(s)\, dt\, d\xi(s)$.

c) Let $g'(t)$ be the derivative of $g(t)$, $t \in [a, b]$, and $g_t(s) = g'(t)$ or 0 according as $t \leqq s$ or $t > s$. Then, for F continuous at a and b

$$\int_a^b \int_a^b g_t(s)\, dt\, d\xi(s) = (g(b) - g(a))\xi[a, b) - \int_a^b \xi[a, t)g'(t)\, dt.$$

What if $F(s)$ is not continuous at a and b?

8. Let $\xi(s)$ be a martingale with

$$E(|\xi[s, t)|^2, \xi_r, r \leqq s) = E|\xi[s, t)|^2 = F[s, t) \text{ a.s.}$$

a) The family of r.f.'s $G(s)$ with $\int E|G(s)|^2\, dF(s) < \infty$ contains those which are measurable with respect to the $ds\, dP$-measure with every r.v. $G(s)$ being $\mathcal{B}(\xi_r, r \leqq s)$-measurable.

If $\xi(s) = \eta(s) - s$ where $\eta(s)$ is a Poisson r.f. with $\lambda = 1$, then $E\, d\xi(s) = 0$, $E(d\xi(s))^2 = dt$, and

$$\int_a^b \xi[a, s)\, d\xi(s) = \tfrac{1}{2}(\xi[a, b))^2 - \tfrac{1}{2}\xi[a, b).$$

If $\xi(s)$ is a Brownian r.f. with $\sigma^2 = 1$, then

$$\int_a^b \xi[a, s)\, d\xi(s) = \tfrac{1}{2}(\xi[a, b))^2 - \tfrac{1}{2}(b - a).$$

b) R.f.'s $X_T = (X_t, t \geqq a)$, $X_t = \int_a^t G(s)\, d\xi(s)$ are square-integrable martingales whose almost all sample functions have one-sided limits at all points and the fixed discontinuities are discontin ities of $F(s)$. If almost all sample functions of $\xi(s)$ are continuous so are those of X_T.

c) If X_T, $T = [a, b]$ is a square-integrable martingale whose almost all sample functions are continuous and $E(|X_{s,t}|^2 | X_r, r \leqq s) = E\left(\int_s^t |G(s)|^2\, ds \mid X_r, r \leqq s\right)$ a.s., then there exists a Brownian r.f. $\xi(s)$, $a \leqq s \leqq b$, such that $X_{a,t} = \int_a^t G(s)\, d\xi(s)$ a.s.; however, the pr. space may have to be enlarged to a product of the given pr. space with an appropriate one unless $G(\omega, s)$ vanishes almost nowhere on (ω, t)-space. What about a converse?

9. Zero-one law. Let X_T, $T = [0, \infty)$, be decomposable. If ξ is a tail r.v. on the family of increments $X_t - X_s$, $s, t \geqq 0$, then ξ is degenerate.

10. Strong law of large numbers. Let X_T, $T = [0, \infty)$, be decomposable with stationary increments and $E(X_t - X_0) = 0$. Then $E(X_1 - X_0 | X_s - X_0, s \geqq t) = E(X_1 - X_0 | X_t - X_0) = (X_t - X_0)/t$ a.s. and $X_t/t \xrightarrow{\text{a.s.}} 0$. Extend as in Section 29.4III.

11. Let $X^{(n)} = (X_t^{(n)}, t \geqq 0)$, $n = 1, 2, \cdots$, be independent Poisson r.f.'s with parameter $\lambda > 0$. The pr. that all $X^{(n)}$ be constant in an interval of length h is $e^{-\lambda hn}$. Let $c_n > 0$, $\sum c_n < \infty$. The series $X_t = \sum c_n X_t^{(n)}$ is a.s. convergent. Are the following statements true? $X = (X_t, t \geqq 0)$ is a.s. continuous, strictly increasing, and its moving discontinuity sets form an everywhere dense set in $[0, \infty)$.

12. Let X_{st} be the number of occurrences in $[s, t)$ of ***purely random events:*** the X_{st} are a.s. finite for finite time-intervals and independent for disjoint ones with $EX_{st} = \lambda(t - s)$, the pr. of occurrence at any specified time and the pr. of two simultaneous occurrences are zero. The X_{st} form a Poisson process, and conversely.

13. Poisson distribution of particles. Let X_i, $i = \cdots -1, 0, +1, \cdots$ be r.v.'s and let ν_I be the number of $X_i \in I$; I, with or without affixes, are intervals of length $|I|$ and X_i and ν_I have same affixes if any. The X_i represent (positions of) particles and the ν_I are their numbers in I. The "distribution of particles" is the family of joint distributions of $\nu_{I_1}, \cdots, \nu_{I_n}$ for all $I_1, \cdots, I_n$, $n = 1, 2, \cdots$. The distribution is "Poissonian" if the ν_I are independent for disjoint intervals I and Poissonian with parameter $\lambda|I|$.

Let the particles $X_i(t)$, $t \geqq 0$, move "independently": for every fixed t, the increments $X_i[0, t)$ are mutually independent, identically distributed, and independent of all $X_i(0)$. Let $P_t[a, b] = P[X_i[0, t) \in [a, b]]$.

a) The Poisson distribution of particles is invariant in time (that is, if their distribution is Poissonian at time 0 then it is Poissonian with same λ at any time t).

b) If, as $t \to \infty$, for every $h > 0$,

$$\sum_{n=-\infty}^{+\infty} | P_t[nh, (n + 1)h] - P_t[(n - 1)h, nh] | \to 0$$

and $E|(\nu_I(0)/|I|) - \lambda| \to 0$ as $|I| \to \infty$, uniformly in all intervals I of length I, for some constant λ, then the distribution of particles converges to the Poisson distribution (the distribution "converges" to the distribution of particles $\{\bar{X}_i\}$ if the $\mathcal{L}(\nu_{I_1}, \cdots, \nu_{I_n})$ converge to $\mathcal{L}(\bar{\nu}_{I_1}, \cdots, \bar{\nu}_{I_n})$ for all $I_1, \cdots, I_n$, $n = 1, 2, \cdots$). What if λ is a r.v.? What if there is no convergence as $|I| \to \infty$?

14. Let $(X_t, \mathcal{B}_t, t \geqq 0)$ with $X_0 = 0$ be a Poisson r.f. with parameter $\lambda > 0$.

a) Set $Y_t = X_t - \lambda t$, $Z_t = Y_t^2 - \lambda t$, $U_t = \exp\{-cX_t + \lambda t(1 - e^{-c})\}$, $c \in R$. Then $(Y_t, \mathcal{B}_t, t \geqq 0)$, $(Z_t, \mathcal{B}_t, t \geqq 0)$, $(U_t, \mathcal{B}_t, t \geqq 0)$ are martingales.

b) Given a positive integer m, let τ_m be the first time X_t reaches m. Then $X_{\tau_m} = m$,

$$\lambda E\tau_m = m, \quad \sigma^2(\tau_m) = m/\lambda^2,$$

and, setting $\alpha = -\lambda(1 - e^{-c})$,

$$Ee^{-\alpha\tau_m} = e^{mc} = (\lambda/(\lambda + \alpha))^m;$$

$\mathcal{L}(\tau_m)$ is a Γ-law with parameters m and λ.

Chapter XIII

BROWNIAN MOTION AND LIMIT DISTRIBUTIONS

Brownian motion, born from and used in Physics, is of ever increasing importance not only in Probability theory but also in classical Analysis. Its fascinating properties and its far-reaching extension of the simplest normal limit theorems to functional limit distributions acted, and continue to act, as a catalyst in random Analysis.

It is recommended that at least portions of this chapter be read as soon as possible. At the cost of some repetitions its dependence upon other chapters of this part is minimized and properties, which obtain from deeper ones—established herein or elsewhere in this part, are also established directly on a relatively elementary level.

§ 41. BROWNIAN MOTION

41.1. Origins. In Physics, the ceaseless and extremely erratic dance of microscopic particles suspended in a liquid or gas, is called "Brownian motion." It was systematically investigated by Brown (1828, 1829) —a botanist, from movement of grains of pollen in water to a drop of water in oil. He was not the first to mention this phenomenon and had many predecessors, starting with Leeuwenhoek in the 17th century. However, Brown's investigation brought it to the attention of the scientific community, hence "Brownian." Brownian motion was frequently explained as due to the fact that particles were alive. Poincaré thought that it contradicted the second law of Thermodynamics. Today we know that this motion is due to the bombardment of the particles by the molecules of the medium. In a liquid, under normal conditions, the order of magnitude of the number of these impacts is of 10^{20} per second!

It is only in 1905 that kinetic molecular theory led Einstein to the first mathematical model of Brownian motion. He began by deriving its possible existence and then only learned that it had been observed. Let us consider the x-component only of the motion—the one-dimensional Brownian motion. Using formal analytic arguments and more or less explicit probabilistic ones, such as stationary independent increments, he derived a partial differential equation

$$\frac{\partial p}{\partial t} = D\frac{\partial^2 p}{\partial x^2}$$

for the pr. density $p = p_t(x)$ that the particle be at x at time t; note that by change of scale, this equation reduces to the "heat equation" $\frac{\partial p}{\partial t} = \frac{1}{2}\frac{\partial^2 p}{\partial x^2}$. If the particle starts at $x = 0$ at time 0, then

$$p_t(x) = \frac{1}{\sqrt{2\pi Dt}} e^{-x^2/Dt}.$$

Using physical arguments, Einstein showed that the "diffusion coefficient" $D = 2RT/Nf$, where R is the ideal gas constant, T the absolute temperature, N the Avogadro number, and f the friction coefficient which depends upon viscosity and upon the particle properties. Soon thereafter, Perrin in a series of experiments based upon Einstein's model obtained a 19% approximation of the Avogadro number. This led to the final acceptance of the kinetic molecular theory even by skeptics such as Mach and Oswald. In fact, Einstein's model was later replaced by a dynamic one of Ornstein and Uhlenbeck, initiated by Langevin's first "stochastic differential equation"; the interested reader is referred to Nelson and to Wax. Perrin mentioned that Einstein's model produces nowhere differentiable continuous functions which, before this model, were considered by most mathematicians as special and somewhat artificial constructs without much mathematical value as counterexamples. Thus, these "monsters" from whom Hermite "turned away with horror," became "natural beings" of importance in Mathematics as well as in Physics.

A completely different origin of mathematical Brownian motion is a game theoretic model for fluctuations of stock prices due to Bachelier. In his thesis, he hinted that it could apply to physical Brownian motion. Therein, and in his subsequent works, he used the heat equation and, proceeding by analogy with "heat propagation" he found, albeit formally, distributions of various functionals of mathematical Brownian motion. Heat equations and related parabolic type equations were used rigorously by Kolmogorov, Petrovsky, and Khintchine (see §42).

Rigorous definition and study of (mathematical) Brownian motion requires measure theory. Some 20 years after Lebesgue's thesis, Wiener (1923) gave its first satisfactory construction and proved its almost sure sample continuity. In 1933, together with Paley and Zygmund, he proved nowhere differentiability of almost all Brownian sample functions. Meanwhile, Khintchine (1924) found the Brownian Law of the Iterated Logarithm. But it is in 1939 that P. Lévy proceeded to an analysis in depth, so exhaustive, that since only improvements of details were obtained. In later works, he investigated multi-dimensional Brownian motion and then Brownian motions where the time interval was replaced by abstract space, especially by Hilbert space.

It is strongly recommended that the interested reader attempt to peruse the chapters on Brownian motion in P. Lévy's book (1964) which are extraordinarily rich in ideas and results. The interested reader is also referred to the remarkable monographs by Freedman and by Ito and McKean.

In this chapter, we shall study the basic case: one-dimensional Brownian motion.

Because of its importance and its intrinsic value, we assign a special notation to Brownian motion W_T or W, with random values W_t or $W(t)$, according to convenience. The symbol "W" comes from "Wiener." Other symbols also in use are "w" and "B."

41.2. Definitions and relevant properties. There are several definitions of (one-dimensional) "Brownian" or "Wiener" or "Wiener-Levy" process or random function or motion. We shall proceed by successive refinements, based on required and relevant properties, to reach the final definition.

A process $W_T = (W_t, t \in T \subset R)$ is "Brownian distributed" if it is decomposable, that is, its increments $W_{st} = W_t - W_s$, $s < t$, $s, t \in T$, on disjoint intervals are independent, and if

$$\mathfrak{L}(W_{st}) = \mathfrak{N}(\alpha(t-s), \sigma^2(t-s)), \sigma^2 > 0;$$

α is called the *drift* and σ^2 is called the *diffusion coefficient of* W_T. A more restrictive definition obtains by adding the requirements

$$W_0 = 0 \text{ a.s. and } T = [0, \infty),$$

equivalently:

A random function (r.f.) $W_T = (W_t, t \geqq 0)$ is "Brownian distributed" if it is decomposable and, for $t \geqq 0$,

$$\mathcal{L}(W_t) = \mathfrak{N}(\alpha t, \sigma^2 t), \sigma^2 > 0,$$

that is, the ch. f.'s are given by

$$f_{W_t}(u) = Ee^{iuW_t} = \exp\{i(\alpha t)u - \sigma^2 u^2/2\}.$$

To simplify the writing and with no real loss, from now on we restrict "Brownian distributed" to the "normalized" or "standard" form obtained by taking $\alpha = 0$ and $\sigma^2 = 1$:

A r.f. $W_T = (W_t, t \geqq 0)$ is *Brownian distributed* if its disjoint increments are independent and

$$\mathcal{L}(W_t) = \mathfrak{N}(0, t), t \geqq 0;$$

thus, $\mathcal{L}(W_t/\sqrt{t}) = \mathfrak{N}(0, 1)$, that is,

$$P(W_t/\sqrt{t} > a) = \frac{1}{\sqrt{2\pi}} \int_a^\infty e^{-v^2/2} dv, \quad a \in R.$$

A. Brownian covariance criterion. *A r.f.* W_T *is Brownian distributed if and only if it is centered normal with covariance given by* $EW_sW_t = s \wedge t$.

Proof. Let W_T be Brownian distributed so that all $EW_t = 0$ and, for any finite section $(W_{t_1}, \cdots, W_{t_m})$ with $t_1 < \cdots < t_m$, by decomposability,

$$\begin{aligned} &E\exp\{iu_1 W_{t_1} + \cdots + iu_m W_{t_m}\} \\ &\quad = E\exp\{i(u_1 + \cdots + u_m)W_{t_1 t_2}\} \times \cdots \times E\exp\{iu_m W_{t_{m-1}, t_m}\} \end{aligned}$$

hence, the increments $W_{t_k, t_{k+1}}$ being normal, so is $(W_{t_1}, \cdots, W_{t_m})$. Thus, W_T is centered normal and $EW_sW_t = s \wedge t$ since for, say, $s \leqq t$,

$$EW_sW_t = EW_{0s}(W_{0s} + W_{st}) = EW_s^2 = s = s \wedge t.$$

Conversely, let W_T be centered normal with $EW_sW_t = s \wedge t$. Then, all W_t are normal with $EW_t = 0$ and $EW_t^2 = t$ so that $\mathcal{L}(W_t) = \mathfrak{N}(0, t)$. Since laws of finite sections are normal so are those of finite sets of increments on disjoint intervals. But such increments, say,

$$W_{st}, W_{s't'}, s \leqq t \leqq s' \leqq t',$$

are orthogonal, for

$$E(W_t - W_s)(W_t' - W_s') = t + s - s - t = 0.$$

Therefore they are independent, and the proof is terminated.

We establish now, in a rather elementary way, a few quite useful properties of Brownian distributed r.f. W_T. Unless otherwise stated, from now on, *we set* $\mathcal{B}_t = \mathcal{B}(W_s, s \leq t)$.

a. Brownian martingales property. *If a r.f.* W_T *is Brownian distributed then*

(i) $$(W_t, \mathcal{B}_t, t \geq 0) \text{ is a martingale}$$

and

(ii) $$(W_t^2 - t \mid \mathcal{B}_t, t \geq 0) \text{ is a martingale}$$

or, equivalently, under (i), *a.s.*

(ii′) $$E(W_{st}^2 \mid \mathcal{B}_s) = t - s, \quad 0 \leq s < t < \infty.$$

Proof. The equivalence assertion is immediate: Let (i) hold. Then, for $s < t$, a.s.

$$E(W_s W_t \mid \mathcal{B}_s) = W_s E(W_t \mid \mathcal{B}_s) = W_s^2$$

so that, a.s.

$$E((W_t - W_s)^2 - (t - s) \mid \mathcal{B}_s) = E(W_t^2 - t \mid \mathcal{B}_s) - (W_s^2 - s).$$

When (ii) holds then the right side vanishes a.s. and when (ii′) holds then the left side vanishes a.s. The equivalence assertion is proved.

Let W_T be Brownian distributed so that its increments W_{st} on disjoint intervals are independent and normal $\mathfrak{N}(0, t - s)$. Then (i) holds since a.s.

$$E(W_t \mid \mathcal{B}_s) = E(W_{st} \mid \mathcal{B}_s) + W_s = EW_{st} + W_s = W_s$$

and (ii′) holds since a.s.

$$E(W_{st}^2 \mid \mathcal{B}_s) = EW_{st}^2 = t - s.$$

The proof is terminated.

Throughout this chapter, we set

$$M_t = \sup_{0 \leq s \leq t} W_s, \; m_t = \inf_{0 \leq s \leq t} W_s.$$

B. Brownian extrema on $(0, \infty)$. *If a r.f.* W_T *is Brownian distributed then* $M_\infty = \lim_{t \to \infty} M_t$ *and* $m_\infty = \lim_{t \to \infty} m_t$ *exist, and a.s.*

(i) $$m_t < 0 < M_t \text{ on } (0, \infty],$$

(ii) $$-\infty = m_\infty = \liminf_{t \to \infty} W_t < \limsup_{t \to \infty} W_t = M_\infty = +\infty.$$

Proof. Since M_t and m_t are monotone in t, M_∞ and m_∞ exist and assertion (ii) has meaning.

1°. We prove (i). Since M_t does not decrease when t increases, it suffices to examine its behavior near $t = 0$. Let $t_n \downarrow 0$ and note that, by symmetry,

$$PA_n = P(W_t > 0) = 1/2$$

hence

$$P(\limsup A_n) \geqq \limsup PA_n = 1/2.$$

But $\limsup A_n$ is a tail event on the sequence $(W_{t_n} - W_{t_{n-1}})$ of independent r.v.'s, hence, by Kolmogorov zero-one law, its pr. is either 0 or 1. It follows that $P(\limsup A_n) = 1$. Since

$$\limsup A_n = [W_{t_n} > 0 \text{ i.o.}] \subset [M_t > 0 \text{ for all } t > 0],$$

the last event has pr. 1. By Brownian symmetry, the same is true for the event $[m_t < 0 \text{ for all } t > 0]$, and (i) is proved.

2°. We prove (ii). By symmetry, $\liminf_{t\to\infty} W_t = -\infty$ is implied by $\limsup_{t\to\infty} W_t = +\infty$ and, since

$$M_\infty = \sup_t W_t \geqq \limsup_{t\to\infty} W_t \geqq \limsup_{t\to\infty} W_n,$$

it suffices to show that $\limsup W_n = +\infty$ a.s. But, given m,

$$P(W_n > m) = \frac{1}{\sqrt{2\pi}} \int_{m/\sqrt{n}} e^{-v^2/2} dv \to 1/2$$

hence

$$P(W_n > m \text{ i.o.}) = P(\limsup_n [W_n > m]) \geqq \limsup P(W_n > m) = 1/2.$$

Therefore, letting $m - \infty$,

$$p = P(\limsup W_n = +\infty) \geqq 1/2.$$

Since the sequence (W_n) is that of successive sums of iid r.v.'s $W_n - W_{n-1}$, the Hewitt Savage zero-one law applies to the exchangeable event $[\limsup W_n = +\infty]$ so that $p = 0$ or 1. But $p \geqq 1/2$ hence $p = 1$, and (ii) is proved.

So far our definition and properties were in terms of Brownian law only. In fact, the most frequent definition of "Brownian motion" contains one basic supplementary condition: W_T is sample continuous, that is, its sample functions $(W_T(\omega),\ \omega \in \Omega)$ are continuous. The definition is then followed by a laborious existence proof. On the other hand, we know that in order to proceed to sample Analysis we need separability: As usual, *our r.f.'s are, or are selected to be, separable.* A Brownian distributed r.f., determined by the preceding Brownian covariance criterion **A** exists because of the Daniell-Kolmogorov consistency theorem 4.3**A.** Furthermore, we know from 38.2 that every r.f. has equivalent separable versions—with the same law.

Clearly, sample continuity implies separability and we shall show that, conversely, we can choose separable versions of Brownian distributed r.f.'s which are sample continuous. In fact, this results at once from 40.3**B**($\mathfrak{N}$) which was obtained by a technically involved analysis of separable decomposable processes. Here we shall give a direct and relatively elementary proof by means of some properties of independent usefulness.

First note that a Brownian distributed r.f. W_T is continuous in law since, as $s \to t$,

$$Ee^{iuW_s} = e^{-su^2/2} \to e^{-tu^2/2} = Ee^{iuW_t}.$$

Moreover, it is continuous in q.m. hence in pr. since, as $s \to t$,

$$E|W_s - W_t|^2 = |s - t| \to 0.$$

But when a r.f. W_T is continuous in pr. we can take for the separating set any countable set dense in T, such as the subset of all rationals or of all dyadic nmubers; we shall do it frequently.

REMARK. Note that, by 40.3**a,** a separable Brownian distributed r.f. is also continuous a.s. and all these types of continuity are uniform on every finite subinterval of T.

b. NORMAL APPROXIMATION LEMMA. *If* $\mathcal{J}_a = \int_a^\infty e^{-v^2/2}\,dv$, *then, as* $a \to \infty$,

$$\mathcal{J}_a \sim e^{-a^2/2}/a$$

and, more precisely, for $a > 0$,

$$\frac{a^2}{1 + a^2} e^{-a^2/2}/a \leq \mathcal{J}_a \leq e^{-a^2/2}/a.$$

Proof. Upon integrating by parts we have, for $a > 0$,

$$(1) \qquad \mathcal{J}_a = \int_a^\infty -\frac{1}{v}\,d(e^{-v^2/2}) = e^{-a^2/2}/a - \int_a^\infty e^{-v^2/2}\frac{dv}{v}$$

where, for $a \to \infty$,

$$(2) \qquad \int_a^\infty e^{-v^2/2}\frac{dv}{v} \leqq \mathcal{J}_a/a^2 = o(\mathcal{J}_a).$$

It follows that, as $a \to \infty$,

$$\mathcal{J}_a \sim e^{-a^2/2}/a.$$

Also (1) and (2) yield, for $a > 0$,

$$\frac{a^2}{1+a^2}e^{-a^2/2}/a \leqq \mathcal{J}_a \leqq e^{-a^2/2}/a.$$

The proof is terminated.

The following inequalities are trivial extensions of P. Lévy inequalities 18.1**C**: the same argument applies. We repeat it.

c. *Let $\xi_1, \cdots, \xi_n$ be independent r.v.'s and set $U_0 = V_n = 0$*

$$U_k = \xi_1 + \cdots + \xi_k,\ V_k = \xi_{k+1} + \cdots + \xi_n,\ k = 0, \cdots, n.$$

If $P(V_k \geqq b) \geqq p$ for all k, then

$$pP(\max_k U_k > a) \leqq P(U_n > a + b).$$

If $P(|V_k| \leqq b) \geqq p$ for all k, then

$$qP(\max|U_k| > a+b) \leqq P(|U_n| > a).$$

Proof. Upon setting

$$A_k = [U_0 \leqq a, \cdots, U_{k-1} \leqq a, U_k > a] \text{ with } A_0 = [U_0 > a]$$

and

$$B_k = [V_k \geqq b],$$

the first hypothesis yields the asserted inequality by

$$P(U_n > a+b) \geqq \sum_k PA_kB_k$$
$$= \sum_k PA_kPB_k \geqq p\sum_k PA_k = pP(\max U_k > a).$$

Since $|U_n| \geqq |U_k| - |V_k|$, upon setting

$$A_k = [|U_0| \leqq a+b, \cdots, |U_{k-1}| \leqq a+b,$$
$$|U_k| > a+b] \text{ with } A_0 = [|U_0| > a+b]$$

and

$$B_k = [|V_k| \leqq b],$$

the second hypothesis yields the asserted inequality by

$$P(|U_n| > a) \geqq \sum_k PA_kB_k = \sum_k PA_kPB_k \geqq qP(\max|U_k| > a + b).$$

d. Extrema equalities. *If W_T is a separable Brownian distributed r.f. then, for $a \geqq 0$,*

$$P(M_t > a) \leqq 2P(W_t > a), P(M_t < -a) \leqq 2P(W_t < -a),$$

$$P(\sup_{0 \leqq s \leqq t} |W_s| > a) \leqq 2P(|W_t| > a).$$

Proof. It suffices to prove the first inequality. For, the second follows by Brownian symmetry and the third one results by adding them. Because of separability and continuity in pr., it suffices to show that the first inequality holds for s varying over a countable set $\{s_1, s_2, \cdots\}$ dense in $[0, t]$; we can and do include therein $s_1 = 0$ and $s_2 = t$. Let $0 = t_0 < \cdots < t_n = t$ be the first n terms of this set reordered by increasing values. Since increments on disjoint intervals are independent

$$W_{t_k} = W_{0,t_1} + \cdots + W_{t_{k-1},t_k}, W_{t_n} - W_{t_k} = W_{t_k,t_{k+1}} + \cdots + W_{t_{k-1},t_k},$$

and the laws of $W_{t_n} - W_{t_k}$ are symmetric hence have zero medians, **c** applies with $b = 0$ and $p = 1/2$. Thus,

$$P(\sup_{0 \leqq k \leqq n} W_{t_k} > a) \leqq 2P(W_t > a)$$

and, letting $n \to \infty$, the asserted inequality follows.

C. Brownian sample continuity. *Almost all sample functions of a separable Brownian distributed r.f. W_T are continuous. More precisely, for every $p < 1/2$, there is a r.v. ν_p such that $n > \nu_p$ implies that a.s.*

$$|W_s - W_t| < 3/n^p \text{ for } |s - t| < 1/n, \qquad s, t \in [0, n].$$

Proof. The continuity assertion results from the last one since then almost all sample functions are uniformly continuous on every finite interval.

Thus, let

$$Y_n = \sup_{|s-t| < 1/n} |W_s - W_t|, s, t \in [0, n]$$

and set

$$Z_k = \sup_{t \in J_k} |W_t - W_{(k-1)/n}|, J_k = [(k-1)/n, k/n], k = 1, \cdots, n^2.$$

By triangle inequality

$$Y_n \leqq 3 \max_k Z_k.$$

Since the Z_k are iid r.v.'s, for every given $\epsilon > 0$,

$$p_n = P(\max Z_k > \epsilon) = P \bigcup_k | Z_k > \epsilon | \leqq \sum_k P(Z_k > \epsilon) \leqq n^2 P(Z_1 > \epsilon).$$

But, by **d**,

$$P(Z_1 > \epsilon) \leqq 2P(| W_{1/n} | > \epsilon) = \sqrt{2/\pi} \int_{\epsilon\sqrt{n}}^{\infty} e^{-v^2/2}\, dv$$

so that, by **b**,

$$p_n \leqq 2n^2 P(| W_{1/n} | > \epsilon) \sim 2\sqrt{2/\pi}\, n^{3/2}\, e^{-n\epsilon^2/2}/\epsilon.$$

Upon taking $\epsilon = \epsilon_n = 1/n^p$ with $p < 1/2$, it follows that $\sum p_n < \infty$. Therefore, by Borel-Cantelli lemma, there is a r.v. ν_p such that a.s. $n > \nu_p$ implies $Y_n \leqq 3/n^p$ on $[0, n]$, and the proof is terminated.

The a.s. sample continuity permits to complete **a** but we will not need the "if" assertion of the criterion below.

D. BROWNIAN MARTINGALES CRITERION. *W_T is a separable Brownian distributed r.f. if and only if W_T is a.s. sample continuous and*

(i) $(W_t, \mathcal{B}_t, t \geqq 0)$ *is a martingale*

with

(ii) $(W_t^2 - t, \mathcal{B}_t, t \geqq 0)$ *is a martingale*

or with

(iii) $E(W_{st}^2 \mid \mathcal{B}_s) = t - s, \quad 0 \leqq s < t < \infty.$

For, by **a**, the "only if" assertion results from Brownian law and the "if" assertion is 39.2**B**.

We are now ready for the final

BROWNIAN MOTION DEFINITION. A r.f. $W_T = (W_t, t \geqq 0)$ is said to be a *Brownian motion*, or *Brownian* for short, if
(i) W_T is *Brownian distributed*, that is, it is centered normal with covariance defined by $EW_sW_t = s \wedge t$ or, equivalently, it is decomposable with $\mathcal{L}(W_t) = \mathfrak{N}(0, t)$, and

(ii) *all sample functions of* W_T *are continuous, unbounded of both signs,* that is, $\liminf_{t\to\infty} W_t = -\infty$, $\limsup_{t\to\infty} W_t = +\infty$, and start at 0, that is, $W_0 = 0$.

Restrictions of Brownian X_T to subintervals of T will also be said to be Brownian.

Such versions exist: Brownian distributed W_T exists. Take a separable version also denoted by W_T so that, outside a null event N_1, the sample functions are continuous. By **B**, outside a null event N_2,

$$\liminf_{t\to\infty} W_t = -\infty, \ \limsup_{t\to\infty} W_t = +\infty$$

and, say, by **A**, $W_0 = 0$ outside a null event N_3.

Thus, (i) and (ii) hold outside the null event $N = N_1 \cup N_2 \cup N_3$. Then, either throw out N or, for all $t \geqq 0$, set, say,

$$W_t = t^{1/2} \sin t \text{ on } N.$$

Because of sample continuity

$$M_t = \max_{0\leqq s\leqq t} W_t, \ m_t = \min_{0\leqq s\leqq t} W_t,$$

and combining **B** with sample continuity we obtain at once

e. *Brownian sample functions have infinity of zeros in every neighborhood of* $t = 0$ *and of* $t = \infty$.

Recall that a *zero* of a function $W_T(\omega)$ is a value of t such that $W_t(\omega) = 0$.

While continuous, Brownian sample functions are not differentiable a.s. at any fixed t, that is,

f. *The* $W_T(\omega)$ *are not differentiable at fixed* t *for* $\omega \in N_t^c$, *where* N_t *is a null event.*

For, with arbitrary $a > 0$, as $h \to 0$,

$$P\left(\left|\frac{W_{t+h} - W_t}{h}\right| < a\right) = \frac{1}{\sqrt{2\pi}} \int_{-a/h}^{a/h} e^{-v^2/2}\, dv \to 0.$$

While N_t may depend upon t, we shall see in the next subsection that it may be replaced by N, so that almost all Brownian sample functions are "Hermitian monsters."

For the forthcoming "time inversion" we need

g. *If* W_T *is Brownian then*

$$tW_{1/t} \xrightarrow{a.s.} 0 \text{ as } t \to 0, \text{ equivalently, } W_t/t \xrightarrow{a.s.} 0 \text{ as } t \to 0.$$

Proof. The equivalence is immediate: Change t into $1/t$ in either assertion, and we prove the first one.
Let $W_t' = tW_{1/t}$ for $t > 0$ and $W_0' = 0$. It follows at once that $EW_s'W_t' = st(1/s, 1/t) = s \wedge t$ for $s, t > 0$ and also for $s = 0$ or $t = 0$. Thus, W_T is Brownian distributed. Since W_T is sample continuous on $[0, \infty)$, W_T' is sample continuous, hence separable but on $(0, \infty)$. Take a separating set in $(0, \infty)$ and adjoin to it $t = 0$. This enlarged separating set yields a separable version on $[0, \infty)$ with the same Brownian law that we continue to denote by W_T'. But separable Brownian distributed r.f. W_T' is a.s. sample continuous on $T = [0, \infty)$ hence $W_t' \xrightarrow{\text{a.s.}} W_0 = 0$, and the assertion is proved.

By refining as above the separable Brownian distributed r.f., we obtain a Brownian motion that we shall still denote by

$$(W_0' = 0, W_t' = tW_{1/t}).$$

The following transformations will be very useful:

E. BROWNIAN INVARIANCE THEOREM. *If $W_T = (W_t, t \geqq 0)$ is Brownian then so are*

$(-W_t, t \geqq 0)$	*symmetry*
$(W_{s+t} - W_s, t \geqq 0)$	*origin change*
$(c^{-1}W_{c^2t}, t \geqq 0), c > 0$	*scale change*
$(W_0' = 0, W_t' = tW_{1/t}, t \geqq 0)$	*time inversion*
$(W_u - W_{u-t}, 0 \leqq t \leqq u)$	*time reversal*

For, clearly Brownian law (compute the covariances) and sample continuity hold as well as the start at 0, and unboundedness holds except for time reversal where it has no content.

CONVENTION. *From now on, $W_T = (W_t, t \geqq 0)$ denotes Brownian motion*, unless otherwise stated. However, we shall mention it in statements of propositions.

41.3. Brownian sample oscillations. The preceding subsection was centered about the search for the most refined version of Brownian r.f.'s by means of direct and rather elementary arguments. The basic achievement was sample continuity. We examine now this property in much more detail and describe the extremely irregular oscillations on intervals and locally in terms of Lipschitz conditions, establish sample nowhere differentiability and study sample variations on intervals. We shall use 41.2 without further comment.
Let s, t belong to $[0, t_0]$.

Kolmogorov a.s. sample continuity condition on $[0, t_0]$ proved in the Corollary of 38.3**B**, is

$$E|W_s - W_t|^a \leqq c_0|s - t|^{1+b},\ a, b, c_0 \text{ positive.}$$

Since for separable Brownian distributed $W_{[0,t_0]}$

$$E|W_s - W_t|^{2n} = c_n|s - t|^n, \quad n = 1, 2, \cdots$$

where $c_n = (2n)!/2^n n!$ (see 13.4.1°), Brownian a.s. sample continuity obtains by taking $n > 1$ and t_0 as large as desired. Furthermore, we proved in the above proposition that, in fact, the Kolmogorov condition yields more than a.s. continuity: For every $c > 1$, there is a r.v. τ_c such that, for $|s - t| < \tau_c$ and $0 < p < b/a$, a.s.

$$|W_s - W_t| < c|s - t|^p.$$

Since $a = 2n$ and $b = n - 1$ where n can be taken as large as desired, $p \leqq (n - 1)/2n \uparrow 1/2$ and it follows that for every $c > 1$ there is a r.v. τ_c such that, for $|s - t| < \tau_c$ and p as close to $1/2$ (from below) as desired

$$|W_s - W_s| < c|t - s|^p.$$

Thus, the first assertion below holds:

a. BROWNIAN LIPSCHITZ p-PROPERTY. *Almost all Brownian sample functions are* Lip_p *for* $p < 1/2$ *but are not* $\text{Lip}_{1/2}$, *on any given interval* $[0, t_0]$.

In fact, both a.s. sample continuity and **a** follow at once from the much more precise

A. BROWNIAN SAMPLE CONTINUITY MODULUS. *If* W_T *is Brownian then, in any given interval* $[0, t_0]$, *a.s.*

$$\limsup_{s-t\to 0} |W_s - W_t| / \sqrt{2|s - t| \log 1/|s - t|} = 1, \tag{1}$$

equivalently,

when $c > 1$, *there is a r.v.* $\tau_c > 0$ *such that for* $|s - t| < \tau_c$, *a.s.,*

$$|W_s - W_t| < c\sqrt{|s - t| \log 1/|s - t|} \tag{1'}$$

when $c < 1$, *there are arbitrarily small values of* $|s - t|$ *such that, a.s.*

$$|W_s - W_t| > c\sqrt{2|s - t| \log 1/|s - t|}. \tag{1''}$$

Proof. The equivalence assertion results from the definition of "limsup," and we prove (1′) and (1″).

To simplify the writing we take $t_0 = 1$. Let

$$h = |s - t| \text{ with } 0 < h < 1, g(h) = \sqrt{2h \log 1/h}.$$

Since $W_s - W_t$ is normal $\mathfrak{N}(0, h)$, we obtain

$$q(h) = P(|W_s - W_t| > cg(h)) = \sqrt{2/\pi} \int_{c\sqrt{2\log 1/h}} e^{-v^2/2}\, dv$$

$$\sim h^{c^2}/c\sqrt{\pi \log 1/h}.$$

When $c > 1$, elementary computations show that all conditions of 38.2**B** hold, and (1′) obtains.

When $c < 1$, set

$$A_{nk} = [|W_{kh_n} - W_{(k-1)k_n}| \geqq cg(h_n)],$$

$$h_n = 1/2^n, k = 1, \cdots, 2^n.$$

The A_{nk} are independent in k so that

$$P \bigcup_k A_{nk} = (1 - q(h_n))^n \to 1.$$

Thus, with pr. as close to 1 as desired, at least one of A_{nk} occurs for n sufficiently large, that is, for h_n sufficiently small, depending upon the sample function, and (1″) obtains. The proof is terminated.

While **A** describes Brownian sample oscillations in fixed finite intervals in terms of a uniform Lipschitz condition, a more precise one—as is to be expected, describes local sample oscillations, that is, in neighborhoods of fixed points in $[0, \infty]$. As background to LIT, we prove

b. *If W_T is Brownian and g is a Borel function on $(0, \infty)$ to $(0, \infty)$, then*

$$\liminf_{t\to 0} W_t\, g(t) \text{ and } \limsup_{t\to 0} W_t/g(t) \text{ are degenerate}$$

and

$$P(\lim_{t\to 0} W_t/g(t) \text{ exists}) = 0 \text{ or } 1.$$

For, setting

$$\mathfrak{B}_{t+} = \bigcap_{u>t} \mathfrak{B}_u = \bigcap_{u>t} \mathfrak{B}(W_s, s \leqq u),$$

the assertions result at once from the fact that the limits therein are $\mathfrak{B}_{0+}$-measurable, by the Blumenthal zero-one law in the Brownian case:

b₀. *The σ-field $\mathfrak{B}_{0+}$ is degenerate.*

Proof. $\mathfrak{B}_{0+} \subset \mathfrak{B}_t$ for all $t > 0$ while, by Brownian decomposability for all $t > s > 0$, $\mathfrak{B}_t$ and $\mathfrak{B}(W_{st})$, hence $\mathfrak{B}_{0+}$ and $\mathfrak{B}(W_{st})$, are independent. But, by Brownian sample continuity (at 0),

$$W_{st} = W_t - W_s \to W_t \text{ as } s \to 0.$$

It follows that $\mathcal{B}_{0+}$ and $\mathcal{B}_t$ are independent for all $t > 0$ hence $\mathcal{B}_{0+}$ and $\bigcap_{t>0} \mathcal{B}_t = \mathcal{B}_{0+}$ are independent, and the assertion obtains.

B. LAWS OF THE ITERATED LOGARITHM (LIT). *Let $W_T = (W_t, t \geqq 0)$ be Brownian. Then, a.s.*

Local LIT: For every fixed $s \geqq 0$

$$\limsup_{t\to 0} (W_{s+t} - W_s)/\sqrt{2t\log\log 1/t} = 1$$

$$\liminf_{t\to 0} (W_{s+t} - W_s)/\sqrt{2t\log\log 1/t} = -1$$

$$\limsup_{t\to 0} |W_{s+t} - W_t|/\sqrt{2t\log\log 1/t} = 1.$$

Asymptotic LIT:

$$\limsup_{t\to\infty} W_t/\sqrt{2t\log\log t} = 1$$

$$\liminf_{t\to\infty} W_t/\sqrt{2t\log\log t} = -1$$

$$\limsup_{t\to\infty} |W_t|/\sqrt{2t\log\log t} = 1.$$

Proof. Local LIT reduces to those at $s = 0$ by Brownian origin change invariance. In turn, these ones and asymptotic LIT reduce to each other by Brownian time inversion invariance. Thus, it suffices to prove, say, local LIT at $s = 0$. But the third one obtains from the first two and the second one obtains from the first one by Brownian symmetry. Thus, it remains only to show that a.s.

(1) $$\limsup_{t\to 0} W_t/\sqrt{2t\log\log 1/t} = 1.$$

Let

$$g(t) = \sqrt{2t\log\log 1/t}$$

and let $t^n = q^n$ with $0 < q < 1$ to be selected suitably.
We have to prove (1) or equivalently, that a.s.

(1′) $$\limsup_{t\to 0} W_t/g(t) \leqq c \text{ for every } c > 1$$

and

(1″) $$\limsup_{t\to 0} W_t/g(t) \geqq c' \text{ for every } c' < 1.$$

1°. To prove (1′), it suffices to show that $P(A_n \text{ i.o.}) = 0$ where

$$A_n = [W_t > cg(t) \text{ for some } t \in [t_{n+1}, t_n]]$$

with given $c > 1$ and some $q \in (0, 1)$.
Since $g(t)$ decreases as t decreases,

$$A_n \subset [M_{t_n} > cg(t_{n+1})].$$

But, for every $\epsilon > 0$,

$$P(M_{t_n} > \epsilon\sqrt{t_n} \leqq 2P(W_{t_n} > \epsilon\sqrt{t_n})$$
$$= \sqrt{2/\pi}\int_\epsilon^\infty e^{-v^2/2}\,dv \leqq \sqrt{2/\pi}\, e^{-\epsilon^2/2}/\epsilon$$

so that, taking

$$\epsilon = \epsilon_n = cg(t_{n+1})/\sqrt{t_n} = c\sqrt{2q\log\{(n+1)\log 1/q\}},$$

it follows that

$$p_n = PA_n \leqq P(M_n > cg(t_n)) \sim a/(n+1)^{c^2q}\sqrt{\log(n+1)}$$

where $a = a(c, q)$ is independent of n.
Since $c > 1$, we can select q such that $1/c^2 < q < 1$ so that $c^2q < 1$. Then $\sum p_n < \infty$ hence, by the Borel-Cantelli lemma, $P(A_n \text{ i.o.}) = 0$, and (1′) is proved.

2°. To prove (1″), it suffices to show that

$$P(W_{t_n} > c'g(t_n) \text{ i.o.}) = 1$$

for given $c' < 1$ and some $q \in (0, 1)$.

Since the increments $Y_n = W_{t_n} - W_{t_{n-1}}$ are normal $\mathfrak{N}(0, t_n - t_{n-1})$, for every $\epsilon > 0$,

$$P(Y_n > \epsilon\sqrt{t_n - t_{n-1}}) = \frac{1}{\sqrt{2\pi}}\int_\epsilon^\infty e^{-v^2/2}\,dv \sim \frac{1}{\sqrt{2\pi}}\, e^{-\epsilon_n^2/2}/\epsilon.$$

Let

$$\epsilon = \epsilon_n = c_1g(t_n)/\sqrt{t_n} = c_1\sqrt{2\log(n\log 1/q)}$$

with $0 < c_1 < c'(<1)$. Then

$$p_n = P(Y_n > c_1(1-q)^{1/2}g(t_n)) \sim a_1/n^{c_1^2}\sqrt{\log n}$$

where $a_1 = a_1(c_1, q)$ is independent of n. Since $c_1^2 < 1$ we have $\sum p_n = \infty$ and, the Y_n being independent, the Borel zero-one law applies, hence

$$P(Y_n > c_1(1-q)^{1/2}g(t_n) \text{ i.o.}) = 1.$$

On the other hand, by Brownian symmetry, (1′) with, say, $c = 2$, yields a.s.

$$-W_{t_{n+1}} < 2g(t_n) \text{ i.o. hence } W_{t_{n+1}} \geqq \frac{-2g(t_{n+1})}{g(t_n)} g(t_n) \text{ i.o.}$$

It follows that a.s.

$$W_{t_n} = Y_n + W_{t_{n+1}} \geqq (c_1(1-q)^{1/2} - 2g(t_{n+1})/g(t_n))g(t_n) \text{ i.o.},$$

where

$$g(t_{n+1})/g(t_n) \to q^{1/2},$$

hence, a.s.,

$$W_{t_n} \geqq (c_1(1-q)^{1/2} - 4q^{1/2})g(t_n) \text{ i.o.}$$

But

$$c_1(1-q)^{1/2} - 4q^{1/2} \to c_1 < c' \text{ as } q \to 0$$

hence we can select $q \in (0, 1)$ sufficiently small so that a.s.

$$W_{t_n} \geqq c'g(t_n) \text{ i.o.},$$

and (1″) is proved.

In preparation for the a.s. sample nowhere differentiability, we remind the reader that a real-valued function on, say, $[0, \infty)$ is differentiable at t if and only if the four Dini derivatives at t have common finite value. The right ones are defined by

$$D^+g(t) = \limsup_{0<h\to 0} \frac{g(t+h)-g(t)}{h} \quad D_+g(t) = \liminf_{0<h\to 0} \frac{g(t+h)-g(t)}{h},$$

and the left ones obtain upon replacing $h > 0$ by $h < 0$. If g is continuous then we can let $h \to 0$ along rationals.
Recall that a property holds a.s. if it holds outside some null event; what happens on this "exceptional" event is irrelevant.

The ingenious proof below is due to Dvoretsky, Erdös and Kakutani.

C. BROWNIAN SAMPLE NOWHERE DIFFERENTIABILITY. *Almost all Brownian sample functions $W_T(\omega) = (W_t(\omega),\ t \geqq 0)$ are nowhere differentiable.*

Proof. It suffices to show that the assertion holds on every finite interval $J \subset T$; to simplify the notation we take $J = [0, 1]$.

Let

$$A = [D^+W_t \text{ and } D_+W_t \text{ are finite for some } t \in [0, 1]].$$

Then

$$A = \bigcup_{m=1}^{\infty} \bigcup_{n=1}^{\infty} A_{mn}$$

where

$$A_{mn} = [|W_{t+h} - W_t| < mh \text{ for all } h \in [0, 1/n] \text{ for some } t \in [0, 1]].$$

It suffices to prove that A is contained in a null event or, equivalently, that so is every A_{mn}.
On A_{mn}, if $t \in [(k-1)/n, k/n]$ then, for $n \geqq 4m$,

$$|W_{k/n} - W_t| < m/n; \ |W_{(k+1)/n} - W_t| < 2m/n,$$
$$|W_{(k+2)/n} - W_t| < 3m/n, \ |W_{(k+3)/n} - W_t| < 4m/n.$$

It follows, by triangle inequality, that

$$A_{mn} \subset B = \bigcap_{n=4m}^{\infty} B_n \text{ with } B_n = \bigcup_{k=1}^{n} B_{nk},$$

where

$$B_{nk} = [|W_{k+1)/n} - W_{k/n}| < 3m/n, \ |W_{(k+2)/n} - W_{(k+1)/n}| < 5m/n,$$
$$W_{(k+3)/n} - W_{(k+2)/n}| < 7m/n].$$

But the Brownian increments on the B_{nk} are independent and normal $\mathfrak{N}(0, 1/n)$. Therefore, for $n \geqq 4m$,

$$PB_{nk} \leqq 3.5.7m^3/n^{3/2}$$

so that

$$PB \leftarrow PB_n \leqq nPB_{nk} \leqq 3.5.7m^3/n^{1/2} \to 0,$$

and the assertion is proved.

COROLLARY. *Almost all Brownian sample functions are not monotone and not of bounded variation on nondegenerate bounded intervals J.*

For continuous functions on $[0, \infty)$ are bounded on J, so that if these functions were monotone they would be of bounded variation on J and such functions are λ-a.e. differentiable (where λ is the Lebesgue measure), contradicting sample nowhere differentiability.

Contrary to this Corollary, since $E(\Delta W_t)^2 = \Delta t$, we may expect that the "quadratic variation" on an interval J, to be defined, would be its length $|J|$: Let $J = [0, t]$ and, for $k = 1, \cdots, k_n$, let

$$W_{nk} = W_{t_{nk}} - W_{t_{n,k-1}}, \qquad s_{nk} = t_{nk} - t_{n,k-1}$$

correspond to subdivisions

$$D_n = \{0 = t_{n0} < \cdots < t_{nk_n} = t\}$$

with

$$D_n \subset D_{n+1} \text{ and } \max_k s_{nk} \to 0.$$

We say that the *quadratic variation* of $W_{[0,t]}$ is

$$\int_{[0,t]} (dW_s)^2 \equiv \text{a.s.} \lim \sum_k W_{nk}^2.$$

D. BROWNIAN QUADRATIC VARIATION. *If $W_{[0,t]}$ is Brownian then its quadratic variation* $\int_{[0,t]} (dW_t)^2 = t$, *in fact,*

$$V_n = \sum_k W_{nk}^2 \xrightarrow[q.m.]{a.s.} t.$$

Proof. The $Y_{nk} = W_{nk}/\sqrt{s_{nk}}$ are iid in $k = 1, \cdots, k_n$ with common law $\mathfrak{N}(0, 1)$, so that

$$E(Y_{nk}^2 - 1) = 0, EY_{nk}^4 = 3$$

hence

$$E(Y_{nk}^2 - 1)^2 = 2.$$

Since the $Y_{nk}^2 - 1$ are independent in k and $t = \sum_k s_{nk}$, it follows that

$$E(V_n - t)^2 = \sum_k E(Y_{nk}^2 - 1)^2 = 2s_{nk}^2 \leqq 2(\max_k s_{nk})\, t \to 0$$

hence

$$V_n \xrightarrow{q.m.} t.$$

To complete the proof, it suffices to show that the V_n form a martingale reversed sequence. For then, by 32**A**(ii), $V_n \xrightarrow{q.m.} t$ implies $V_n \xrightarrow{a.s.} t$. Without loss of generality, we can take $k_n = n$ by adding subdivisions if necessary. Then, for every n there is a $k \leqq n$ such that

(1) $$V_n = V_{n+1} + 2W_{n+1,k}W_{n+1,k+1}.$$

Let

$$\mathcal{C}_n = \mathcal{B}(V_{n+1}, V_{n+2}, \ldots)$$

and

$$\mathcal{D}_n = \mathcal{B}(W_t, t \leqq t_{n+1,k}; |W_s - W_t|, s, t \geqq t_{n+1,k}).$$

Since $\mathcal{D}_n$ conditions symmetrically the symmetrically distributed r.f. $(W_t, t_{n+1,k} \leqq t \leqq t_{n+1,k+1})$, a.s.,

$$E(W_{n+1,k}W_{n+1,k+1} \mid \mathcal{D}_n) = W_{n+1,k}E(W_{n+1,k+1} \mid \mathcal{D}_n) = 0$$

so that, conditioning (1) by $\mathcal{C}_n \subset \mathcal{D}_n$, a.s.

$$E(V_n \mid V_{n+1}, V_{n+2}, \cdots) = V_{n+1},$$

and the proof is completed.

Note that the preceding Corollary to **C** follows also from **D**: For, Brownian sample functions being uniformly continuous on $[0, t]$, $\max_k |W_{nk}| \to 0$ so that, if $W_{(0,t)}$ were a.s. of bounded variation for $t > 0$ then a.s.

$$0 < t \leftarrow V_n \leqq \max_k |W_{nk}| . \sum_k |W_{nk}| \to 0.$$

41.4. Brownian times and functionals. In what follows, $W_T = (W_t, t \geqq 0)$ is a Brownian motion, s and t with or without affixes belong to $T = [0, \infty)$, and

$$\mathcal{B}_t = \mathcal{B}(W_s, s \leqq t),\ \mathcal{B}_T = (\mathcal{B}_t, t \geqq 0);\ \mathcal{B}_\infty = \bigvee \mathcal{B}_t.$$

A $\mathcal{B}_T$-time (or W_T-time) will be called a "Brownian time" and we recall the definition and various properties of such "times" in 38.4 that we shall use without further comment.

A measurable function τ on Ω to $[0, \infty)$ is a *Brownian time* if $[\tau \leqq t] \in \mathcal{B}_t$ for every t, and then so are all $\tau \wedge t$ and $\tau + t$. Every t is a degenerate Brownian time.

To Brownian times τ are associated σ-fields

$$\mathcal{B}_\tau = \{B_\infty \in \mathcal{B}_\infty : B_\infty[\tau \leqq t] \in \mathcal{B}_t \text{ for all } t \geqq 0\},$$

with respect to which they are measurable. If $\sigma \leqq \tau$ are Brownian then $\mathcal{B}_\sigma \subset \mathcal{B}_\tau$. Since our W_T is sample continuous, for every finite Brownian τ, the W_τ defined by $W_\tau(\omega) = W_{\tau(\omega)}(\omega)$ are $\mathcal{B}_\tau$-measurable r.v.'s. Thus, we write "$\mathcal{B}_\tau = \mathcal{B}(W_t, t \leqq \tau)$" and speak about events $B_\tau \in \mathcal{B}_\tau$ as "defined on $(W_t, t \leqq \tau)$" or "on W_T up to time τ" (included).

Extremely useful Brownian times are finite elementary ones:

$$\tau = \sum_j t_j I_{B_j}, \quad t_1 < t_2 < \ldots, \quad B_j \in \mathcal{B}_{t_j}$$

to which correspond r.v.'s

$$W_\tau = \sum W_{t_j} I_{B_j}.$$

For, given a finite Brownian time τ, the elementary ones

$$\tau_n = \sum_{k=1}^{\infty} kh_n I_{[(k-1)h_n \leqq \tau < kh_n]}, \quad h_n = 1/2^n,$$

are such that

$$\tau_n \downarrow \tau, \quad \mathcal{B}_{\tau_n} \supset \mathcal{B}_\tau$$

and, by sample (right) continuity, the r.v.'s

$$W_{\tau_n} = \sum_{k=1}^{\infty} W_{kh_n} I_{[(k-1)h_n \leqq \tau < kh_n]} \to W_\tau.$$

Among the most important Brownian times are *first exit* times τ_U from open sets U (or *first hitting* times of closed sets U^c): $\tau_U(\omega)$ is defined to be the infimum of all those t for which the distance of the sets $\{W_s(\omega)$: $s \leqq t\}$ and U^c is zero; if such $\tau_U(\omega)$ does not exist, we set $\tau_U(\omega) = +\infty$. *The τ_U are Brownian times.* For, setting $S_n = \{x: d(x, U^c) < 1/n\}$ and taking all rationals $r < t$,

$$[\tau_U \leqq t] = [W_t \in U^c] \cup (\bigcap_n \bigcup_{r<t} [W_r \in S_n] \in \mathcal{B}_t.$$

We shall be using two such times and, to simplify the writing, set

$$\tau_{a,b} = \tau_U \text{ when } U = (a, b) \text{ with } a < 0 < b,$$

$$\tau_c = \tau_U \text{ when } U \text{ is the complement of } \{c\}.$$

Since Brownian motion starts at 0 and is sample continuous and unbounded (of both signs), *$\tau_{a,b}$ and τ_c are finite.*

When $\tau = \tau_{a,b}$, we can describe completely $\mathcal{L}(W_\tau)$ and find $E\tau$ in a rather elementary way. We require a very simple case of 39.2**A** that we prove directly.

a. Martingale lemma. *Let $(X_s, s, 0 \leqq s \leqq t)$ be a sample rightcontinuous martingale. If $Y_t = \sup_{0 \leqq s \leqq t} |X_t|$ is integrable then $EX_\sigma = EX_0$ or every* time σ ($\leqq t$) *of this martingale.*

Proof. Let $h_n = 1/2^n$, $k = 1, \cdots, 2^n$, and set

$$B_{nk} = [(k-1)h_n t < \sigma \leqq kh_n t]$$

so that

$$\sigma_n = \sum_k kh_n t I_{B_{nk}} \downarrow \sigma$$

and

$$X_{\sigma_n} = \sum_k X_{kh_n t} I_{B_{nk}} \to X_\sigma.$$

By martingale property, $kh_n t \leqq t$ implies that

$$EX_{\sigma_n} = \sum_k \int_{B_{nk}} X_t = EX_t = EX_0.$$

Since all

$$|X_{\sigma_n}| \leqq Y_t \text{ integrable,}$$

the dominated convergence theorem applies so that

$$EX_0 = EX_{\sigma_n} \to EX_\sigma,$$

and the lemma is proved.

b. *If τ is a Brownian time then, for every t,*

(1) $EW_{\tau \wedge t} = 0$, (2) $EW_{\tau \wedge t}^2 = E(\tau \wedge t)$.

For, by the Brownian martingales property 41.2**a**, $(W_s, \mathcal{B}_s, 0 \leqq s \leqq t)$ and $(W_s^2 - s, \mathcal{B}_s, 0 \leqq s \leqq t)$ are martingales while, 41.2**a** implies that $\sup_{0 \leqq s \leqq t} |W_s|$ and $\sup_{0 \leqq s \leqq t} W_s$ are integrable, so that the above lemma with $\sigma = \tau \wedge t$ applies, and $EW_{\tau \wedge t} = 0$

$$EW_{\tau \wedge t} - \tau \wedge t = EW_0^2 - 0 = 0.$$

A. *If $\tau = \tau_{a,b}$ then W_τ takes either value a or value b,*

(1) $$P(W_\tau = a) = \frac{b}{|a| + b}, \quad P(W_\tau = b) = \frac{|a|}{|a| + b}$$

and

(2) $$E\tau = EW_\tau^2 = |a|b.$$

Proof. We use the above proposition without further comment. Since Brownian sample functions start at $0 \in (a, b)$ and are unbounded of both signs, τ is finite and $W_\tau(\omega)$ leaves (a, b) either at a or at b with pr. p and $1\text{-}p$, respectively. Therefore,

$$|W_t| \leqq c = |a| + b \text{ on } [\tau > t]$$

so that, as $t \to \infty$,

$$\int_{[\tau > t]} |W_t| \leqq cP(\tau > t) \to 0$$

and, by the dominated convergence theorem, letting $t \to \infty$ along integers,

$$0 = EW_{\tau \wedge t} = I_{[\tau \leqq t]} W_\tau + I_{[\tau > t]} W_t \to EW_\tau.$$

Thus, $ap + b(1 - p) = 0$ and (i) obtains.
The same argument applied to W_t^2 in lieu of W_t yields

$$E\tau \uparrow E(\tau \wedge t) = EW_{\tau \wedge t}^2 = I_{[\tau \leqq t]} W_\tau^2 + I_{[\tau > t]} W_t^2 \to EW_\tau^2,$$

and (2) obtains using (1). The proof is terminated.

By proving first that $\tau_{a,b}$ is integrable, the preceding proposition would result from the much more general

A'. BROWNIAN WALD RELATIONS. *If τ is an integrable Brownian time then*

$$EW_\tau = 0 \textit{ and } EW_\tau^2 = E\tau < \infty.$$

Proof. According to **b**

$$(1)\ EW_{\tau \wedge t} = 0 \quad (2)\ EW_{\tau \wedge t}^2 = E(\tau \wedge t).$$

Since $E(\tau \wedge t) \leqq E\tau < \infty$, (2) yields (3) $\sup_t EW_{\tau \wedge t}^2 < \infty$.

Thus, letting t take integer values, the $W_{\tau \wedge t}$ are uniformly integrable hence we can interchange "$\lim_{t\ \infty}$" and "E" in (1) and, applying the monotone convergence theorem, the first relation obtains by

$$EW_\tau \leftarrow EW_{\tau \wedge t} = 0.$$

As for the second relation, since the W_t form a martingale, by (3) and 39.1**C**,

$$E \sup_t W_{\tau \wedge t}^2 \leqq 4 \sup_t W_{\tau \wedge t}^2 < \infty$$

so that, by (2), applying the dominated and monotone convergence theorem, as $t \to \infty$ along integers,

$$EW_\tau^2 \leftarrow EW_{\tau\wedge t}^2 = E(\tau \wedge t) \uparrow E\tau < \infty,$$

and the second relation obtains.

Perhaps the most important general Brownian property is the one proved below. It was used systematically by P. Lévy and rigorously proved by Hunt and by Blumenthal. It will be applied without further comment in the remainder of this subsection, especially with $\tau = \tau_c$.

B. BROWNIAN TIMES ORIGIN INVARIANCE. *Brownian motion starts anew at every finite Brownian time: If τ is time of Brownian $W_T = (W_t, t \geqq 0)$ then the r.f.*

$$W_T^\tau = W_{\tau+T} - W_\tau = (W_{\tau+t} - W_\tau, t \geqq 0)$$

is Brownian and independent of $W_{[0,\tau]} = (W_t, t \leqq \tau)$.

Proof. Clearly, W_T^τ, like W_T, starts at 0 and its sample functions are continuous and unbounded of both signs. Thus, it remains to prove that W_T^τ is Brownian distributed and independent of $W_{[0,\tau]}$.
Let $(W_{t_1}, \cdots, W_{t_m})$ and $(W_{t_1}^\tau, \cdots, W_{t_m}^\tau)$ be arbitrary finite sections of W_T and W_T^τ. Let $g_1, \cdots, g_m$ be arbitrary, real- or complex-valued, bounded continuous functions on R, and let an arbitrary event $B \in \mathcal{B}_\tau = (X_t, t \leqq \tau)$. It suffices to prove that

(C) $E\{I_B g(W_{t_1}^\tau) \times \cdots \times g(W_{t_m}^\tau)\} = PBE\{g_1(W_{t_1}) \times \cdots \times g(W_{t_m})\}.$

For, say, take $g_j(a) = e^{iu_ja}$ so that $g_j(W_t) = e^{iuW_t}$, $g_j(W_t^\tau) = e^{iuW_t^\tau}$ and recall that mutually consistent ch. f.'s of finite sections determine consistent distributions of these sections which, in turn, determine the distribution of the corresponding r.f.'s. Or, take $g_j(a) = I_{[0,a_j)}$ so that, say, $g_j(W_t) = I_{[W_t<a]}$ and proceed from there.

Let

$$\tau_n = \sum_{k=1}^{\infty} kh_n I_{[(k-1)h_n \leqq \tau < kh_n]}, \quad h_n = 1/2^n,$$

so that

(1) $$\tau_n \downarrow \tau \text{ and } X_{\tau_n+t} \to X_{\tau+t}$$

and

(2) $$B[\tau_n = kh_n] \in \mathcal{B}_{kh_n} = \mathcal{B}(X_s, s \leqq kh_n).$$

As usual, we first prove (C_n), that is, (C) with τ_n in lieu of τ. Then, letting $n \to \infty$, (C) obtains by (1) and the dominated convergence theorem. But, by (2) and Brownian decomposability and invariance under change of origin to kh_n,

$$E\{I_{B[\tau_n=kh_n]}\, g_1(W_{kh_n+t_1} - W_{kh_n}) \times \cdots \times g_m(W_{kh_n+t_m} - W_{kh_n})\} = P(B[\tau_n = kh_n])E\{g_1(W_{t_1} \times \cdots \times g_m(W_{t_m})\}$$

so that, summing over k, (C_n) obtains and the proof is completed.

This powerful proposition makes the properties of W_T and of W_T^{τ} the same. For example,

COROLLARY. *Let τ be a finite Brownian time. Then*
(i) *Almost surely*

$$\limsup_{t\to 0}(W_{\tau+t} - W_\tau)/\sqrt{2t\log\log 1/t} = 1$$

and

$$\liminf(W_{\tau+t} - W_\tau)/\sqrt{2t\log\log 1/t} = -1$$

(i') *Almost surely*

$$\limsup_{t\to \infty}(W_{\tau+t} - W_\tau)/\sqrt{2t\log\log} = 1$$

and

$$\liminf_{t\to \infty}(W_{\tau+t} - W_\tau)/\sqrt{2t\log\log t} = -1$$

(ii) *Almost all sample functions $W_T^{\tau}(\omega)$ have an infinity of values of t in every neighborhood $V(\omega)$ of $\tau(\omega)$ such that $W_{\tau+t}(\omega) = W_\tau(\omega)$.*

(ii') *Almost all sample functions $W_T^{\tau}(\omega)$ have an infinity of values in every neighborhood of $t = \infty$ such that $W_{\tau+t}(\omega) = W_\tau(\omega)$.*

Note that, by sample continuity, (i) implies (ii) and (i') implies (ii').

C. BROWNIAN LEVEL SETS STRUCTURE. For almost all ω, the *c-level sets*

$$\Lambda(\omega) = \{t: W_t(\omega)\} = c$$

are perfect, unbounded and λ-null.

By definition, a set is *perfect* if it is closed and dense in itself (hence uncountable). Recall that λ is Lebesgue measure and $\tau_c = \{t: W_t = c\}$ is a finite Brownian time.

Proof. It suffices to prove the assertion for the "zero" level sets $\Lambda_0(\omega)$ since, for $\tau = \tau_c$,

$$\Lambda_c(\omega) = \{t: W_{\tau_c+t}(\omega) = W_{\tau_c}(\omega)\} = \{t: W_t^{\tau_c}(\omega) = 0\}.$$

The $\Lambda_0(\omega)$ are closed since $W_T(\omega)$ are continuous, and they are unbounded by Corollary (ii'). For almost all ω, $\lambda(\Lambda_0(\omega)) = 0$ since W_T is Borelian hence $(\lambda \times P)$-measurable and $P(W_t = 0)$ for every t, so that, by Fubini theorem and $P(W_t = 0)$ for every t,

$$E(\lambda(\Lambda_0)) = E\int_0^\infty I_{[W_t=0]}\,dt = \int_0^\infty P(W_t = 0)\,dt = 0.$$

It remains to prove that almost all $\Lambda_0(\omega)$ are dense in themselves. For every rational $r \geqq 0$, let

$$\tau(r) = \inf\{t \geqq r: W_t = 0\}.$$

Clearly, $\tau(r)$ is a finite Brownian time and $W_{\tau(r)} = 0$. By Corollarys (ii), $W(0) = 0$ is a.s. rightlimit of zeros of W_T, that is, 0 is limit point of $\Lambda_0(\omega)$ for almost all ω. The same is true with $\tau(r)$ for Brownian $W_T^{\tau(r)} = 0$. Thus, if $A(r)$ is the set of those w for which $r(r)$ is limit point of $\Lambda_0(\omega)$ then $P(\bigcap_r A(r)) = 1$. But $\omega \in \bigcap_r A(r)$ if and only if $\Lambda_0(\omega)$ is dense in itself, and the proof is completed.

The following application of **B** is an extremely useful and a far-reaching generalization of the "symmetry" or "reflection principle" of Desire Andre for the ballot problem studied in Intuitive Background CDIII.

D. REFLECTION PRINCIPLE. *Brownian motion reflected at a finite Brownian time τ is Brownian: If W_T is Brownian so is $\rho_\tau W_T$ defined by*

$$\rho_\tau W_t = W_t \text{ or } 2W_\tau - W_t \text{ according as } t \leqq \tau \text{ or } t > \tau.$$

Proof. Clearly, $\rho_\tau W_T$ starts at 0 and is sample continuous and unbounded of both signs. It remains to prove that it is Brownian distributed.

Let $Y_t = W_T$ for $t \leqq \tau$ and $Y_t = W_\tau$ for $t > \tau$ be W_T "stopped at τ" and, as before, let $W_T^\tau = W_{\tau+T} - W_\tau$ be W_T "starting anew at τ." Since τ and Y_τ are $\mathcal{B}_\tau$-measurable, the Brownian W_T^τ is independent of (τ, Y_T) and, by symmetry, so is $-W_T^\tau$. Therefore setting

$$U_t = V_t = W_t \text{ for } t \leqq \tau$$

and

$$U_t = W_\tau + W'_{t-\tau},\ V_t = W_\tau - W'_{t-\tau} \text{ for } t \geqq \tau,$$

U_T and V_T have the same distribution. But, for $t \geqq \tau$,

$$W'_{t-\tau} = W_t - W_\tau$$

so that

$$U_t = W_\tau + (W_t - W_\tau) = W_t$$

and

$$V_t = W_\tau - (W_t - W_\tau) = 2W_\tau - W_t = \rho_\tau W_t.$$

Thus,

$$U_T = W_T,\ V_T = \rho_\tau W_T,$$

and the asserted principle obtains.

In the remainder of this subsection we derive distributions of various Brownian functionals using without further comment **B** and **C** with $\tau = \tau_c$—the *c-level reflection*: The changes of W_T after c-level time τ_c are independent of its changes before τ_c and their pr.'s are not changed by reflecting W_T at c-level.

First we transform extrema inequalities 41.2**d** into

c. EXTREMA EQUALITIES. *For* $c > 0$

$$P(M_t > c) = P(+m_t < -c) = P(|W_t| > c) = \sqrt{2/\pi} \int_c^\infty e^{-v^2/2}\, dv.$$

For,

$$P(M_t > c) = P(M_t > c, W_t > c) + P(M_t > c, W_t = c) + P(M_t > c, W_t < c)$$

where

$$P(M_t > c, W_t > c) \leqq P(W_t = c) = 0,$$

$$P(M_t > c, W_t > c) = P(W_t > c)$$

and, by c-level reflection,

$$\rho_{\tau_c} W_t = W_t \text{ or } 2c - W_t \text{ according as } t \leqq \tau_c \text{ or } t > \tau_c$$

hence, setting $\rho_{\tau_c} M_t = \max_{0 \leqq s \leqq t} \rho_{\tau_c} W_s$,

$$P(M_t > c, W_t < c) = P(\rho_{\tau_c} M_t > c, 2c - W_t < c)$$
$$= P(M_t > c, W_t > c) = P(W_t > c).$$

More generally

E. Distribution of (M_t, W_t). *Let* $c > 0$ *and* $J = [c_1, c_2]$. *Then*

$$P(M_t > c, W_t \in J) = \frac{1}{\sqrt{2\pi t}} \int_{c_1 \vee c}^{c_2 \vee c} e^{-v^2/2t}\, dv + \frac{1}{\sqrt{2\pi t}} \int_{(2c-c_2)\vee c}^{(2c-c_1)\vee c} dv.$$

Proof. Since $P(W_t = 0)$,

$$P(M_t > c, W_t \in J) = P(W_t \in J[c, \infty)) + P(M_t > c, W_t \in (-\infty, c]).$$

Let p_1 and p_2 be the first and the second right side terms, respectively. Thus,

$$p_1 = \frac{1}{\sqrt{2\pi t}} \int_{c_1 \vee c}^{c_2 \vee c} e^{-v^2/2t}\, dv$$

while, using $P(M_t = c) = 0$,

$$p_2 = P(\rho_c M_t > c, \rho W_t \in J \cup (-\infty, c])$$
$$= P(M_t \geqq c), W_t \in [2c - c_2, 2c - c_1] \quad [c, \infty))$$
$$= P(M_t \geqq c) = \frac{1}{\sqrt{2\pi t}} \int_{(2c-c_2)\vee c}^{(2c-c_1)\vee c} e^{-v^2/2t}\, dv$$

where the last but one equality obtains by

$$W_t \in [2c - c_2, 2c - c_1] \cap [c, \infty) \subset [M_t \geqq c].$$

Note that **E** contains **c**: Take $J = (-\infty, +\infty)$. Also, taking $c = x$ and $c_1 = -\infty$, $c_2 = y$, **E** becomes

$$P(M_t > x, W_t < y) = 2P(W_t > 2x - y)$$

and yields

Corollary 1. *The joint distribution of* (M_t, W_t) *has pr. density*

$$p(x,y) = \sqrt{2/\pi t}\, \frac{2x - y}{t} \exp\{-(2x - y)^2/2t\}$$

for $x > 0$ *and* $y \leqq x$ *and is* 0 *otherwise.*

Upon setting $y = x - z$ so that z is a value of $M_t - W_t$, Corollary 1 yields

COROLLARY 2. *The joint distribution of* $(M_t, M_t - W_t)$ *has pr. density*

$$p_1(x, z) = \sqrt{2/\pi t}\, \frac{x + z}{t} \exp\{(x + z)^2/2t\}$$

for $x > 0$, $z > 0$ *and is* 0 *otherwise.*

Since $p_1(x, a)$ is symmetric in x and z, upon taking **c** into account, we obtain

COROLLARY 3. *The r.v.'s* $M_t, -m_t, |W_t|, M_t - W_t, W_t - m_t$ *have the same d.f.*

$$F(x) = \sqrt{2/\pi t} \int_0^x e^{-v^2/2t}\, dv$$

for $x > 0$ *and* 0 *otherwise.*

Yet, there is a glaring contrast between, on the one hand, M_T and $-m_T$ which are nondecreasing, positive on $(0, \infty)$ and vanish only at $t = 0$ and, on the other hand, the identically distributed $|W_T|$, $M_T - W_T$ and $W_T - m_T$ which oscillate a.s. as t varies and have a.s. an infinity of zeros.

§ 42. LIMIT DISTRIBUTIONS

The far-reaching extension of limit laws of sums of random variables and of random vectors to convergence of laws of random functions and of their functionals began with Kolmogorov (1931). He was followed by Petrovsky and by Khintchine (1936). Their main concern was with events related to sequences of sums of independent or Markov dependent r.v.'s remaining within given boundaries determined by continuously differentiable functions, and with the relevant heat equation and more general parabolic type ones.

Some twenty years later, Erdös and Kac reopened the investigation of functional limit distributions in a new and direct probabilistic way. They brought to bear their "invariance principle" idea: To find limit distributions for functionals of random walks, say, $\xi(0,1)$ with common expectation 0 and variance 1, do it first for simple random walks, say, the coin-tossing one which then remain valid or "invariant" for all $\xi(0,1)$.

In 1952, Donsker obtained his crucial invariance principle for $\xi(0,1)$. An unpublished extension of Donsker's result is given by L. Le Cam in a note at the end of this chapter. His argument is direct and extends to (separable) Banach-valued r.v.'s. Within a few years, a flurry of contributions, especially those of Prohorov and Skorobod, state and solve the problem in full generality in terms of convergence of laws *of r.f.'s* and of their functionals. Prohorov (1936) builds the foundations (see § 12) and, in particular, solves the problem for r.f.'s which are sample continuous; Donsker's principle becomes a special case. Simultaneously, Skorohod investigates the case of r.f.'s whose sample functions are continuous except for jumps.

Five years later, Skorohod created a new approach, especially, for Brownian convergence he introduced his "Brownian embedding." Soon thereafter, Strassen used it in an unexpected direction—for a.s. convergence, and obtains new laws of iterated logarithm.

Besides the relevant works by the above-mentioned authors, the interested reader is referred to books rich in ideas and results, by Billingsley, by Gikhman and Skorohod and, for the Skorohod approach, by Breiman and by Freedman.

42.1. Pr.'s on $\mathcal{C}$. Let $\mathcal{C} = \mathcal{C}\,[0, 1]$ be the family of real-valued continuous functions on $[0, 1]$. The t-th *coordinate* of its "point" x is the value $x(t)$ of the function x at t, and $\mathcal{C}$ is *linear*:

$$x, y \in \mathcal{C},\ a, b \in R \Longrightarrow ax + by \in \mathcal{C}.$$

Since $[0, 1]$ is compact, these functions are bounded and uniformly continuous. Since uniform limits of continuous functions are continuous, it follows that, under the *uniform norm*

$$||x|| = \sup_t |x(t)| = \max_t |x(t)|,$$

the space $\mathcal{C}$ with the corresponding metric $d(x, y) = ||x - y||$ is complete. Thus, $\mathcal{C}$ becomes a Banach space.

Furthermore, this space is separable. For, its members x can be approximated uniformly by polygonal functions x_n linear between $(k - 1)$ and k/n with values $x(k/n)$ at the vertices $(k = 1, \cdots, n)$ and these ones can be approximated uniformly by polygonal ones with rational values at the vertices, which form a countable subset of $\mathcal{C}$.

The *continuity modulus* or *δ-oscillation* $\gamma_x(\delta)$ of $x \in \mathcal{C}$ is defined by

$$\gamma_x(\delta) = \sup_{|s-t|<\delta} |\, x(s) - x(t)\,|, \quad 0 < \delta < 1.$$

a. *$\gamma_x(\delta)$ is uniformly continuous in x for every fixed δ.*

For, triangle inequality yields

$$\gamma_x(\delta) \leqq \| x - y \| + \gamma_y(\delta)$$

so that upon interchanging x and y,

$$| \gamma_x(\delta) - \gamma_y(\delta) | \leqq 2 \| x - y \|.$$

b. *$\gamma_x(\delta)$ is nondecreasing in δ, and $\gamma_x(\delta) \to 0$ as $\delta \to 0$, uniformly in $x \in K$ compact.*

For clearly, $\gamma_x(\delta)$ is nondecreasing in δ, $\gamma_x(\delta) \to 0$ as $\delta \to 0$, and the Dini lemma for uniform convergence on compact $K \subset \mathcal{C}$ applies.

Since the Arzéla-Ascoli theorem is crucial for tightness criteria of pr.'s on $\mathcal{C}$, we give its proof, using 2.3 and especially **b** and **B** therein without further comment.

A. ARZÉLA-ASCOLI THEOREM. *$A \in \mathcal{C}$ has compact closure $\bar{A}$ if and only if*

(i) *A is uniformly equicontinuous:* $\sup_{x \in A} \gamma_x(\delta) \to 0$ *as* $\delta \to 0$

and

(ii) *A is uniformly bounded:* $\sup_t \sup_{x \in A} | x(t) < \infty$

or

(ii′) *A is uniformly bounded at* 0: $\sup_{x \in A} | x(0) | < \infty$.

Proof. Clearly, (ii) implies (ii′) and, under (i), (ii′) implies (ii) since, by taking n sufficiently large so that $\sup_{x \in A} \gamma_x(1/n) < \infty$,

$$|x(t)| \leqq |x(0)| + \sum_{k=1}^{n} |(x(kt/n) - x((k-1)t/n)|$$

and (ii) follows.

If $\bar{A}$ is compact then it is totally bounded hence bounded and, a fortiori, so is A, that is, (ii) holds and (i) follows by **b**.

Conversely, let (i) and (ii) hold. If $x_n \in A$ then, by (ii) and use of the diagonal process, we obtain a subsequence $(x_{n'})$ which, on the set of all rationals $r \in [0, 1]$, converges to some x with $x \in \bar{A}$. But, by (i), for every $\epsilon > 0$ there is a $\delta = \delta(\epsilon) > 0$ such that $| s - t | < \delta$ implies $| x_n(s) - x_n(t) | < \epsilon, n = 1, 2, \cdots$. Thus, for every t there is a rational $r = r(\epsilon)$ such that

$$
\begin{aligned}
|x_{n'}(t) - x_{m'}(t)| &\\
\leq |x_{n'}(t) - x_{n'}(r)| &+ |x_{n'}(r) - x_{m'}(r)| + |x_{m'}(r) - x_{m'}(t)| \\
\leq 2\epsilon &+ |x_{n'}(r) - x_{m'}(r)|
\end{aligned}
$$

and, letting $m, n \to \infty$ then $\epsilon \to 0$, $\|x_{n'} - x_{m'}\| \to 0$ hence $x_{n'} \to x \in \bar{A}$. If $y_n \in \bar{A}$ then there are $x_n \in A$ with $d(x_n, y_n) < 1/n$ and $x_{n'} \to x \in \bar{A}$ so that $y_{n'} \to x \in \bar{A}$ hence $\bar{A}$ is compact.

Projections of $\mathcal{C}$ are defined by

$$pr_{t_1, \ldots, t_m}(x) = x(t_1, \cdots, t_m), x \in \mathcal{C},$$

and are continuous mappings of $\mathcal{C}$ onto $R_{t_1} \times \cdots \times R_{t_m}$;
Borel cylinders in $\mathcal{C}$ are defined by $pr^{-1}_{t_1, \ldots, t_m}(A)$ with *base* A a Borel set in $R_{t_1} \times \cdots \times R_{t_m}$.

c. *The σ-field $\mathcal{S}$ of Borel sets in $\mathcal{C}$, generated by the class of open (closed) sets in $\mathcal{C}$, is generated by the class of Borel cylinders in $\mathcal{C}$, that is, $\mathcal{S}$ is the smallest σ-field for which all projections are measurable.*

For, $\mathcal{C}$ being separable, the open sets in it are countable unions of open spheres, and

$$\{x: \|x - x_0\| < c\} = \bigcup_{n=1}^{\infty} \bigcap_{r} \{x: |x(r) - x_0(r) < c - 1/n\}$$

where r varies over all rationals in $[0, 1]$.

In what follows, P with or without affixes denotes pr.'s on $\mathcal{S}$ and the *finite restrictions* of P are *sections* $Ppr^{-1}_{t_1, \ldots, t_m}$ of P to Borel cylinders with bases in $R_{t_1} \times \cdots \times R_{t_m}$; they are pr.'s on the class of such cylinders. We use constantly the concepts and results in Section 12.

d(i). *Weak convergence of sequences of pr.'s on $\mathcal{S}$ entails weak convergence of their finite sections: For every finite subset $(t_1, \cdots, t_m)$ of* $[0, 1]$,

$$P_n \xrightarrow{w} P_0 \Rightarrow P_n pr^{-1}_{t_1, \ldots, t_m}$$

(ii) *If finite sections of pr.'s P_n converge weakly then they converge weakly to finite sections of some pr. P_0.*

For, (i) is immediate and (ii) obtains by the Consistency theorem.

While weak convergence of finite sections of the P_n determines P_0, it does not imply weak convergence of the P_n to P_0. For this we require moreover tightness of the P_n.

We say that h is *P_0-admissible on* $\mathcal{C}$ if h is a P_0-a.e. continuous Borel function on $(\mathcal{C}, \mathcal{S})$ to a metric space with its σ-field of Borel sets.

B. WEAK CONVERGENCE CRITERION. $P_n \xrightarrow{w} P_0$ *if and only if*

(i) *the finite sections of the* P_n *converge weakly to those of* P_0 *and the sequence* P_n *is tight or, equivalently, relatively compact*

or

(ii) $P_n h^{-1} \xrightarrow{w} P_0 h^{-1}$ *for every P_0-admissible* h on $\mathcal{C}$.

Proof. We prove (i): The equivalence assertion results from 12.3**A** since $\mathcal{C}$ is complete and separable. The "only if" assertion results from $P_n \xrightarrow{w} P_0$ implying relative compactness of the sequence P_n and from **d**(i). The "if" assertion obtains as follows: Relative compactness means that every subsequence of the sequence P_n contains a weakly convergent subsequence. But by finite sections hypothesis, d(ii) applies and the weak pr. limit P_0 is the same for all such subsequences hence $P_n \xrightarrow{w} P_0$.

As for assertion (ii), the "only if" part obtains by 12.1**A** Corollary and the "if" part obtains by taking h to be the identity function on $\mathcal{C}$; $h(x) = x$.

For $\mathcal{C}$, we have the most useful

C. TIGHTNESS CRITERION. *The sequence* (P_n) *is tight if and only if*

(i) $$\sup_n P_n(x: |x(0)| > c) \to 0 \text{ as } c \to \infty$$

and, for every $\epsilon > 0$, *as* $\delta \to 0$,

(ii) $$\limsup_n P_n(\gamma_x(\delta) > \epsilon) \to 0$$

or

(ii′) $$\sup P_n(\gamma_x(\delta) > \epsilon) \to 0.$$

Note that (i) says that the sequence $(P_n pr_0^{-1})$ is tight.

Proof. Since finite families of pr.'s on $\mathcal{C}$ are tight, (ii) and (ii′) are equivalent.

Let (P_n) be tight. Then, given $\epsilon > 0$ and $\eta > 0$ there is a compact $K \subset \mathcal{C}$ with $P_n K^c < \eta$ for all n, so that for c sufficiently large

$$\{x: |x(0)| > c\} \subset K^c,$$

for δ sufficiently small,

$$\{x: \gamma_x(\delta) > \epsilon\} \subset K^c,$$

and (i) and (ii) follow.

Conversely, let (i) and (ii) hold and set

$$A = \{x: |x(0)| > c\}, \quad B_m = \{x: \gamma_x(\delta_m) > 1/m\}.$$

Then, given $\epsilon > 0$ and $\eta > 0$, we can choose c and δ_m so that, for all n,

$$P_n A < \eta/2, \quad P_n B_m < \eta/2^{m+1}.$$

Thus, $P_n K^c < \eta$ for all n, where the closure K of $A \cup \bigcap_m B_m{}^c$ is compact by **A**, and the sequence (P_n) is tight.

42.2. Limit distributions on $\mathcal{C}$. Let $X = (X(t), 0 \leqq t \leqq 1)$ be a sample continuous r.f. so that *X is a measurable* function on some pr. space $(\Omega, \mathcal{A}, P)$ to $(\mathcal{C}, \mathcal{S})$. *Finite sections* of X are of the form $(X(t_1), \cdots, X(t_m))$. The law $\mathcal{L}(X)$ is given by the *distribution* P_X on $\mathcal{S}$ defined by

$$P_X(B) = P(X \in B), \quad B \in \mathcal{S}.$$

Let $X_n = (X_n(t), 0 \leqq t \leqq 1)$ be also sample continuous r.f.'s. *Convergence of laws $\mathcal{L}(X_n)$ to $\mathcal{L}(X)$ means that $P_{X_n} \xrightarrow{w} P_X$.* We are concerned primarily with limit laws, that is, limit distributions of functionals $h(X_n)$. By 42.1 **B**(ii), this problem reduces to

A. FUNCTIONAL LIMIT LAWS CRITERION. $\mathcal{L}(X_n) \to \mathcal{L}(X)$ *if and only if* $\mathcal{L}(h(X_n)) \to \mathcal{L}(h(X))$ *for every P_X-admissible h.*

By 42.1**B**(i) and **c**, the above criterion becomes

B. RANDOM FUNCTIONS LAWS CONVERGENCE CRITERION. $\mathcal{L}(X_n) \to \mathcal{L}(X)$ *if and only if*

(i) *the laws of finite sections of the r.f.'s X_n converge to those of the r.f. X: For every finite subset $(t_1, \cdots, t_m)$ of $[0, 1]$,*

$$\mathcal{L}(X_n(t_1), \cdots, X_n(t_m)) \to \mathcal{L}(X(t_1), \cdots, X(t_m))$$

and

(ii) *the laws of the sequence (X_n) of r.f.'s are tight:*

$$\text{(ii}_1\text{)} \qquad P(|X_n(0)| > c) \to 0 \quad \text{as} \quad c \to \infty$$

and, for every $\epsilon > 0$,

$$\text{(ii}_2\text{)} \qquad \limsup_n P(\sup_{|s-t|<\delta} |X_n(s) - X_n(t)| > \epsilon) \to 0 \quad \text{as} \quad \delta \to 0.$$

From now on, W denotes Brownian motion on [0, 1] and we call *Brownian convergence* convergence of laws of r.f.'s to $\mathfrak{L}(W)$; s and t with or without affixes belong to [0, 1].

Let ξ_{nk}, $k = 1, \cdots, k_n \to \infty$ be independent r.v.'s with d.f.'s F_{nk}, zero expectations, and finite variances $\sigma_{nk}^2 = \sigma_{nk}^2(X_{nk}) = EX_{nk}^2$ such that

$$\sum_k \sigma_{nk}^2 = 1, \quad \max_k \sigma_{nk}^2 \to 0.$$

Set

$$t_{nk} = \sum_{j=1}^{k} \sigma_{nj}^2, \quad S_{nk} = \sum_{j=1}^{k} \xi_{nj}, \quad t_{n0} = 0, \quad S_{n0} = 0.$$

Thus,

$$0 = t_{n0} \leqq \cdots \leqq t_{nk_n} = 1,$$

and we define the corresponding sample continuous r.f. $X_n = (X_n(t), 0 \leqq t \leqq 1)$ to be linear between t_{nk} and $t_{n,k+1}$ with vertices (t_{nk}, S_{nk}) for $k = 0, \cdots, k_n$, that is,

$$X_n(t) = S_{nk} + \frac{t - t_{nk}}{t_{n,k+1} - t_{n,k}} \xi_{n,k+1}, \quad t \in [t_{nk}, t_{n,k+1}).$$

We use the normal convergence criterion in 22.2 without comment to obtain its very far-reaching generalization:

C. BROWNIAN CONVERGENCE CRITERION. *Let r.v.'s ξ_{nk} and r.f.'s X_n be as above and let W be Brownian on* [0, 1].
Then

$$\mathfrak{L}(X_n) \to \mathfrak{L}(W), \quad \textit{in fact} \quad \mathfrak{L}(h(X_n)) \to \mathfrak{L}(h(W))$$

for every P_W-admissible h on $\mathfrak{C}$
if and only if Lindeberg condition holds: for every $\epsilon > 0$,

$$g_n(\epsilon) = \sum_{k=1}^{k_n} \int_{|a|>\epsilon} a^2 dF_{nk}(a) \to 0.$$

Note that Lindeberg condition implies $\max_k \sigma_{nk}^2 \to 0$.

Proof. For the "only if" assertion, take h on $\mathfrak{C}$ to R to be defined by $h(x) = x(1)$ so that h is continuous hence admissible, $h(X_n) = S_{nk_n}$ and $h(W) = W(1)$ hence $\mathfrak{L}(S_{nk_n}) \to \mathfrak{N}(0, 1)$ and Lindeberg condition holds.

It remains to prove the "if" assertion. By **A**, it suffices to show that $\mathfrak{L}(X_n) \to \mathfrak{L}(W)$. Only proof of **B**(i) and **B**(ii$_2$) is required since $X_n(0) = W(0)$ hence **B**(ii$_1$) holds.

1°. We prove **B**(i). Let

$$X_n'(t) = \sum_{t_{nj}<t} \xi_{nj}$$

so that, for every t,

$$|X_n'(t) - X_n(t)| \leqq \max_k |\xi_{nk}|$$

hence, for every $\epsilon > 0$,

$$P(X_n'(t) - X_n(t)| \leqq \sum_k P(|\xi_{nk}| > \epsilon)$$

$$\leqq \sum_k \int_{k|a|>\epsilon} a^2 dF_{nk}(x) \leqq g_n(\epsilon)/\epsilon^2 \to 0.$$

Thus, it suffices to prove **B**(i) for the X_n' in lieu of the W_n. But, the r.f.'s X_n' and W being decomposable and starting at 0, we have only to show that for all s, t,

$$\mathfrak{L}(X_n'(t) - X_n'(s)) \to \mathfrak{L}(W(t) - W(s)).$$

Since the Lindeberg condition holds for $\sum_k \xi_{nk}$, it holds a fortiori for

$$X_n'(t) - X_n'(s) = \sum_{s \leqq t_{nj} < t} \xi_{nj}$$

while

$$\left| \sum_{s \leqq t_{nj} < t} \sigma_{nj}^2 - (t - s) \right| \leqq \max_k \sigma_{nk}^2 \to 0.$$

Thus

$$\mathfrak{L}(X_n'(t) - X_n'(s)) \to \mathfrak{N}(0, t - s) = \mathfrak{L}(W_t - W_s),$$

and **B**(i) is proved.

2°. We prove **B**(ii). For $|s - t| < \delta$, s and t belong to some interval of the form $[k\delta, (k + 1)\delta]$ or to two adjacent such intervals. By triangle inequality, it follows that

$$\gamma_{X_n}(\delta) = \sup_{|s-t|<\delta} |X_n(t) - X_n(s)|$$

$$\leqq 2 \sup_k \sup_{k\delta<t<(k+2)\delta} |X_n(t) - X_n(k\delta)|$$

$$\leqq 4 \sup_k \sup_{k\delta < t < (k+1)\delta} | X_n(t) - X_n(k\delta) |$$

$$\leqq 8 \sup_{j_{nk} \leqq i < j_{n,k+1}} \left| \sum_{j=j_{nk}}^{i} \xi_{nj} \right|$$

where

$$j_{nk} = \max\{j : t_{nj} \leqq k\delta\}.$$

Since, for $j_{nk} < i < j_{n,k+1}$,

$$\limsup_n \sup_i P\left(\left| \sum_{j=1}^{j_{n,k+1}} \xi_{nj} \right| > \epsilon/16 \right) \leqq (16/\epsilon)^2 \delta,$$

by the second part of 41.2**c** with $a = b = \epsilon/8$, it follows that for δ sufficiently small

$$\limsup_n P(\gamma_{X_n}(\sigma) > \epsilon)$$
$$\leqq \frac{1}{1 - (16/\epsilon)^2\delta} \limsup_n P\left(\left| X_n(t_{n,j_{n,k+1}}) - X_n(t_{n,j_{n,k}}) \right| > \frac{\epsilon}{16} \right).$$

But, by 1°, the right side "limsup" is

$$p(\epsilon, \delta) = \frac{1}{\sqrt{2\pi}} \int_{|v| > \epsilon/16\sqrt{\delta}} e^{-v^2/2}\, dv$$

and, by normal approximation 41.2**b**, $p(\epsilon, \delta)/\delta \to 0$ as $\delta \to 0$. Therefore, as $\delta \to 0$,

$$\limsup_n P(\gamma_{X_n}(\delta) > \epsilon) = O(\sum_{k\delta < t} p(\epsilon, \delta)) = O(p(\epsilon, \delta)/\delta) \to 0$$

and **B**(ii) holds.

Upon setting $\zeta_{nk} = \zeta_k/\sqrt{n}$, **C** yields

D. Donsker invariance principle on $\mathfrak{C}$. *Let $\xi = (\xi_1, \xi_2, \cdots)$ be a centered second-order random walk with common exp 0 and variance 1. Let X_n be polygonal r.f.'s with vertices $(k/n, S_n/\sqrt{n})$ where $S_k = \xi_1 + \cdots + \xi_k$, $S_0 = 0$ and $k = 0, 1, \cdots, n$.*

Then

$\mathfrak{L}(X_n) \to \mathfrak{L}(W)$, *in fact* $\mathfrak{L}(h(X_n)) \to \mathfrak{L}(h(W))$ *for every P_W-admissible h.*

42.3. Limit distributions; Brownian embedding. For Brownian convergence there is a direct approach based upon "Brownian embedding" of random walks into Brownian motion at Brownian times.

a. BROWNIAN EMBEDDING LEMMA. *For every r.v.* ξ *with* $E\xi = 0$ *and* $0 < \sigma^2\xi < \infty$ *there is a Brownian time* $\tau_{U,V}$ *such that*

$$\mathcal{L}(W(\tau_{U,V})) = \mathcal{L}(\xi), \quad E\tau_{U,V} = \sigma^2\xi.$$

Proof. Let Brownian $W = (W(t), t \geqq 0)$ be on a pr. space $(\Omega_0, \mathcal{A}_0, P_0)$. Let (U,V) with $U \leqq 0 \leqq V$ be a random vector on a pr. space $(\Omega_1, \mathcal{A}_1, P_1)$ with d.f. G defined by

$$dG(u, v) = \frac{1}{\alpha}(v - u)dF_\xi(u)dF_\xi(v), \quad u \leqq 0 \leqq v,$$

where

$$\alpha = E\xi^+ = E\xi^- > 0$$

since $E\xi = E\xi^+ - E\xi^- = 0$ and ξ is not degenerate.
On the product pr. space

$$(\Omega, \mathcal{A}, P) = (\Omega_0 \times \Omega_1, \mathcal{A}_0 \times \mathcal{A}_1, P_0 \times P_1)$$

W and (U, V) are independent, that is, $\mathcal{B}(W(t), t \geqq 0)$ and $\mathcal{B}(U, V)$ are independent σ-fields, and

$$\tau_{U,V} = \inf\{t: W(t) \not\subset (U, V)).$$

is Brownian since

$$[\tau_{U,V}] \leqq t = [\min_{0 \leqq s \leqq t} W(s) \leqq]U \cup [\max_{0 \leqq s \leqq t} W(s) \leqq V].$$

If $a > 0$ then

$$P(W(\tau_{U,V}) > a) = EP(W(\tau_{U,V}) > a \mid U, V)$$

with, by **41.4A**

$$P(W(\tau_{U,V}) > a \mid U = u, V = v) = 0 \text{ or } \frac{-u}{v - u}$$

according as $v < a$ or $v \geqq a$. It follows that

$$P(W(\tau_{U,V}) > a) = \int_{u=-\infty}^{0} \int_{v=0}^{+\infty} \frac{-u}{v - u} dG(u, v) = P(\xi > a);$$

similarly, if $a < 0$ then

$$P(W(\tau_{U,V}) < a) = P(\xi < a).$$

Thus

$$\mathfrak{L}(W(\tau_{U,V})) = \mathfrak{L}(\xi),$$

by the same theorem,

$$E\tau_{U,V} = E(E(\tau_{U,V})|\ U,V)) = E|\ UV\ | = E(W(\tau_{U,V}))^2 = E\xi^2$$

and the lemma is proved.

From now on, *let* $\bar{\xi}(0, 1)$ *denote second-order centered random walk of iid r.v.'s* $\xi_1, \xi_2, \cdots$ *with common expectation* 0 *and variance* 1 *and denote by* $S_0 = 0, \cdots, S_n = \xi_1 + \cdots + \xi_n, \cdots$ *their successive sums.*

A. BROWNIAN EMBEDDING THEOREM. *Given* $\bar{\xi}(0, 1)$, *there is a Brownian r.f.* $W = (W(t), t \geqq 0)$ *and an independent of* W *sequence of iid Brownian times* $\tau_1, \tau_2, \cdots$ *with common expectation* 1 *such that*

(i) $W(\tau_1), W(\tau_1 + \tau_2) - W(\tau_1), W(\tau_1 + \tau_2 + \tau_3) - W(\tau_1 + \tau_2), \cdots$ *are iid r.v.'s with common law* $\mathfrak{L}(\xi_1)$, *equivalently*

(ii) $\mathfrak{L}(W(\tau_1), W(\tau_1 + \tau_2), W(\tau_1 + \tau_2 + \tau_3), \cdots) = \mathfrak{L}(S_1, S_2, S_3, \cdots)$.

Proof. The equivalence assertion is immediate, and we prove (i). Let Brownian W be on a pr. space $(\Omega_0, \mathfrak{a}_0, P_0)$ and let $(U_1, V_1), (U_2, V_2), \cdots$ be random vectors on pr. spaces $(\Omega_1, \mathfrak{a}_1, P_1), (\Omega_2, \mathfrak{a}_2, P_2), \cdots$ with common d.f. G introduced in the above lemma. On the product space

$$(\Omega, \mathfrak{a}, P) = (\Omega_0 \times \Omega_1 \times \Omega_2 \times \cdots, \mathfrak{a}_0 \times \mathfrak{a}_1 \times \mathfrak{a}_2 \times \cdots, P_0 \times P_1 \times P_2 \times \cdots),$$

W and the iid random vectors $(U_1, V_1), (U_2, V_2), \cdots$ are independent. According to the lemma, there is a Brownian first exit time $\tau_1 = \tau_{U_1, V_1}$ of W from (U_1, V_1) such that

$$\mathfrak{L}(W(\tau_1)) = \mathfrak{L}(\xi_1), \quad E\tau_1 = E\xi_1^2 = 1,$$

$W^{(1)} = (W(\tau_1 + t) - W(\tau_1), t \geqq 0)$ is independent of $(W_s, s \leqq \tau_1, U_1, V_1)$ and, by Brownian times origin invariance **41.4B** $W^{(1)}$ is a Brownian r.f. independent of $W(\tau_1)$. Proceeding similarly with first exit time τ_2 of $W^{(1)}$ from (U_2, V_2) and so on, we obtain iid r.v.'s

$$W(\tau_1), W(\tau_1 + \tau_2) - W(\tau_1), W(\tau_1 + \tau_2 + \tau_3) - W(\tau_1 + \tau_2), \cdots$$

with common law $\mathfrak{L}(W(\tau_1)) = \mathfrak{L}(\xi_1)$, and the proof is concluded.

b. BREIMAN LEMMA. Let $\tau_1, \tau_2, \cdots$ be the Brownian exit times in **A.** *Then*

$$Y_n = \sup_{0 \leqq t \leqq 1} \left| \frac{\tau_1 + \cdots + \tau_{[nt]}}{n} - t \right| \xrightarrow{P} 0.$$

Proof. Let $\sigma_k = (\tau_1 - 1) + \cdots + (\tau_k - 1)$ where the summands have mean 0 since the means of the τ's are 1 and set $\bar{\sigma}_k = \sigma_k/k$. Then

$$T_n \leqq \sup_{0 \leqq t \leqq 1} |\sigma_{[nt]}/n| + \frac{1}{n}$$

so that

$$T_n \leqq \sup_{0 \leqq t \leqq 1} t|\bar{\sigma}_{[nt]}| + \frac{1}{n}.$$

Therefore, for $0 < \epsilon < 1$, setting $\bar{\sigma} = \sup_{k \geqq 1} |\bar{\sigma}_k|$,

$$(1) \qquad T_n \leqq \epsilon\bar{\sigma} + \sup_{k \geqq [\epsilon n]} |\bar{\sigma}^k| + \frac{1}{n}$$

since

$$T_n \leqq T_n' + T_n'' + \frac{1}{n}$$

where

$$T_n' = \sup_{0 \leqq t \leqq \epsilon} t|\bar{\sigma}_{[nt]}| \leqq \epsilon \sup_{0 \leqq t \leqq \epsilon} |\bar{\sigma}_{[nt]}| \leqq \epsilon \sup_{k \geqq 1} |\bar{\sigma}_k| = \epsilon\bar{\sigma}$$

and

$$T_n'' = \sup_{\epsilon \leqq t \leqq 1} t|\bar{\sigma}_{[nt]}| \leqq \sup_{\epsilon \leqq t \leqq 1} |\bar{\sigma}_{[nt]}| \leqq \sup_{k \geqq [\epsilon n]} |\bar{\sigma}_k|.$$

But the Kolmogorov strong law of large numbers applies to successive sums

σ_n of iid r.v.'s $\tau_1 - 1, \tau_2 - 1, \cdots$ with mean 0, that is, $\bar{\sigma}_k \xrightarrow{a.s.} 0$ as $k \to \infty$. Thus, for any $a > 0$, upon letting in (1) $n \to \infty$ then $\epsilon \to 0$,

$$\limsup P(T_n > a) \leqq P(\epsilon\bar{\sigma} > a) \to 0$$

and $T_n \xrightarrow{P} 0$ obtains.

By using Brownian embedding one could prove directly the invariance principle on $\mathfrak{C}$ (42.2**D**). The same procedure yields also a modified invariance principle which sometimes is more convenient: The one on $\mathfrak{C}$ was established by first interpolating, linearly, successive sums so as to deal with sample continuous r.f.'s X_n. But this approach is somewhat arbitrary and may be avoided working with the more "direct" r.f.'s Y_n defined by $Y_n(t) = S_{[nt]}/\sqrt{n}$ for $0 \leqq t \leqq 1$. However, their sample functions are not continuous and, thus, the space $\mathfrak{C}$ cannot be used. In fact, these sample functions belong to the *linear space* D (containing $\mathcal{C}$) of functions $x = (x(t), 0 \leqq t \leqq 1)$ which are right continuous and have left limits:

$$x(t) = x(t+) = \lim_{s \downarrow t} x(s) \text{ for } 0 \leqq t < 1,$$

$$x(t-) = \lim_{s \uparrow t} x(s) \text{ exists for } 0 < t \leqq 1,$$

and we set $x(1) = x(1-)$; note that these functions are bounded.

The space D can be given (Skorohod) a metrizable topology which makes it complete and separable. Then the Prohorov approach applies: An analogue of the Arzéla-Ascoli theorem is obtained for the space D and yields an analogue of 42.2**B** for r.f.'s with sample functions in D, which then yields Brownian convergence with an invariance principle *in* D. However, we are concerned only with conditions for $\mathcal{L}(Y_n) \rightarrow \mathcal{L}(W)$ and those can be established directly, using Brownian embedding, as follows.

Let x, y belong to D and introduce on it the uniform norm $\| x \| = \sup_{0 \leqq t \leqq 1} | x(t) |$ ($< \infty$ since every x is bounded) with corresponding metric $\| x - y \|$ so that its subspace $\mathcal{C}$ has the same metric as previously. (Note that with this metric D is not separable!)
Extend P_W, which lives on $\mathcal{C}$, to D by assigning to every Borel set in D the P_W-measure of its trace on $\mathcal{C}$ (which trace is a Borel set in $\mathcal{C}$ and in D). We are now ready to treat the problem:

Given $\xi(0, 1)$, let $Y_n = (Y_n(t), 0 \leqq t \leqq 1)$ be defined by $Y_n(t) = S_{[nt]}/\sqrt{n}$. Let $W_n = (W_n(t), 0 \leqq t \leqq 1)$ be defined by $W_n(t) = \sqrt{n}\, W(t/n)$; by Brownian scale change invariance 41.1**D** the r.f.'s W_n are Brownian. Replace W by W_n in **A** and denote by $\tau_{n1}, \tau_{n2}, \cdots$ the corresponding iid r.v.'s.

Thus,

$$\mathcal{L}(W_n(\tau_{n1}), W_n(\tau_{n1} + \tau_{n2}), \cdots) = \mathcal{L}(S_1, S_2, \cdots)$$

that is, returning to W,

$$\mathcal{L}\left(W(\tau_{n1}), \cdots, W\left(\frac{\tau_{n1} + \cdots + \tau_{n,[nt]}}{n}\right), \cdots\right),$$
$$= \mathcal{L}(S_1, \cdots, S_{[nt]}/\sqrt{n}, \cdots)$$

hence, setting $Y_n = (Y_n(t), 0 \leqq t \leqq 1)$ with

$$Y_n(t) = W\left(\frac{\tau_{n1} + \cdots + \tau_{n,[nt]}}{n}\right)$$

we have $\mathfrak{L}(\tilde{Y}_n) = \mathfrak{L}(Y_n)$. It follows that to prove the theorem stated below, we can replace Y_n by $\tilde{Y}_n$ and show that $\mathfrak{L}(\tilde{Y}_n) \to \mathfrak{L}(W)$ and $\mathfrak{L}(h(\tilde{Y}_n)) \to \mathfrak{L}(h(W))$ where h are P_W*-admissible* on D, that is, P_W-a.e. continuous Borel functions on D to metric spaces with their σ-fields of Borel sets.

Since W_n and W are both Brownian, the iid r.v.'s $\tau_{n1}, \tau_{n2}, \cdots$ have the same common distribution as the iid r.v.'s $\tau_1, \tau_2, \cdots$, so that, by **b** with W_n in lieu of W,

$$T_{nn} = \sup_{0 \leq t \leq 1} \left| \frac{\tau_{n1} + \cdots + \tau_{n,[nt]}}{n} - t \right| \xrightarrow{\text{P}} 0.$$

B. INVARIANCE PRINCIPLE ON D. *Given* $\xi(0, 1)$, *let* Y_n *on* $[0, 1]$ *be r.f.'s with* $Y_n(t) = S_{[nt]}/\sqrt{n}$. *Then* $\mathfrak{L}(Y_n) \to \mathfrak{L}(W)$, *in fact* $\mathfrak{L}(h(Y_n)) \to \mathfrak{L}(h(W))$ *for every* P_W*-admissible* h *on* D.

Proof. According to what precedes, it suffices to prove this principle with the Y_n replaced by the equivalent r.f.'s $\tilde{Y}_n$ with

$$\tilde{Y}_n(t) = W\left(\frac{\tau_{n1} + \cdots + \tau_{n,[nt]}}{n}\right).$$

By 12.1**A** and its Corollary 1, we have only to show that, given a bounded continuous function g on D, $\mathfrak{L}(g(Y_n)) \to \mathfrak{L}(g(W))$, equivalently that every subsequence of integers n contains a sequence (n') such that $\mathfrak{L}(g(Y_n')) \to \mathfrak{L}(g(W))$.
But $T_{nn} \xrightarrow{\text{P}} 0$ implies that every subsequence of integers n contains a sequence (n') such that $T_{n'n'} \xrightarrow{\text{a.s.}} 0$. Since Brownian sample functions on $[0, 1]$ are uniformly continuous, it follows that

$$\sup_{0 \leq t \leq 1} | W(\tau_{n'1} + \cdots + \tau_{n'[n't]})/n' - W(t) | \xrightarrow{\text{a.s.}} 0,$$

that is, almost all sample functions of $\tilde{Y}_n'$, converge to those of W uniformly on $[0, 1]$. Therefore $g(\tilde{Y}_n') \xrightarrow{\text{a.s.}} g(W)$, a fortiori

$$\mathfrak{L}(g(\tilde{Y}_{n'})) \to \mathfrak{L}(g(W)),$$

and the proof is concluded.

In a completely different direction—from convergence of laws to almost sure convergence, Brownian embedding yields the theorem below which is essentially the start of Strassen's LIT invariance investigation.

C. LIT COMPARISON THEOREM. *Given* $\xi(0, 1)$, *as* $t \to \infty$

$$(S_{[t]} - W(t))/\sqrt{2t\log\log t} \xrightarrow{\text{a.s.}} 0.$$

Proof. When $t \to \infty$, $[t]/t \to 1$ and, by the strong law of large numbers, $\sigma_{[t]}/t \xrightarrow{\text{a.s.}} 1$. Thus, given $q > 1$, there is an a.s. finite r.v. τ_q such that, for $t > \tau_q$,

$$t/q \leqq \sigma_{[t]} \leqq tq$$

hence

$$| W(\sigma_{[t]}) - W(t) | \leqq Z(t)$$

where

$$Z(t) = \sup\{| X(s) - X(t) |: t/q \leqq s \leqq tq\}.$$

For $q^n \leqq t \leqq q^{n+1}$, by triangle inequality,

$$Z(t) \leqq \sup\{| X(s) - X(t) |: q^{n-1} \leqq s \leqq q^{n+2}\} \leqq 2Z_n$$

where

$$Z_n = \sup\{| X(s) - X(q^{n-1}) |: q^{n-1} \leqq s \leqq q^{n+2}\},$$

while $g(t) \geqq g(q^n)$ so that

$$Z(t)/g(t) \leqq 2Z_n/g(q^n)$$

and, $q > 1$ being arbitrary, the assertion will follow by proving that, as $q \to 1$,

$$\limsup_n Z_n/g(q^n) \to 0.$$

Let $q^{n+2} - q^{n-1} = 2\epsilon^2 q^n$ so that, as $q \to 1$,

$$\epsilon = \sqrt{(q^2 - 1/q)2} \to 0.$$

We are now on usual grounds. By normal approximation and extrema bounds in 41.2,

$$P(Z_n/g(q^n) > \epsilon)$$

$$\leqq 2P(| W(q^{n+2}) - W(q^{n-1}) | > 2\sqrt{(q^{n+2} - q^{n-1})\log\log q^n})$$

$$\leqq \sqrt{2/\pi}/n^2(\log q)^2.$$

Thus, by the Borel-Cantelli lemma,

$$P(Z_n/g(q^n) > \epsilon \text{ i.o.}) = 0$$

that is, a.s.

$$\limsup_n Z_n/g(q^n) \leqq \epsilon$$

and, letting $q \to 1$, the proof is concluded.

The (Hartman-Wintner) LIT for $\bar{\xi}(0, 1)$ is that a.s.

$$\limsup S_n/g(n) = 1, \quad \liminf S_n/g(n) = -1, \quad \limsup |S_n|/g(n) = 1.$$

The LIT comparison theorem **C** yields at once

COROLLARY. *The* LIT *for* $\bar{\xi}(0, 1)$ *and for Brownian motion entail each other and follow from* LIT *for any specific* $\bar{\xi}(0, 1)$, *say, the coin-tossing random walk.*

42.4. Some specific functionals. So far we emphasized, as usual, ideas and methods. We shall now give some applications to limit distributions of specific functionals.

Recall that $\bar{\xi}(0, 1)$ is a random walk $(\xi_1, \xi_2, \cdots)$ with common expectation 0 and variance 1, and successive sums, $S_0 = 0, \cdots, S_n = \xi_1 + \cdots + \xi_n, \cdots$.

The r.f.'s $X_n = (X_n(t), 0 \leqq t \leqq 1)$ are defined by

$$X_n(t) = \frac{1}{\sqrt{n}} S_{[nt]} + (nt - [nt]) \frac{1}{\sqrt{n}} \xi_{[nt]+1}$$

and their sample functions belong to $\mathcal{C}$.

The r.f.'s $Y_n = (Y_n(t), 0 \leqq t \leqq 1)$ are defined by

$$Y_n(t) = S_{[nt]}/\sqrt{n}$$

and their sample functions belong to D. We use invariance principles on $\mathcal{C}$ and on D:

$\mathcal{L}(h(X_n)) \to \mathcal{L}(h(W)$ *for every* P_W*-a.e. continuous Borel function* h *on* $\mathcal{C}$ *to a metric space.*

$\mathcal{L}(h(Y_n)) \to \mathcal{L}(h(W))$ *for every* P_W*-a.e. continuous function* h *on* D *to a metric space.*

Two approaches are available: Given a functional h on $\mathcal{C}$ or on D *either* $\mathcal{L}(h(W)$ is known or relatively easy to find so that the limit distribution of $\mathcal{L}(h(X_n)$ or $\mathcal{L}(h(Y_n))$ expressed in terms of the S_n obtains *or* a suitable random walk for which $\mathcal{L}(h(X_n))$ or $\mathcal{L}(h(Y_n))$ is known or relatively easy to find so that, by passage to the limit, $\mathcal{L}(h(W))$ obtains. Among such random walks is the *coin-tossing* one: the common law assigning pr. 1/2 to values $+1$ and -1.

The following examples will serve to illustrate these approaches.

ARCSINE LAW. We use invariance principle on D. Let $h = \lambda\{t : 0 \leqq t \leqq 1 \text{ and } x(t) > 0\}$, $x \in D$. It is a Borel function on D to R and, by 41.4**C**, it is P_W-continuous. If ν_n is the number of those of the n first successive sums which are positive, then $h(Y_n) = \nu_n/n$. Andersen Arcsine law applies to the coin-tossing walk. Thus, $\mathfrak{L}(\nu_n/n)$ converges to the Arcsine law defined by its pr. density

$$p(a) = 1/\pi\sqrt{a(1-a)}, \quad 0 \in a \in 1,$$

and its d.f. is given by

$$F(a) = \frac{2}{\pi} \text{Arcsine} \sqrt{a}.$$

Since this limit law is $\mathfrak{L}(h(W))$, it is the Arcsine law due to P. Lévy, that of the total amount of time above 0 spent in [0, 1] by Brownian motion.

From now on, we use the invariance principle on $\mathcal{C}$ without further comment.

Let $h(x) = x(1)$ so that $X_n(t) = S_n/\sqrt{n}$ and the *normal convergence theorem obtains* since $\mathfrak{L}(S_n/\sqrt{n}) \to \mathfrak{L}(W(1)) = \mathfrak{N}(0, 1)$. This very particular case of the invariance principle shows how far-reaching it is, since this principle results from the above normal convergence.

EXTREMA LAWS. Let

$$m_n = \min_{1 \leq k \leq n} S_k, \quad M_n = \max_{1 \leq k \leq n} S_k$$

and recall that

$$m(t) = \min_{0 \leq s \leq t} W_s, \quad M(t) = \max_{0 \leq s \leq t} W_s.$$

The functions h_1 and h_2 on $\mathcal{C}$ to R and h_3 on $\mathcal{C}$ to R , defined by

$$h_1(x) = \min_{0 \leq t \leq 1} (t), \quad h_2(x) = \max_{0 \leq t \leq 1} x(t), \quad h_3(x) = (\max_{0 \leq t \leq 1} x(t), x(1))$$

are P_W-continuous hence

$$\mathfrak{L}(m_n/\sqrt{n}) \to \mathfrak{L}(m(1)), \quad \mathfrak{L}(M_n/\sqrt{n}) \to \mathfrak{L}(M(1)),$$

$$\mathfrak{L}(M_n/\sqrt{n}, W(1)) \to \mathfrak{L}(M(1), W(1))$$

where the limit laws obtain by **41.4c** and **E.**

These limit distributions are but particular cases of the *limit distribution described by (I) and (II) below,* obtained by using the coin-tossing walk as indicated in what follows.
The function h on $\mathcal{C}$ to R^3, defined by

$$h(x) = (\min_{0 \leq t \leq 1} x(t), \max_{0 \leq t \leq 1} x(t), x(1))$$

is continuous so that

$$\text{(I)} \qquad \mathfrak{L}(m_n/\sqrt{n}, M_n/\sqrt{n}, S_n/\sqrt{n}) \to \mathfrak{L}(m(1), M(1), W(1)).$$

Let

$$p_n(k) = P(S_n = k), p_n(a, b, c) = P(a < m_n \leqq M_n < b, S_n = c).$$

The relation to be established in the coin-tossing case is

$$\text{(1)} \quad p_n(a, b, c) = \sum_{j=-\infty}^{+\infty} p_n(c + 2j(b - a)) - \sum_{j=-\infty}^{+\infty} p_n(2b - c + 2j(b - a))$$

with integers a, b, c such that

$$a \leqq 0 \leqq b, \quad a < b, \quad a \leqq c \leqq b.$$

Proceed by induction: For $n = 0$ this relation obtains by direct consideration of cases. Suppose it holds for $n - 1$. If $a = 0$ then both sides in (1) vanish—the right side because $p_n(k) = p_n(-k)$; similarly if $b = 0$. Thus (1) holds for n if $a = 0$ or $b = 0$. Since then $a + 1 \leqq 0$ or $b - 1 \geqq 0$, the relation holds with the arguments replaced by $n - 1, a - 1, b - 1, c - 1$ and by $n - 1, a + 1, b + 1, c + 1$. Then (1) obtains by using the immediate recurrence relations

$$p_n(k) = \frac{1}{2} p_{n-1}(k - 1) + \frac{1}{2} p_{n-1}(k + 1)$$

and

$$p_n(a, b, c) = \frac{1}{2} p_{n-1}(a - 1, b - 1, c - 1) + \frac{1}{2} p_{n-1}(a + 1, b + 1, c + 1).$$

By summing over c given $a \leqq 0 \leqq b$ and $a \leqq c_1 < c_2 \leqq b$, (1) yields

$$\begin{aligned}\text{(2)} \quad & P(a < m_n \leqq M_n < b, c_1 < S_n < c_2) \\ & = \sum_{j=-\infty}^{+\infty} P(c_1 + 2j(b - a) < S_n < c_2 + 2j(b - a)) \\ & - \sum_{j=-\infty}^{+\infty} P(2b - c_2 + 2j(b - a) < S_n < 2b - c_1 + 2j(b - a)).\end{aligned}$$

Replace a, b, c_1, c_2 by $[a\sqrt{n}], -[-b\sqrt{n}], [c_1\sqrt{n}], -[c_2\sqrt{n}]$, respectively. It is possible to pass to the limit termwise in (2) so that

(II) $P(a < m(1) \leqq M(1) < b, c_1 < W(1) < c_2)$

$$= \sum_{j=-\infty}^{+\infty} P(c_1 + 2j(b-a) < W(1) < c_2 + 2j(b-a))$$

$$- \sum_{j=-\infty}^{+\infty} P(2b - c_2 + 2j(b-a) < W(1) < 2b - c_1 + 2j(b-a),$$

where $\mathfrak{L}(W(1)) = \mathfrak{N}(0, 1)$.

COMPLEMENTS AND DETAILS

All r.f.'s are separable. Unless otherwise stated, $W_T = (W_t, t \geqq 0)$ *or* $(W(t), t \geqq 0)$, $T = [0, \infty)$, *is a Brownian motion with σ-fields* $\mathfrak{B}_T = (\mathfrak{B}_t, t \geqq 0)$, *and notation is that of the chapter.*

1. *Some transformations.* a) Let $Y(t) = e^{-t}W(e^{2t}), t \geqq 0$,

$$Z(t) = W(t) - tW(1), \quad 0 \leqq t \leqq 1.$$

Then

$$EY(s)Y(t) = e^{-|s-t|}, \quad EZ(s)Z(t) = (1-s)t \quad \text{for} \quad t \leqq s.$$

b) Let $U(t) = \int_0^t W(s)\,ds$, $V(t) = \exp\left\{c\int_0^t g(s)W(s)\,ds\right\}$, $c \in R$, g continuous on $\langle 0, \infty)$. Then

$$EU(t) = 0, \quad EU^2(t) = t^3/3,$$

$$E\exp\left\{c\int_0^t W(s)\,ds\right\} = e^{c^2t^3/6}, \quad E\exp\left\{c\int_0^t sW(s)\,ds\right\} = e^{c^2t^5/15}$$

and, in the general case,

$$EV(t) = \exp\left\{c^2\int_0^t g(v)\left(\int_0^v ug(u)\,du\right)dv\right\}.$$

c) $|W_T|$ is the *reflected at the origin* Brownian W_T, with

$$E|W(t)| = \sqrt{2t/\pi}, \quad \sigma^2(|W(t)|) = (1 - 2/\pi)t$$

d) $(e^{W(t)}, \ t \geqq 0)$ is called *geometric Brownian*. Then

$$Ee^{W(t)} = e^{t/2}, \quad Ee^{2W(t)} = e^{2t}, \quad \sigma^2(e^{W(t)}) = e^{t/2}(e^t - 1).$$

Find its so-called "diffusion coefficients" (for $e^{W(t)} = y$):

$$\lim_{h\downarrow 0} E(e^{W(t+h)} - e^{W(t)}|e^{W(t)})/h, \quad \lim_{h\downarrow 0} E\{(e^{W(t+h)} - e^{W(t)})^2|e^{W(t)}\}/h.$$

2. *Brownian second-order properties.* W_T is a r.f. of second order. Use systematically every applicable to W_T result in Chapter XI, and to Complements and Details, in 38.3**C** and in Complements and Details 5 to 8 of Chapter XII. Examples:

a) Let $L_2 = L_2(\Omega, \mathfrak{A}, P)$ be the space of equivalence classes of square integrable r.v.'s on Ω. The L_2-norm is $\| W(s) - W(t) \| = \sqrt{|s - t|}$, W_T is a continuous curve in L_2, and the subspace $L_2(W_T)$ spanned by the W_t is separable.

b) Let $s < t < u$. $E(W(t) \mid W(s), W(u))$ is the projection of $W(t)$ on the plane in $L_2(W_T)$ spanned by $W(s)$ and $W(u)$. In fact, a.s.

$$W(t) = E(W(t) \mid W(s), W(u)) + \sigma(t)\xi(t)$$

where $\xi(t)$ and $(W(s), W(t))$ are independent, $\mathfrak{L}(\xi(t)) = \mathfrak{N}(0, 1)$,

$$\sigma(t) = ((t - s)(u - t)/(u - s))^{1/2},$$

and $E(W(t) \mid W(s), W(u))$ is obtained by the linear interpolation of $W(s)$ and $W(u)$ given by

$$E(W(t) \mid W(s), W(u)) = \frac{u - t}{u - s} W(s) + \frac{t - s}{u - s} W(u) \quad \text{a.s.},$$

$\xi(t)$ is independent of $(W(r), r \notin (s, u))$ and for disjoint intervals with $s < u < s' < u'$, $(\xi(t), s < t < u)$ and $(\xi(t'), s' < t' < u')$ are independent. Furthermore, for $s \leqq t < t' \leqq u$, the square of the covariance

$$E\xi(t)\xi(t') = ((t - s)(u - t')/(t' - s)(u - t))^{1/2}$$

is the anharmonic ratio of s, u, t, t'. It is invariant under projective transformations of $[0, \infty)$ and so is $\mathfrak{L}(\xi(t), t \geqq 0)$, provided $[0, \infty)$ is transformed into itself. In particular, $\mathfrak{L}(W(t)/\sqrt{t}, 0 < t < \infty) = \mathfrak{L}(\sqrt{t}\, W(1/t), 0 < t < \infty)$.

c) Construct $W_{[0,\pi]}$: Use 37.5**B** and verify that

$$s \wedge t = \frac{st}{2} + \frac{2}{\pi} \sum_{n=1}^{\infty} \frac{\sin ns \sin nt}{n^2}$$

where the series converges uniformly and absolutely on $[0, \pi] \times [0, \pi]$. Then

$$X(t) = \frac{t}{\sqrt{\pi}} \xi_0 + \sqrt{2/\pi} \sum_{n=1}^{\infty} \frac{\sin nt}{n} \xi_n,$$

where $\xi_0, \xi_1, \cdots$ are iid with common law $\mathfrak{N}(0, 1)$, is $W(t)$ on $[0, \pi]$ provided sample continuity is proved: take partial sums, show that they converge uniformly on $[0, \pi]$ outside a null event, and then eliminate it, as usual.

d) Examine the r.f.'s in 1 in L_2-terms.

3. *Martingales.* $(W_t, \mathfrak{B}_t, t \geqq 0)$ and $(W^2(t) - t, \mathfrak{B}_t, t \geqq 0)$ are martingales. Also

$$(Y_c(t, \mathfrak{B}_t, t \geqq 0) \quad \text{where} \quad Y_c(t) = \exp\left\{cW(t) - \frac{1}{2} c^2\right\}$$

is a martingale for every $c > 0$ and so is $(Y_c'(t), \mathfrak{B}_t, t \geqq 0)$ where $Y_c'(t) = Y_c(t) + Y_{-c}(t)$.

a) The Kolmogorov inequality for Brownian motion is

$$P(\sup_{0 \leqq s \leqq t} | W(s) | \geqq \epsilon) \leqq t/\epsilon^2$$

and yields

$$W(t)/t \xrightarrow{a.s.} 0 \quad \text{as} \quad t \to +\infty:$$

Take $t = 2^n$, $\epsilon = 2^{2n/3}$ and use Borel-Cantelli.

b) Let $\tau = \tau_{a,b}$. Then

$$E(W^2(\tau \wedge n) - \tau \wedge n) = E(W^2(0) - 0) = 0$$

hence

$$E(\tau \wedge n) = E(W^2(\tau \wedge n)) \leqq (| a | + b)^2$$

and

$$E\tau = \lim_n \int_0^n P(\tau > t)\, dt = \lim_n E(\tau \wedge n) \leqq (| a | + b)^2 < \infty.$$

c) Let $W_T{}^\alpha$ be a Brownian r.f. *with drift* $\alpha \neq 0$. The martingale with $Y_c(t)$ becomes a martingale with

$$Y_c{}^\alpha(t) = \exp\left\{cW^\alpha(t) - \left(cx + \frac{1}{2}c^2t\right)\right\}$$

Take $c_0 = -2\alpha$ so that $Y^\alpha(t) = \exp\{c_0W^\alpha(t)\}$. It follows, using the corollary of 39.2**A**, that

$$1 = E(Y_{c_0}{}^\alpha(\tau)) = \exp\{c_0a\}P(\tau = a) + \exp\{c_0b\}P(\tau = b).$$

Compute $P(\tau = a)$ and $P(\tau = b)$.

Let $\alpha < 0$ and $M_\infty{}^\alpha = \max_{0 \leqq t < \infty} W^\alpha(t)$. Then

$$P(M_\infty{}^\alpha \geqq c) = e^{-2|\alpha|c}, \quad c \geqq 0.$$

Let $\tau = \tau_c$ with $c > 0$. Then the pr. density of τ_c is given by

$$\frac{c}{\sqrt{2\pi t^3}} \exp\{-c^2/2t\}, \quad t > 0.$$

Let $\tau = \tau_{-a,\, +a}$, $a > 0$. Use the martingale $Y_c'(t)$ with $c = \sqrt{2\lambda}$ to prove that

$$Ee^{-\lambda\tau} = 1/\cosh(a\sqrt{2\lambda}).$$

4. *Cogburn and Tucker quadratic variation.* The Brownian quadratic variation is generalized for continuous decomposable processes X_T, say, $T = [0, 1]$ with i.d. law (α_T, Ψ_T); it is assumed that α_T is of bounded variation. Let the limits be taken along sequences of partitions $0 = t_{n0} < \cdots < t_{nk_n} = 1$ ordered by refinement with $\max_{k \leqq k_n} (t_{n,k} - t_{n,k-1}) \to 0$. Then

a) $$\int_T (dX_t)^2 = \lim_{n\to\infty} \sum_k X_{nk}^2 = \sigma^2 + \sum \mathcal{J}_t^2 \quad \text{a.s.}$$

where σ^2 is the variance of the normal component and the sum is that of squares of the jumps of X_T.

b) If g on R is continuous with $g(0) = 0$ and has a second derivative at 0, then a.s.

$$\int_T g(dX_t) = \lim_{n\to\infty} \sum_k g(X_{nk}) = g'(0)X_T + \frac{1}{2}g''(0)\sigma^2 + \sum\{g(\mathcal{J}_t) - g'(0)\mathcal{J}_t\}.$$

What if $X_T = W_T$ is a Brownian motion?

5. $\mathcal{L}(m(t), M(t), W(t))$. Let $a < 0 < b$, $c = b - a$, $t > 0$,

a) Let $S \subset [a, b]$ be a Borel set, and let

$$g(v) = \frac{1}{\sqrt{2\pi t}} \sum_{n=-\infty}^{+\infty} (\exp\{(-v - 2nc)^2/2t\} - \exp\{-(v - 2a + 2nc)^2/2t\}).$$

Then

$$P(a < m(t) \leqq M(t) < b,\ \ W(t) \in S) = \int_S g(v)\, dv.$$

The proof is based upon the reflection principle used repeatedly.

b) Deduce the individual laws of these 3 r.v.'s, then their two by two joint laws, and compare with the results in 41.4.

c) Find the conditional law of $M(t)$ given $M(t) - W(t) = 0$ and show that

$$P(M(t) > a |\, M(t) = W(t)) = e^{-a^2/2t}.$$

Let $W'(t) = W(1) - W(1 - t), 0 \leqq t \leqq 1; W'_{[0,1]}$ is a Brownian motion. Then

$$P(W(1) \leqq a |\, W(t) \geqq 0,\ \ 0 \leqq t \leqq 1) = P(W'(1) \leqq a |\, W'(1) = M'(1)) = 1 - e^{-a^2/2}.$$

6. *Passage times*. The passage times are times $\tau_c = \min\{t: W(t) = c\}, c \in R$. The r.f. $(\tau_c, c \geqq 0)$ is one-sided stable r.f. with "exponent" 1/2 and "rate" $\sqrt{2}$, that is, it is decomposable with

$$P(\tau_{c+h} - \tau_h \leqq t) = P(\tau_c \leqq t) = \frac{c}{\sqrt{2\pi}} \int_0^t s^{-3/2} e^{-c^2/2s}\, ds.$$

To prove, show that what follows holds.

a) $P(\tau_c \leqq t) = P(M(t) \geqq c)$ is the above probability: use change of variable in the integral giving $P(M(t) \geqq c)$ in the text. Also

$$P(\tau_c < c^2 u) = \frac{1}{\sqrt{2\pi}} \int_0^u e^{-1/2v}\, v^{-3/2}\, dv,\ c > 0,\quad u > 0,$$

and $\mathfrak{L}(\tau_c/c^2) = \mathfrak{L}(\tau_1)$ is independent of c. This obtains *a priori* from the fact that the nature of the r.f. $W(t)$, $t \geqq 0$, hence that of $M(t)$, $t \geqq 0$ and τ_c, $c \geqq 0$, does not change when simultaneously c and t are changed into λc and $\lambda^2 t$, respectively.

b) $(\tau_c, c \geqq 0)$ is decomposable with

$$\psi(u, c) = \log Ee^{iu\tau_c}$$
$$= (-1 + i)c\sqrt{u} = \frac{c}{\sqrt{2\pi}} \int_0^\infty (e^{iuc} - 1)u^{-3/2}\, du, \quad c > 0, \quad u > 0:$$

Show that $\mathfrak{L}(\tau_{c+h} - \tau_h) = \mathfrak{L}(\tau_c)$, use what precedes, note that τ_c does not decrease as c increases and that the decomposable r.f. has only positive jumps.

c) $(\tau_c, c \geqq 0)$ is a sum of positive jumps. For every interval of length c, the number $\nu(h)$ of jumps whose height exceeds h is a Poisson r.v. with expectation

$$\frac{c}{\sqrt{2\pi}} \int_h^\infty u^{-3/2}\, du = c\sqrt{2/\pi h}$$

and the total length Λ of jumps exceeded by h has for expectation

$$\frac{c}{\sqrt{2\pi}} \int_0^h \frac{du}{\sqrt{u}} = c\sqrt{2h/\pi}.$$

Furthermore, setting $c\sqrt{2/\pi} = s$, the Law of Large Numbers yields

$$\nu(h)/\sqrt{h} \xrightarrow{a.s.} s, \quad \Lambda/\sqrt{h} \xrightarrow{a.s.} s.$$

7. *Zeros.* Let $s \leqq t \leqq u$.

a) The conditional pr. $P(W(t)$ has at least one zero between s and $u \mid X_s = c) =$
$\dfrac{|c|}{\sqrt{2\pi}} \displaystyle\int_0^{u-s} e^{-c^2/2v}\, v^{-3/2}\, dv.$

It follows that

$$P(W(t) \text{ has at least one zero between } s \text{ and } u) = \frac{2}{\pi} \operatorname{Arccos}\sqrt{s/u}.$$

b) Let τ be the largest zero of W_T not exceeding t, and let τ' be the smallest zero of W_T exceeding t. Then

$$P(\tau < s) = \frac{2}{\pi} \operatorname{Arcsin}\sqrt{s/t},\ P(\tau' < u) = \frac{2}{\pi} \operatorname{Arccos}\sqrt{t/u},$$

$$P(\tau < s, \tau' > u) = \frac{2}{\pi} \operatorname{Arcsin} \sqrt{s/u}.$$

c) $P(M(t) - W(t)$ has at least one zero in $(s, u)) = (2/\pi)\operatorname{Arccos}\sqrt{s/u}$. If $\bar{\tau}$ is the largest zero of $M_T - W_T$ not exceeding t and $\bar{\tau}'$ is the smallest zero of $M_T - W_T$ exceeding t, then $\mathfrak{L}(\tau) = \mathfrak{L}(\tau)$ and $\mathfrak{L}(\tau') = \mathfrak{L}(\bar{\tau}')$.

AN EXTENSION OF DONSKER'S THEOREM

NOTE by L. Le Cam

Donsker's theorem, Section 42.2, may be extended to cover cases where variances do not exist. Since passage from the symmetric cases to nonsymmetric ones present absolutely no difficulty, we shall describe only the symmetric situation.

Consider a double array ξ_{nk}; $k = 1, 2, \cdots, k_n$, $n = 1, 2, \cdots$ as in Section 42.2 but subject to the following assumptions:

a) For each n the ξ_{nk} are independent

b) $\mathfrak{L}(\xi_{nk}) = \mathfrak{L}(-\xi_{nk})$

c) If $\sigma_{nk}^2 = E\min(1, \xi_{nk}^2)$, then $\sup_k \sigma_{nk}^2 \to 0$ and

d) $\sum_k \sigma_{nk}^2 = 1$.

Construct a process $X_n(t)$ by linear interpolation exactly as in Section 42.2 but with times t_{nk} computed with the present definition of the σ_{nk}^2. Let S_n be the last sum $S_n = S_{nk_n}$ and let W be the Wiener process on $[0, 1]$.

THEOREM. Assume that the conditions a, b, c, d *above are satisfied. Then the following statements are equivalent.*

i) *There is a tight sequence* $\{G_n\}$ *of Gaussian distributions such that the Lévy distance between* $\mathfrak{L}(S_n)$ *and* G_n *tends to zero.*

ii) $\mathfrak{L}(X_n) \to \mathfrak{L}(W)$.

The proof uses a rougher interpolation as follows. For each integer n construct times $\tau(n, r, j)$ starting with $\tau(n, r, 0) = 0$. The value $\tau(n, r, j)$ will be a certain t_{nk} for $k = k(j)$.

If $\tau(n, r, j)$ has been chosen, equal to some $t_{n\,k(i)}$ then $\tau(n, r, j + 1)$ is equal to t_{nk} for the first $k > k(j)$ for which $t_{nk} \geq t_{n\,k(j)} + (1/r)$.

Define a process $Y_{nr}(t)$, $t \in [0, 1]$ by taking $Y_{nr}(t) = X_n(t)$ if t is one of the times $\tau(n, r, j)$. Interpolate linearly between the values so obtained.

The essential result is as follows.

LEMMA 1. Let conditions a, b, c, d *be satisfied and assume that statement* (i) *of the theorem holds. Then, for every* $\epsilon > 0$, *there is an integer* r *and an integer* $n(\epsilon)$ *such that*

$$P\{\sup_t | X_n(t) - Y_{nr}(t) | > \epsilon\} < \epsilon.$$

Proof. Let $\tau_1 = \tau(n, r, j)$ and $\tau_2 = \tau(n, r, j + 1)$ be two successive points of the rough division. Note that by symmetry

$$P\{\sup_{\tau_1 \leq t \leq \tau_2} | X_n(t) - Y_{nr}(t) | > \epsilon\} \leq 2P\{| X_n(\tau_2) - X_n(\tau_1) | > \epsilon\}.$$

The difference involved here is a certain block sum $B(n, r, j) = \sum_k \{\xi_{nk};$ $k(j) < k \leq k(j + 1)\}$. It will be sufficient to show that one can select r so that eventually $\sum_j P\{|B(n, r, j)| > \epsilon\} < \frac{\epsilon}{2}$. If not, there would exist an $\epsilon_0 > 0$ such that for each r there is some $n(r) \geq r$ for which

$$\sum_j P\{|B(n(r), r, j)| \geq \epsilon_0\} \geq \epsilon_0.$$

Now for the block in question the sum of truncated variances $\sum_k \{\sigma_{nk}^2; k(j) < k \leq k(j+1)\}$ is at most $r^{-1} + \sup_k \sigma_{nk}^2$.

Thus the variables $B(n(r), r, j)$ are still uniformly asymptotically negligible. The existence of the $\epsilon_0 > 0$ would then violate the Normal Convergence Criterion (page 328, Vol. 1). This completes the proof of the lemma.

The lemma may be restated noting that the interpolation procedure used to pass from X_n to Y_{nr} is a certain linear transformation of $C[0, 1]$ into itself, say M_r. The conclusion of the Lemma 1 can then be restated as the assumption (i) in the following lemma.

LEMMA 2. Let X_n, $n = 0, 1, \cdots$ *and* W *be random elements with values in* $C[0, 1]$. *Assume*

i) *There are linear transformations* M_r *of* $C[0, 1]$ *into itself such that for each* $\epsilon > 0$ *there is an* $r > \epsilon^{-1}$ *and an* $n(\epsilon)$ *such that* $n \geq n(\epsilon)$ *implies* $P\{\| (I - M_r)X_n \| > \epsilon\} < \epsilon$.

ii) *For each r the Prohorov distance between* $\mathfrak{L}(M_rX_n)$ *and* $\mathfrak{L}(M_rW)$ *tends to zero.*

iii) $\lim_r \mathfrak{L}(M_rW) = \mathfrak{L}(W)$.

Then $\mathfrak{L}(X_n) \to \mathfrak{L}(W)$.

This is easily proven by looking at differences of expectations $Eh(M_rX_n) - Eh(X_n)$ where h is a bounded uniformly continuous function defined on $C[0, 1]$. Condition (i) makes these differences small. Condition (ii) makes the differences $Eh(M_rX_n) - Eh(M_rW)$ small and condition (iii) makes the differences $Eh(M_rW) - Eh(W)$ small.

This is sufficient to imply convergence since each bounded continuous function is at the same time the pointwise supremum of a countable set of uniformly continuous functions and the pointwise infimum of another such set.

Here we have just proved that (i) of the Lemma holds. That (ii) holds follows from the usual finite dimensional central limit theorem. Part (iii) may be proved by noting that the argument of Lemma 1 applies just as well to the Wiener process itself.

With appropriate modifications the Theorem admits extensions to the case where the variables ξ_{nk} take values in a separable Banach space or its weak second dual. Statements of this nature can be found in "Remarks on a theorem of Donsker and Kolmogorov" by A. P. Araujo and L. Le Cam.

Chapter XIV

MARKOV PROCESSES

Markov dependence was introduced by Markov (1906) as a natural extension of independence such that the asymptotic properties of sums of r.v.'s, say the law of large numbers and the normal convergence, may be expected to continue to hold under reasonable restrictions. As the restrictions were gradually removed and the setup expanded, new types of behavior and problems appeared. Independently, similar ones appeared in physics (Chapman, Fokker, Planck, etc.).

In his fundamental paper (1931) Kolmogorov introduces, rigorously, Markov dependence in the continuous case and shows that the transition pr.'s satisfy certain differential or integro-differential equations under various restrictions (primarily of the Lindeberg type for normal convergence—on the conditional moments—or of the continuity type $P[X_{t+h} = X_t] \to 1$ as $h \to 0$). The *leit-motif* is the search for local characteristics. Feller explores this new field of research in two series of basic papers essentially of a purely analytical character. First (1936, 1940) he pursues Kolmogorov's approach through local characteristics of transition pr.'s, investigating conditions under which there do exist transition pr.'s with various local characteristics. Then (1952 on) he uses and expands the semi-group theory (created by Hille and Yosida and immediately applied by them to Markov evolution in time) and introduces and analyzes "Feller" or "stable Markov" processes; his work is at the root of novel developments in Markov processes.

Meanwhile Doblin (1938–39) proceeds to a direct sample analysis under a uniform continuity condition on transition pr.'s which leads to step sample functions; Doob (1945), upon removing the uniformity restriction, discovers sample discontinuities more complicated than jumps; P. Lévy (1951) flushes the then monstrous sample possibilities into the open; and Kinney (1953) investigates sample continuity properties.

The purely analytical and the sample analysis gradually merge. Fortet early (1943) examines sample continuity in connection with the Kolmogorov-Feller approach. Later (1955) in connection with the Hille-Yosida-Feller semi-group approach, Neveu combines the two lines of attack. At the same time, Dynkin begins a series of basic papers in which he establishes and extends Feller's results and pursues an investigation of various Markov evolutions in time by means of an intimate blend of sample and semi-group analysis; for this purpose Dynkin and Yushkevich isolate and analyze the concept of strong Markov dependence, first mentioned by Doob (1945), and which is essential for the sample analysis of Markov processes. Hunt (1957) in a basic series of papers connects potential theory with Markov processes.

Let us only mention another line of attack by difference, differential, and integral stochastic equations in the r.f.'s themselves, now in the process of growth, thanks to the works of Bernstein, P. Lévy, Doob, Maruyama, and especially Itō (1951).

§43. MARKOV DEPENDENCE

43.1. Markov property. In order to describe and analyze Markov dependence in intuitive yet precise terms and also to simplify the writing, we recall and add some terminology and notation.

Let r, s, t with or without affixes denote the elements of a linearly ordered index set T. To fix the ideas, we take T to be a set of reals with the usual ordering, say, the set of all reals or of all integers or of all nonnegative reals or of all nonnegative integers; in fact, we shall end up by taking $T = [0, \infty)$. It is convenient to give the indices a phenomenological meaning: they will represent moments of "time." Thus, an ordered triplet $U < V < W$ of disjoint index subsets (that is, every element of U precedes every element of V, which precedes every element of W) represents a "past" U, a "present" V, and a "future" W.

R.v.'s X, Y, $\cdots$ with or without affixes are measurable functions on a pr. space $(\Omega, \mathcal{A}, P)$ to a measurable space $(\mathcal{X}, \mathcal{S})$ or *state space*. In general, the r.v.'s take their values in a topological space $\mathcal{X}$, and $\mathcal{S}$ is the σ-field of topological Borel sets, that is, the σ-field generated by the class of open sets. Most of the considerations in this chapter remain valid for more general state spaces and especially for locally compact separable metric state spaces and, thus, are frequently couched in general terms. However, to fix the ideas we assume that unless otherwise stated our r.v.'s are numerical: the state space is a Borel set of the (extended) real line together with the σ-field of its Borel sets. In general, the extension to the N-dimensional real spaces will be trivial. But the reader is invited to transpose what follows to more general state spaces.

$\mathcal{B}$ with or without affixes will denote sub σ-fields of events B with same affixes if any, unless otherwise stated. *In this subsection only* we denote by $\mathcal{B}_T'$ the union σ-field of the $\mathcal{B}_t$, $t \in T'$, that is, the smallest σ-field containing all of them. Note that the finite sums of finite intersections $B_{t_1} \cap \cdots \cap B_{t_n}$ of their events over all finite subsets of T' form a field $\mathcal{C}$ which generates $\mathcal{B}_{T'}$—the smallest monotone field containing $\mathcal{C}$. In connection with this construction we shall use the properties of c.pr.'s and c.exp.'s without further comment. $\mathcal{B}_{T'}$-measurable functions whose expectation exists will be denoted by $Y_{T'}$, unless otherwise stated.

Finally, in accordance with 25.3**A**, we introduce the following equivalent terminologies for conditional independence

$\mathcal{B}_1$ *and* $\mathcal{B}_3$ *are c.ind. given* $\mathcal{B}_2$: $P(B_1B_3 \mid \mathcal{B}_2) = P(B_1 \mid \mathcal{B}_2)P(B_3 \mid \mathcal{B}_2)$ *a.s.*

$\mathcal{B}_1$ *is c.ind. of* $\mathcal{B}_3$ *given* $\mathcal{B}_2$: $P(B_1 \mid \mathcal{B}_{23}) = P(B_1 \mid \mathcal{B}_2)$ *a.s.*

$\mathcal{B}_3$ *is c.ind. of* $\mathcal{B}_1$ *given* $\mathcal{B}_2$: $P(B_3 \mid \mathcal{B}_{12}) = P(B_3 \mid \mathcal{B}_2)$ *a.s.*

A family $\mathcal{B}_t$, $t \in T$, of σ-fields of events is said to be *Markovian* or a *Markov family* if the following equivalent forms of *Markov property* hold:

For any past, present, and future

(M_{future}) *The future of the family is c.ind. of its past given its present:*

$$P(B_{\text{future}} \mid \mathcal{B}_{\text{present+past}}) = P(B_{\text{future}} \mid \mathcal{B}_{\text{present}}) \text{ a.s.}$$

(M_{past}) *The past of the family is c.ind. of its future given its present:*

$$P(B_{\text{past}} \mid \mathcal{B}_{\text{present+future}}) = P(B_{\text{past}} \mid \mathcal{B}_{\text{present}}) \text{ a.s.}$$

($M_{\text{past, future}}$) *The past and the future of the family are c.ind. given its present:*

$$P(B_{\text{past}}B_{\text{future}} \mid \mathcal{B}_{\text{present}}) = P(B_{\text{past}} \mid \mathcal{B}_{\text{present}})P(B_{\text{future}} \mid \mathcal{B}_{\text{present}}) \text{ a.s.}$$

It will be convenient to have at our disposal seemingly weaker and seemingly stronger yet equivalent forms of Markov property. The lemma below permits the passage from events to functions.

a. *Let* $\mathcal{B}_1$, $\mathcal{B}_2$, $\mathcal{B}_3$ *be* σ*-fields of events and let* $\mathcal{B}_3$ *be generated by a field* $\mathcal{C}$ *of finite sums of events* D *belonging to a class* $\mathfrak{D}$.

If $P(D \mid \mathcal{B}_{12}) = P(D \mid \mathcal{B}_2)$ *a.s. then* $P(B_3 \mid \mathcal{B}_{12}) = P(B_3 \mid \mathcal{B}_2)$ *a.s. and, more generally,* $E(Y_{23} \mid \mathcal{B}_{12}) = E(Y_{23} \mid \mathcal{B}_2)$ *a.s.*

Proof. The class of events B such that $P(B \mid \mathcal{B}_{12}) = P(B \mid \mathcal{B}_2)$ a.s. is closed under countable summations and contains $\mathcal{D}$ hence contains $\mathcal{C}$. It is also closed under monotone passages to the limit hence it contains the monotone field $\mathcal{B}_3$ generated by the field $\mathcal{C}$. This proves the particular assertion.

Since the family of functions Y (whose exp.'s exist) such that $E(Y \mid \mathcal{B}_{12}) = E(Y \mid \mathcal{B}_2)$ a.s. is closed under multiplications by the indicators I_{B_2} and we just proved that it contains the I_{B_3}, it follows that it contains the $I_{B_2B_3} = I_{B_2}I_{B_3}$. Therefore, by the particular assertion, it contains the $I_{B_{23}}$. But this family is also closed under linear combinations whose exp.'s exist and under passages to the limit by nondecreasing sequences of its nonnegative elements. Thus it contains the simple $\mathcal{B}_{23}$-measurable functions then the nonnegative $\mathcal{B}_{23}$-measurable functions and finally all the $\mathcal{B}_{23}$-measurable functions whose exp.'s exist. The proof is terminated.

The next lemma permits the extension of the futures.

b. *Let $\mathcal{B}_1$, $\mathcal{B}_2$, $\mathcal{B}_3$, $\mathcal{B}_4$ be σ-fields of events. If* (i) $P(B_3 \mid \mathcal{B}_{12}) = P(B_3 \mid \mathcal{B}_2)$ *a.s. and* (ii) $P(B_4 \mid \mathcal{B}_{123}) = P(B_4 \mid \mathcal{B}_3)$ *a.s., then* $P(B_{34} \mid \mathcal{B}_{12}) = P(B_{34} \mid \mathcal{B}_2)$ *a.s. and, more generally,* $E(Y_{34} \mid \mathcal{B}_{12}) = E(Y_{34} \mid \mathcal{B}_2)$ *a.s.*

Proof. According to **a**, (i) yields

(1) $$E(Y_{23} \mid \mathcal{B}_{12}) = E(Y_{23} \mid \mathcal{B}_2) \text{ a.s.}$$

while, upon multiplying by I_{B_3} and conditioning by $\mathcal{B}_{23}$, (ii) yields

(2) $$E(I_{B_3}I_{B_4} \mid \mathcal{B}_{23}) = E(I_{B_3B_4} \mid \mathcal{B}_3) = E(I_{B_3B_4} \mid \mathcal{B}_{123}) \text{ a.s.}$$

But we always have

(3) $$E(I_{B_3B_4} \mid \mathcal{B}_{12}) = E\{E(I_{B_3B_4} \mid \mathcal{B}_{123}) \mid \mathcal{B}_{12}\} \text{ a.s.}$$

where, by (2), the right side reduces to $E\{E(I_{B_3B_4} \mid \mathcal{B}_{23}) \mid \mathcal{B}_{12}\}$ a.s. and the c.exp. within the last expression is $\mathcal{B}_{23}$-measurable. Therefore, by (1), the right side of (3) reduces to

$$E\{E(I_{B_3B_4} \mid \mathcal{B}_{23}) \mid \mathcal{B}_2\} = E(I_{B_3B_4} \mid \mathcal{B}_2) \text{ a.s.}$$

and (3) becomes

$$P(B_3B_4 \mid \mathcal{B}_{12}) = P(B_3B_4 \mid \mathcal{B}_2) \text{ a.s.}$$

The lemma follows on account of **a**, and the proof is terminated.

We are primarily interested in the Markovian evolution and thus explore the future as time increases. Therefore, we shall use primarily the corresponding form (M_{future}) of the Markov property and be content with stating its equivalent formulations. Besides, the equivalent formulations of the other forms will then follow at once on account of 25.3**A.**

A. MARKOV EQUIVALENCE THEOREM. *The following equivalent relations characterize the Markov property:*

(M) $P(B_{\text{future}} \mid \mathcal{B}_{\text{present+past}}) = P(B_{\text{future}} \mid \mathcal{B}_{\text{present}})$ *a.s. or*

(M′) $E(Y_{\text{present+future}} \mid \mathcal{B}_{\text{present+past}}) = E(Y_{\text{present+future}} \mid \mathcal{B}_{\text{present}})$ *a.s.*

for

(i) *any past, present, and future or*

(ii) *any instant present s, the whole past* $[r: r < s]$, *and the whole future* $[t: t > s]$ *or*

(iii) *any finite past, instant present, and instant future.*

Note that it suffices that

(M″) *the c.pr.'s of future events given the present and the past depend only upon the present:*

$$P(B_{\text{future}} \mid \mathcal{B}_{\text{present+past}}) \text{ is } \mathcal{B}_{\text{present}}\text{-measurable.}$$

For, this property is implied by (M) and, upon conditioning it by $\mathcal{B}_{\text{present}}$, it yields (M).

Proof. Since (M) $\Rightarrow$ (M′) on account of **a** and (M′) $\Rightarrow$ (M) as a particular case, it suffices to consider (M). Since (i) $\Rightarrow$ (ii) $\Rightarrow$ (iii), it remains only to prove that (iii) $\Rightarrow$ (i). Thus, let $U \prec V \prec W$ and

$$\text{(M)} \qquad P(B_W \mid \mathcal{B}_{U+V}) = P(B_W \mid \mathcal{B}_V) \text{ a.s.}$$

for every finite U, singleton V, and singleton W. Since, by **b**, we can increase the future point by point, (M) extends to all finite subsets W_n of any given future, whence, by **a**, to W.

Since for a singleton V and for all finite subsets U_n of any given past U

$$\int_B P(B_W \mid \mathcal{B}_{U+V}) = \int_B P(B_W \mid \mathcal{B}_{U_n+V}) = \int_B P(B_W \mid \mathcal{B}_V)$$

for every $B \in \mathcal{B}_{U_n+V}$ and since the $\mathcal{B}_{U_n+V}$ generate $\mathcal{B}_{U+V}$, it follows that the equality between the finite measures on every $\mathcal{B}_{U_n+V}$ defined by the extreme terms extends on $\mathcal{B}_{U+V}$. Therefore, the integrands coincide a.s. for any given U.

Finally, (M) extends to any given present V, as follows. If V has a last element t, then take $U + V - \{t\}$ as the past, $\{t\}$ as the present, and W as the future, so that

$$P(B_W \mid \mathcal{B}_{U+V}) = P(B_W \mid \mathcal{B}_t) \text{ a.s.}$$

It follows, upon conditioning by $\mathcal{B}_V$ that

$$P(B_W \mid \mathcal{B}_V) = P(B_W \mid \mathcal{B}_t) = P(B_W \mid \mathcal{B}_{U+V}) \text{ a.s.}$$

If V has no last element, decompose V into $V' + V''$ where $V' < V''$ and V' has a last element. Then take U as the past, V' as the present, and $V'' + W$ as the future. According to what precedes

$$P(B_{V''+W} \mid \mathcal{B}_{V'+U}) = P(B_{V''+W} \mid \mathcal{B}_{V'}) \text{ a.s.}$$

and, interchanging the past U and the future $V'' + W$, it follows that

$$P(B_U \mid \mathcal{B}_{V+W}) = P(B_U \mid \mathcal{B}_{V'}) \text{ a.s.,}$$

Finally, conditioning by $\mathcal{B}_V$ so that

$$P(B_U \mid \mathcal{B}_V) = P(B_U \mid \mathcal{B}_{V'}) = P(B_U \mid \mathcal{B}_{V+W}) \text{ a.s.}$$

and interchanging the past U and the future W so that

$$P(B_W \mid \mathcal{B}_V) = P(B_W \mid \mathcal{B}_{U+V}) \text{ a.s.,}$$

the extension of (M) to any given past is proved and the proof is terminated.

The time set T consists most frequently of all $t \geqq t_0$ with $t_0 = 0$ and the Markov evolution is analyzed from instant presents s to the futures $t \geqq s$. Thus, it is primarily the form (M_{future}) which is used. However, difficulties arise at once: The c.pr.'s in (M_{future}) may have no regular versions, so that we cannot treat them as measures; and even if they have, we may not be able to select the regular versions so as to transform all the a.s. equalities into strict equalities, and then we are faced with too many exceptional null events since T is uncountable. If none of these difficulties arises, so that by a suitable choice of c.pr.'s the equalities (M_{future}) become strict equalities and we have no exceptional null events, then we say that the Markov property is *regular* and further analysis of a regular Markov evolution becomes possible. If, moreover, the evolution is independent of the instant present, that is, the c.pr.'s are invariant under translations in time, then we say that the Markov property is *stationary* (or *homogeneous in time*). We shall be mostly concerned with the stationary Markov property.

43.2. Regular Markov processes. A r.f. $X_T = (X_t, t \in T)$ is said to be *Markovian* or a *Markov r.f.* if the family of σ-fields $\mathcal{B}_t = \mathcal{B}(X_t) = X_t^{-1}(\mathcal{S})$ of events induced by the X_t is Markovian. A process is said to be *Markovian* or a *Markov process* if it consists of Markov r.f.'s with same induced σ-fields. Since $\mathcal{B}(X_t) = \mathcal{B}(Y_t)$ means that there exists an invertible Borel g_t such that $Y_t = g_t(X_t)$ and $X_t = g_t^{-1}(Y_t)$, we may consider a Markov process as a Markov r.f. defined up to an invertible scale function g_T.

In this connection, it is sometimes convenient to modify the definitions so as not to require invertibility. A Markov r.f. becomes a family $(X_t, \mathcal{B}_t, t \in T)$ where the X_t are $\mathcal{B}_t$-measurable and the σ-fields $\mathcal{B}_t$ form a Markov family so that, *a fortiori*, $(X_t, t \in T)$ is a Markov r.f. according to our definition. Then, a Markov process becomes the family of all r.f.'s X_T with $\mathcal{B}_t$-measurable X_t's, where the given $\mathcal{B}_t$ form a Markov family.

For r.f.'s X_T, the Markov property (Mii), equivalently, (M'ii) which emphasizes the Markov evolution from any instant present becomes

$$P(B \mid X_r, r \leqq s) = P(B \mid X_s) \text{ a.s., equivalently,}$$
$$E(Y \mid X_r, r \leqq s) = E(Y \mid X_s) \text{ a.s.}$$

where B and Y are defined on $(X_t, t \geqq s)$. In order to examine regularity possibilities, we take its most particular case. First, we take an instant past $\{r\}$ so that, upon conditioning by (X_r, X_s), we have

$$P(B \mid X_r, X_s) = P(B \mid X_s) \text{ a.s.}$$

On account of the general smoothing property $P(B \mid X_r) = E\{P(B \mid X_r, X_s) \mid X_r\}$, it yields the relation

$$P(B \mid X_r) = E\{P(B \mid X_s) \mid X_r\} \text{ a.s.}$$

Next, we take an instant future $\{t\}$ so that the relation becomes

$$P(X_t \in S \mid X_r) = E\{P(X_t \in S \mid X_s) \mid X_r\}, \quad r < s < t, \quad S \in \mathcal{S},$$

outside a null event of the form $[X_r \in S_0]$, where the Borel set S_0 may vary with r, s, t, and with S.

If there are regular versions of the c.pr.'s which figure in the last relation, then its right side can be written as an integral. If moreover the regular versions may be selected so as to make disappear the exceptional sets S_0 for all r, s, t and all S, then we have regularity at least in the particular case of instant present, past, and future. Thus, the requirement that it be possible to transform this relation into strict equality is a minimal one for regularity. We then give to the relation the name of (Chapman-)Kolmogorov equation and to the c.pr.'s therein the name of transition pr.'s. More precisely, a function $P_{st}(x, S)$ defined for all pairs $(s, t) \in T \times T$ with $s < t$ and for all $x \in \mathfrak{X}$, $S \in \mathcal{S}$, is a *transition pr.* (*tr.pr.*) if it is $\mathcal{S}$-measurable in x and a pr. in S and if the *Kolmogorov equation* holds: for all r, s, t with $r < s < t$, all x, and all S,

$$\text{(K)} \qquad P_{rt}(x, S) = \int P_{rs}(x, dy) P_{st}(y, S).$$

We add the *convention:* $P_{ss}(x, S) = I(x, S) (= 1$ or 0 according as $x \in S$ or $x \notin S)$ so that the tr.pr.'s are now defined for all pairs (s, t) with $s \leqq t$, and then the Kolmogorov equation clearly holds with $r \leqq s \leqq t$.

The Kolmogorov equation can be given a seemingly stronger form which parallels the passage of (M) to (M′), that is, from events to functions. Let G be the space of all bounded Borel functions on $\mathfrak{X}$ (so that all integrals below exist and are finite). The relation

$$T_{uv}g(x) = \int P_{uv}(x, dy) g(y), \quad u \leqq v,$$

defines a function $T_{uv}g$ which clearly also belongs to G. In particular, for $g(x) = I(x, S)$ we have $T_{uv}g(x) = P_{uv}(x, S)$, and $T_{uu} = I$—the identity operator on G. Thus, it defines a *tr. operator* T_{uv} on G (to G) which is an extension of the transformation of indicators $I(x, S)$ into tr.pr.'s $P_{uv}(x, S)$. Furthermore, according to the Kolmogorov equation, for $r \leqq s \leqq t$,

$$(T_{rt}g)(x) = \int P_{rt}(x, dz) g(z) = \int \left\{ \int P_{rs}(x, dy) P_{st}(y, dz) \right\} g(z)$$

$$= \int P_{rs}(x, dy) \left\{ \int P_{st}(y, dz) g(z) \right\} = \{ T_{rs}(T_{st}g) \} (x).$$

Therefore, the family of transformations T_{uv} on G has the so-called *generalized semi-group property:*

$$\text{(K}'\text{)} \qquad T_{rt} = T_{rs} T_{st}, \quad r \leqq s \leqq t.$$

Conversely, upon applying (K') to indicators $g(x) = I(x, S)$, Kolmogorov equation follows. Thus

a. *The Kolmogorov equation for tr.pr.'s and the generalized semi-group property for corresponding tr. operators are equivalent.*

What precedes leads to the following definition: A r.f. X_T is a *regular Markov r.f.* if there exists a tr.pr. $P_{st}(x, S)$ such that for all pairs (s, t) with $s \leqq t$

$$P(X_t \in S \mid X_r, r \leqq s) = P_{st}(X_s, S) \text{ a.s.}$$

The r.f. is Markovian, since conditioning by X_s yields

$$P(X_t \in S \mid X_s) = P_{st}(X_s, S) = P(X_t \in S \mid X_r, r \leqq s) \text{ a.s.};$$

by omitting "a.s.", we can and do select regular versions of its c.pr.'s so as to have a regular Markov property. Thus, the definition of a regular Markov r.f. X_T and the choice of regular versions of c.pr.'s coalesce into *the strict equality*

$$\text{(MR)} \qquad P(X_t \in S \mid X_r, r \leqq s) = P(X_t \in S \mid X_s)$$

with the identity

$$P(X_t \in S \mid X_s) \equiv P_{st}(X_s, S),$$

where the values $P_{st}(x, S)$ of the right side are those of a tr.pr.

Note that for $t = s$, the identity reduces to, hence justifies, our convention for tr.pr.'s.

From now on and unless otherwise stated the time set T has a first element t_0, $P_{st}(x, S)$ is a tr.pr. and the "initial distribution" P_{t_0} is a pr. on $\mathcal{S}$.

Let $(\mathfrak{X}_T, \mathcal{S}_T) = \prod_{t \in T} (\mathfrak{X}_t, \mathcal{S}_t)$ be the sample space of a regular Markov r.f. On account of (MR) the tr.pr. $P_{st}(x, S)$ determines its conditional distributions $P_T{}^x$ given $X_{t_0} = x$. For, according to the Tulcea theorem 8.3**A** and to 27.2**b** with $g(x_1, \cdots, x_n) = I_{S_{t_1}}(x_1) \times \cdots \times I_{S_{t_n}}(x_n)$, $t_1 < \cdots < t_n$, $S_t \in \mathcal{S}_t$, $P_T{}^x$ is determined by

$$\text{(CD)} \quad P_T{}^x C(S_{t_1} \times \cdots \times S_{t_n})$$
$$= \int_{S_{t_1} \cdots S_{t_n}} P_{t_0 t_1}(x, dx_1) \cdots P_{t_{n-1} t_n}(x_{n-1}, dx_n).$$

It follows that, for the law of X_0 given by an initial distribution P_{t_0}, the law of X_T is determined by its distribution defined by

$$\text{(D)} \qquad P_T S_T = \int_{\mathfrak{X}} P_{t_0}(dx) P_T{}^x S_T, \quad S_T \in \mathcal{S}_T.$$

Thus, the c.pr. P^x given $X_{t_0} = x$ on $\mathcal{B}(X_T)$—the σ-field of events $B = [X_T \in S_T]$, is given by

$$\text{(CP)} \qquad P^x[X_T \in S_T] = P_T{}^x S_T, \quad S_T \in \mathcal{S}_T.$$

On account of the integration theorem 26.1**A**, the c.exp. E^x given $X_{t_0} = x$ on the $\mathcal{B}(X_T)$-measurable functions ξ (defined on X_T) whose expectations exist, is given by

$$\text{(CE)} \qquad E^x\xi = \int P^x(d\omega)\xi(\omega),$$

and, for an initial pr. P_{t_0}, hence for an initial distribution $P_{t_0}(S) = P[X_{t_0} \in S]$,

$$\text{(P)}: \quad PB = \int P_{t_0}(dx)P^xB, \quad \text{(E)}: \quad E\xi = \int P_{t_0}(dx)E^x\xi.$$

What precedes leads us to the following definition: A *regular Markov process* is a family of regular Markov r.f.'s with common tr.pr. and arbitrary initial distributions. The tr.pr. $P_{st}(x, S)$ determines by means of (CD), (CP), and (CE) the common conditional distributions $P_T{}^x$, c.pr.'s P^x, and c.exp.'s E^x, given $X_{t_0} = x$. The choice of the initial distribution P_{t_0} determines the law of the corresponding r.f. of the process.

Since the regular Markov property (MR) is in terms of c.pr.'s only, we may also define a regular Markov process as follows. Let X_T be a family of measurable functions X_t on a measurable space $(\Omega, \mathcal{A})$ to a measurable space $(\mathfrak{X}, \mathcal{S})$ (with all singletons $\{x\} \in \mathcal{S}$) and let $(P^x, x \in \mathfrak{X})$ be a family of pr.'s on the induced σ-field $\mathcal{B} = \mathcal{B}(X_T)$ with

$$P^x[X_{t_0} = x] = P^x(X_{t_0}{}^{-1}\{x\}) = 1.$$

If for every $x \in \mathfrak{X}$ there exists a tr.pr. $P_{st}(x, S)$ such that (CD) holds for $P_T{}^x$ defined by (CP), then we may say that X_T is a *regular Markov process*. For every choice of an initial distribution P_{t_0} on $\mathcal{S}$, X_T becomes a regular Markov r.f. on the pr. space $(\Omega, \mathcal{B}, P)$ where P on $\mathcal{B}$ is determined by (P).

The two definitions correspond to the two ways of viewing regular processes and, as long as we are concerned with one regular Markov process with a given tr.pr., we may use either of these definitions according to convenience. However, this raises a basic question: whether given an arbitrary tr.pr. there always exists a corresponding Markov process. The answer is in the affirmative, as follows.

A. Regular Markov existence theorem. *To any tr.pr. there corresponds a regular Markov process with a determined law of evolution.*

To any tr.pr. and any initial distribution there corresponds a regular Markov r.f. with a determined law.

Proof. The assertions relative to r.f.'s follow from those relative to processes on account of the necessary condition (P). The assertion relative to the law of evolution follows from the existence assertion on account of the necessary condition (CD). It remains to prove the existence assertion for processes. The necessary conditions (CD) and (CP) show the way: Take $(\Omega, \mathfrak{A}) = (\mathfrak{X}_T, \mathfrak{S}_T)$ with points $x_T = (x_t, t \in T)$. Define $X_T = (X_t, t \in T)$ by $X_t(x_T) = x_t$ and using the given tr.pr. $P_{st}(x, S)$, by means of (CD) and (CP), construct the family of pr.'s $P^x = P_T^x$, $x \in \mathfrak{X}$. The process so defined is a regular Markov process with the given tr.pr., provided we can show that for every finite index set $t_1 < \cdots < t_n < s < t$ and every $S_t \in \mathfrak{S}_t$ it is possible to select versions of c.pr.'s such that

$$P(X_t \in S_t \mid X_{t_1}, \cdots, X_{t_n}, X_s) = P_{st}(X_s, S_t)$$

or, equivalently,

$$P^x C(S_{t_1} \times \cdots \times S_{t_n} \times S_s \times S_t) = \int_{C(S_{t_1} \times \cdots \times S_s)} P^x(dx_T) P_{st}(x_s, S_t).$$

But by the construction of P^x, both sides reduce to

$$\int_{S_{t_1} \times \cdots \times S_{t_n} \times S_s \times S_t} P_{t_0 t_1}(x, dx_{t_1}) \cdots P_{t_n s}(x_{t_n}, dx_s) P_{st}(x_s, dx_t),$$

and the proof is terminated.

There is a one-to-one correspondence between tr.pr.'s $P_{st}(x, S)$ and *tr.d.f.'s* F_{st}^x defined by $F_{st}^x(y) = P_{st}(x, (-\infty, y))$ (when $\mathfrak{X} = R$). The Kolmogorov equation in terms of tr.d.f.'s is

$$F_{rt}^x(z) = \int F_{rs}^x(dy) F_{st}^y(z).$$

Tr.d.f.'s F_{st}^x are d.f.'s with a parameter x in which they are Borelian, and there is a one-to-one correspondence between tr.d.f.'s F_{st}^x and *tr.ch.f.'s* f_{st}^x defined by $f_{st}^x(u) = \int e^{iuy} F_{st}^x(dy)$. However, in the study of local characteristics under some continuity condition we are primarily interested in the Markov behavior in the neighborhood of given states x at times s. This leads to the centering at x of the tr.d.f. F_{st}^x and of the tr.ch.f. f_{st} hence to the introduction of

$$\bar{F}_{st}^x(y) = P(X_t - X_s < y \mid X_s = x) = F_{st}^x(x + y), \quad \bar{f}_{st}^x(u) = e^{-iux} f_{st}^x(u).$$

A trivial example of regular Markov processes and of tr.d.f.'s is provided by processes $(X_t, t \geqq 0)$ with independent r.v.'s; then $F_{st}^x = F_t$ and Kolmogorov's equation reduces to an identity. An important and suggestive example of regular Markov processes and of tr.d.f.'s is provided by decomposable processes with increments X_{st}; then $\overline{F}_{st}^x = F_{st}$ and Kolmogorov's equation becomes the composition relation $F_{rt} = F_{rs} * F_{st}$. Concepts and problems relative to decomposable processes suggest similar ones for regular Markov processes. In particular, the concept of law derivative will lead us to the historically important local characterizations of tr.d.f.'s, as follows.

We say that a ch.f. $\dot{f}_t^x$ represents the *tr. law (right) (left) derivative* given x at t of a regular Markov process X_T if $(\overline{f}_{t-h,t+k}^x)^{[1/(h+k)]} \to \dot{f}_t^x$ as $h + k \to 0$ with $h, k \geqq 0$ $(h \equiv 0)$ $(k \equiv 0)$, $h + k > 0$. According to the central convergence theorem, if the limit ch.f. exists then it is necessarily an i.d. ch.f. $\dot{f}_t^x = e^{\psi_t^x}$, $\psi_t^x = (\alpha_t^x, \Psi_t^x)$, and the process is tr. law (right) (left) continuous given x at t: $F_{t-h,t+k}^x(y) \to 0$ or 1 according as $y < x$ or $y > x$. While in the decomposable case we could use the convergence theorem for i.d. laws here we need the more general central convergence criterion but for identically distributed summands, that is, with $F_{nk} = F_n$:

$$f_n^n \to f = e^{\psi}, \psi = (\alpha, \Psi) \textit{ if and only if } \Psi_n \xrightarrow{c} \Psi \textit{ and } \int d\Psi_n(x)/x \to \alpha$$

$$\textit{with } \Psi_n(x) = n\int_{-\infty}^{x} \frac{y^2}{1+y^2}\, dF_n.$$

It suffices to note that because of $F_{nk} \equiv F_n$, condition (iii) of the criterion yields $n\left(\int_{|x|<\epsilon} x\, dF_n\right)^2 \leqq \left|\int_{|x|<\epsilon} x\, dF_n\right| \times \left|\int_{|x|<\epsilon} n\, dF_n\right| \to 0$ so that in its condition (ii) the left side can be replaced by $n\int_{|x|<\epsilon} \frac{x^2}{1+x^2}\, dF_n$, and the assertion follows by elementary computations.

Upon applying this particular form of the central convergence criterion to tr. law derivatives, we have

b. TR. LAW DERIVATIVE EXISTENCE CRITERION. *The tr. law (right) (left) derivative $e^{\psi_t^x}$, $\psi_t^x = (\alpha_t^x, \Psi_t^x)$ exists if and only if*

$$\Psi_{t-h,t+k}^x \xrightarrow{c} \Psi_t^x \textit{ and } \int d\Psi_{t-h,t+k}^x(y)/y \to \alpha_t^x$$
$$\textit{as } h + k \to 0\ (h \equiv 0)\ (k \equiv 0)$$

where

$$\Psi_{t-h,t+k}^x(x') = \frac{1}{h+k}\int_{-\infty}^{x'} \frac{y^2}{1+y^2}\, d\overline{F}_{t-h,t+k}^x$$

Next we note that, as for $g(x) = e^{iux}$, $x \in R$,

If a function g on R is bounded and twice differentiable then the function h_g on $R \times R$, given by

$$h_g(x, y) = \left(g(x + y) - g(x) - \frac{y}{1 + y^2} g'(x)\right) \frac{1 + y^2}{y^2}$$

and defined by continuity at $y = 0$ to be $h_g(x, 0) = g''(x)/2$, is bounded and continuous in y for every fixed x.

The two foregoing remarks yield

c. *If $f_n^n \to f = e^{\psi}$, $\psi = (\alpha, \Psi)$, then for every bounded and twice differentiable function g on R and every fixed $x \in R$*

$$n\left\{\int g(x + y)\, dF_n(y) - g(x)\right\} \to \alpha g'(x) + \int h_g(x, y)\, d\Psi(y).$$

It suffices to apply the Helly-Bray theorem to the left side expressed as

$$n\int (g(x + y) - g(x))\, dF_n(y) = g'(x)\int d\Psi_n(y)/y + \int h_g(x, y)\, d\Psi_n(y).$$

B. TR. LAW DERIVATIVES THEOREM. *Let $e^{\psi_s^x}$, $\psi_s^x = (\alpha_s^x, \Psi_s^x)$, $t \geq 0$ represent the (right) (left) tr. law derivatives corresponding to a tr.pr. $P_{st}(x, S)$, and let g on R be a bounded twice differentiable function. Then as $h + k \to 0$ ($h \equiv 0$) ($k \equiv 0$)*

$$\frac{1}{h + k}\left\{\int P_{s-h,s+k}(x, dy)g(y) - g(x)\right\} \to \alpha_s^x g'(x) + \int h_g(x, y)\, d\Psi_s^x(y).$$

In particular, for left tr. law derivatives, if $P_{st}(x, S)$ is twice differentiable in x then its left derivative $\frac{\partial^-}{\partial s} P_{st}(x, S) = \lim_{h \to 0} \frac{1}{h}\left\{P_{s-h,t}(x, S) - P_{st}(x, S)\right\}$ exists and

$$\frac{\partial^-}{\partial s} P_{st}(x, S) = \alpha_s^x \frac{\partial}{\partial x} P_{st}(x, S)$$

$$+ \int\left\{P_{st}(x + y, S) - P_{st}(x, S) - \frac{y}{1 + y^2}\frac{\partial}{\partial x} P_{st}(x, S)\right\} \frac{1 + y^2}{y^2}\, d\Psi_s^x(y).$$

Proof. The general assertion results from **c**, and the particular one follows upon setting $k = 0$, $g(x) = P_{st}(x, S)$, and using Kolmogorov's equation

$$P_{s-h,t}(x, S) = \int P_{s-h,s}(x, dy) P_{st}(y, S).$$

Upon introducing the P. Lévy form $\psi_s^x = (\alpha_s^x, (\beta_s^x)^2, L_s^x)$, the foregoing integro-differential equation is explicited into

$$\frac{\partial^-}{\partial s} P_{st}(x, S) = \alpha_s^x \frac{\partial}{\partial x} P_{st}(x, S) + \tfrac{1}{2}(\beta_s^x)^2 \frac{\partial^2}{\partial x^2} P_{st}(x, S)$$

$$+ \int \left\{ P_{st}(x + y, S) - P_{st}(x, S) - \frac{y}{1 + y^2} \frac{\partial}{\partial x} P_{st}(x, S) \right\} dL_s^x(y).$$

Kolmogorov's "continuous case" corresponds to vanishing P. Lévy functions L_s^x hence to normal tr. law derivatives $\psi_s^x(u) = i\alpha_s^x(u) - (\beta_s^x)^2 \frac{u^2}{2}$. Feller's "purely discontinuous" and "mixing" cases correspond to vanishing functions α_s^x, $(\beta_s^x)^2$ and to finite $L_s^x(\pm 0)$, respectively. In the first purely analytical approach the study of Markov processes was centered about the questions of existence, unicity, and tr.pr. properties of solutions of these equations. Itō, to whom the foregoing theorem is due, answers these questions by solving stochastic integral equations under somewhat stringent restrictions. As we shall see later, the semigroup approach leads to answers under weaker restrictions.

43.3. Stationarity. To discuss stationarity, that is, invariance under translations in time, it is convenient to take *once and for all* $T = [0, \infty)$, $u, v, r, s, t \geqq 0$ and to use the terminology and notation of Section 33 for translates. Let $\xi = g(X_T)$ denote Borel functions g of X_T and let B denote events defined on X_T; the ξ and the I_B are $\mathcal{B}(X_T)$-measurable functions. The translate X_{s+T} of $X_T = (X_t, t \geqq 0)$ by s is the family $(X_{s+t}, t \geqq 0)$ of the translates X_{s+t} of the X_t by s. The *translate* ξ_s of ξ by s is defined by $\xi_s = g(X_{s+T})$ and the *translate* B_s of B by s is defined by $I_{B_s} = (I_B)_s$; the ξ_s and the I_{B_s} are $\mathcal{B}(X_T)$-measurable, in fact, $\mathcal{B}(X_{s+T})$-measurable functions. To avoid ambiguities, it suffices to take for pr. space the sample space of X_T (see 33.2).

A tr.pr. $P_{u,v}(x, S)$ is *stationary* if it is invariant under translations in time: $P_{s+u,s+v}(x, S) = P_{u,v}(x, S)$ for all u, v, x, S. Thus, a tr.pr. $P_{u,v}(x, S)$ is stationary if and only if its dependence upon the time arguments u, v reduces to dependence upon their differences $t = v - u$ only, so that a stationary tr.pr. may be denoted by $P_t(x, S)$. Its complete definition is then as follows: A *stationary tr.pr.* $P_t(x, S)$, $t \geqq 0$, $x \in \mathfrak{X}$, $S \in \mathcal{S}$, is measurable in x, a pr. in S, with $P_0(x, S) = I(x, S)$, and it satisfies the *stationary Kolmogorov equation*

$$(\mathrm{K_{st}}) \qquad P_{s+t}(x, S) = \int P_s(x, dy)P_t(y, S).$$

The corresponding generalized semi-group property becomes then the *semi-group property*

$$(\mathrm{K'_{st}}) \qquad T_{s+t} = T_s T_t$$

of *stationary tr. operators* or *Markov endomorphisms* T_t on the space G of bounded Borel functions g on $\mathfrak{X}$, defined by

$$(T_t g)(x) = \int P_t(x, dy)g(y), \quad T_0 = I,$$

and 38.2**a** becomes

a. *The stationary Kolmogorov equation for stationary tr.pr.'s and the semi-group property for corresponding stationary tr. operators (Markov endomorphisms) are equivalent.*

A regular Markov process on $T = [0, \infty)$ is *stationary* if its law of evolution is invariant under translations in time. On account of 38.2(MR) the process is stationary if and only if its tr.pr. is stationary.

Let the c.pr. P^ξ be defined by $P^\xi = P^x$ when $\xi = x$, where P^x is defined by 38.2(CD) and (CP); similarly for E^ξ. Thus, P^ξ (E^ξ) is the c.pr. (c.exp.) on a regular Markov process given the initial random value $X_0 = \xi$. The Markov equivalence and existence theorems, together with (CD) and (CP) and integral definitions of c.pr.'s and c.exp.'s, yield without any difficulty the following theorem where the functions under exp. signs are limited to those whose exp.'s exist.

A. MARKOV STATIONARITY THEOREM. *To every stationary tr.pr. $P_t(x, S)$ there corresponds a stationary regular Markov process X_T. The following equivalent relations characterize regular Markov stationarity:*

$$\text{(i)} \qquad P(X_{s+t} \in S \mid X_r, r \leq s) = P_t(X_s, S)$$

or

$$P(X_{s+t_1} \in S_1, \cdots, X_{s+t_n} \in S_n \mid X_r, r \leq s) = P^{X_s}[X_{t_1} \in S_1, \cdots, X_{t_n} \in S_n]$$

for all s, all finite index sets and all Borel sets.

(ii) $$P(B_s \mid X_r, r \leqq s) = P(B_s \mid X_s) = P^{X_s}B$$

or

$$E(\xi_s \mid X_r, r \leqq s) = E(\xi_s \mid X_s) = E^{X_s}\xi$$

for all events B and measurable functions ξ defined on X_T and their translates B_s, ξ_s.

(iii) $$P^x AB_s = \int_A P^x(d\omega)P^{X_s(\omega)}B$$

or

$$E^x\eta\xi_s = E^x(\eta E^{X_s}\xi)$$

for all $x \in \mathfrak{X}$, all events A and measurable functions η defined on $(X_r, r \leqq s)$, all events B and measurable functions ξ and their translates B_s, ξ_s.

Note that, in particular, stationarity implies that

$$P(B_s \mid X_s = x) = P(B \mid X_0 = x), \quad E(\xi_s \mid X_s = x) = E(\xi \mid X_0 = x).$$

A r.f. X_T is *stationary* if its law is invariant under translations in time, while the r.f.'s of a *stationary process* have only a stationary law of evolution. In order that X_T be stationary it is necessary but, in general, not sufficient that the initial distribution be stationary: $P_s = P_0$ for all s, where P_s is the distribution of X_s. However, if the law of evolution is stationary then this condition is also sufficient, since then, for all events B on X_T and their translates B_s,

$$PB_s = \int P_s(dx)P(B_s \mid X_s = x) = \int P_0(dx)P(B \mid X_0 = x) = PB.$$

COROLLARY. *A Markov r.f. with a stationary tr.pr. is stationary if and only if the initial distribution is stationary.*

For stationary tr.pr.'s the study of tr. law derivatives becomes as follows: The *stationary tr.d.f.* $F_t^x(y)$ is defined by $F_t^x(y) = P_t(x, (-\infty, y))$ (when $\mathfrak{X} = R$) and the corresponding stationary tr.d.f. f_t^x is its ch.f. Upon centering at x, we have

$$\bar{F}_t^x(y) = F_t^x(x + y), \quad \bar{f}_t^x(u) = e^{-iux}f_t^x(u).$$

The representation of tr. law derivatives, when they exist, reduces to the limit ch.f.'s $(\bar{f}_h^x)^{[1/h]}$ as $h \to 0$ (independent of t), necessarily of the form e^{ψ^x}, $\psi^x = (\alpha^x, \Psi^x)$. Note that if e^{ψ^x} exists, then $P_h(x, (x - \epsilon, x + \epsilon)^c) \to 0$ for every $\epsilon > 0$ as $h \to 0$. The corresponding existence criterion becomes

b. *The stationary tr. law derivative* e^{ψ^x}, $\psi^x = (\alpha^x, \Psi^x)$ *exists if and only if*

$$\Psi_h^x \xrightarrow{c} \Psi^x \quad and \quad \int d\Psi_h^x(y)/y \to \alpha^x \quad as \quad h \to 0,$$

$$\Psi_h^x(x') = \frac{1}{h}\int_{-\infty}^{x'} \frac{y^2}{1+y^2}\, dF_h^x(y).$$

The corresponding theorem becomes

B. *Let* e^{ψ^x}, $\psi^x = (\alpha^x, \Psi^x)$, *represent the tr. law derivative corresponding to a stationary tr.pr.* $P_t(x, S)$ *and let* g *on* R *be a bounded twice differentiable function. Then, as* $h \to 0$,

$$\frac{1}{h}\left\{\int P_h(x, dy)g(y) - g(x)\right\} \to \alpha^x g'(x) + \int h_g(x, y)\, d\Psi^x(y).$$

In particular, if $P_t(x, S)$ *is twice differentiable in* x *then its right derivative* $\frac{\partial^+}{\partial t} P_t(x, S) = \lim_{h\to 0} \frac{1}{h}\{P_{t+h}(x, S) - P_t(x, S)\}$ *exists and, upon introducing the P. Lévy form* $\psi^x = (\alpha^x, (\beta^x)^2, L^x)$,

$$\frac{\partial^+}{\partial t} P_t(x, S) = \alpha^x \frac{\partial}{\partial x} P_t(x, S) + \tfrac{1}{2}(\beta^x)^2 \frac{\partial^2}{\partial x^2} P_t(x, S)$$

$$+ \int \left\{P_t(x + y, S) - P_t(x, S) - \frac{y}{1+y^2}\frac{\partial}{\partial x} P_t(x, S)\right\} dL^x(y).$$

The Kolmogorov continuous case (L^x vanishes) becomes then the Fokker-Planck original case.

From now on and unless otherwise stated, all our Markov processes will be regular and stationary on $T = [0, \infty)$ *so that, whenever convenient, we will drop "regular" and "stationary."*

43.4. Strong Markov property. The central problem of random analysis is that of the sample functions behavior. As soon as this problem arises, random times appear, say, the time of appearance of the first discontinuity of sample functions or the time when the r.f. takes a given value. In the Markov case, we might expect the Markov property to hold under conditionings with a random time τ as "present," since it holds when every one of its values is used as "present." In fact, for a long time this possibility was not even questioned. Yet, let X_T be a stationary regular Markov r.f. with tr.pr.

$$P_t(x, S) = \frac{1}{\sqrt{2\pi t}} \int_S e^{-(x-y)^2/2t}\, dy, \quad x \neq x_0, \quad P_t(x_0, S) = I(x_0, S).$$

Let τ be the time X_T first reaches x_0: $X_t(\omega) \neq x_0$ for $t < \tau(\omega)$ and $X_t(\omega) = x_0$ for $t = \tau(\omega)$. For almost every one of its sample functions, if we know that $X_T(\omega)$ is in a state $x \neq x_0$ at a time t_0, then this sample function is a Brownian continuous one, and, when it passes through x_0, it does not stop there; while if we know that it is in the state x_0 at time t_0, then it stays there forever. It follows that the Markov property of X_T is no more true with τ as "present." Thus, the Markov property is to be strengthened if we want to be able to investigate the sample functions behavior. In other words, we restrict ourselves to those Markov r.f.'s, always separable, for which, *formally*,

$$P(B_\tau \mid X_r, r \leqq \tau) = P^{X_\tau} B,$$

where $B \in \mathcal{B}(X_T)$ and B_τ is its translate by a random time τ—a "random present." However, the introduction of random "presents" raises immediate difficulties. Let, say, τ be the first time X_T takes a given value x_0. To begin with, $\tau(\omega)$ does not exist for those sample functions $X_T(\omega)$ which never take the value x_0. If τ exists it may not be measurable; even if it is measurable it may take infinite values $\tau(\omega) = \infty$, and $X_{\tau(\omega)}(\omega)$ does not exist unless the point at infinity is added to the time interval. If τ is measurable and finite, $X_\tau = (X_{\tau(\omega)}(\omega), \omega \in \Omega)$ may not be measurable; even if all the $X_{\tau+t}$ are measurable, the formal Markov property above has still to be given a meaning. Thus, at first we have to consider and eliminate these difficulties.

A *random time* τ will be a nonnegative measurable function, not necessarily finite but not a.s. infinite: $P\Omega^\tau > 0$, $\Omega^\tau = [\tau < \infty]$. For random times corresponding to sample properties, existence and measurability are to be proved. However, if a random time exists only outside some event, we take it to be infinite on the exceptional event. This convention is acceptable as long as we consider sample functions on the given time interval $[0, \infty)$ only. In other words, *we limit ourselves to the restricted pr. space* $(\Omega^\tau, \mathcal{A}^\tau, P^\tau)$ where $\mathcal{A}^\tau = (\Omega^\tau A, A \in \mathcal{A})$ and $P^\tau = P/P\Omega^\tau$; this is one reason for excluding the possibility of $P\Omega^\tau = 0$. Since we seek sample properties which, at best, are those of almost all sample functions, that is, are valid outside a null event, the exclusion of $P\Omega^\tau = 0$ is not a restriction on random times. Note that whenever relations between pr.'s are homogeneous in P^τ *we may and will replace* P^τ *by* P (upon multiplying throughout by a suitable power of $P\Omega^\tau$).

Translates by τ of events and, more generally, random variables defined on X_T are defined as in 30.2: The translate of $A_t = [X_t \in S]$ is $A_{\tau+t} = [X_{\tau+t} \in S]$ where $X_{\tau+t}(\omega) = X_{\tau(\omega)+t}(\omega)$, the translate of X_t is $X_{\tau+t}$ and, in general, the translate of $\xi = g(X_T)$ is $\xi_\tau = g(X_{\tau+T})$ where $X_{\tau+T} = (X_{\tau+t}, t \in T)$. Translates by a random time τ are considered on Ω^τ only, and *we drop "on* Ω^τ*."* If a random time τ is elementary $\tau = \sum t_j I_{A_j}$ (on Ω^τ), then, for any r.f. $X_T = (X_t, t \geqq 0)$, the translates $X_{\tau+t}$ by τ are r.v.'s $X_{\tau+t} = \sum X_{t_j+t} I_{A_j}$ so that the translate $X_{\tau+T} = (X_{\tau+t}, t \geqq 0)$ is a r.f. If τ is not elementary, then we have to assume or to prove that $X_{\tau+T}$ is a r.f. It is a r.f. when X_T is Borelian. In particular, X_T is Borelian when it is sample right continuous. For then, it is Borelian as limit of Borel r.f.'s $X_T{}^{(n)} = (X_t{}^{(n)}, t \geqq 0)$ where

$$X_t{}^{(n)} = \sum_{k=0}^{\infty} X_{\frac{k+1}{2^n}} I_{\left[\frac{k}{2^n} \leqq t < \frac{k+1}{2^n}\right]}.$$

To summarize

a. *The translates of arbitrary r.f.'s by elementary random times are r.f.'s and so are the translates by arbitrary random times of Borel r.f.'s, in particular, of sample right continuous r.f.'s.*

A random time τ will be a *time of* X_T if $[\tau \leqq t] \in \mathcal{B}(X_s, s \leqq t)$ for every t; in other words, if we know what happened up to time t inclusive, that is, if we know the sample values $X_s(\omega)$, $s \leqq t$, then we know whether $\tau(\omega) \leqq t$ or not. In particular, every "degenerate" time t is time of any r.f. and if τ is a time of X_T so is every $\tau + t$. In fact, since the inverse images under a time τ of X_T of Borel sets in $[0, t]$ belong to $\mathcal{B}(X_s, s \leqq t)$, we have

b. *If* τ *is a time of* X_T *so are the random times* $\tau + t$ *and, in fact, so are the random times* $g(\tau) \geqq \tau$*, where the functions* g *are Borelian.*

For, by hypothesis, $[g(\tau) \leqq t] = \tau^{-1}g^{-1}[0, t] \subset \mathcal{B}(X_s, s \leqq t)$.

A trivial example of a time of X_T is any elementary time $\tau = \sum t_j I_{A_j}$, $t_1 < t_2 < \cdots$, where every $A_j \in \mathcal{B}(X_r, r \leqq t_j)$.

A nontrivial and important example is as follows. Let U be an open state set and let $\tau_U(\omega)$ be the infimum of all t such that the distance of the sets $(X_s(\omega), s \leqq t)$ and U^c be zero; if $\tau_U(\omega)$ does not exist we take it to be infinite. If X_T is sample right continuous, then τ_U is the time X_T

first *hits* U^c, and if X_T is sample continuous, then τ_U is the time X_T first *reaches* U^c.

c. *If X_T is sample continuous (from the right), then the time X_T first reaches (hits) U^c is a time of X_T.*

For, setting $S_n = [x: d(x, U^c) < 1/n]$ and letting $r < t$ vary over the rationals,

$$[\tau_U \leqq t] = [X_t \in U^c] \cup (\bigcap_n \bigcup_{r<t} [X_r \in S_n]).$$

If τ is a time of X_T, the events $A \in \mathfrak{A}^\tau$ such that $A[\tau \leqq t] \in \mathfrak{B}(X_s, s \leqq t)$ for all t form the σ-field $\mathfrak{B}(X_s, s \leqq \tau)$—in Ω^τ—of *events defined on* $(X_s, s \leqq \tau)$ or *on X_T up to time τ* (inclusive). Since the definitions of times τ of X_T and of events defined on X_T up to time τ are in terms of events on X_T, it follows that in the Markov case they pertain to all the r.f.'s of the Markov process simultaneously, that is, to the process as a whole. The same is true for what follows and, therein, a "stationary Markov X_T" will mean the process X_T as well as any r.f. belonging to it.

A time τ of a stationary Markov X_T with stationary Borel tr.pr. $P_t(x, S)$ is *Markovian* or *a Markov time of X_T* if all $X_{\tau+t}$ are r.v.'s and, given X_τ, the Markov evolution starts anew:

$$P(X_{\tau+t} \in S \mid X_r, r \leqq \tau) = P(X_{\tau+t} \in S \mid X_\tau)$$

with the same Markov law of evolution:

(TM) $$P(X_{\tau+t} \in S \mid X_{\tau+r}, r \leqq s) = P_{t-s}(X_{\tau+s}, S), \quad s \leqq t;$$

it may and will happen that X_τ is to be replaced by $X_{\tau+0}$ for these relations and hence for the following ones but, to simplify, we shall still speak of a Markov time τ; as usual, we write $\mid\cdot$ in lieu of $\mid \mathfrak{B}(\cdot)$.

Upon setting $s = 0$ in the second relation, the first becomes equivalent to the relation

(SM) $$P(X_{\tau+t} \in S \mid X_r, r \leqq \tau) = P_t(X_\tau, S).$$

Thus Markov times τ are also characterized by (SM) and (TM). Furthermore, since our Markov evolution is stationary in terms of its degenerate Markov times t, it is natural to require that the same be true in terms of its random Markov times τ: we say that τ is a *stationary Markov time of X_T* if it is and remains a Markov time under translations in time, that is, if all $\tau + s$, $s \geqq 0$, are Markov times of X_T. Thus, stationary Markov times τ are characterized by (SM) and (TM) with τ replaced by any

$\tau + s$ or, equivalently, by

$$(\mathrm{SM_{st}})\quad P(X_{\tau+t} \in S \mid X_r, r \leqq \tau + s) = P_{t-s}(X_{\tau+s}, S), \quad s \leqq t.$$

For, (SM) with $\tau + s$ in lieu of τ yields $(\mathrm{SM_{st}})$ with $t + s$ in lieu of t, while $(\mathrm{SM_{st}})$, with $s + t$ and $s + s'$ in lieu of t and s, conditioned by $(X_{\tau+s+r}, r \leqq s')$, yields (TM) with $\tau + s$ and s' in lieu of τ and s..

A very useful example is that of elementary times:

d. *Elementary times of an arbitrary stationary Markov X_T are stationary Markov times of X_T.*

Proof. Let $\tau = \sum t_j I_{A_j}$ (on Ω^τ), $t_1 < t_2 < \cdots$, $A_j \in \mathcal{B}(X_r, r \leqq t_j)$, be an elementary time of X_T. The integral form of $(\mathrm{SM_{st}})$ is

$$P^x A[X_{\tau+t} \in S] = \int_A P^x(d\omega) P_{t-s}(X_{\tau(\omega)+s}(\omega), S), \quad A \in \mathcal{B}(X_r, r \leqq \tau + s).$$

We have to prove that this relation holds for the elementary time τ, namely that

$$\sum_j P^x A A_j [X_{t_j+t} \in S] = \sum_j \int_{AA_j} P^x(d\omega) P_{t-s}(X_{t_j+s}(\omega), S).$$

Since $AA_j = A[\tau = t_j] \in \mathcal{B}(X_r, r \leqq t_j + s)$, the ordinary Markov property of X_T applies, that is,

$$P(X_{t_j+t} \in S \mid X_r, r \leqq t_j + s) = P_{t-s}(X_{t_j+s}, S),$$

and its integral form is the equality between the terms with same j of both sums. Therefore, these sums are equal and the assertion is proved.

Note that it would have sufficed to prove that elementary times of X_T are Markovian since their translates are also elementary times, hence Markovian.

If all the times of a stationary Borel Markov X_T with a Borel tr.pr. are Markovian, we say that X_T is (*stationary*) *strongly Markovian* or has the *strong Markov property*. If τ is a time of X_T so are all $\tau + s$, hence all the times of a stationary strongly Markovian X_T are stationary Markov times of X_T. Therefore, (SM) holds for all times of X_T or $(\mathrm{SM_{st}})$ holds for all times of X_T if and only if X_T is strongly Markovian.

If τ' varies over all the times of X_T so does $\tau = \tau' I_A + \infty I_{A^c}$ with A varying over all the events defined on $(X_r, r \leqq \tau')$. For, τ is a Borel function of τ' at least equal to τ' and $\tau = \tau'$ when $A = \Omega^{\tau'}$. The integral form of (SM) with τ' in lieu of τ is

$$P^x A[X_{\tau'+t} \in S] = \int_A P^x(d\omega) P_t(X_{\tau'(\omega)}(\omega), S)$$

and, in terms of τ, becomes

$$P^x[X_{\tau+t} \in S] = \int_{\Omega^\tau} P^x(d\omega) P_t(X_{\tau(\omega)}(\omega), S).$$

Set

$$P_\tau(x, S) \equiv P^x[X_\tau \in S]$$

by analogy with the identity $P_t(x, S) = P^x[X_t \in S]$. Then the above relation becomes: for every x

$$(\mathrm{SK}_{st}) \qquad P_{\tau+t}(x, S) = \int P_\tau(x, dy) P_t(y, S).$$

This equality reduces to the stationary Kolmogorov equation for degenerate times and will be called (*stationary*) *strong Kolmogorov equation.* The same arguments as in 43.2 and 43.3 lead to the equivalent semi-group property

$$T_{\tau+t} = T_\tau T_t$$

of the Markov endomorphisms on G with T_τ defined as T_t by

$$(T_\tau g)(x) = \int P_\tau(x, dy) g(y), \quad g \in G.$$

In fact, the arguments in the preceding subsections remain valid when, therein, s is replaced by τ and, together with what precedes, yield

A. STRONG MARKOV EQUIVALENCE THEOREM. *Let $P_t(x, S)$ be a Borel stationary tr.pr., where x varies over $\mathfrak{X}$, S over $\mathcal{S}$, and let $s \leqq t$ vary over* $[0, \infty)$. *Let τ vary over all the times of Borel r.f. X_T.*

(i) *The following equivalent relations characterize the* (*stationary*) *strong Markov property of X_T with tr.pr. $P_t(x, S)$*

$$P(X_{\tau+t} \in S \mid X_r, r \leqq \tau) = P_t(X_\tau, S)$$

or

$$P^x AB_\tau = \int_A P^x(d\omega) P^{X_{\tau(\omega)}} B \text{ or } E^x \eta \xi_\tau = E^x(\eta E^{X_\tau} \xi)$$

for all events B and measurable functions ξ (whose exp.'s exist) on X_T, for all $x \in \mathfrak{X}$ and all events A *and measurable η on $(X_r, r \leqq \tau)$.*

(ii) *The following equivalent relations characterize the strong Markov property of a stationary Markov X_T with tr.pr. $P_t(x, S)$*

$$P_{\tau+t}(x, S) = \int P_\tau(x, dy)P_t(y, S) \quad or \quad T_{\tau+t} = T_\tau T_t.$$

An important example of strong Markov X_T is as follows (Dynkin and Ushkevitch).

COROLLARY. *Let X_T be stationary Markovian with Borel tr.pr. If X_T is sample right continuous and the Markov endomorphisms T_t transform bounded continuous functions into bounded continuous functions, then X_T is strongly Markovian.*

Proof. Let τ be a time of X_T and note that the elementary times

$$\tau_n = \sum_{k=1}^{\infty} \frac{k}{2^n} I_{\left[\frac{k-1}{2^n} \leqq \tau < \frac{k}{2^n}\right]}$$

converge to τ from the right (on Ω^τ). Let $s \geqq 0$ be arbitrary. Since X_T is sample right continuous, it follows by **a** that it is Borelian so that $X_{\tau+s}$ are r.v.'s, and $X_{\tau_n+s} \to X_{\tau+s}$. Therefore, if $g \in G$ is continuous then $g(X_{\tau_n+s}) \to g(X_{\tau+s})$ and

$$E^x g(X_{\tau_n+s}) \to E^x g(X_{\tau+s}) \quad \text{or} \quad T_{\tau_n+s} g \to T_{\tau+s} g.$$

Since, by **d**, elementary times τ_n of X_T are Markov times, hence, by **A**(ii), $T_{\tau_n+t}g = T_{\tau_n}T_t g$ and $T_t g \in C$, we have $T_{\tau+t}g = T_\tau T_t g$. Thus, the family of bounded functions g on $\mathfrak{X}$ for which this relation holds contains the continuous ones. It is closed under passages to the limit by bounded sequences. Therefore, by the Baire definition of Borel functions, it contains the family G of all bounded Borel functions on $\mathfrak{X}$ and, by **A**(ii), τ is a strong Markov time of X_T. The proof is terminated.

§ 44. TIME-CONTINUOUS TRANSITION PROBABILITIES

Let $P_t(x, S)$ be a stationary tr.pr., that is, a Borel function in x and a pr. in S with $P_0(x, S) = I(x, S)$, obeying the Kolmogorov equation:

$$P_{s+t}(x, S) = \int P_s(x, dy)P_t(y, S), \quad s, t \geq 0.$$

Unless otherwise stated, the time arguments r, s, t, with or without affixes, vary over $[0, \infty)$, x varies over the state space $\mathfrak{X}$, and S varies over the σ-field of state sets $\mathcal{S}$ generated by the class of open sets in $\mathfrak{X}$. As usual, to fix the *ideas, we take* $(\mathfrak{X}, \mathcal{S})$ *to be the Borel line and take the limits in* h *as* $0 < h \to 0$.

However, the properties of the tr.pr. to be established and the proofs given are valid for all state spaces $\mathfrak{X}$ such that the diagonal (which consists of all points (x, x)) in $\mathfrak{X} \times \mathfrak{X}$ belong to the product σ-field and, consequently, all singletons $\{x\}$ belong to $\mathcal{S}$ as sections of the diagonal. Also they remain valid when the tr.pr. $P_t(x, S)$ is only a measure in S bounded by 1 in lieu of being a pr. in S. We leave the search for corresponding extensions of properties in 44.2 to the reader.

Denote by $\dot{P}_s(x, S)$ the derivative of $P_t(x, S)$ with respect to t at $t = s$, provided it exists. Note that the derivative $\dot{P}_0(x, S)$ at $t = 0$ is necessarily a right derivative. Set

$$q(x) = \lim_{h \to 0} \frac{1 - P_h(x, \{x\})}{h}, \quad q(x, S) = \lim_{h \to 0} \frac{P_h(x, S)}{h}, \quad x \notin S,$$

whenever the limits exist, and make the convention that

$$q(x, S) = q(x, S\{x\}^c);$$

then

$$\dot{P}_0(x, S) = \lim_{h \to 0} \frac{P_h(x, S) - I(x, S)}{h} = q(x, S) - q(x)I(x, S).$$

Formal differentiation of Kolmogorov's equation with respect to s at $s = 0$ and with respect to t at $t = 0$ (followed by the change of s into t) yields the *backward* and the *forward* equations

$$\text{(B)} \qquad \dot{P}_t(x, S) = \int_{\{x\}^c} q(x, dy)P_t(y, S) - q(x)P_t(x, S)$$

$$\text{(F)} \qquad \dot{P}_t(x, S) = \int P_t(x, dy)q(y, S) - \int_S P_t(x, dy)q(y)$$

Formal solutions (by formal substitution) of these equations are given by

$$\bar{P}_t(x, S) = \sum_{n=0}^{\infty} P_t^{(n)}(x, S)$$

where

$$P_t^0(x, S) = e^{-q(x)t}I(x, S)$$

and

$$P_t^{(n+1)}(x, S) = \int_0^t ds \int_{\{x\}^c} e^{-q(x)s} q(x, dy) P_{t-s}^{(n)}(y, S)$$

or, alternatively,

$$P_t^{(n+1)}(x, S) = \int_0^t ds \int P_s^{(n)}(x, dy) \int_{S\{x\}^c} q(y, dz) e^{-q(z)(t-s)}.$$

In probabilistic language, $\bar{P}_t(x, S)$ is the pr. of transition from x into S in time t in finitely many steps (see the probabilistic interpretation of $q(x)$ and $q(x, S)$ at the end of the section). Thus, whenever there is a possibility of such a transition but not in finitely many steps, it may be expected that $\bar{P}_t(x, S)$ will be a measure in S smaller than 1 in lieu of a pr. in S.

The foregoing formal discussion brings into light the problems to be considered: to begin with, the problem of existence and properties of tr.pr. derivatives, hence of $q(x)$ and $q(x, S)$, and of the corresponding sample properties. Since existence of $\dot{P}_0(x, S)$ implies the *continuity condition*

$$\text{(C)} \qquad P_h(x, S) \to I(x, S),$$

this condition will be assumed at the start. This is the Doblin-Doob approach to the analysis of Markov evolution. The analytical problem of existence, unicity, and tr.pr. properties of solutions of the backward and forward equations—the Kolmogorov-Feller approach—will be left out for it fits within the wider and more powerful Hille-Yosida-Feller-Dynkin semi-group approach.

44.1. Differentiation of tr.pr.'s. The Doblin-Doob results under condition (C) were improved by Kolmogorov who established the existence and finiteness of the function $q(x, S)$ for countable state spaces and his result was extended by Kendall to noncountable state spaces of the general type described above, under a parallel restriction of "σ-uniformity."

Note that condition (C) is equivalent to $P_h(x, \{x\}) \to 1$or, on account of lemma **c** below, to the continuity in t of the tr.pr.

a. *For every section D_x of $D \in \mathcal{S} \times \mathcal{S}$ and in particular for sections $\{x\}$ of the diagonal, the function $P_t(x, D_x)$ is Borelian in x.*

Proof. The class of sets D for which the assertion holds is closed under finite summations and monotone passages to the limit by sequences. It contains all rectangles $S \times S' \in \mathcal{S} \times \mathcal{S}$ since the function $P_t(x, (S \times S')_x) = I(x, S)P_t(x, S')$ has the asserted property. Therefore, it contains the minimal field $\mathcal{S} \times \mathcal{S}$ generated by the rectangles.

b. *The function $P_t(x, \{x\})$ is supermultiplicative in t.*

For, by Kolmogorov's equation,

$$P_{s+t}(x, \{x\}) \geqq \int_{\{x\}} P_s(x, dy)P_t(y, \{x\}) = P_s(x, \{x\})P_t(x, \{x\}).$$

c. *Under* (C), *the function* $g(t) = -\log P_t(x, \{x\})$ *exists and is finite and subadditive in* t, *and the function* $P_t(x, S)$ *is uniformly continuous in* t *uniformly in* S.

Proof. Let $P_h(x, \{x\}) \to 1$. Then, by **b**, $P_t(x, \{x\}) \geqq P_{t/n}{}^n(x, \{x\}) > 0$ for n sufficiently large. Therefore, $g(t)$ exists and is finite, and super-multiplicativity in **b** becomes subadditivity of $g(t)$. The first assertion is proved. By Kolmogorov's equation

$$\begin{aligned}\Delta &= P_{t+h}(x, S) - P_t(x, S)\\ &= \int_{\{x\}^c} P_h(x, dy)P_t(y, S) - (1 - P_h(x, \{x\})P_t(x, S)\end{aligned}$$

so that

$$-(1 - P_h(x, \{x\}) \leqq \Delta \leqq P_h(x, \{x\}^c) = 1 - P_h(x, \{x\}).$$

Therefore,

$$| P_s(x, S) - P_t(x, S) | \leqq 1 - P_{|s-t|}(x, \{x\}) \to 0$$

uniformly in t and in S as $s - t \to 0$, and the second assertion is proved.

d. POINT DIFFERENTIATION LEMMA. *Under* (C),

$$\frac{1 - P_h(x, \{x\})}{h} \to q(x) \leqq \infty$$

where the function $q(x)$ *is Borelian, and* $P_t(x, \{x\}) \geqq e^{-q(x)t}$.

Proof. If $q(x)$ exists, then it is Borelian as limit of sequences of Borel functions corresponding to $h = h_n \to 0$.
Fix x, set $g(t) = -\log P_t(x, \{x\})$ and note that, by **c**,

$$0 \leqq g(t) < \infty, \quad g(s + t) \leqq g(s) + g(t), \quad g(+0) = g(0) = 0.$$

Given $t > 0$ and $h > 0$, take $n = [t/h]$ so that $t = nh + \theta$, $0 \leqq \theta < h$ and, by subadditivity,

$$\frac{g(t)}{t} \leqq \frac{ng(h)}{t} + \frac{g(\theta)}{t} = \frac{g(h)}{h}\frac{nh}{t} + \frac{g(\theta)}{t}.$$

Therefore, $g(t)/t \leqq \liminf_{h\to 0} g(h)/h$ and

$$\limsup_{h\to 0} \frac{g(h)}{h} \leqq \sup_{t>0} \frac{g(t)}{t} \leqq \liminf_{h\to 0} \frac{g(h)}{h}.$$

Thus, $q(x) = \lim_{h\to 0} \frac{g(h)}{h}$ exists with

$$q(x) = \sup_{t>0} \frac{g(t)}{t}.$$

The second assertion follows by

$$P_t(x, \{x\}) = e^{-\frac{g(t)}{t}t} \geqq e^{-q(x)t},$$

and the first one follows by

$$\frac{1 - P_h(x, \{x\})}{h} = \frac{1 - e^{-g(h)}}{h} = (1 + o(1))\frac{g(h)}{h} \to q(x).$$

The proof is terminated.

We say that a state x is *absorbing*, *instantaneous*, or *steady* according as $q(x) = 0$, $q(x) = \infty$, or $0 < q(x) < \infty$. We say that a set $U = [x: q(x) \leqq c]$ with c finite is *q-bounded*. Since the function $q(x)$ is Borelian, q-bounded sets are state sets and, since $1 - P_h(x, \{x\}) \leqq 1 - e^{-ch} \to 0$ uniformly in $x \in U$, the continuity condition $P_h(x, \{x\}) \to 1$ holds uniformly on every q-bounded set. The same is true on every finite state set even if it has instantaneous states. In general, let $\mathfrak{U}$ be the class of all *uniform continuity state sets*—on each of which (C) holds uniformly. Clearly $\mathfrak{U}$ is closed under taking finite unions and state subsets of its sets and we shall use these closure properties without further comment. Unless otherwise stated, *we denote the sets of* $\mathfrak{U}$ *by* U, *with or without affixes*.

e. *Under* (C), *for every x and every uniform continuity state set* $U \not\ni x$,

$$\frac{P_h(x, U)}{h} \to q(x, U)$$

finite bounded by $q(x)$, *Borelian in x and a measure in* U.

Proof. If the function $q(x, U)$ exists, then it is Borelian in x as limit of sequences of Borel functions corresponding to $h = h_n \to 0$ and it is bounded by $q(x)$ since

$$0 \leqq \frac{1 - P_h(x, \{x\})}{h} - \frac{P_h(x, U)}{h} \to q(x) - q(x, U).$$

Let $V \in \mathfrak{U}$ be such that $U + \{x\} \subset V$, (say, $V = U + \{x\}$). Given a positive $\epsilon < \frac{1}{3}$ there exists a positive $t_0 = t_0(V, \epsilon)$ such that, for $t \leqq t_0$, $P_t(y, \{y\}) > 1 - \epsilon$ for all $y \in V$ hence $P_t(x, U) < \epsilon$. Set

$$P'_h(x, S) = P_h(x, S), \quad P'_{(j+1)h}(x, S) = \int_{U^c} P'_{jh}(x, dy) P_h(y, S).$$

In probabilistic language, $P'_{(j+1)h}(x, S)$ is the pr. of transition from x into S in time $(j + 1)h$ avoiding U at times $h, \cdots, jh$. From this probabilistic interpretation, or directly by induction, it follows that

$$(1) \qquad P_{kh}(x, S) = \sum_{j=1}^{k-1} \int_U P'_{jh}(x, dy) P_{(k-j)h}(y, S) + P'_{kh}(x, S).$$

Given h and $t \leqq t_0$, let $n = [t/h]$ so that $nh = t - \theta \leqq t_0$, $0 \leqq \theta < h$. Since $P'_{kh}(x, S) \geqq P'_{(k-1)h}(x, \{x\}) P_h(x, S)$ and, by (1) with $S = U$,

$$P_{nh}(x, U) = \sum_{k=1}^{n} \int_U P'_{kh}(x, dy) P_{(n-k)h}(y, U),$$

it follows that

$$P_{nh}(x, U) \geqq \sum_{k=1}^{n} P'_{(k-1)h}(x, \{x\}) \int_U P_h(x, dy) P_{(n-k)h}(y, U)$$

hence

$$(2) \qquad P_{nh}(x, U) \geqq \sum_{k=1}^{n} P'_{(k-1)h}(x, \{x\})(1 - \epsilon) P_h(x, U).$$

It also follows that

$$\epsilon > P_{nh}(x, U) \geqq (1 - \epsilon) \sum_{k=1}^{n} P'_{kh}(x, U)$$

hence

$$(3) \qquad \sum_{k=1}^{n} P'_{kh}(x, U) \leqq \frac{\epsilon}{1 - \epsilon}.$$

On account of (3) and (1) with $S = \{x\}$ and $k \leqq n$,

$$1 - \epsilon < P_{kh}(x, \{x\}) \leqq \sum_{j=1}^{k-1} P'_{jh}(x, U) + P'_{kh}(x, \{x\})$$

$$\leqq \frac{\epsilon}{1 - \epsilon} + P'_{kh}(x, \{x\})$$

hence

$$P'_{kh}(x, \{x\}) \geqq \frac{1 - 3\epsilon}{1 - \epsilon}. \tag{4}$$

Similarly, on account of (2) and (4),

$$P_{nh}(x, U) \geqq n \left(\frac{1 - 3\epsilon}{1 - \epsilon}\right) (1 - \epsilon) P_h(x, U)$$

or

$$\frac{P_h(x, U)}{h} \leqq \frac{1}{1 - 3\epsilon} \frac{P_{nh}(x, U)}{nh}.$$

Therefore, on account of **c**, letting $h \to 0$ then $t \to 0$,

$$\limsup_{h \to 0} \frac{P_h(x, U)}{h} \leqq \frac{1}{1 - 3\epsilon} \liminf_{t \to 0} \frac{P_t(x, U)}{t}$$

and letting $\epsilon \to 0$,

$$q(x, U) = \lim_{h \to 0} \frac{P_h(x, U)}{h}$$

exists with

$$q(x, U) \leqq \frac{1}{1 - 3\epsilon} \frac{P_t(x, U)}{t}, \quad t \leqq t_0(V, \epsilon), \quad U + \{x\} \subset V. \tag{5}$$

It follows that $q(x, U) < \infty$ is a measure in U since it is clearly finitely additive and, as $U_n \downarrow \emptyset$, for $t_0 = t_0(U_1 + \{x\}, \epsilon)$,

$$q(x, U_n) \leqq \frac{1}{1 - 3\epsilon} \frac{P_{t_0}(x, U_n)}{t_0} \to 0.$$

The proof is terminated.

Note that given x and $V \in \mathfrak{U}$, for $x \notin U \subset V$, $\dfrac{P_h(x, U)}{h} \to q(x, U)$ uniformly in $U \subset V$. For, if $h \leqq t_0(V \cup \{x\}, \epsilon)$ then, by (5),

$$\frac{P_h(x, U')}{h} - q(x, U') \geqq -3\epsilon q(x, U') \geqq -3\epsilon q(x, V), \quad U' \subset V,$$

and

$$\frac{P_h(x, U)}{h} - q(x, U)$$

$$= \frac{P_h(x, V)}{h} - q(x, V) + q(x, V - U) - \frac{P_h(x, V - U)}{h}$$

$$\leqq \frac{P_h(x, V)}{h} - q(x, V) + 3\epsilon q(x, V),$$

so that

$$\sup_{U \subset V} \left| \frac{P_h(x, U)}{h} - q(x, U) \right| \leqq 3\epsilon q(x, V) + \left| \frac{P_h(x, V)}{h} - q(x, V) \right| \to 0.$$

What precedes, together with our convention $q(x, S) = q(x, S\{x\}^c)$, yields

A. TR.PR.'S DIFFERENTIATION THEOREM. *Under the continuity condition, the derivative $\dot{P}_0(x, U)$ at $t = 0$ of the tr.pr. $P_t(x, U)$ exists for every state and every uniform continuity state set U. In fact, for every state x and every uniform continuity state set V,*

$$\frac{P_h(x, U) - I(x, U)}{h} \to q(x, U) - q(x)I(x, U) = \dot{P}_0(x, U)$$

uniformly in $U \subset V$; the nonnegative function $q(x) \leqq \infty$ is Borelian; the function $q(x, U) = q(x, U\{x\}^c) < \infty$ bounded by $q(x)$ is Borelian in x and a finite measure in U, with

$$q(x, U) \leqq \frac{1}{1 - 3\epsilon} \frac{P_t(x, U\{x\}^c)}{t}$$

for all $U + \{x\} \subset V$ and $t \leqq t_0(V, \epsilon)$, $0 < \epsilon < \frac{1}{3}$.

If the function $q(x)$ is bounded then $\mathfrak{X}$ is q-bounded and the continuity condition is *uniform* (for all $x \in \mathfrak{X}$). Conversely, if the continuity condition is uniform, that is, $\mathfrak{X} \in \mathfrak{U}$ then, by **A**, $q(x) = q(x, \mathfrak{X}) \leqq 1/(1 - 3\epsilon)t_0$ for $t_0 = t_0(\mathfrak{X}, \epsilon)$ and the function $q(x)$ is bounded. Thus

COROLLARY. *The function $q(x)$ is bounded if and only if the continuity condition is uniform. Then the tr.pr.'s differentiation theorem is valid with $U \in \mathfrak{U}$ replaced by $S \in \mathcal{S}$, and $q(x, \mathfrak{X}) = q(x)$.*

If the function $q(x)$ is finite, hence $\mathfrak{X} = [x: q(x) < \infty]$, then there exists a countable partition of $\mathfrak{X}$ into uniform continuity state sets U_j

(say, $U_j = [x: j - 1 \leqq q(x) < j]$). In general, whenever there exists such a countable partition $\mathfrak{X} = \sum U_j$, we say that the continuity condition is *σ-uniform*. For example, if the set of instantaneous states is countable $[x: q(x) = \infty] = (x_1, x_2, \cdots)$, then the continuity condition is σ-uniform; it suffices to take $U_j = [x: j - 1 \leqq q(x) < j] + \{x_j\}$. In particular, if the state space is countable then the continuity condition is σ-uniform. Thus, we may consider σ-uniformity as a "natural" transposition of this property of countable state spaces to general state spaces.

Let $\bar{q}(x) = \sup_{U \in \mathfrak{U}} q(x, U)$ and note that there is always a sequence $V_n \uparrow V = \bigcup V_n$ and contained in $\mathfrak{U}$ such that $q(x, V_n) \uparrow \bar{q}(x)$.

f. EXTENSION LEMMA. *If $\bar{q}(x) < \infty$ then, for this x, the measure $q(x, U)$ in U extends to a measure $q(x, S)$ in S, and the extension is finite with $q(x, \mathfrak{X}) = \bar{q}(x)$.*

If the continuity condition is σ-uniform then, for every x, the measure $q(x, U)$ in U extends to a measure $q(x, S)$ in S, and the extension is unique.

Proof. We use, without further comment the closure properties of $\mathfrak{U}$ and the fact that a nondecreasing sequence of measures $\mu_n \uparrow \mu$ on $\mathcal{S}$ converges to a measure μ on $\mathcal{S}$; it suffices to note that, as $n \to \infty$ then $m \to \infty$,

$$\sum \mu(S_j) \geqq \mu_n(\sum S_j) \to \mu(\sum S_j),$$

while

$$\mu(\sum S_j) \geqq \mu_n(\sum S_j) \geqq \sum_{j=1}^{m} \mu_n(S_j) \to \sum \mu(S_j)$$

Since

$$q(x, V_n) \uparrow \bar{q}(x) < \infty, \quad V_n \uparrow V, \quad V_n \in \mathfrak{U},$$

it follows, by **A**, that

$$q(x, UV^c) = q(x, UV^c + V_n) - q(x, V_n) \leqq \bar{q}(x) - q(x, V_n) \to 0$$

and

$$q(x, U) = q(x, UV) + q(x, UV^c) = q(x, UV) = \lim_n q(x, UV_n).$$

Therefore, the measure $q(x, S)$ in S defined by

$$q(x, SV_n) \uparrow q(x, S) = q(x, SV)$$

is an extension of the measure $q(x, U)$ in U, with

$$q(x, \mathfrak{X}) = q(x, V) = \bar{q}(x) < \infty.$$

The first assertion is proved.

If the continuity condition is σ-uniform, then $\mathfrak{X} = \sum U_n$, or $V_n = \sum_{k=1}^{n} U_k \uparrow \mathfrak{X}$, $V_n \in \mathfrak{U}$. Therefore, $q(x, UV_n) \uparrow q(x, U)$ and the extension of the measure $q(x, U)$ in U to a measure $q(x, S)$ in S is determined by the necessary condition $q(x, SV_n) \uparrow q(x, S)$. The second assertion is proved, and the proof is terminated.

B. EXTENDED TR.PR.'S DIFFERENTIATION THEOREM. *Under the continuity condition, if $\bar{q}(x) = q(x) < \infty$ (or under the σ-uniform continuity condition, if $q(x, \mathfrak{X}) = q(x) < \infty$) then, for this x, the derivative $\dot{P}_0(x, S)$ exists for every S. In fact,*

$$\frac{P_h(x, S) - I(x, S)}{h} \to q(x, S) - q(x)I(x, S) = \dot{P}_0(x, S)$$

uniformly in S and $q(x, S)$ is a finite measure in S.

Proof. It suffices to prove the assertion for $x \notin S$ so that the term in $q(x)$ disappears. According to the hypothesis and the extension lemma, the measure $q(x, U)$ in U extends to a measure $q(x, S)$ in S with

$$q(x, \mathfrak{X}) = \bar{q}(x) = q(x) < \infty$$

and

$$q(x, SV_n) \uparrow q(x, S) = q(x, SV), \quad V_n \uparrow V, \quad V_n \in \mathfrak{U}.$$

It follows that

$$\left| \frac{P_h(x, S)}{h} - q(x, S) \right| \leqq \left| \frac{P_h(x, SV_n)}{h} - q(x, SV_n) \right| + \left| \frac{P_h(x, SV_n^c)}{h} - q(x, SV_n^c) \right| \to 0$$

uniformly in S. For, by the tr.pr.'s differentiation theorem, for n fixed, the first term on the right converges to zero uniformly in SV_n hence in S as $h \to 0$, and the upper bound of the second term below

$$\frac{P_h(x, V_n^c\{x\}^c)}{h} + q(x, V_n^c) = \frac{1 - P_h(x, \{x\})}{h} - \frac{P_h(x, V_n\{x\}^c)}{h} + q(x, V_n^c)$$

contains no S and converges to $q(x) - q(x, V_n)$ then to zero as $h \to 0$ then $n \to \infty$.

What precedes is valid with $V = \mathfrak{X}$ under the σ-uniform continuity condition and $q(x, \mathfrak{X}) = q(x)$. The proof is terminated.

COROLLARY. *If the two functions $\bar{q}(\cdot)$ and $q(\cdot)$ coincide and there are no instantaneous states, then the derivative $\dot{P}_t(x, S)$ exists and is finite and continuous in $t \geqq 0$ for every x and every S, and the backward equation holds:*

$$\dot{P}_t(x, S) = \int_{\{x\}^c} q(x, dy) P_t(y, S) - q(x) P_t(x, S).$$

For, by Kolmogorov's equation,

$$\frac{P_{t+h}(x, S) - P_t(x, S)}{h} = \int_{\{x\}^c} \frac{1}{h} P_h(x, dy) P_t(y, S)$$

$$- \frac{1 - P_h(x, \{x\})}{h} P_t(x, S) \to \int_{\{x\}^c} q(x, dy) P_t(y, S) - q(x) P_t(x, S)$$

upon using the following propositions:

g. *If finite measures μ_n on $\mathcal{S}$ converge to a finite measure μ on $\mathcal{S}$ and g on $\mathfrak{X}$ is a bounded Borel function, then $\int g\, d\mu_n \to \int g\, d\mu$.*

It suffices to note that g can be approximated up to any given $\epsilon > 0$ by simple functions $g' = \sum_{j=1}^{m} x_j I_{S_j}$ uniformly bounded by some finite constant c so that, as $n \to \infty$ then $\epsilon \to 0$,

$$\left| \int g\, d\mu - \int g\, d\mu_n \right| \leqq \int | g - g' |\, d\mu + \left| \int g'\, d\mu - \int g'\, d\mu_n \right|$$

$$+ \int | g' - g |\, d\mu_n \leqq \epsilon\mu(\mathfrak{X}) + c \sum_{j=1}^{m} | \mu S_j - \mu_n S_j | + \epsilon\mu_n(\mathfrak{X}) \to 0.$$

According to this proposition, the foregoing passage to the limit as $h \to 0$ is valid. Furthermore

If for $t \geqq 0$, the function $g(t)$ is continuous and its right derivative $\dot{g}^+(t)$ exists and is continuous, then the derivative exists (and coincides with the right derivative).

For, setting $h(t) = \int_0^t \dot{g}^+(s)\, ds$ so that $\dot{h}(t) = \dot{g}^+(t)$, the assertion reduces to the classical one that a continuous function $g(t) - h(t)$ whose right derivative vanishes is a constant, hence has a vanishing derivative.

Since after the foregoing passage to the limit as $h \to 0$, the existence of a right derivative $\dot{P}_t^+(x, S)$ equal to a continuous function in t is established, the term "right" may be omitted on account of the proposition just established. The proof of the corollary is completed.

44.2. Sample functions behavior. Let $X_T = (X_t, t \geqq 0)$ be a separable r.f. and let $\mathcal{B}(X_T)$ be the σ-field of events defined on X_T. We denote by $X_{s+T} = (X_{s+t}, t \geqq 0)$ the translate by s of X_T. Let g be a numerical Borel function on the sample space of X_T such that the exp. of $g(X_T)$ exists. The translate by s of $g(X_T)$ is $g(X_{s+T})$ and B_s—the translate by s of an event $B \in \mathcal{B}(X_T)$—is defined by $I_{B_s} = (I_B)_s$. We denote by P_t the distribution of X_t: $P_t(S) = P[X_t \in S]$.

Throughout this subsection, we assume that X_T is a stationary regular Markov r.f. with tr.pr. $P_t(x, S)$, unless otherwise stated. To be precise: there exists a family $(P^x, x \in \mathfrak{X})$ of pr.'s on $\mathcal{B}(X_T)$ and regular versions of c.pr.'s and c.exp.'s below—the only ones we shall use—such that, for every $x \in \mathfrak{X}$, $S \in \mathcal{S}$, $B \in \mathcal{B}(X_T)$, $0 \leqq s \leqq t < \infty$, the Markov property holds:

(M) $$P(B_s \mid X_r, r \leqq s) = P(B_s \mid X_s)$$

and is stationary:

(S) $$P(B_s \mid X_s = x) = P(B \mid X_0 = x) = P^x B$$

with tr.pr. $P_t(x, S)$:

(Tr.) $$P(X_t \in S \mid X_0 = x) = P^x[X_t \in S] = P_t(x, S).$$

Upon denoting by E^x the exp. which corresponds to P^x and approximating Borel functions by simple Borel ones, it follows that

$$E(g(X_{s+T}) \mid X_r, r \leqq s) = E(g(X_{s+T}) \mid X_s),$$
$$E(g(X_{s+T}) \mid X_s = x) = E^x g(X_T).$$

Note that upon conditioning by $X_{s_1}, \cdots, X_s$, $s_1 \leqq \cdots \leqq s$, we may replace in what precedes X_r, $r \leqq s$, by $X_{s_1}, \cdots, X_s$.

We also assume that, unless otherwise stated, the continuity condition holds:

(C) $$P_h(x, S) \to I(x, S),$$

equivalently, $P_h(x, \{x\}) \to 1$ or, by 39.1**c**, $P_t(x, S)$ is uniformly continuous in t uniformly in S.

a. *X_T is continuous in pr., in fact, for every x, $P^x[X_{t+h} \neq X_t] \to 0$ for every $t \geqq 0$ and $P^x[X_{t-h} \neq X_t] \to 0$ for every $t > 0$.*

Proof. The first assertion follows from the second one since, by the dominated convergence theorem,

$$P[|X_{t\pm h} - X_t| \geqq \epsilon]$$
$$= EP(|X_{t\pm h} - X_t| \geqq \epsilon \mid X_0) \leqq EP(X_{t\pm h} \neq X_t \mid X_0) \to 0.$$

Since, by Markov property,

$$P(X_{t+h} \neq X_t \mid X_0) = E\{P(X_{t+h} \neq X_t \mid X_t, X_0) \mid X_0\}$$
$$= E\{P(X_{t+h} \neq X_t \mid X_t) \mid X_0\},$$

it follows that, for $X_0 = x$,

$$P^x[X_{t+h} \neq X_t] = E^x P(X_{t+h} \neq X_t \mid X_t),$$

and, by stationarity,

$$P(X_{t+h} \neq X_t \mid X_t = y) = P(X_h \neq X_0 \mid X_0 = y) = P_h(y, \{y\}^c).$$

Therefore, by the continuity condition which says that $P_h(y, \{y\}^c) \to 0$ and by the dominated convergence theorem,

$$P^x[X_{t+h} \neq X_t] = \int P_t(x, dy) P_h(y, \{y\}^c) \to 0.$$

Similarly, replacing t by $t - h$, by the continuity condition and its implication $P_{t-h}(x, S) \to P_t(x, S)$,

$$P^x[X_t \neq X_{t-h}] = \int P_{t-h}(x, dy) P_h(y, \{y\}^c) \to 0,$$

and the assertion is proved.

The last passage to the limit is based upon the following proposition.

b. *If on $\mathcal{S}$ finite measures μ_n converge to a finite measure μ and on $\mathcal{X}$ uniformily bounded Borel functions g_n converge to a Borel function g, then*

$$\int g_n \, d\mu_n \to \int g \, d\mu.$$

For, by Egorov's theorem, given $\epsilon > 0$ there exists S with $\mu S^c < \epsilon$ and $s_n = \sup_{x \in S} |g_n(x) - g(x)| \to 0$ and, by 39.1**g**, as $n \to \infty$ then $\epsilon \to 0$,

$$\left| \int g_n \, d\mu_n - \int g \, d\mu \right| \leqq \left| \int g(d\mu_n - d\mu) \right| + s_n \mu_n S + 2c\mu_n S^c \to 0.$$

Note that the assumptions "measure" μ and "Borel function" g may be omitted, for they follow from the convergence assumptions.

c. Duration of stay lemma. *The pr. that starting from x at time s, X_T stays in x during time t is given by*

$$P(X_{s+r} = x, 0 \leqq r \leqq t \mid X_s = x) = e^{-q(x)t}.$$

Proof. Since X_T is separable and continuous in pr., we can replace $[0, \infty)$ by a countable set S dense in $[0, \infty)$, say, the set of dyadic numbers jh_n, $h_n = (\frac{1}{2})^n$. Thus, the sets

$$[X_{s+r} = x, 0 < r < t] = [X_{s+r} = x, 0 < r < t, r \in S]$$

are events. Because of stationarity, it suffices to prove the asserted relation for $s = 0$.

Let $k_n = [t/h_n]$ so that $k_n h_n \to t$ and, to simplify the writing, drop the subscripts n so that $h = h_n \to 0$ and $k = k_n \to \infty$ as $n \to \infty$. By Markov property and continuity condition $(P(X_t = x \mid X_{kh} = x) \to 1)$.

$$p_h = \prod_{j \leqq k} P(X_{jh} = x \mid X_{(j-1)h} = x) \to P(X_s = x, 0 \leqq s \leqq t \mid X_0 = x)$$

and, by stationarity and 44.1 **d** (see its proof),

$$p_h = P_h^k(x, \{x\}) = \exp\left\{\frac{\log P_h(x, \{x\})}{h} kh\right\} \to \exp\{-q(x)t\}.$$

The proof is terminated.

Upon introducing the *duration of stay* $\tau(\omega)$ of $X_T(\omega)$ in some state and considering the three cases $q(x) = 0$, $q(x) = \infty$, and $0 < q(x) < \infty$, **c** yields

A. Duration of stay theorem. *The duration of stay τ is a random time, not necessarily finite, with $P^x[\tau > t] = e^{-q(x)t}$.*

In particular, outside P^x-null events, when at some fixed time X_T takes the value x, then it stays in x forever when x is absorbing, or leaves it at once

when x is instantaneous, or stays in it for some (random) time and then leaves it when x is steady.

The last statement explains the classification of states x into absorbing, instantaneous, or steady, according as $q(x) = 0$, $q(x) = \infty$, or $0 < q(x) < \infty$.

The sets $B_n = [X_{s+t} = X_s, 0 \leqq t < 1/n] \uparrow B = \bigcup B_n$ are events since X_T is separable, and so is their limit B—the set to which correspond the sample functions remaining constant for some positive time after s. By **A** and the dominated convergence theorem,

$$PB \leftarrow PB_n = \int P_s(dx) e^{-q(x)/n}$$

$$= \int_{[q(\cdot) < \infty]} P_s(dx) e^{-q(x)/n} \rightarrow P_s[q(\cdot) < \infty].$$

Similarly, $C_n = [X_{s+t} \neq X_s \text{ for some } t \leqq n] \uparrow C$—the set to which correspond sample functions having a discontinuity some time after s, and

$$PC \leftarrow PC_n = \int P_s(dx)(1 - e^{-q(x)n})$$

$$= \int_{[q(\cdot) > 0]} P_s(dx)(1 - e^{-q(x)n}) \rightarrow P_s[q(\cdot) > 0].$$

In particular

Corollary. *If there are no instantaneous states then, after any given time s, almost all sample functions are constant for some positive times, finite or infinite.*

If there are no absorbing states then, after any given time s almost all sample functions have a discontinuity at some finite time, positive or not.

If there are only steady states then, after any given time s, almost all sample functions are constant for some finite positive time and then have a discontinuity.

At first sight, if there are no instantaneous states, then we expect almost all sample functions to stay constant for some positive times, then jump and remain constant for some positive times, and so on, unless they get into some absorbing state and then stay there forever. Thus, we expect that almost all the sample functions will have a finite or infinite sequence of *isolated jumps*, that is, preceded and followed by time intervals of constancy. Yet, more complicated discontinuities may occur unless some restrictions are imposed. To begin, we shall study those sample functions $X_T(\omega)$ whose first discontinuity time τ after any given s is of the type: $X_T(\omega)$ is constant up to a finite positive time $\tau(\omega)$, is discontinuous at $\tau(\omega)$, and is constant with a different value after $\tau(\omega)$ for some time $h(\omega) > 0$. Because of the separability implications 38.2(3°), these simple discontinuities are isolated jumps, provided we neglect a null set of sample functions. From now on, we assume $q(\cdot)$ finite and we complete our measures so that subsets of null sets are null.

Fix $t > 0$, a steady state x, and a uniform continuity state set $U \not\ni x$. Let $D_{n,h}$ be the set of ω's such that $X_r(\omega) = x$ for $0 \leqq r \leqq \frac{kt}{n}$ and $X_r(\omega) = y$ for $\frac{(k+1)t}{n} \leqq r \leqq \frac{(k+1)t}{n} + h$, for some $k < n$ and some $y \in U$. According to **A**,

$$P^x D_{n,h} = \sum_{k=1}^{n-1} e^{-q(x)kt/n} \int_U P_{t/n}(x, dy) e^{-q(y)h}$$

$$= \frac{e^{-q(x)t/n} - e^{-q(x)t}}{(1 - e^{-q(x)t/n})/(t/n)} \int_U \frac{1}{t/n} P_{t/n}(x, dy) e^{-q(y)h}.$$

Therefore, by 44.1**A** and 44.1**g**, as $n \to \infty$,

$$P^x D_{n,h} \to p_h = \frac{1 - e^{-q(x)t}}{q(x)} \int_U q(x, dy) e^{-q(y)h},$$

then, as $h \to 0$,

$$p_h \to p = (1 - e^{-q(x)t}) \frac{q(x, U)}{q(x)}.$$

Let $\underline{D}_h = \liminf_n D_{n,h}$, $\overline{D}_h = \limsup_n D_{n,h}$ and note that $\underline{D}_h \uparrow \underline{D} = \bigcup_n D_{1/n}, \overline{D}_h \uparrow \overline{D} = \bigcup_n \overline{D}_{1/n}$ as $h \to 0$. Then, by the Fatou-Lebesgue theorem, from $P^x \underline{D}_h \leqq p_h \leqq P^x \overline{D}_h$, it follows that

$$P^x \underline{D} \leqq p \leqq P^x \overline{D}.$$

Let D_h be the set of ω's such that $X_r(\omega) = x$ for $0 \leqq r < \tau(\omega) < t$ and $X_r(\omega) = y$ for $\tau(\omega) < r < \tau(\omega) + h$ for some $y \in U$. Then $D_h \uparrow D$ as $h \to 0$ and D is the set of all such ω with some $h = h(\omega) > 0$. Thus, D corresponds to the set of those sample functions which have an isolated jump from x into U at some time less than t. According to the above definitions, if $\omega \notin D_h$ then $\omega \in D_{n,h}$ for finitely many values of n only, hence $\omega \notin \bar{D}_h$. Thus, $\bar{D}_h \subset D_h$ and $\bar{D} \subset D$. Similarly, if $\omega \in D_h$ then, for $n > 2t/h$ sufficiently large, there exists a $k < n$ such that $\frac{kt}{n} \leqq \tau(\omega) < \frac{(k+1)t}{n}$ and $X_r(\omega) = y$ for $\frac{(k+1)t}{n} \leqq r \leqq \frac{(k+1)t}{n} + h - \frac{t}{n}$ for some $y \in U$, hence $\omega \in D_{n,h/2}$ for n sufficiently large. Thus, $D_h \subset D_{n,h/2}$ and $D \subset \underline{D}$. It follows from $\underline{D} \subset \bar{D}$ that $\bar{D} = D = \underline{D}$. Therefore, $P^x D = p$ since $P^x \underline{D} \leqq p \leqq P^x \bar{D}$.

The preceding discussion remains valid when U is replaced by S and $q(x, S) = \lim_{h \to 0} P_h(x, S)/h$ exists for all $S \not\ni x$; by 44.1**B**, it is so when $\bar{q}(x) = \sup_{U \in \mathfrak{U}} q(x, U) = q(x)$ since $q(x) < \infty$. Thus, *when* $q(\cdot)$ *is finite*

d. *The sample functions starting from a steady state x at time s, which remain constant for some positive times less than t then jump into a uniform continuity state set $U \not\ni x$ and remain constant for some times, correspond to a set D with $P^x D = (1 - e^{-q(x)t}) q(x, U)/q(x)$. If, moreover, $\bar{q}(x) = q(x)$ then what precedes is valid with S in lieu of U.*

We recall that S denotes any state set while U denotes only the uniform continuity ones.

We make the following convention: $q(x, S)/q(x) = 0$ when $q(x) = 0$.

B. ISOLATED JUMPS THEOREM. *Let X_T be in a state x at time s and $q(\cdot)$ be finite. Then*

The pr. that there be a sample discontinuity in the finite or infinite interval $(s, s + u)$ and the first one be an isolated jump into $U \not\ni x$ is given by $(1 - e^{-q(x)u}) q(x, U)/q(x)$. If there is a sample discontinuity in $(s, s + u)$, (and when x is steady, a.s. there is at least one after s), then the pr. that the first one be an isolated jump into U is given by $q(x, U)/q(x)$.

If, moreover, $\bar{q}(x) = q(x)$ then what precedes holds with S in lieu of U, and, when x is steady, a.s. there is a first discontinuity which is an isolated jump.

Proof. The first assertion and the one with S in lieu of U replaces **d** together with the convention about absorbing states, and the second assertion follows by **A**. The assertion about an isolated jump without specifying into which set means that the pr. of an isolated jump from x into $\{x\}^c$ is one and results from $q(x, \{x\}^c) = q(x)$. The proof is terminated.

At first sight, once isolated jumps occur, the same stationary Markov evolution starts anew. However, this means that we can use the random time τ, to be precise $\tau + 0$, as a present or past moment in the relations (M), (S), and (Tr.) at the beginning of this subsection. Since these relations pertain to all states and all state sets, we shall have to assume not only that there are no instantaneous states but also that the two functions $q(x)$ and $\bar{q}(x)$ coincide. It will suffice to prove that $\tau + 0$ is a stationary Markov time of X_T, that is, the relation

$$(\mathrm{SM_{st}})\quad P(X_{\tau+t} \in S \mid X_r, r \leqq \tau + s) = P_{t-s}(X_{\tau+s}, S),\quad 0 \leqq s < t < \infty$$

has meaning and is valid. Thus, we shall have first to show that all $X_{\tau+t}$, $t \geqq 0$ are r.v.'s for τ finite, that is, on $\Omega^\tau = [\tau < \infty]$. The conditioning by $(X_r, r \leqq \tau + s)$ means then conditioning by the σ-field in Ω^τ of all events $A \subset \Omega^\tau$ such that $A[\tau \leqq t] \in \mathcal{B}(X_r, r \leqq t + s)$ The proof will be based upon 43.4**d**—the only result we require in Section 43.

e. ISOLATED JUMP TIME LEMMA. *Let the functions $q(x)$ and $\bar{q}(x)$ coincide and be finite. Then the first isolated jump time τ ($\tau + 0$ to be precise) is a stationary Markov time of X_T.*

Proof. Assume that there are only steady states so that, by **A**, τ is a r.v. with $P^x[\tau > t] = e^{-q(x)t}$. By its definition, τ is a time of X_T, that is, $[\tau \leqq t] \in \mathcal{B}(X_r, r \leqq t)$. For, if we know a sample function $(X_r(\omega), r \leqq t)$ up to time t inclusive, then we know whether it left the state $X_0(\omega)$ or not during this time interval.

To prove the assertion, we subdivide $[0, \infty)$ into intervals of length $h_n = (\frac{1}{2})^n$, denote by $\tau_n(\omega)$ the first of the subdivision points which follows $\tau(\omega)$, approximate functions of τ by functions of τ_n and let $n \to \infty$. The following immediate properties will be used without further comment: $\tau < \tau_n \leqq \tau + h_n$, $\tau_n \to \tau + 0$, $[\tau_n + t - h_n, \tau_n + t] \downarrow \{\tau + t\}$, and, knowledge of $\tau(\omega)$ implying that of $\tau_n(\omega)$, τ_n is an elementary time of X_T hence, by 38.4**d**, is a stationary Markov time of X_T. The property to be established and to play a central role is that, for every $\tau + t$, $t \geqq 0$, almost all sample functions $X_T(\omega)$ have a time interval of constancy at $\tau(\omega) + t$ that is, on $[\tau(\omega) + t, \tau(\omega) + t + h(\omega)]$, $h(\omega) > 0$.

The constancy property at $\tau + 0$ is immediate. For, by definition of an isolated jump almost all sample functions have a time interval of constancy from $\tau + 0$. Therefore, the limit $X_{\tau(\omega)+0}(\omega) = X_{\tau_n(\omega)}(\omega)$ from some $n = n(\omega)$ on exists, and $X_{\tau+0}$ is a r.v.

Given $t > 0$, we take n sufficiently large so that $t - h_n > 0$. The event $B_n = [X_{\tau_n+r} = z, r \in [t - h_n, t], z \in S]$ corresponds to the set of those sample functions $X_T(\omega)$ which are constant in S during the time interval $[\tau_n + t - h_n, \tau_n + t]$. Thus,

$$P(B_n \mid X_{\tau_n} = y) = \int_S P_{t-h_n}(y, dz)e^{-q(z)h_n} \to P_t(y, S)$$

and, moreover, $P(B_n \mid X_{\tau_n}(\omega)) = P(B_n \mid X_{\tau(\omega)+0}(\omega))$ from some $n = n(\omega)$ on, because of the time interval of constancy at $\tau + 0$. Since $B_n \downarrow B = \bigcap B_n$, it follows that

$$PB \leftarrow PB_n = \int P(B_n \mid X_{\tau(\omega)+0}(\omega))P(d\omega) + o(1) \to \int P_{\tau+0}(dy)P_t(y, S).$$

Similarly, if the event C_n corresponds to the set of all sample functions $X_T(\omega)$ which at time $\tau_n(\omega) + t - h_n$ are either in S or in some state $z \in S^c$ and leave z within time h_n, then

$$P(C_n \mid X_{\tau_n} = y)$$
$$= P_{t-h_n}(y, S) + \int_{S^c} P_{t-h_n}(y, dz)(1 - e^{-q(z)h_n}) \to P_t(y, S)$$

and, setting $C = \liminf C_n$,

$$PC \leqq \liminf PC_n = \lim PC_n = PB.$$

Since $B_n \subset [X_{\tau+t} \in S] \subset C_n$ so that $B \subset [X_{\tau+t} \in S] \subset C$, it follows that $[X_{\tau+t} \in S]$ differs from B by a null event; furthermore, $PB = 1$ when $S = \mathfrak{X}$. Thus, $X_{\tau+t}$ is a r.v., almost all sample functions have a time interval of constancy at $\tau + t$, and $P(B_n \mid X_{\tau_n(\omega)}(\omega)) \to P(X_{\tau+t} \in S \mid X_{\tau(\omega)+0}(\omega))$. Therefore, there is a regular version of c.pr. $P(X_{\tau+t} \in S \mid X_{\tau+0}) = P_t(X_{\tau+0}, S)$. This result is valid for every $X_{\tau+s}$, $s < t$, and what precedes applies with $X_{\tau+s}$ in lieu of $X_{\tau+0}$; in particular, we can take $P(X_{\tau+t} \in S \mid X_{\tau+s}) = P_{t-s}(X_{\tau+s}, S)$.

Since τ_n is a stationary Markov time of X_T and $A \in \mathcal{B}(X_r, r \leqq \tau + s) \subset \mathcal{B}(X_r, r \leqq \tau_n + s)$, it follows that

$$PA[X_{\tau+t} \in S] \leftarrow PAB_n$$
$$= \int_A P(B_n \mid X_{\tau_n+s})\, dP \to \int_A P(X_{\tau+t} \in S \mid X_{\tau+s})\, dP.$$

Thus,

$$PA[X_{\tau+t} \in S] = \int_A P_{t-s}(X_{\tau+s}, S)\, dP, \quad A \in \mathcal{B}(X_r, r \leqq \tau + s),$$

that is, the integral form of $(\mathrm{SM_{st}})$, hence $(\mathrm{SM_{st}})$, are valid. So far, we assumed that there were only steady states so that $\Omega^\tau = [\tau < \infty]$ was an a.s. event. If there are also absorbing states and $P\Omega^\tau > 0$, then what precedes applies upon replacing all events by their intersections with Ω^τ. If $P\Omega^\tau = 0$, then what precedes has no content but almost all sample functions remain constant forever from time $\tau + 0$ on, and we may consider this property as the trivial degenerate form of the proposition. The proof is terminated.

A *q-pair* of functions $q(x)$, $q(x, S)$ will be called *regular* if they are finite, nonnegative, Borelian in x, and $q(x, S)$ is a measure in S with $q(x, \{x\}) = 0$ and $q(x, \mathfrak{X}) = q(x)$. We say that such a pair is *bounded* if the functions are bounded. We say that such a pair *derives* from a tr.pr. $P_t(x, S)$ if the function $\dot{P}_0(x, S) = q(x, S) - q(x)I(x, S)$. In fact, then, by the corollary of 44.1**B**, the derivative $\dot{P}_t(x, S)$ exists and is continuous in t, the backward equation holds, and the tr.pr. obeys the σ-uniform continuity condition; if, moreover, the q-pair is bounded then, by the corollary of 44.1**A**, the tr.pr. obeys the uniform continuity condition.

C. Sample step functions theorem. *Let a q-pair of functions $q(x)$, $q(x, S)$ be regular.*

If the q-pair derives from the tr.pr. $P_t(x, S)$ of a separable stationary Markov r.f. $X_T = (X_t, t \geqq 0)$, then there exists a random time τ_θ of accumulation of isolated jumps, not necessarily infinite, and almost all sample functions $X_T(\omega)$ are step functions in $[0, \tau_\theta(\omega))$. If, moreover, the q-pair is bounded, then almost all sample functions are step functions.

Conversely, the q-pair derives from at least one tr.pr. $P_t(x, S)$ of a separable stationary Markov r.f. $X_T = (X_t, t \geqq 0)$ with a corresponding random time τ_θ. If, moreover, τ_θ is a.s. infinite, in particular, if the q-pair is bounded, then the tr.pr. is unique.

Proof. We use without further comment the isolated jumps theorem and the strong Markov jump time lemma.

1° If there are no absorbing states (that is, if the q-functions are positive), then there is a sequence of finite positive random times $\tau_1, \tau_2, \cdots$ such that almost all sample functions $X_T(\omega)$ are constant on $[0, \tau_1(\omega)), (\tau_1(\omega), \tau_1(\omega) + \tau_2(\omega)), \cdots$, with different values in any two consecutive intervals; we set $\tau_0(\omega) + 0 = 0$. If there are absorbing states, then, whenever $X_T(\omega)$ is in such a state for the first time—at some $\tau_{n-1}(\omega) + 0$, then it stays there forever so that $\tau_n(\omega) = \infty$ and we set $\tau_{n+1}(\omega) = \tau_{n+2}(\omega) = \cdots = \infty$.

In either case, the sum of the series $\tau_\theta(\omega) = \sum \tau_n(\omega)$ of positive terms exists and is finite or infinite. If $\tau_\theta(\omega) = \infty$, then the sample function is a step function. If $\tau_\theta(\omega) < \infty$, then we know only that $X_T(\omega)$ is a step function on $[0, \tau_\theta(\omega))$. Thus, the sample functions which correspond to the set $[\tau_\theta = \infty]$ are step functions.

In particular, if the q-pair is bounded by $c < \infty$, then $P[\tau_\theta = \infty] = 1$ so that almost all sample functions are step functions. For, then

$$P[\tau_n \geqq t] = \int P_{\tau_{n-1}+0}(dx)P^x[\tau_n \geqq t] \geqq e^{-ct}$$

for every n and every $t > 0$, so that $P(\limsup [\tau_n \geqq t]) \geqq e^{-ct}$, hence infinitely many τ_n are $\geqq t$ with pr. $\geqq e^{-ct}$ and, thus, $P[\tau_\theta = \infty] \geqq e^{-ct} \to 1$ as $t \to 0$. The direct assertions are proved.

2° Conversely, given a regular q-pair, we construct a separable stationary Markov r.f. X_T with a tr.pr. from which the q-pair derives, upon following the pattern set by what precedes.

We select $\tau_0 = 0$, $\xi_0 = X_0$, τ_1, $\xi_1 = X_{\tau_1}$, $\cdots$, as follows: the r.v. ξ_0 is chosen arbitrarily and for $n > 0$, given the preceding choices, we choose τ_{n+1} and ξ_{n+1} so that

$$P(\tau_{n+1} \geqq t \mid \tau_0, \xi_0, \cdots, \tau_n, \xi_n) = e^{-q(\xi_n)t}$$

$$P(\xi_{n+1} \in S \mid \tau_0, \xi_0, \cdots, \tau_n, \xi_n, \tau_{n+1}) = q(\xi_n, S)/q(\xi_n)$$

and, whenever $q(\xi_n(\omega)) = 0$, we take $\tau_{n+1}(\omega) = \tau_{n+2}(\omega) = \cdots = \infty$, $\xi_{n+1}(\omega) = \xi_{n+2}(\omega) = \cdots = \xi_n(\omega)$; we set $X_t(\omega) = \xi_n(\omega)$ for $t \in \left[\sum_{k=1}^{n} \tau_k(\omega), \sum_{k=1}^{n+1} \tau_k(\omega)\right)$ and $\tau_\theta(\omega) = \sum \tau_n(\omega)$. Thus X_t is defined for all $t \in [0, \tau_\theta)$.

If $P[\tau_\theta = \infty] = 1$, then $X_T = (X_t, t \geqq 0)$ is so defined. If $P[\tau_\theta = \infty] < 1$, then we continue the construction with τ_θ as with τ_0: we choose an arbitrary r.v. ξ_θ independent of the ξ_n, τ_n, choose $\tau_{\theta+1}$ with $P(\tau_{\theta+1} \geqq t \mid \xi_0, \tau_0, \cdots, \tau_\theta) = e^{-q(\xi_\theta)}$, set $X_t(\omega) = \xi_\theta(\omega)$ for $t \in [\tau_\theta(\omega), \tau_{\theta+1}(\omega))$, and so on, starting over, if necessary, at the new accumulation points of jump times with r.v.'s with distribution P_θ of ξ_θ. It is intuitive that this defines $X_T = (X_t, t \geqq 0)$, but we shall not prove it, for the proof requires the use of ordinals.

X_T is a stationary Markov r.f. and the q-pair derives from its tr.pr., as follows: Note that $P[\tau > t] = e^{-qt}$ implies that $P(\tau > s + t \mid \tau > s) = e^{-qt}$ for any $s > 0$. This means that if we stop the construction when we reach a $\tau > s$ and start it anew at s in lieu of at 0 but use X_s in lieu of ξ_0, then the τ and ξ which follow have the same distribution as when the construction was not stopped. Thus, $P(X_{s+t} \in S \mid X_r, r \leqq s) =$

$P(X_{s+t} \in S \mid X_s)$ and is independent of s, and the above assertion is true.

If $P[\tau_\theta = \infty] = 1$ then, up to the choice of $\xi_0 = X_0$, the r.f. X_T is the only one which conforms to what precedes 2°. Therefore, the tr.pr. is unique and the q-pair derives from it. If $P[\tau_\theta < \infty] < 1$, then the constructed X_T and its tr.pr. depend upon the choice of P_θ.

The converse assertions are proved, and the proof is terminated.

§ 45. MARKOV SEMI-GROUPS

45.1. Generalities. Markov semi-groups on G characterize stationary Markov laws of evolution. Their analysis requires introduction of analytical concepts (limits, continuity, integration, differentiation) in G.

We recall the notation to be used throughout. Unless otherwise stated, times r, s, t, with or without affixes, are points of $T = [0, \infty)$, states x, y, z, with or without affixes, are points of a locally compact separable metric state space $\mathfrak{X}$, sets S, with or without affixes, are topological Borel sets—sets of the σ-field $\mathcal{S}$ generated by the class of open sets in $\mathfrak{X}$, and $V_x(\epsilon)$ are open spheres of radius ϵ centered at x.

To fix the ideas, *we take the state space $\mathfrak{X}$ to be a Borel set in R with the usual topology in it.* What follows extends at once to the general case.

The space G is the Banach space of all bounded Borel f.'s g on $\mathfrak{X}$ with the uniform norm $\| g \| = \sup_x | g(x) |$ for every $g \in G$. The space Φ is the Banach space of all bounded signed measures φ on $\mathcal{S}$ with the variation norm $\| \varphi \| = \operatorname{Var} \varphi = \varphi^+(\mathfrak{X}) + \varphi^-(\mathfrak{X})$; in particular, all pr. measures δ_x which degenerate at x belong to Φ. The elements φ of Φ may be considered as linear functionals on G:

$$\varphi(g) = (\varphi, g) = \int \varphi(dx) g(x), \quad \varphi \in \Phi.$$

However, Φ is not the adjoint space of G; it is only a "reciprocal" subspace of it, that is, such that $\| g \| = \sup_{\|\varphi\| \leq 1} (\varphi, g)$ for every $g \in G$ (follows upon using the δ_x). It ought to be noted that Φ is the adjoint of the subspace $C_0 (\subset G)$ of bounded continuous functions on $\mathfrak{X}$ vanishing at infinity.

We introduce two concepts of limit or types of convergence in G. Let $t \to t_0$.

Strong convergence means uniform pointwise convergence: If $g_t(x) \to g(x)$ uniformly in $x \in \mathfrak{X}$, we say that g_t *converges strongly* to g or that g is *strong limit* of g_t, and we write $g_t \xrightarrow{s} g$. Thus, $g_t \xrightarrow{s} g$ is equivalent to $\| g_t - g \| \to 0$, that is, to convergence in norm.

Φ-*weak convergence* means bounded pointwise convergence: If $g_t(x) \to g(x)$ for every $x \in \mathfrak{X}$ and the g_t are uniformly bounded, we say that g_t *converges* Φ-*weakly* to g or that g is Φ-*weak limit* of g_t, and we write $g_t \xrightarrow{w} g$. Clearly, strong convergence implies weak convergence, and it is easily seen that $g_t \xrightarrow{w} g$ is equivalent to $(\varphi, g_t) \to (\varphi, g)$ for every $\varphi \in \Phi$ (use the Banach-Steinhaus uniform boundedness theorem). Thus, if we limit ourselves to the subspace C_0 (so that Φ is its adjoint space), then Φ-weak convergence becomes the usual "weak" convergence in C_0. This explains the "Φ-weak" terminology.

To each of the foregoing concepts of limit correspond concepts of continuity, differentiation, and integration. Let g_t be a function in $t \in [a, b]$ R with values in G. Let $t' \to t$ and $0 < h \to 0$. The function g_t is strongly continuous at t if $g_{t'} \xrightarrow{s} g_t$, and it is *strongly differentiable* at t if $(g_{t'} - g_t)/(t' - t)$ converges strongly, necessarily to an element of G—to be called *strong derivative* of g_t at t and to be denoted by Dg_t. If $t' = t + h$ (or $t' = t - h$), then the derivative is from the right (or left) and denoted by D^+g_t (or D^-g_t); the derivative at a (or b) is necessarily from the right (or left). We drop "at t" when the foregoing properties hold for every t. The same definitions apply upon replacing "strong" by "Φ-weak," "s" by "w" and "D" by "D'." Clearly, strong (Φ-weak) differentiability implies strong (Φ-weak) continuity. In fact

a. Φ-*weak differentiability implies strong continuity.*

For, if $(g_{t'} - g_t)/(t' - t)$ converges weakly hence boundedly, then $\| g_{t'} - g_t \| \leqq c \, | t' - t | \to 0$.

The function g_t is *strongly integrable* on a bounded interval $[c, d)$ if its Riemann sums converge strongly in the usual way. The limit is then necessarily an element of G to be called the *strong integral* of g_t on $[c, d)$ and denoted by $\int_c^d g_t \, dt$

Strong integrals on unbounded intervals are defined by strong passages to the limit exactly as for the improper Riemann integrals. The usual properties of Riemann integrals remain valid: change of variables, additivity, integrability of strongly continuous functions on bounded intervals and also on unbounded intervals when these functions are bounded in norm by numerical functions integrable on these intervals. Similarly, the inequality $\left\| \int_c^d g_t \, dt \right\| \leqq \int_c^d \| g_t \| \, dt$, remains valid, and the convergence property: as $h \to 0$,

$$\frac{1}{h}\int_c^{c+h} g_t\,dt \xrightarrow{s} g_c \quad \text{when} \quad g_t \xrightarrow{s} g_c \quad \text{as} \quad t \to c+0.$$

The Φ-*weak integral* $g_{[c,d)} \in G$ is defined by $g_{[c,d)}(x) = \int_{[c,d)} g_t(x)\,dt$ for functions g_t measurable in (x, t) and bounded in norm by numerical Lebesgue integrable functions. The convergence property holds: as $h \to 0$, $g_{[c,c+h)} \xrightarrow{w} g_c$ when $g_t \xrightarrow{w} g_c$ as $t \to c + 0$, and the Fubini theorem with finite measures μ on $\mathfrak{X}$ applies:

$$\int_{\mathfrak{X}} \mu(dx)\left\{\int_{[c,d)} g_t(x)\,dt\right\} = \int_{\mathfrak{X}\times[c,d)} \mu(dx) g_t(x)\,dt$$
$$= \int_{[c,d)}\left\{\int_{\mathfrak{X}} \mu(dx) g_t(x)\right\} dt.$$

Let T, with or without affixes, denote an *endomorphism on G*— a linear bounded mapping on G to G:

$$T(ag + a'g') = aTg + a'Tg', \quad \|Tg\| \leqq c\|g\|, \quad c < \infty.$$

The smallest possible value of c as g varies, is $\|T\| = \sup_{\|g\|\leqq 1}\|Tg\|$ or the *norm* of T. If $g \geqq 0 \Rightarrow Tg \geqq 0$ we say that T is nonnegative, and if $\|T\| \leqq 1$ we say that T is a *contraction*. Multiplication by scalars, addition, and multiplication of endomorphisms, defined by

$$(aT)g = a(Tg), \quad (T + T')g = Tg + T'g, \quad (TT')g = T(T'g)$$

yield endomorphisms. It follows that the space $\mathcal{E}$ of our endomorphisms is linear and with the foregoing norm becomes a Banach space. Furthermore, multiplication of endomorphisms commutes with their multiplication by scalars, is distributive with respect to addition, and $TI = IT$ where I is the identity mapping. Thus $\mathcal{E}$ is an "algebra with unit I" and since $\|TT'\| \leqq \|T\| \cdot \|T'\|$ it is a "Banach algebra:"

b. *The Banach space of endomorphisms on a Banach space is a Banach algebra.*

Let $t \to t_0$. In $\mathcal{E}$ on our space G we have at our disposal the usual convergence in norm $\|T_t - T\| \to 0$ or *uniform convergence* and the types of convergence induced by those in G: *strong convergence* $T_t \xrightarrow{s} T$ meaning $T_t g \xrightarrow{s} Tg$ for every $g \in G$ and Φ-*weak convergence* $T_t \xrightarrow{w} T$

meaning $T_t g \xrightarrow{w} Tg$ for every $g \in G$. To each of these types of convergence correspond types of limits hence of continuity and of integrals. For example: If T_t is uniformly continuous in $t \in [c, d]$, then the uniform integral $\int_c^d T_t\,dt$ defined as uniform limit of corresponding Riemann sums exists and is an endomorphism. The strong and Φ-weak integrals are induced by those for $g \in G$: the strong integral is defined by $\left(\int_c^d T_t\,dt\right) g = \int_c^d T_t g\,dt$ for all $g \in G$, and the Φ-weak integral is defined by

$$\left(\int_{[c,d)} T_t\,dt\right) g = \int_{[c,d)} T_t g\,dt$$

for all $g \in G$.

If an endomorphism T on G and an endomorphism U on Φ are such that, for all $g \in G$, $\varphi \in \Phi$,

$$(\varphi, Tg) = \int \varphi(dx)\, Tg(x) = \int U\varphi(dx) g(x) = (U\varphi, g),$$

then we shall set $U\varphi = \varphi T$, write the above relation $(\varphi, Tg) = (\varphi T, g)$, and say that T is Φ-*adjoint*—its adjoint on the adjoint space of G leaves Φ invariant. Clearly

c. *Endomorphisms T and strong passages to the limit commute.*

But generally this is not true of Φ-weak passages to the limit. However

c′. *Φ-adjoint endomorphisms T commute with Φ-weak passages to the limit.*

For, $g_t \xrightarrow{w} g \Rightarrow (\varphi, Tg_t) = (\varphi T, g_t) \to (\varphi T, g) = (\varphi, Tg) \Rightarrow Tg_t \xrightarrow{w} Tg$.

We are now ready for the introduction of Markov endomorphisms. Let $P(x, S)$ denote Borel functions in $x \in \mathfrak{X}$ and pr.'s in $S \in \mathcal{S}$ or, more generally, measures in S bounded by 1. To every $P(x, S)$ there corresponds an endomorphism T on G and an endomorphism U on Φ, to be called *Markov endomorphisms*, defined by

$$Tg(x) = \int P(x, dy) g(y), \quad U\varphi(S) = \int \varphi(dx) P(x, S).$$

Clearly, T and U are nonnegative contractions, and when $P(x, S)$ is a pr. in S so that $P(x, \mathfrak{X}) = 1$ for all x then $\| T \| = \| U \| = 1$. Either T or U determines $P(x, S)$; it suffices to take $g(\cdot) = I(\cdot, S)$ or $\varphi(\cdot) =$

$I(x, \cdot)$. In fact, T is Φ-adjoint (to U) since

$$(\varphi, Tg) = \int \varphi(dx) P(x, dy) g(y) = (U\varphi, g).$$

We shall concentrate on Markov endomorphisms T on G and have

A. MARKOV ENDOMORPHISMS CRITERION. *An endomorphism T on G is Markovian if and only if it is a nonnegative Φ-adjoint contraction.*

Proof. The "only if" assertion is contained in what precedes. As for the "if" assertion, let T be a nonnegative contraction Φ-adjoint (to U). Set $P(x, S) = \Delta_x(S) = I(x, S)T$ (where the last term stands for $U\varphi(S)$ with $\varphi(\cdot) = I(x, \cdot)$). Since every φT is a measure so is Δ_x and, from $(\varphi, Tg) = (\varphi T, g)$ it follows that

$$Tg(x) = \int I(x, dy) Tg(y) = \int P(x, dy) g(y).$$

In particular, $TI(x, S) = P(x, S) \in G$ and, since our Tg is a nonnegative contraction and $I(x, S)$ is bounded by 1, it follows that $P(x, S)$ is a nonnegative Borel function in x bounded by 1. The proof is terminated.

Let $P_t(x, S)$ be a stationary tr.pr. except that in lieu of $P_t(x, \mathfrak{X}) = 1$ we assume only that $P_t(x, \mathfrak{X}) \leqq 1$, unless otherwise stated. In terms of Markov r.f.'s this assumption may mean that its r.v.'s when numerical may take infinite values with positive pr.'s. According to what precedes, $P_t(x, S)$ as a Borel function in x and a measure in S bounded by 1 determines and is determined by a Markov endomorphism T_t on G defined by

$$T_t g(x) = \int P_t(x, dy) g(y), \quad g \in G.$$

There remains the stationary Kolmogorov equation, which links the values of the tr.pr. for different values of t and which, by

$$T_{s+t} g(x) = \int P_{s+t}(x, dz) g(z) = \int P_s(x, dy) P_t(y, dz) g(z) = T_s T_t g(x),$$

is equivalent to the *semi-group property* $T_{s+t} = T_s T_t$.

We say that this family of Markov endormorphisms, which is in a one-to-one correspondence with a stationary tr.pr. hence with the corresponding law of evolution of a stationary Markov process, is a *Markov semi-group*. Unless otherwise stated, semi-groups are semi-groups of endomorphisms on G and all endomorphisms are on G. If a semi-group consists of Φ-adjoint (nonnegative) (contraction) endomorphisms we say that it is a Φ-*adjoint* (*nonnegative*) (*contraction*) semi-group. If, moreover, all functions $T_t g(x)$ are Borelian in (x, t) we say that the semi-group is *Borelian*.

B. MARKOV SEMI-GROUPS CRITERION. *A semi-group is a Markov semi-group if and only if it is a nonnegative Φ-adjoint contraction semi-group.*

A Markov semi-group is Borelian if and only if the corresponding stationary tr.pr. is Borelian.

Proof. The first assertion follows from **A**. The second assertion follows from the integral form of $T_t g(x)$, the "only if" part upon taking $g(\cdot) = I(\cdot, S)$ and the "if" part upon approximating g by simple functions.

45.2. Analysis of semi-groups. While our concern is with Markov semi-groups, the concepts and general properties below are valid for more general contraction semi-groups of endomorphisms on Banach spaces G, and they are stated accordingly. As usual, the limits in h are taken as $0 < h \to 0$, and if a property holds at all t, *we drop* "at t".

Let $(T_t, t \geqq 0)$ be a contraction semi-group:

$$T_{s+t} = T_s T_t, \quad T_0 = I, \quad \| T_t \| \leqq 1, \quad s, t \geqq 0.$$

A set G_0 in G is *invariant* (by the semi-group) if all $T_t G_0 \subset G_0$; in other words there is a restriction of the semi-group on G_0 to G_0. The whole space G and the singleton which consists of the origin of G are trivially invariant.

We denote by G_c the set on which our semi-group is strongly continuous at $t = 0$: $g \in G_c \Leftrightarrow T_h g \xrightarrow{s} g$.

a. STRONG CONTINUITY LEMMA. *G_c is the set on which the contraction semi-group is strongly continuous, and it is an invariant Banach subspace.*

Proof. Clearly G_c contains the strong continuity set of the semi-group. But G_c is also contained in this set and is invariant. For, $(T_h - I)T_t = T_t(T_h - I)$ and, for every $g \in G_c$,

$$\| T_{t+h}g - T_t g \| = \| T_t(T_h - I)g \| \leqq \| (T_h - I)g \| \to 0,$$

$$\| T_t g - T_{t-h}g \| = \| T_{t-h}(T_h - I)g \| \leqq \| (T_h - I)g \| \to 0.$$

G_c is obviously closed under linear combinations. It is also closed under strong passages to the limit by sequences $g_n \xrightarrow{s} g$, $g_n \in G_c$, hence is a Banach space. For, as $h \to 0$ then $n \to \infty$,

$$\| T_h g - g \| \leqq \| T_h(g - g_n) \| + \| (T_h - I)g_n \| + \| g_n - g \|$$

$$\leqq 2\| g - g_n \| + \| (T_h - I)g_n \| \to 0.$$

The proof is terminated.

We denote by G_d the set on which the semi-group is strongly differentiable at $t = 0$: $g \in G_d \Leftrightarrow D_h g \xrightarrow{s} Dg$; $D_h = (T_h - I)/h$. Clearly $G_d \subset G_c$. The (*strong*) *differentiation operator* D on G_d to G is also called the (*strong*) *infinitesimal operator* or the *generator* of the semi-group. For, it generates the semi-group at least on G_c, as will be seen later. Clearly, D is a linear operator on the obviously linear space G_d. But in general, D is not bounded and G_d is not a Banach subspace. Note that if $g \in G_c$ and a, b are finite then the strong integral $g_a{}^b = \int_a^b T_t g\, dt$ exists. For, the function $T_t g$ is then strongly continuous in t. Furthermore $g_a{}^b \in G_d$ since

$$(T_h - I)g_a{}^b = \int_a^b T_{t+h} g\, dt - \int_a^b T_t g\, dt = \int_b^{b+h} T_t g\, dt - \int_a^{a+h} T_t g\, dt$$

implies that

$$D_h g_a{}^b \xrightarrow{s} (T_b - T_a)g = Dg_a{}^b;$$

in particular

$$G_d \ni \frac{1}{h}\int_0^h T_t g\, dt \xrightarrow{s} g \in G_c.$$

b. Strong differentiation lemma. *G_d is the set on which the contraction semi-group is strongly differentiable and it is an invariant set dense in G_c with $DT_t = T_t D$ on it.*

Proof. Clearly, G_d contains the strong differentiability set of the semi-group. But G_d is also contained in this set and is invariant with $DT_t = T_t D$ on it. For, if $g \in G_d$, then

$$D_h(T_t g) = (T_{t+h} - T_t)g/h = T_t(D_h g) \xrightarrow{s} T_t(Dg) = D(T_t g),$$

hence $T_t g \in G_d$ and $D(T_t g) = T_t(Dg)$, and

$$(T_t - T_{t-h})g/h = T_{t-h}(D_h g) = T_{t-h}(D_h g - Dg) + T_{t-h}Dg \xrightarrow{s} T_t(Dg).$$

Finally, G_d is dense in G_c since every $g \in G_c$ is strong limit of elements $\frac{1}{h}\int_0^h T_t g\,dt$ of G_d. The proof is terminated.

We replace now strong limits by Φ-weak limits. If we try to proceed as for the strong continuity, we find that left limits have to be left out and we need the functional form $(\varphi, g_n) \to (\varphi, g)$ of $g_n \xrightarrow{w} g$ hence, to use 40.1**c′**, we assume that the semi-group is Φ-adjoint. If we try to proceed as for the strong differentiability, we find that in order to introduce Φ-weak integrals to establish the closure property, we have to assume that the semi-group is Borelian. We are then led without any difficulty to the following definitions and propositions. Denote by G'_c the set on which the semi-group is Φ-weakly continuous at $t = 0$: $g \in G'_c \Leftrightarrow T_h g \xrightarrow{w} g$.

a′. Φ-WEAK CONTINUITY LEMMA. *G'_c is the set on which the Φ-adjoint contraction semi-group is Φ-weakly rightcontinuous, and it is an invariant Banach subspace.*

Denote by G'_d the set on which the semi-group is Φ-weakly differentiable at $t = 0$: $g \in G'_d \Leftrightarrow D_h g \xrightarrow{w} D'g$. The *$\Phi$-weak differentiation operator* D' is also called the *Φ-weak infinitesimal operator* or the *Φ-weak generator* of the semi-group. Then

b′. Φ-WEAK DIFFERENTIATION LEMMA. *G'_d is the set on which the Φ-adjoint contraction semi-group is Φ-weakly right differentiable, and it is an invariant set with $D'T_t = T_tD'$ on it.*

If, moreover, the semi-group is Borelian, then the Φ-weak closure of G'_d contains G'_c.

The four spaces so distinguished are related by

c. INCLUSION AND CLOSURE LEMMA. *The invariant continuity and differentiation sets of a Φ-adjoint contraction semi-group are ordered by the inclusions $G_d \subset G'_d \subset G_c \subset G'_c$, and G_c is the strong closure of G_d.*

If, moreover, the semi-group is Borelian, then these four sets have common Φ-weak closure G'_c.

Proof. The first part follows from the preceding lemmas except for $G'_d \subset G_c$ which follows from 40.1**a**. The second part follows from the asserted ordering and the last assertion in **b′**.

Instead of replacing "strong" by "Φ-weak" we may replace it by "uniform." Then much more is true as follows. Let E be an endomorphism on a Banach space and let E^n be its n-th iterate ($E^0 = I$).

Clearly, the exponentials e^{tE} defined by $e^{tE} = \sum_{n=0}^{\infty} \frac{t^n}{n!} E^n$ are endomorphisms bounded by $e^{t\|E\|}$ and form a semi-group—*the exponential semi-group generated by E.* This semi-group is uniformly continuous and uniformly differentiable with "uniform differential operator" E, since

$$e^{(t+h)E} - e^{tE} = e^{tE}(e^{hE} - I), \quad e^{tE} - e^{(t-h)E} = e^{(t-h)E}(e^{hE} - I)$$

and

$$\left\| \frac{e^{hE} - I}{h} - E \right\| \leqq (e^{h\|E\|} - 1 - h\| E \|)/h \to 0.$$

d. Uniform continuity and differentiation lemma. *If the contraction semi-group is uniformly continuous at $t = 0$ on an invariant Banach subspace G_u, then the semi-group is exponential on it.*

Proof. G_u is invariant; and the semi-group is uniformly continuous on it since

$$\| T_{t+h} - T_t \| \leqq \| T_h - I \| \to 0 \text{ and } \| T_t - T_{t-h} \| \leqq \| T_h - I \| \to 0$$

on G_u. From here on we consider the semi-group on the invariant Banach space G_u only. Clearly, the uniform integral $I_h = \int_0^h T_s\, ds$ exists and $I_h/h \xrightarrow{u} I$. Therefore, for $h = h_0$ sufficiently small, I_{h_0} has an inverse $I_{h_0}{}^{-1}$. Let $E = (T_{h_0} - I)I_{h_0}{}^{-1}$. It follows from

$$(T_t - I)\int_0^{h_0} T_s\, ds = \int_t^{t+h_0} T_s\, ds - \int_0^{h_0} T_s\, ds = (T_{h_0} - I)\int_0^t T_s\, ds$$

that

$$T_t - I = E\int_0^t T_s\, ds,$$

hence $D_h = EI_h/h \xrightarrow{u} E$. Also, proceeding by induction,

$$T_t - I = \sum_{k=1}^{n} \frac{t^k}{k!} E^k + E^{n+1}\int_0^t \frac{(t-s)^n}{n!} T_s\, ds$$

where the norm of the right side summation is bounded by $e^{t\|E\|}$ and that of the remaining term is bounded by $t^{n+1}\| E \|^{n+1}/(n+1)! \to 0$. Thus $T_t = \sum_{n=0}^{\infty} \frac{t^n}{n!} E^n = e^{tE}$, and the proof is terminated.

LOCAL CHARACTERIZATION. The purely analytical problem relative to Markov processes (regular and stationary) corresponding to stationary tr.pr.'s, hence to Markov semi-groups, can now be stated as follows: Characterize Markov semi-groups in terms of their local properties. It is a specialization of the same problem for general contraction semi-groups to nonnegative Φ-adjoint endomorphisms on our function space G. Thus, we first treat the general case (then the results extend at once to general Banach spaces G), and specialize it later. The problem may be decomposed as follows: The *existence problem* is that of characterizing infinitesimal operators of contraction semi-groups, the *unicity problem* is that of characterizing those infinitesimal operators which determine their semi-group, and the *generation problem* is that of constructing the corresponding semi-groups. However, at best we may hope for answers on the strong or Φ-weak closures of the domains of the infinitesimal operator. Thus appears the *extension problem:* find conditions under which extensions of semi-groups exist and are unique on domains containing the space G with which we are concerned.

In the numerical case, where the contractions are endomorphisms on R, the semi-group property reduces to the classical functional equation $f(s+t) = f(s)f(t)$ with $f(0) = 1, |f(t)| \leqq 1$. The only continuous and, in fact, the only measurable solutions are exponential: $f(t) = e^{td}$ where $d \leqq 0$. In the general case, the *formal* solution of the equation $\frac{d}{dt} T_t = DT_t$ with $T_0 = I$ is similarly exponential $T_t = e^{tD}$ which would require that the infinitesimal operator D be an endomorphism. Yet, an infinitesimal operator D, while not necessarily an endomorphism, is always the limit of endomorphisms $D_h = (T_h - I)/h$. This fact, as well as the numerical case and the formal approach, lead us to expect that, at least on the strong closure G_c of the domain G_d of D, the corresponding semi-group would be determined by D and could be represented as limit of exponential semi-groups. We shall show that these expectations are justified. First, we have to introduce the necessary tool—the Laplace transform or "resolvent" of a contraction semi-group.

The (*strong*) *resolvent* R_λ, on the (strong) continuity subspace G_c of the contraction semi-group $(T_t, t \geqq 0)$ is defined by the strong integrals

$$R_\lambda g = \int_0^\infty e^{-\lambda t} T_t g \, dt, \quad g \in G_c, \quad \lambda \in (0, \infty).$$

$R_\lambda g$ exists and belongs to the Banach subspace G_c, since the integrand is strongly continuous and is bounded in norm by $e^{-\lambda t}\| g \|$ whose integral on $[0, \infty)$ is $\| g \|/\lambda$. Clearly R_λ is linear. Thus R_λ is an endomorphism on G_c with $\| \lambda R_\lambda \| \leqq 1$. The basic role of the resolvent is due to

e. RESOLVENT LEMMA. *The resolvent R_λ of the contraction semi-group with infinitesimal operator D on G_d is the inverse of the one-to-one mapping $\lambda I - D$ of G_d onto G_c.*

Proof. If $g \in G_c$ then, from

$$T_h R_\lambda g = \int_0^\infty e^{-\lambda t} T_{t+h} g \, dt$$

$$= e^{\lambda h} \int_h^\infty e^{-\lambda t} T_t g \, dt = e^{\lambda h} \left(R_\lambda g - \int_0^h e^{-\lambda t} T_t g \, dt \right),$$

it follows that

$$D_h R_\lambda g = \frac{e^{\lambda h} - 1}{h} R_\lambda g - \frac{e^{\lambda h}}{h} \int_0^h e^{-\lambda t} T_t g \, dt \xrightarrow{s} (\lambda R_\lambda - I) g,$$

hence $R_\lambda g \in G_d$. If $g \in G_d (\subset G_c)$ then, moreover,

$$D_h R_\lambda g = R_\lambda D_h g \xrightarrow{s} R_\lambda D g.$$

Therefore,

$$D R_\lambda = \lambda R_\lambda - I \text{ on } G_c, \quad R_\lambda D = \lambda R_\lambda - I \text{ on } G_d,$$

equivalently,

$$(\lambda I - D) R_\lambda = I \text{ on } G_c, \quad R_\lambda (\lambda I - D) = I \text{ on } G_d.$$

The proof is terminated.

The theorem which follows will lead us to the "natural" appearance of spaces of continuous functions

A. UNICITY THEOREM. *Let D with its domain G_d be the infinitesimal operator of a contraction semi-group.*

(i) *D determines the semi-group on its strong continuity space G_c and determines the semi-group on the Φ-weak closure G' of G_c when the contractions are Φ-adjoint.*

(ii) *The unique solution of the equation* $\frac{df_t}{dt} = Df_t, f_t \in G_d$, *which is bounded, reduces to* $g \in G_d$ *for* $t = 0$ *and has a strongly continuous strong derivative, is* $f_t = T_t g$, $g \in G_d$.

Proof 1° Given D on the strong differentiability set G_d, the strong continuity space G_c is determined as strong closure of G_d and then, by **e**, R_λ on G_c is determined. Therefore, by the classical unicity theorem for numerical Laplace transforms $R_\lambda g(x) = \int_0^\infty e^{-\lambda t} T_t g(x)\, dt$ of continuous functions $T_t g(x)$ in t, these functions are determined. The G_c-assertion is proved, and the G'-assertion follows since Φ-adjoint endomorphisms commute with Φ-weak passages to the limit.

2° According to **b**, $f_t = T_t g$, $g \in G_d$ is a solution of the stated equation with the asserted properties. The "unicity" assertion will follow if we show that such a solution, which vanishes for $t = 0$ (in lieu of reducing to g), vanishes for all t. Let $g_t = e^{-\lambda t} f_t$ and note that, according to the hypotheses made, $g_0 = 0, \frac{dg_t}{dt} \in G_c$ and $\| g_s \| \to 0$ as $s \to \infty$. On account of the stated equation, we have $\frac{dg_t}{dt} = (D - \lambda I)g_t$ hence, by **e**, $g_t = -R_\lambda \frac{dg_t}{dt}$. It follows that for all $\lambda(> 0)$, as $s \to \infty$,

$$\int_0^\infty e^{-\lambda t} f_t\, dt \overset{s}{\leftarrow} \int_0^s g_t\, dt = -R_\lambda g_s \overset{s}{\rightarrow} 0.$$

Therefore, by the unicity theorem recalled above, $f_t = 0$. The proof is terminated.

Upon replacing "strong" by "Φ-weak," parallel definitions and slightly more involved but similar arguments yield without difficulty

e′. Φ-WEAK RESOLVENT LEMMA. *The Φ-weak resolvent R'_λ of a Φ-adjoint contraction Borel semi-group is the inverse of the one-to-one mapping* $\lambda I - D'$ *of* G'_d *onto* G'_c.

A′. Φ-WEAK UNICITY THEOREM. *Let D' with its domain G'_d be the Φ-weak infinitesimal operator of a Φ-adjoint contraction Borel semi-group.*

(i) *D' determines the semi-group on the common Φ-weak closure G' of its continuity and differentiation spaces.*

(ii) *The unique solution of the equation* $\frac{d^+ f_t}{dt} = D' f_t, f_t \in G'_d$, *which is bounded and reduces to* $g \in G'_d$ *for* $t = 0$ *is measurable in* (x, t) *and continuous in* t *and has a right continuous right derivative bounded on every finite interval of values of* t, *is* $f_t = T_t g$, $g \in G'_d$.

The foregoing unicity theorems lead at once to the following question: when is the closure of a subset of G the whole of G? For then, the semi-group is completely determined—on G—and the extension problem is solved in the affirmative. An answer is immediately available if we recall the Baire definition of Borel functions: The class of Borel functions on a Euclidian space $\mathfrak{X}$ is the closure of its subclass of continuous functions under pointwise passages to the limit of sequences. The space C of bounded continuous functions on $\mathfrak{X}$ is a Banach subspace of the Banach space G of bounded Borel functions on $\mathfrak{X}$. Thus, G is the Φ-weak closure of its Banach subspace C.

B. C-EXTENSION THEOREM. *If the* Φ*-weak closure of the strong* (Φ*-weak*) *continuity subspace of a* Φ*-adjoint* (*Borel*) *contraction semi-group contains the subspace* C *then the semi-group is completely determined by its strong* (Φ*-weak*) *infinitesimal operator. In particular, the semi-group is completely determined when* C *is invariant and the semi-group is strongly* (*weakly*) *continuous on* C.
Moreover, if the contraction Borel semi-group leaves C_0 *invariant and is* Φ*-weakly continuous on* C_0, *then it is* Φ *adjoint, strongly continuous on* C_0, *and, in fact, "weak" and "strong" concepts coincide in* C_0.

Proof. The determination assertion results from the unicity theorems and the Φ-weak closure property of C. The infinitesimal operators assertion results from the continuity assertion since then, by the resolvent lemma, R_λ is a one-to-one mapping of $C \cap G_c$ onto $C \cap G_d$ and of $C \cap G'_c$ onto $C \cap G'_d$.
Since Φ is the adjoint of C_0, the "Φ-adjoint" assertion is immediate. The "weak"–"strong" assertion follows from 40.3**d**. The proof is terminated.

The stated problems are answered in terms of (strong) infinitesimal operators as follows.

Let $(T_{\lambda,t}, t \geqq 0)$ be a family in $\lambda > 0$ of strongly continuous contraction semi-groups on a common invariant Banach space H. Take all limits as $\lambda, \mu \to \infty$, unless otherwise stated. We say that these *semi-groups converge strongly* if on H there exist endomorphisms T_t, necessarily contractions, such that $T_{\lambda,t} \xrightarrow{s} T_t$ uniformly in $t \leqq a$ for any finite a: $\| T_{\lambda,t}\, g - T_t g \| \to 0$ uniformly in $t \leqq a$, for every $g \in H$. It is easily seen that such convergence implies and is implied by the corresponding mutual convergence $T_{\lambda,t} - T_{\mu,t} \xrightarrow{s} 0$ uniformly in $t \leqq a$, and that if either convergence holds on a set dense in H then it holds on H. Furthermore, the limits T_t form a contraction semi-group. Finally, because of the uniformity condition, it is strongly continuous on H and, for $\nu > 0$, $g \in H$, as $\lambda \to \infty$,

$$R_{\nu,\lambda}g = \int_0^\infty e^{-\nu t} T_{\lambda,t} g\, dt \xrightarrow{s} \int_0^\infty e^{-\nu t} T_t g\, dt = R_\nu g.$$

f. Semi-groups convergence lemma. *On a Banach space H, if endomorphisms D_λ commute, $\| e^{tD_\lambda} \| \leqq 1$, and $D_\lambda \xrightarrow{s} D$ on a set H' dense in H, then the strongly (in fact, uniformly) continuous exponential contraction semi-groups $(e^{tD_\lambda}, t \geqq 0)$ converge to a strongly continuous contraction semi-group $(T_t, t \geqq 0)$ whose infinitesimal operator is D on $H_d \supset H'$.*

Proof. We use two elementary relations applicable to commuting endomorphisms:

$$|\alpha^n - \beta^n| \leqq n|\alpha - \beta|, \quad |\alpha|, \quad |\beta| \leqq 1,$$

$$\frac{1}{h}(e^{h\alpha} - 1)x \to \alpha x, \ h \to 0.$$

Let $g \in H'$, $t \leqq a$, and exclude the trivial case $D_\lambda g = D_\mu g$. As $n \to \infty$,

$$\| (e^{tD_\lambda} - e^{tD_\mu})g \| \leqq a \left\| \frac{e^{(t/n)(D_\lambda - D_\mu)} - I}{t/n} g \right\| \to a \|(D_\lambda - D_\mu)g\|.$$

As λ, $\mu \to \infty$, the last expression converges to 0, and the convergence assertion follows.

For any λ, let R_λ^ν be the resolvent of the semi-group of endomorphisms e^{tD_ν} and let R_λ be that of the limit semi-group, so that $R_\lambda^\nu \xrightarrow{s} R_\lambda$ as $\nu \to \infty$. Since

$$R_\lambda^\nu(\lambda I - D)g = R_\lambda^\nu(\lambda I - D_\nu)g + R_\lambda^\nu(D_\nu - D)g,$$

where on the right side the first term is g and the norm of the second term is bounded by $\| (D_\nu - D)g \|/\lambda \to 0$ as $\nu \to \infty$, it follows that $R_\lambda(\lambda I - D)g = g$ and the infinitesimal operator assertion follows by the resolvent lemma. The proof is terminated.

We are now ready for the basic Hille-Yosida

C. INFINITESIMAL OPERATORS CRITERION. *A linear transformation D on a linear subset H_d of a Banach space H is the infinitesimal operator of one and only one strongly continuous contraction semi-group $(T_t, t \geqq 0)$ on H, if and only if*

(i) *The inverse R_λ of the transformation $\lambda I - D$ exists and is an endomorphism on H with $\| \lambda R_\lambda \| \leqq 1$ for every $\lambda > 0$.*

(ii) $\lambda R_\lambda \xrightarrow{s} I$ *on H as $\lambda \to \infty$ or* (ii′) *H_d is dense in H.*
Then the exponential semi-groups $(e^{tD_\lambda}, t \geqq 0)$, where $D_\lambda = \lambda(\lambda R_\lambda - I)$, converge strongly to the semi-group $(T_t, t \geqq 0)$ as $\lambda \to \infty$.

Proof. Let $\lambda \to \infty$. Properties (ii) and (ii′) are equivalent under (i): Let $\lambda R_\lambda g \xrightarrow{s} g$ for every $g \in H$. Then, from $\lambda R_\lambda g \in H_d$ it follows that H_d is dense in H. Conversely, let H_d be dense in H. Since $\| \lambda R_\lambda \| \leqq 1$ implies that for every $g \in H_d$

$$\| (\lambda R_\lambda - I)g \| = \| R_\lambda D g \| \leqq \| Dg \| / \lambda \to 0,$$

hence the contractions $\lambda R_\lambda \xrightarrow{s} I$ on H_d, it follows that $\lambda R_\lambda \xrightarrow{s} I$ on H.

The "only if" and the "unicity" assertions result from **a** and **A**. The "if" assertion then results from the convergence one which we prove by showing that, because of (i), (ii), and (ii′), the semi-groups convergence lemma applies: Since, by (i), the R_λ hence the D_λ commute and $\| \lambda R_\lambda \| \leqq 1$ hence $\| e^{tD_\lambda} \| \leqq e^{-\lambda t} e^{\lambda t \| \lambda R_\lambda \|} \leqq 1$, on account of (ii′) it suffices to show that $D_\lambda \xrightarrow{s} D$ on H_d. But, by (i), $R_\lambda(\lambda I - D) = I$ on H_d hence, by (ii), $D_\lambda = \lambda R_\lambda D \xrightarrow{s} D$ on H. The proof is terminated.

COROLLARY. *Let H be an invariant subspace of G. Under the conditions of the Hille-Yosida theorem, $T_t \geqq 0$ for all $t > 0 \Rightarrow \lambda R_\lambda \geqq 0$ for all $\lambda > 0 \Rightarrow e^{tD_\lambda} \geqq 0$ for all $t, \lambda > 0 \Rightarrow T_t \geqq 0$ for all $t > 0$.*

For, the first implication results from the definition of R_λ, the second one follows from $e^{tD_\lambda} = e^{-\lambda t} \sum_{n=0}^{\infty} \frac{(\lambda t)^n}{n!} (\lambda R_\lambda)^n$, and the third one is obtained by letting $\lambda \to \infty$.

The definitions and properties in this subsection apply to semi-groups restricted to any given *invariant subspace* $H \subset G$, provided they are relativized accordingly, that is, $G, G_c, G'_c, G'_d, G', \cdots$ are replaced by their intersections with H: $H, H \cap G_c$, and so on. However, there is a difficulty in the Φ-weak case: Φ-weak integrals, supposed to exist, of functions $g_t \in H$ may not belong to H, and some restriction is needed to eliminate this difficulty. For example, it suffices that H be the space

C. This brings out once more the advantage of invariant C. In fact, our study became fast restricted to invariant subspaces selected according to the semi-group continuity requirements. Thus we are led to classify our semi-groups according to their invariant subspaces selected according to suitable requirements. What precedes, and the C-unicity theorem show the convenience of the class for which C is invariant.

45.3. Markov processes and semi-groups. We apply what precedes to the Markov case, searching for a probabilistic interpretation of the concepts and properties.

Let X_T, $T = [0, \infty)$, be a stationary regular Markov process. There is a one-to-one correspondence between the stationary law of evolution $P^{X_0} = (P^x, x \in \mathfrak{X})$ of X_T, the tr.pr. $P_t(x, S) = P(X_{s+t} \in S \mid X_s = x)$, the tr.d.f. $F_t^x(y) = P(X_{s+t} < y \mid X_s = x)$, the tr.ch.f. $f_t^x(u) = E(e^{iuX_{s+t}} \mid X_s = x)$, and the Markov semi-group $(T_t, t \geqq 0)$ with

$$T_t g(x) = \int P_t(x, dy) g(y) = E^x g(X_t), \quad g \in G,$$

or, to emphasize stationarity,

$$T_t g(X_s) = E\{g(X_{s+t}) \mid X_s\}.$$

The Markov endomorphisms are adjoint to endomorphisms on the space Φ, defined by

$$\varphi T_t(S) = \int \varphi(dx) P_t(x, S) = \int \varphi(dx) P^x[X_t \in S]$$

hence translating distributions $\varphi = P_s$ of X_s according to

$$P_s T_t = P_{s+t}.$$

Let the tr.pr., equivalently the semi-group, be Borelian. Form the function

$$\lambda R_\lambda(x, S) = \lambda \int_0^\infty e^{-\lambda t} P_t(x, S)\, dt, \quad \lambda > 0.$$

It is a Borel function in x and a pr. in S. Therefore, the endomorphism λR_λ defined by

$$\lambda R_\lambda g(x) = \int \lambda R_\lambda(x, dy) g(y), \quad g \in G,$$

is Markovian. Since by interchanging the integrations

$$\lambda R_\lambda g(x) = \lambda \int_0^\infty e^{-\lambda t} T_t g(x)\, dt,$$

endomorphisms R_λ appear as an extension on G of the resolvents of $(T_t, t \geqq 0)$. Furthermore, the iterates $(\lambda R_\lambda)^n$ are Markov endomorphisms defined by

$$(\lambda R_\lambda)^n g(x) = \int (\lambda R_\lambda)^{(n)}(x, dy) g(y), \quad g \in G,$$

with

$$(\lambda R_\lambda)^{(n)}(x, S) = \int (\lambda R_\lambda)^{(n-1)}(x, dy) \lambda R_\lambda(y, S).$$

It follows that every e^{tD_λ} with $D_\lambda = \lambda(\lambda R_\lambda - I)$ is a Markov endomorphism defined by

$$e^{tD_\lambda} g(x) = \int (e^{tD_\lambda})(x, dy) g(y), \quad g \in G,$$

with

$$(e^{tD_\lambda})(x, S) = e^{-\lambda t}\left(1 + \frac{\lambda t}{1!} \lambda R_\lambda(x, S) + \frac{\lambda^2 t^2}{2!} (\lambda R_\lambda)^{(2)}(x, S) + \cdots\right).$$

Thus, the exponential Markov semi-groups $(e^{tD_\lambda}, t \geqq 0)$ appear as extensions of the exponential semi-groups of the infinitesimal operators criterion. Clearly, what precedes applies to the relativization on an invariant subspace H.

Finally, $D_h = (T_h - I)/h$ given by

$$D_h g(x) = \int (P_h(x, dy) - I(x, dy)) g(y)/h = E^x(g(X_h) - g(x))/h, \quad h > 0,$$

or, to emphasize stationarity,

$$D_h g(X_s) = E\{g(X_{s+h}) - g(X_s) \mid X_s\}/h$$

may be thought of as a "mean speed" operator. If

$$E\{g(X_{s+h}) - g(X_s) \mid X_s\}/h \to Dg(X_s), \quad g \in G',$$

the "speed" operator D on G' appears as an extension of the infinitesimal operators of $(T_t, t \geqq 0)$.

We pursue our specialization to Markov semi-groups. We recall that C denotes the subspace of G formed by bounded continuous functions on the state space $\mathfrak{X}$. *We denote by* C_u *the subspace of uniformly continuous functions on* $\mathfrak{X}$ *and by* C_0 *that of continuous functions on* $\mathfrak{X}$ *vanishing at infinity*, and have

$$C \supset C_u \supset C_0.$$

For any locally compact space $\mathfrak{X}$, $g \in C_0$ means that for every $\epsilon > 0$ there is a compact $K_\epsilon \subset \mathfrak{X}$ such that $|g(x)| < \epsilon$ for $x \notin K_\epsilon$; we write $g(x) \to 0$ as $x \to \infty$. (Note that if $\mathfrak{X}$ is compact there is no "point at infinity.") If $\mathfrak{X}$ is compact, the above condition is void and, also, continuity becomes uniform. Thus

If $\mathfrak{X}$ is compact, then $C = C_u = C_0$.

The restriction of a Markov semi-group on $H \subset G$ is defined by

$$T_t g(x) = \int P_t(x, dy) g(y), \quad g \in H$$

and will be called *Markovian on H*.

a. *The restriction of a Markov semi-group on $H \supset C_0$ determines the semi-group.*

Proof. Upon letting g vary over G in the above integral representation of T_t on H, the semi-group so extended is Markovian. This Markovian extension is unique because T on C_0 determines the tr.pr. $P_t(x, S)$ as follows. Let $[a, b) \subset \mathfrak{X}$ be a bounded interval and take $g_n(x) = 1$ for $a \leqq x < b - 1/n$, $= 0$ for $x \leqq a - 1/n$ and $x \geqq b$, and linear for $a - 1/n \leqq x \leqq a$ and for $b - 1/n \leqq x \leqq b$. Thus, $g_n \in C_0$, $g_n \to I_{[a,b)}$ and, by the dominated convergence theorem,

$$T_t g_n(x) = \int P_t(x, dy) g_n(y) \to P_t(x, [a, b)).$$

Therefore, T_t on C_0 determines $P_t(x, S)$ on the class of all bounded intervals $S = [a, b)$ hence on the Borel field $\mathcal{S}$. The lemma is proved.

A. MARKOV INFINITESIMAL OPERATORS CRITERION. *A linear transformation D on a subset H_d of a Banach subspace $H \supset C_0$ to H is the infinitesimal operator of one and only one Markov semi-group $(T_t, t \geqq 0)$ on G strongly continuous on invariant H, if and only if on H*

(i) *The inverse R_λ of the transformation $\lambda I - D$ exists and λR_λ is a Markov endomorphism for every $\lambda > 0$*

(ii) $\lambda R_\lambda \xrightarrow{s} I$ *on H as $\lambda \to \infty$ or* (ii') *H_d is dense in H.*
Then, on G, the Markov extension λR_λ is determined and the exponential Markov semi-groups $(e^{tD_\lambda}, t \geqq 0)$, where $D_\lambda = \lambda(\lambda R_\lambda - I)$, converge strongly to the semi-group $(T_t, t \geqq 0)$ as $\lambda \to \infty$.

Proof. According to the infinitesimal operators criterion, the proposition holds on H when we drop "Markov" and "$\supset C_0$" therein. Since the semi-group $(T_t, t \geqq 0)$ on H is strongly continuous, hence Borelian, by the foregoing discussion and lemma, from $H \supset C_0$ it follows that we have:

Markov $(T_t, t \geqq 0)$ on $G \Rightarrow$ Markov λR_λ on $H \Rightarrow$ Markov $(e^{tD_\lambda}, t \geqq 0)$ on $H \Rightarrow$ Markov $(e^{tD_\lambda}, t \geqq 0)$ on $G \Rightarrow$ Markov $(T_t, t \geqq 0)$ on G.

The proof is terminated.

The spaces C, C_0, C_u of continuous functions appear in the convergence of laws criteria:

Let $F, f, \hat{f}$, with same affixes if any, denote corresponding d.f.'s, ch.f.'s, and integral ch.f.'s. According to the complete (weak) convergence criterion and the Helly-Bray theorem (extended lemma), we have the equivalences (convergence of d.f.'s is, as usual, up to additive constants);

$$F_n \xrightarrow{\text{c}} F \Leftrightarrow f_n \to f \Leftrightarrow \int g\,dF_n \to \int g\,dF, \quad g \in C$$

$$F_n \xrightarrow{\text{w}} F \Leftrightarrow \hat{f}_n \to \hat{f} \Leftrightarrow \int g\,dF_n \to \int g\,dF, \quad g \in C_0.$$

Note that the integral ch.f.'s correspond to the subset of functions $g \in C_0$ defined by $g(x) = \dfrac{e^{iux} - 1}{ix}$, $u \in R$, and the ch.f.'s correspond to the subset of functions $g \in C$ defined by $g(x) = e^{iux}$, $u \in R$. Since the last subset is also in C_u, we may replace C by C_u in what precedes.

The last equivalent forms of convergence of laws leads to the general concept of *H-convergence* of laws: $F_n \xrightarrow{\text{H}} F$ meaning that $\int g\,dF_n \to \int g\,dF$ for every $g \in H$, and $F_n^x \xrightarrow{\text{H}} F^x$ *uniformly in* x meaning that $\int g\,dF_n^x \to \int g\,dF^x$ uniformly in x for every $g \in H$. But for a stationary regular Markov process X_T with c.pr. $(P^x, x \in \mathfrak{X})$, hence c.d.f.'s $F_t^x(y) = P^x[X_t < y]$, we have

$$\text{(G)} \qquad T_t g(x) = E^x g(X_t) = \int g\,dF_t^x, \quad g \in G.$$

Therefore, to H-convergence of laws correspond continuity properties of its semi-group on H, as follows. Let the limits in h be taken as $0 < h \to 0$ and note that $F_0^x(y) = 0$ or 1 according as $y \leqq x$ or $y > x$. $F_h^x \xrightarrow{\text{H}} F_0^x$ corresponds to $T_h g \xrightarrow{\text{w}} g$ for every $g \in H$, that is, $H \subset G'_c$—

the weak continuity space of our semi-group. Similarly, $F_h^x \xrightarrow{H} F_0^x$ uniformly in x corresponds to $T_h g \xrightarrow{s} g$ for every $g \in H$, that is, $H \subset G_c$—the strong continuity space of our semi-group. This emphasizes the distinguished role of spaces C_0, C_u, C, when the usual weak or complete convergence of laws are considered.

The spaces of continuous functions also appear under "natural" requirements for semi-groups: stability and continuity of evolution. According to the celebrated Hadamard principle, a "well set" evolution problem is stable, in the sense that small variations of initial data lead to small variations in the evolution. The weakest probabilistic interpretation would be in terms of individual laws given the initial state. To be precise, let X_T, $T = [0, \infty)$, be a regular process. We say that the process is *stable* if the laws $\mathfrak{L}(X_t \mid X_0 = x)$ are continuous in x for every t, equivalently, if for every x and every t, as $x' \to x$

$$F_t^{x'} \xrightarrow{c} F_t^x \quad \text{or} \quad f_t^{x'} \to f_t^x \quad \text{or} \quad \int g\, dF_t^{x'} \to \int g\, dF_t^x, \quad g \in C.$$

For a stationary regular Markov process X_T and its semi-group $(T_t, t \geqq 0)$, it suffices to use (G) to obtain

b. *A Markov semi-group leaves C invariant if, and only if, the corresponding Markov process is stable.*

For, if $x' \to x$ and $g \in C$, then stability implies that

$$T_t g(x') = \int g\, dF_t^{x'} \to \int g\, dF_t^x = T_t g(x),$$

and invariance of C implies that

$$\int g\, dF_t^{x'} = T_t g(x') \to T_t g(x) = \int g\, dF_t^x.$$

What precedes remains valid with "w" and "C_0" in lieu of "c" and "C" except that we obtain $T_t g \in C$ in lieu of $T_t g \in C_0$. Thus, in order that C_0 be invariant, we have to add the requirement $T_t g(x') = \int g\, dF_t^{x'} \to 0$ as $x' \to \pm\infty$ for $g \in C_0$, equivalently $F_t^{x'}[a, b) \to 0$ as $x' \to \pm\infty$ for every bounded $[a, b)$. We may interpret the conditions so obtained as *weak stability including infinity* of the process, in the following sense: $F_t^{x'} \xrightarrow{w} F_t^x$ for every t and x including $x = \pm\infty$, upon setting $\operatorname{Var} F_t^{\pm\infty} = P^{\pm\infty}[|X_t| < \infty] = 0$. Thus

b′. *A Markov semi-group leaves* C_0 *invariant if, and only if, the corresponding process is weakly stable including infinity.*

The other "natural" requirement is that of continuous evolution. The weakest probabilistic interpretation would be that of continuity of laws as $h \to 0$: $F_h^x \to F_0^x$, ($F_0^x(y) = 0$ or 1 according as $y < x$ or $y > x$), that is, $P_h(x, (x - \epsilon, x + \epsilon)^c) \to 0$ or $P_h(x, (x - \epsilon, x + \epsilon)) \to 1$ for every $\epsilon > 0$; equivalently $T_h g(x) = \int g \, dF_h^x \to g(x)$, $g \in C$, that is, $C \subset G'_c$. Thus

c. *A Markov semi-group is Φ-weakly continuous on C at t* $= 0$ *if, and only if, the corresponding Markov process is continuous in law at* $t = 0$: $P_h(x, (x - \epsilon, x + \epsilon)) \to 1$ *for every x and every* $\epsilon > 0$.

If we consider C_0 and C_u, then uniform conditions appear. We shall denote by K closed bounded intervals in the state space and set $V_x(\epsilon) = (x - \epsilon, x + \epsilon)$, whenever convenient; in fact, what follows is valid with K interpreted as a compact set.

c′. *A Markov semi-group is strongly continuous on* C_0 *at* $t = 0$ *only if* $P_h(x, (x - \epsilon, x + \epsilon)) \to 1$ *uniformly in* $x \in K$ *for every K and every* $\epsilon > 0$.

Proof. Since every K is compact, it can be covered by a finite number of $V_{x_k}(\epsilon/4)$. Thus every $x \in K$ belongs to some $V_{x_k}(\epsilon/4)$, so that $V_x(\epsilon) \supset V_{x_k}(\epsilon/2)$ and the "only if" assertion reduces to $P_h(x, V_{x_k}(\epsilon/2)) \to 1$ uniformly in $x \in V_{x_k}(\epsilon/4)$. By hypothesis, $T_h g(x) = \int P_h(x, dy) g(y) \to g(x)$ uniformly in x for every $g \in C_0$. Since we can select $g \in C_0$ such that $0 \leqq g \leqq 1$, $g = 1$ on the closure $\bar{V}_{x_k}(\epsilon/4)$ and $g = 0$ on $V_{x_k}^c(\epsilon/2)$, it follows then that, uniformly in $x \in V_{x_k}(\epsilon/4)$,

$$P_h(x, V_{x_k}(\epsilon/2)) \geqq \int P_h(x, dy) g(y) \to g(x) = 1.$$

The assertion is proved.

c″. *A Markov semi-group is strongly continuous on* C_u *at* $t = 0$ *if* $P_h(x, (x - \epsilon, x + \epsilon)) \to 1$ *uniformly in x for every* $\epsilon > 0$.

Proof. Let $g \in C_u$. If $P_h(x, (x - \epsilon, x + \epsilon)) \to 1$ uniformly in x then, from $\| g \| \leqq c < \infty$ and

$$T_h g(x) - g(x)$$

$$= \int_{|y-x|<\epsilon} (g(y) - g(x))\, dF_h^x(y) + \int_{|y-x|\geqq\epsilon} (g(y) - g(x))\, dF_h^x(y)$$
$$+ (\operatorname{Var} F_h^x - 1)g(x),$$

it follows that as $h \to 0$ then $\epsilon \to 0$

$$\sup |\, T_h g(x) - g(x) \,|$$
$$\leqq \sup_{|y-x|<\epsilon} |\, g(y) - g(x) \,| + 3c \sup_x P_h(x, (x - \epsilon, x + \epsilon)^c) \to 0,$$

hence $T_h g \xrightarrow{s} g$.

We combine now the preceding lemmas **b** and **c** into

B. C-INVARIANCE AND CONTINUITY CRITERION. *A Markov semi-group leaves the space C invariant and is Φ-weakly rightcontinuous on it if, and only if, the corresponding process is stable and continuous in law at* $t = 0$: $F_t^{x'} \to F_t^x$ *as* $x' \to x$ *and* $P_h(x, (x - \epsilon, x + \epsilon)) \to 1$ *as* $h \to 0$, *for every x and every* $\epsilon > 0$.

If the state space $\mathfrak{X}$ is compact, then $C = C_u = C_0$ and **b**′, **c**′ and **c**″ apply. In fact,

d. INVARIANCE AND CONTINUITY LEMMA. *Let C be invariant and* $\mathfrak{X}$ *be compact or let* C_0 *be invariant and* $\mathfrak{X}$ *be only locally compact. Then on the invariant subspace, Φ-weak rightcontinuity implies strong continuity.*

Proof. 1° Let C be invariant and $\mathfrak{X}$ be compact. For every $g \in C$, form

$$g_\lambda(x) = \lambda R_\lambda g(x) = \int_0^\infty \lambda e^{-\lambda s} T_s g(x)\, ds, \quad \lambda > 0,$$

and note that $g_\lambda \in C$; for, by C invariance and the dominated convergence theorem, $g_\lambda(x_n) \to g_\lambda(x)$ as $x_n \to x$. In fact, $g_\lambda \in C_c$: By a straightforward computation,

$$T_h g_\lambda(x) = \lambda e^{\lambda h} \int_h^\infty e^{-\lambda s} T_s g(x)\, ds$$

so that, for $g \geqq 0$, $e^{-\lambda h}\, T_h g_\lambda(x) \uparrow g_\lambda(x)$ as $h \downarrow 0$. However, on a compact space monotone pointwise convergence implies uniform convergence (Dini's lemma). Therefore, $T_h g_\lambda \xrightarrow{s} g_\lambda$ for $g \geqq 0$, hence for any $g = g^+ - g^-$.

Since $\mathfrak{X}$ is compact, Φ is adjoint to C. Therefore, by Riesz's representation theorem, any bounded linear functional is of the form

$$\varphi(g) = \int \varphi(dx)g(x), \quad \varphi \in \Phi, \quad g \in C.$$

In particular,

$$\varphi(g_\lambda) = \int_0^\infty \lambda e^{-\lambda s} f(s)\, ds = \int_0^\infty e^{-u} f(u/\lambda)\, du$$

where

$$f(s) = \int \varphi(dx) T_s g(x).$$

When the semi-group is Φ-weakly rightcontinuous, $f(s)$ is right continuous in s so that, as $\lambda \to \infty$,

$$f(s/\lambda) \to f(0) = \varphi(g), \quad \text{hence} \quad \varphi(g_\lambda) \to \varphi(g).$$

Thus, if there exists a $\varphi(\cdot)$ which vanishes on C_c, hence on all g_λ, then it vanishes on all $g \in C$. Therefore, by the Hahn-Banach theorem, $C_c = C$.

2° Let C_0 be invariant and $\mathfrak{X}$ only locally compact. The preceding argument applies upon noting that $g_\lambda(x) \to 0$ as $|x| \to \infty$. Or, to compactify $\mathfrak{X}$ into $\overline{\mathfrak{X}} = \mathfrak{X} + \{\infty\}$, set

$$\overline{P}_t(\infty, \{\infty\}) = 1, \quad \overline{P}_t(\infty, \mathfrak{X}) = 0$$

and, for $x \in \mathfrak{X}$, $S \in \mathcal{S}$, set

$$\overline{P}_t(x, S) = P_t(x, S), \quad \overline{P}_t(x, \{\infty\}) = 1 - P_t(x, \mathfrak{X}).$$

This function is a tr.pr. and the corresponding Markov semi-group $\overline{T}_t$ leaves $\overline{C} = C(\overline{\mathfrak{X}})$ invariant: For, every function $\bar{g} \in \overline{C}$ is of the form $\bar{g} = g + c$ with $g \in C_0$, c constant, and

$$\overline{T}_t g = T_t g, \quad \overline{T}_t c = c.$$

Furthermore, Φ-weak rightcontinuity of T_t on C_0 implies that of $\overline{T}_t$ on $\overline{C}$ and 1° applies. The proof is terminated.

We say that weak stability including infinity is weak stability *uniform at infinity* if at infinity $P_t(x', K) \to 0$ as $x' \to \pm\infty$ for every K uniformly in $t \leqq t_0$ arbitrary but finite.

B′. C_0-INVARIANCE AND CONTINUITY CRITERION. *A Markov semi-group leaves the space C_0 invariant and is strongly continuous on it if, and only if, the corresponding process is weakly stable uniformly at infinity and*

$P_h(x, (x - \epsilon, x + \epsilon)) \to 1$ as $h \to 0$ uniformly in $x \in K$ for every K and every $\epsilon > 0$.

Proof. By **b′** and **c′**, C_0-invariance and weak stability including infinity are equivalent and strong continuity implies the uniform in $x \in K$ condition. Therefore, denoting the strong continuity at $t = 0$ condition by (i), the uniform at infinity condition by (ii), and the uniform in $x \in K$ condition by (iii), it remains to prove that under C_0-invariance, (i) implies (ii) and (ii) and (iii) imply (i). Thus, let C_0 be invariant.

1° Given K, form

$$g_\lambda(x) = \lambda R_\lambda g(x) = \int_0^\infty \lambda e^{-\lambda s} T_s g(x)\, ds, \quad \lambda > 0,$$

where $g \in C_0$ is positive and exceeds 1 on K. If (i) holds, then $g_\lambda \in C_0$ (g_λ is a strong integral),

$$T_t g_\lambda(x') = e^{\lambda t} \int_t^\infty \lambda e^{-\lambda r} T_r g(x')\, dr \leqq e^{\lambda t} g_\lambda(x'),$$

and $g_\lambda(x)$ converges to $g(x)$ uniformly in x as $\lambda \to \infty$ (apply **A** or, directly, set $\lambda s = u$ and note that $T_{u/\lambda} g(x) \to g(x)$ uniformly in x). Therefore, $g_{\lambda_0} \geqq 1$ on K for λ_0 sufficiently large, so that for $t \leqq t_0 < \infty$, as $x' \to \pm\infty$

$$P_t(x', K) \leqq T_t g_{\lambda_0}(x') \leqq e^{\lambda_0 t_0} g_{\lambda_0}(x') \to 0,$$

and (ii) holds.

2° Let (ii) and (iii) hold. To establish strong continuity on C_0 at $t = 0$, it suffices to prove it on the subspace $C_{00} \subset C_0$ of those continuous functions which vanish outside K's. For, given $\delta > 0$, every $g \in C_0$ can be decomposed into $g' + g''$ with $g' \in C_{00}$ and $|g''| < \delta$, so that as $h \to 0$ then $\delta \to 0$,

$$\| T_h g - g \| \leqq \| T_h g' - g' \| + 2\delta \to 0$$

provided $\| T_h g' - g' \| \to 0$. Similarly, because of (ii), $P_h(x, K) \leqq \delta$ for sufficiently small h and for x outside a sufficiently large interval, and we can assume C_{00} invariant under T_h. Thus, we can take every g and $T_h g$ arbitrarily small outside some corresponding K sufficiently large.

Let $x, y \in K$, and let c be a bound of g. Since K is compact, given $\delta > 0$ we can select $\epsilon > 0$ sufficiently small so that $|g(y) - g(x)| < \delta$ for $|y - x| < \epsilon$. It follows, by (iii), that uniformly in x, as $h \to 0$

then $\delta \to 0$,

$$|T_h g(x) - g(x)| \leqq \left| \int_{|y-x|<\epsilon} P_h(x, dy)(g(y) - g(x)) \right|$$
$$+ |g(x)| \{1 - P_h(x, (x - \epsilon, x + \epsilon))\}$$
$$+ \left| \int_{|y-x|\geqq \epsilon} P_h(x, dy) g(y) \right|$$
$$\leqq \delta + 2c\{1 - P_h(x, (x - \epsilon, x + \epsilon))\} \to 0.$$

Thus (i) holds, and the proof is terminated.

We illustrate the foregoing concepts and properties in characterizing the infinitesimal operators of Markov semi-groups which correspond to stationary continuous decomposable processes.

Let $(Z_t, t \geqq 0)$ be a stationary continuous decomposable process. Let F_t and $e^{t\psi}$, $\psi = (\alpha, \Psi)$ be the i.d. d.f.'s and ch.f.'s of the $Y_t = Z_t - Z_0$. Our process is a stationary regular Markov process with $Z_t - Z_0$ independent of Z_0, $P^x = P$ and

$$F_t^x(z) = P(Z_t < z \mid Z_0 = x) = P(Y_t < z - x) = F_t(z - x).$$

It follows that $F_t^{x'} \xrightarrow{c} F_t^x$ as $x' \to x$ and $F_t^x \to 0$ or 1 according as $x \to +\infty$ or $-\infty$, that is, $F_t^x[a, b) \to 0$ for every bounded $[a, b)$; thus, the process is stable and weakly stable at infinity. Also $P_h(x, (x - \epsilon, x + \epsilon)^c) = P[|Y_h| \geqq \epsilon] \to 0$ uniformly in x for every $\epsilon > 0$, for the process is continuous in law. Therefore, according to the preceding propositions, the corresponding semi-group defined by

$$T_t g(x) = \int g(z)\, dF_t(z - x) = \int g(x + y)\, dF_t(y), \quad g \in G$$

leaves the space C_0 (also C) invariant and is strongly continuous on it. Thus, the Markov infinitesimal operators criterion applies, and we can and do limit ourselves to C_0. Let g vary over C_0. The resolvent R_λ is defined by

$$\lambda R_\lambda g(x) = \int_0^\infty \left\{ \int g(x + y)\, dF_t(y) \right\} \lambda e^{-\lambda t}\, dt = \int g(x + y)\, dG_\lambda(y),$$

where the function $G_\lambda = \int_0^\infty \lambda e^{-\lambda t} F_t\, dt$ is a d.f. (weighted by the exponential d.f. family of d.f.'s F_t) with ch.f.

$$g_\lambda(u) = \int_0^\infty \lambda e^{-\lambda t} e^{t\psi(u)}\, dt = \lambda/(\lambda - \psi(u)).$$

The operator D_λ of the infinitesimal operators criterion is given by

$$D_\lambda g(x) = \lambda(\lambda R_\lambda - I)g(x) = \int (g(x+y) - g(x))\lambda\, dG_\lambda(y),$$

and the infinitesimal operator D (in C_0) is the strong limit as $\lambda \to \infty$ of D_λ on its domain of existence. Since, as $\lambda \to \infty$,

$$(\alpha, \Psi) = \psi \leftarrow \lambda\left(\frac{\lambda}{\lambda - \psi} - 1\right) = \int (e^{iuy} - 1)\lambda\, dG_\lambda(y) = \psi_\lambda = (\alpha_\lambda, \Psi_\lambda)$$

where

$$\alpha_\lambda = \int \frac{y}{1+y^2}\lambda\, dG_\lambda(y), \quad d\Psi_\lambda(y) = \frac{y^2}{1+y^2}\lambda\, dG_\lambda(y),$$

the convergence theorem 22.1 **D** applies and

$$\alpha_\lambda \to \alpha, \quad \Psi_\lambda \xrightarrow{c} \Psi.$$

In order to use this convergence property, we make appear α_λ and Ψ_λ in the integral representation of $D_\lambda g(x)$. Proceeding formally, it takes the form

$$D_\lambda g(x) = g'(x)\alpha_\lambda + \int h_g(x, y)\, d\Psi_\lambda(y)$$

where

$$h_g(x, y) = \left(g(x+y) - g(x) - \frac{y}{1+y^2}g'(x)\right)\frac{1+y^2}{y^2}$$

is defined by continuity at $y = 0$ to be $h_g(x, 0) = \frac{1}{2}g''(x)$. Always formally, the convergence property then yields

$$D_\lambda g(x) \to Dg(x) = \alpha g'(x) + \int h_g(x, y)\, d\Psi(y).$$

Thus, we are led to consider the operator D on the set $C''_0 \subset C_0$ of all twice differentiable $g \in C_0$ with $g', g'' \in C_0$. Note that C''_0 is dense in C_0 so that the infinitesimal operator on C''_0 determines the Markov semi-group. We use the preceding notation for the Ito-Neveu theorem

C. *The infinitesimal operators D in C_0 of Markov semi-groups corresponding to stationary continuous decomposable processes are of the form*

$$Dg(x) = \alpha g'(x) + \int h_g(x, y) d\Psi(y), \quad g \in C''_0.$$

Proof. If $g \in C''_0$, then the function $h_g(x, y)$ is defined and is bounded and continuous in x and y with $h_g(x, 0) = \frac{1}{2}g''(x)$ and, by elementary computations,

(i) $h_g(x, y) \to h_g(x, 0)$ uniformly in x, as $y \to 0$
(ii) $h_g(x', y) \to h_g(x, y)$ uniformly in y, as $x' \to x$
(iii) $h_g(x, y) \to 0$ uniformly in $|y| \leqq a$ (arbitrary but finite), as $x \to \pm\infty$.

Using these properties, it suffices to go over the foregoing formal argument to verify that it is valid for $g \in C''_0$.

§46. SAMPLE CONTINUITY AND DIFFUSION OPERATORS

46.1. Strong Markov property and sample rightcontinuity. The basic results to be established here permit us to recognize the strong Markov property and lead to Dynkin's blending of the semi-group and sample analysis, to be performed in the next subsection.

The first basic theorem is a generalized formulation (by Yushkevitch) of the corollary to 38.4**A**.

A. Strong Markov property theorem. *Let $X_T = (X_t, t \geqq 0)$ be stationary Markovian with measurable state space $(\mathfrak{X}, \mathcal{S})$, tr.pr. $P_t(x, S)$, and corresponding semi-group $(T_t, t \geqq 0)$ on the space G of bounded $\mathcal{S}$-measurable functions g on $\mathfrak{X}$.*

If X_T and its tr.pr. are Borelian, then X_T is strongly Markovian whenever there exists a topology in $\mathfrak{X}$ such that the subspace C of functions g continuous in this topology is Φ-weakly dense in G and

(i) *the semi-group $(T_t, t \geqq 0)$ leaves C invariant,*

(ii) *the sample functions $X_T(\omega)$, $\omega \in \Omega$, are rightcontinuous in this topology.*

Proof. Let τ be an arbitrary time of X_T and let $s, t \geqq 0$ be arbitrary degenerate times. Since X_T and its tr.pr. are Borelian, because of the strong Markov equivalence theorem 38.4**A**(ii), it suffices to prove the validity for $\tau < \infty$ of the strong semi-group property

$$T_{\tau+t}g = T_\tau T_t g, \quad g \in G.$$

Thus, in what follows, we restrict ourselves to $\Omega^\tau = [\tau < \infty]$ without further comment.

Let $g \in C$. The elementary times $\tau_n = \sum_{k=1}^{\infty} \frac{k}{2^n} I\left[\frac{k-1}{2^n} \leqq \tau < \frac{k}{2^n}\right]$ converge to τ from the right. Since X_T is sample rightcontinuous (in the selected topology), $X_{\tau_n+s} \to X_{\tau+s}$, so that $g(X_{\tau_n+s}) \to g(X_{\tau+s})$ and, consequently,

$$T_{\tau_n+s}g(x) = E^x g(X_{\tau_n+s}) \to E^x g(X_{\tau+s}) = g(x).$$

The elementary times τ_n are times of X_T, hence, by 38.4**d**, are its Markov times and $T_{\tau_n+t}g = T_{\tau_n}T_t g$. But $T_t g \in C$ hence

$$T_{\tau+t}g \leftarrow T_{\tau_n+t}g = T_{\tau_n}(T_t g) \to T_\tau T_t g.$$

Since C is Φ-weakly dense in G and Markov endomorphisms commute with Φ-weak passages to the limit, it follows that $T_{\tau+t}g = T_\tau T_t g$ for all $g \in G$. The theorem is proved.

Particular cases. The advantage of the foregoing general formulation lies in the freedom of choice of the topology, whether or not there is already one in $\mathfrak{X}$; the state sets $S \in \mathcal{S}$ are not necessarily topological Borel sets in the topology to be selected. The price of freedom is the requirement that C be dense in G and X_T and its tr.pr. be Borelian. This price is reduced when the freedom of choice is restricted, as follows.

1° *If the tr.pr. is Borelian and the state sets are topological Borel sets in a given metric topology in* $\mathfrak{X}$, *then* X_T *is strongly Markovian whenever* (i) *and* (ii) *hold in a topology at least as fine as the given metric one.*

Note that the recalled corollary enters into this particular case upon selecting the given topology.

To pass from the general formulation **A** to the particular case, let C and C' be the subspaces of those functions belonging to G which are continuous in the new topology $\mathcal{O}$ and in the given metric topology $\mathcal{O}'$, respectively. By hypothesis, $\mathcal{O} \supset \mathcal{O}'$ hence $C \supset C'$. But the state sets $S \in \mathcal{S}$ are now topological Borel sets in the metric topology $\mathcal{O}'$ and the functions $g \in G$ are $\mathcal{S}$-measurable. Therefore C' and *a fortiori* C is Φ-weakly dense in G. Also, the sample functions being rightcontinuous in $\mathfrak{O}$ are rightcontinuous in $\mathfrak{O}'$. Right continuity in t and measurability in ω of $X_t(\omega)$ imply that X_T is Borelian

The finest topology in a space is the trivial discrete topology $\mathfrak{D}$—in which all singletons hence all sets are open. With this topology, only condition (ii) remains:

2° *If the sample functions of X_T are rightcontinuous in the discrete topology in the state space, then X_T is strongly Markovian.*

For, in the discrete topology all functions on $\mathfrak{X}$ are continuous, so that the density and invariance conditions are trivially true. Sample right continuity in $\mathfrak{D}$ implies that, for every $x \in \mathfrak{X}$, $P_h(x, \{x\}) \to 1$ as $h \to 0$, hence $P_h(x, S) \to I(x, S)$ and, by the Kolmogorov equation, $P_{t+h}(x, S) = \int P_t(x, dy)P_h(y, S) \to P_t(x, S)$. Rightcontinuity in t and measurability in ω or x imply that X_T and its tr.pr. are Borelian.

The second basic theorem, essentially due to Dynkin, is as follows.

B. MARKOV TIME THEOREM. *Let stationary Markovian X_T and its semi-group $(T_t, t \geqq 0)$ be Borelian, with extended resolvent R_λ on G, and infinitesimal operators D on G_d and D' on G'_d. If τ is a Markov time of X_T then, whatever be $g \in G$,*

$$E^x\{e^{-\lambda\tau}R_\lambda g(X_\tau)\} = E^x\int_\tau^\infty e^{-\lambda u}g(X_u)\,du$$

and, for $g \geqq 0$, $(-e^{-\lambda(\tau+t)}R_\lambda g(X_{\tau+t}),\ t \geqq 0)$ is a submartingale on $(\Omega, \mathfrak{A}, P^x)$.

If, moreover, $E^x\tau < \infty$ then, for $g \in G_d$,

$$E^x g(X_\tau) = g(x) + E^x\int_0^\tau Dg(X_t)\,dt$$

and the same is true with D' and G'_d in lieu of D and G_d.

Proof. We have $T_u g(x) = E^x g(X_u)$ and, upon interchanging the integrations,

$$R_\lambda g(x) = \int_0^\infty e^{-\lambda u}T_u g(x)\,du = E^x\int_0^\infty e^{-\lambda u}g(X_u)\,du.$$

Since a Markov time τ of a stationary Markovian X_T is stationary and

$e^{-\lambda\tau} = 0$ on $[\tau = \infty]$, it follows that

$$E^x\{e^{-\lambda\tau}R_\lambda g(X_\tau)\} = E^x\left\{e^{-\lambda\tau}E\left(\int_0^\infty e^{-\lambda u}g(X_{\tau+u})\,du \,\middle|\, X_\tau\right)\right\}$$

$$= E^x\int_\tau^\infty e^{-\lambda u}g(X_u)\,du,$$

and the first asserted equality is proved. Similarly, for $s \leqq t$,

$$E^x\{e^{-\lambda(\tau+t)}R_\lambda g(X_{\tau+t}) \mid X_{\tau+r}, r \leqq s\}$$

$$= E^x\left\{\int_{\tau+t}^\infty e^{-\lambda u}g(X_u)\,du \mid X_{\tau+r}, r \leqq s\right\}$$

so that, for $g \geqq 0$, the left side is no larger than

$$E^x\left\{\int_{\tau+s}^\infty e^{-\lambda u}g(X_u)\,du \mid X_{\tau+r}, r \leqq s\right\} = e^{-\lambda(\tau+s)}R_\lambda g(X_{\tau+s}),$$

and the submartingale assertion follows.

Finally, by the first asserted equality, for any $g' \in G$,

$$E^x\{e^{-\lambda\tau}R_\lambda g'(X_\tau)\} = R_\lambda g'(x) - E^x\int_0^\tau e^{-\lambda u}g'(X_u)\,du.$$

Therefore, if $g' = (\lambda I - D)g$ with $g \in G_d$ hence $R_\lambda g' = g$, then

$$E^x\{e^{-\lambda\tau}g(X_\tau)\} = g(x) + E^x\int_0^\tau e^{-\lambda u}Dg(X_u)\,du - E^x\int_0^\tau \lambda e^{-\lambda u}g(X_u)\,du.$$

Since $|g| \leqq c < \infty$, the last term is bounded by $cE^x(1 - e^{-\lambda\tau}) \leqq c\lambda E^x\tau$ so that, letting $\lambda \to 0$, if $E^x\tau < \infty$ then

$$E^x g(X_\tau) = g(x) + E^x\int_0^\tau Dg(X_u)\,du.$$

Similarly for D' and G'_d in lieu of D and G_d, and the last assertion is proved. The proof is terminated.

The Markov time theorem applies to all degenerate times $t \geqq 0$ for our stationary Markovian X_T. It applies to all times of X_T, provided X_T satisfies the requirements of theorem **A**. These requirements are of two different kinds: **A**(i) is relative to the corresponding Markov semi-group and, thus, is in terms of the tr.pr. **A**(ii) requires rightcontinuity of sample functions and, thus, is not in terms of the tr.pr. Yet, the primary datum in the investigation and use of Markov property is the tr.pr. This leads us to a search for tr.pr.'s, equivalently, Markov semi-groups, to which correspond sample rightcontinuous X_T.

Given a tr.pr. on a measurable state space $(\mathfrak{X}, \mathcal{S})$, the preliminary question is whether there exists a corresponding X_T. The answer is as follows: According to the existence theorem 43.2**A**, X_T exists when $\mathfrak{X}$ is the Euclidean line and $\mathcal{S}$ is the σ-field of its topological Borel sets. The proof based upon the Tulcea theorem 8.3**A** extends trivially to any finite-dimensional Borel space or Borel subset thereof and, in fact, is valid in the abstract case below.

Let $\mathfrak{X}$ be a separable locally compact metric space with metric "d," and let $\mathcal{S}$ be its σ-field of topological Borel sets; we shall denote by $V_x(\epsilon)$ the sphere $[y: d(x, y) < \epsilon]$. The proof of the separability existence theorem 38.2**B** remains valid and, thus, there exists a separable for closed sets $\tilde{X}_T$ equivalent to X_T, hence Markovian with same tr.pr.

We are now ready for our problem. What follows applies to any state space of the above nature and is couched in corresponding terms. In particular, if x_n goes out of any compact $K \subset \mathfrak{X}$, we write $x_n \to \infty$ and denote by $\mathfrak{X} + \{\infty\}$ the one point compactification of $\mathfrak{X}$ (in the case of a Borel line, "∞" denotes "$\pm\infty$" lumped together). However, as usual, to fix the ideas, we take $(\mathfrak{X}, \mathcal{S})$ to be the Borel line, without further comment. The results are essentially those of Kinney (with modifications due to Blumenthal and Maruyama). We emphasize the methods: the direct method and the martingales method.

From now on, X_T is Borel stationary Markovian and separated for closed sets, with tr.pr. $P_t(x, S)$ and semi-group $(T_t, t \geqq 0)$. As usual, the limits in h are taken as $h \to 0$.

a. SAMPLE LIMITS LEMMA. *Let $T_h g(x) \to g(x)$ for every $x \in \mathfrak{X}$ and every $g \in C_0$.*

If the $T_h g \in C$ or the convergence is uniform in $x \in K$ for every compact K, then almost all sample functions of X_T have at any time at most one left and one right limit value belonging to $\mathfrak{X}$.

If the convergence is uniform in $x \in \mathfrak{X}$, then almost all sample functions of X_T have at any time left and right limits belonging to $\bar{\mathfrak{X}}(= \mathfrak{X}$ or $\mathfrak{X} + \{\infty\}$ according as $\mathfrak{X}$ is compact or not).

Proof. It suffices to give the proof for sample functions which start at any given $x \in \mathfrak{X}$, that is, on the pr. space $(\Omega, \mathcal{A}, P^x)$. Because of separability of X_T, all limits along $[0, \infty)$ may and will be taken along some fixed countable separating set, without further comment. Take $g \geqq 0$.

1° Let $g \in C_0$. By theorem **B**, $(-e^{-\lambda t}R_\lambda g(X_t), t \geqq 0)$ is a semi-martingale. Since it is bounded, the martingales limits theorem (39.1**C**) applies. Thus, neglecting a null event throughout the rest of the proof, at any time all sample functions $R_\lambda g(X_T(\omega))$ have left and right limits for any g and any rational $\lambda > 0$. Since

$$\lambda R_\lambda g(x) = \int_0^\infty \lambda e^{-\lambda t} T_t g(x)\, dt = \int_0^\infty e^{-u} T_{u/\lambda} g(x)\, du,$$

it follows that, as $\lambda \to \infty$, $\lambda R_\lambda g(x) \to g(x)$ for every x or uniformly in $x \in K$, according as $T_h g(x) \to g(x)$ for every x or uniformly in $x \in K$. Let, say, $t' \uparrow t$; what follows applies as well to $t' \downarrow t$.

2° Suppose that $X_{t'}(\omega)$ has distinct limit values $x' \neq x''$ belonging to $\mathfrak{X}$.

Let the $T_h g \in C$. There is a $g \in C_0$ such that $g(x') \neq g(x'')$, hence $\lambda R_\lambda g(x') \neq \lambda R_\lambda g(x'')$ for a sufficiently large rational λ. This contradicts the existence of a unique limit value of $\lambda R_\lambda g(X_{t'}(\omega))$.

Let the convergence be uniform in $x \in K$ for every compact K. There is a $g \in C_0$ with $g = 1$ on K' and $g = 0$ on K'', where K' and K'' are disjoint compact neighborhoods of x' and x'', respectively. Thus, given $\epsilon > 0$, for a sufficient large rational λ independent of $x \in K' + K''$, $|\lambda R_\lambda g(x) - g(x)| < \epsilon$, while there are two sequences $t'_n, t''_n \uparrow t$ such that $g(x'_n) \to 1$, $g(x''_n) \to 0$, where $x'_n = X_{t'_n}(\omega)$, $x''_n = X_{t''_n}(\omega)$. Therefore, as $n \to \infty$, then $\epsilon \to 0$,

$$1 \leftarrow |g(x'_n) - g(x''_n)| \leqq |\lambda R_\lambda g(x'_n) - \lambda R_\lambda g(x''_n)| + 2\epsilon \to 0.$$

and we reach a contradiction.

Finally, let the convergence be uniform in $x \in \mathfrak{X}$. Upon compactifying $\mathfrak{X}$ as in 45.3**d**, the preceding case applies (a fortiori, if $\mathfrak{X}$ is already compact). The proof is terminated.

The first part of **a** yields (use 45.3c for the first hypothesis)

a′. *For every $\epsilon > 0$, let $P_h(x, V_x(\epsilon)) \to 1$ for every $x \in \mathfrak{X}$ and C be invariant, or let $P_h(x, V_x(\epsilon)) \to 1$ uniformly in $x \in K$ for every compact K.*

Then, for almost all $\omega \in \Omega$ and all $t > 0$, as $t' \uparrow t$ and as $t' \downarrow t$, $X_{t'}(\omega)$ converges to some $x \in \mathfrak{X}+\{\infty\}$ or $X_{t'}(\omega)$ has two limit values: $x \in \mathfrak{X}$ and ∞.

When $\mathfrak{X}$ is compact, the point at infinity disappears. Otherwise to eliminate the point at infinity and, thus, have at any time left and right limits belonging to $\mathfrak{X}$, it suffices that the sample functions be bounded. In fact, uniform stability at infinity suffices:

b. *If $P_t(x, K) \to 0$ as $x \to \infty$ uniformly in t on every finite time interval for every compact K, then almost all sample functions of X_T are bounded on every finite time interval.*

If $\mathfrak{X}$ is compact, then the sample functions are necessarily bounded, and we interpret the hypothesis as trivially true.

Proof. Let $t > 0$ and compact $K_n \uparrow \mathfrak{X}$. Suppose there are on $[0, t]$ unbounded sample functions corresponding to an event of pr. δ. It suffices to prove that $P[X_t \in K] \leqq 1 - \delta$ for any given compact K, for then

$$1 - \delta \geqq P[X_t \in K_n] \to P[X_t \in \mathfrak{X}] = 1$$

and $\delta = 0$. Let q_n be the supremum of $P_s(x, K)$ over all $s \leqq t$ and over all $x \notin K_n$; by hypothesis, $q_n \to 0$.

Let $t_0 < \cdots < t_m$ be points of an arbitrary finite set T' in $[0, t]$. Given K_n, let $\tau(\omega) = t_j$ where t_j is the first of these points for which $[X_{t_j}(\omega) \in K_n]$ or $\tau(\omega) = \infty$ if there are no such points. Thus

$$[\tau < \infty] = [X_{t_j} \notin K_n \text{ for some } t_j \in T'],\quad [\tau = \infty] = [X_{t_j} \in K_n \text{ for all } t_j \in T'].$$

The simple time τ is a time of X_T, hence is its Markov time. Therefore,

$$P[\tau < \infty, X_t \in K] = E\{I_{[\tau<\infty]}P_{t-\tau}(X_\tau, K)\} \leqq q_nP[\tau < \infty]$$

and

$$P[X_t \in K] \leqq P[\tau = \infty] + q_n.$$

Apply the standard separability procedure: take a sequence of finite subsets T' of a separating set of $X_{[0,t]}$ converging increasingly to this set. It follows that

$$P[X_t \in K] \leqq P[X_s \in K_n, 0 \leqq s \leqq t] + q_n \leqq 1 - \delta + q_n \to 1 - \delta,$$

and the assertion is proved. The proof is terminated.

REMARK. The above method of proof is direct. But we may also use the martingales method, as follows: For positive $g \in C_0$,

$$T_tg(x) = \int_{K+K^c} P_t(x, dy)g(y) \leqq cP_t(x, K) + \sup_{y \notin K} g(y)$$

so that, by hypothesis, $T_tg(x) \to 0$ as $x \to \infty$. It follows that the unbounded sample functions on $[0, a]$ correspond to the event $A = [\inf_{t \leqq a} R_\lambda g(X_t) = 0]$ and, by 39.1**b**(ii) applied to the submartingale formed by $-e^{-\lambda t}R_\lambda g(X_t)$, $\int_A e^{-\lambda a}R_\lambda g(X_a)dP = 0$. The integrand is positive so that $PA = 0$.

Under the condition of **b** and either one of the conditions of **a′**, almost all sample functions of X_T have at any time left and right limits belonging to $\mathfrak{X}$. Since X_T is separable for closed sets, almost all sample functions are continuous except for countable sets of strict jumps. Furthermore, the weakest condition: $P_h(x, V_x(\epsilon)) \to 1$ for every $x \in \mathfrak{X}$ and every $\epsilon > 0$ implies that X_T (that is, every r.f. belonging to it) is right continuous in pr., since for every x

$$P^x[d(X_t, X_{t+h}) \geqq \epsilon] = \int P_t(x, dy)P_h(y, V_y^c(\epsilon)) \to 0,$$

hence for every initial distribution P_0

$$P[d(X_t, X_{t+h}) \geqq \epsilon] = \int P_0(dx)P^x[d(X_t, X_{t+h}) \geqq \epsilon] \to 0.$$

Thus, if S separates X_T and we set $\tilde{X}_t = X_t$ for $t \in S$ and $\tilde{X}_t = \lim_{t' \downarrow t} X_{t'} (t' \in S)$ for $t \notin S$, ($\tilde{X}_t = \lim_{t' \uparrow t} X_{t'}$ for $t \notin S$), then $\tilde{X}_T$ separated for closed sets by S is equivalent to X_T. Hence, $\tilde{X}_T$ has same tr.pr. and almost all its sample functions are right (left) continuous except perhaps at those points of S which are fixed discontinuity points of X_T (where $X_t(\omega)$ may coincide with $X_{t+0}(\omega)$ or with $X_{t-0}(\omega)$ according to the choice of ω); to eliminate this last obstacle for sample right (left) continuity, we may replace the X_t by the $X_{t+0}(X_{t-0})$ and, in fact, a.s. sample continuity may also be imposed upon X_T using 38.3**A** and the particular cases which follow it. However, it will be more instructive to give a direct answer (which overlaps the preceding lemmas) to the problem of sample right (left) continuity.

C. MARKOV SAMPLE RIGHTCONTINUITY THEOREM. *Let a separated for closed sets X_T be stationary Markovian with tr.pr. $P_t(x, S)$. Let either of the two following conditions hold: as $h \to 0$*

(i) $P_h(x, V_x(\epsilon)) \to 1$ *uniformly in $x \in \mathfrak{X}$ for every $\epsilon > 0$*

(ii) $P_h(x, V_x(\epsilon)) \to 1$ *uniformly in $x \in K$ for every compact K and every $\epsilon > 0$, and $P_t(x, K) \to 0$ as $x \to \infty$ uniformly in t on every finite time interval for every compact K.*

Then, X_T is a.s. continuous, almost all its sample functions are continuous except for countable sets of strict jumps, and there exists a separated for closed sets equivalent X_T with almost all sample functions right (left) continuous and left (right) limits for every t.

Proof. Let T' be a time subset with first element t_0. Let $\epsilon > 0$ and $k = 1, \cdots, n$. Let $\omega \in A_n(T') \Leftrightarrow X_T(\omega)$ has n oscillations greater than ϵ on T', that is, there exist n pairs (t'_k, t''_k) of points $t'_k < t''_k \leqq t'_{k+1}$ of T' such that $d(X_{t'_k}(\omega), X_{t''_k}(\omega)) > \epsilon$. For every $\omega \in \Omega$, let $\tau_0(\omega) = t_0$ and let $\tau_k(\omega)$ be the first point t of T' following $\tau_{k-1}(\omega)$ for which $d(X_t(\omega), X_{\tau_{k-1}(\omega)}(\omega)) > \epsilon/2$ or $\tau_k(\omega) = \infty$ if there is no such point. We have

$$A_n(T') \subset B_n(T') = [\tau_0 < \cdots < \tau_n < \infty]$$

and if T' is a finite set, then the τ_k are simple times of X_T, hence are its Markov times.

1° Suppose (i) holds: Given $\epsilon, \delta > 0$, there exists an $h = h(\epsilon, \delta) > 0$ such that $P_{h'}(x, V_x^c(\epsilon/4)) < \delta$ for all $h' \leqq h$ and all $x \in \mathfrak{X}$. Let $I = [a, b]$ be an interval of length h, set $P_\tau^x C = P(C \mid X_\tau = x)$, and let p_n be the supremum of $P_{\tau_0}^x[\tau_0 < \cdots < \tau_n < \infty]$ over all finite sets $T' \subset I$ and over all $x \in \mathfrak{X}$. For any such T'

$$P_{\tau_0}^x[\tau_0 < \cdots < \tau_n < \infty] = E_{\tau_0}^x\{I_{[\tau_0 < \tau_1 < \infty]} P(\tau_1 < \cdots < \tau_n < \infty \mid X_{\tau_1})\}$$

so that $p_n \leqq p_{n-1} p_1$ and, by induction, $p_n \leqq p_1^n$. Furthermore,

$$P_{\tau_0}^x[\tau_0 < \tau_1 < \infty, X_b \in V_x(\epsilon/4)]$$
$$\leqq E_{\tau_0}^x\{I_{[\tau_0 < \tau_1 < \infty]} P(X_b \notin V_{X_{\tau_1}}(\epsilon/4) \mid X_{\tau_1})\} < \delta,$$

so that

$$P_{\tau_0}^x[\tau_0 < \tau_1 < \infty] < \delta + P_{\tau_0}^x[X_b \notin V_x(\epsilon/4)] < 2\delta,$$

hence $p_1 < 2\delta$ and $p_n < (2\delta)^n$. Therefore, upon applying the standard separability procedure,

$$PA_n(I) \leqq PB_n(I) < (2\delta)^n.$$

Thus, taking $n = 1$,

$$P\,[\sup_{t', t'' \in I} d(X_{t'}, X_{t''}) > \epsilon] < 2\delta,$$

and $a, \epsilon, \delta > 0$ being arbitrary, it follows that X_T is a.s. continuous. On the other hand, taking $\delta < \frac{1}{2}$,

$$\sum_{n=1}^{\infty} PA_n(I) \leqq \sum_{n=1}^{\infty} PB_n(I) < \sum (2\delta)^n < \infty$$

so that, by the Borel-Cantelli lemma, on I almost all sample functions of X_T have only a finite number of oscillations greater than ϵ and of consecutive oscillations greater than $\epsilon/2$. Since every finite time interval $[0, c]$ is covered by a finite number of intervals of positive length h, it follows that on $[0, c]$ almost all sample functions of X_T are bounded and have left and right limits. The equivalence assertion results then from the discussion which precedes the theorem.

2° Suppose (ii) holds. In fact, in lieu of its second part, we suppose only the boundedness conclusion of **b**. The preceding argument applies but with restrictions to suitable compacts. First, given ϵ, $\delta > 0$ and $[0, c]$, take $c' > c$ and select a compact K so that the set of sample functions which do not stay in K on $[0, c']$ corresponds to an event of pr. less than δ (thus $P[X_t \in K] > 1 - \delta$ for all $t \leqq c'$); then take a compact K' containing all $V_x(\epsilon/2)$, $x \in K$ (decreasing ϵ if necessary). There exists an $h = h(\epsilon, \delta, K') > 0$ such that $P_{h'}(x, V_x^c(\epsilon/4)) < \delta$ for all $h' \leqq h$ and all $x \in K'$; decrease h if necessary so that $h < c' - c$ and a finite number of intervals $I \subset [0, c']$ covers $[0, c]$.

Let p_n be the supremum of $P_{\tau_0}^x[\tau_0 < \cdots < \tau_n < \infty, X_{\tau_0} \in K, \cdots, X_{\tau_{n-1}} \in K]$ over all finite sets $T' \subset I$ and over all $x \in K$. Upon proceeding as in (i), the relation $p_n \leqq p_1^n$ is still valid. Similarly, for $x \in K$,

$$P_{\tau_0}^x[\tau_0 < \tau_1 < \infty, X_{\tau_0} \in K, X_{\tau_1} \in K', X_b \in V_x(\epsilon/4)] < \delta$$

so that

$$P_{\tau_0}^x[\tau_0 < \tau_1 < \infty, X_{\tau_0} \in K]$$
$$< \delta + P_{\tau_0}^x[X_{\tau_1} \notin K'] + P_{\tau_0}^x[X_b \notin V_x(\epsilon/4)] < 3\delta,$$

hence $p_1 < 3\delta$ and $p_n < (3\delta)^n$.

Upon proceeding as in (i) and using $P[\tau_0 < \tau_1 < \infty] \leqq P[\tau_0 < \tau_1 < \infty, X_{\tau_0} \in K] + P[X_{\tau_0} \notin K]$ it follows that

$$P[\sup_{t',t'' \in I} d(X_{t'}, X_{t''}) > \epsilon] \leqq p_1 P[X_a \in K] + P[X_a \notin K] < 4\delta,$$

and X_T is a.s. continuous on $[0, c]$; since c is arbitrarily large, X_T is a.s. continuous. Similarly, on $[0, c]$, almost all of those sample functions which stay in K have left and right limits; since the others correspond to an event of pr. less than δ and δ is arbitrarily small, the restriction to K can be removed. The equivalence assertion follows. Quasileftcontinuity, page 383, completes this subsection.

46.2. Extended infinitesimal operator. Let $X_T = (X_t, t \geqq 0)$ be stationary Markovian with Borel tr.pr. $P_t(x, S)$ and corresponding Borel semi-group $(T_t, t \geqq 0)$ on the space G of bounded measurable functions on a measurable state space with metric "d." Let D on G_d and D' on G'_d be the strong and the weak infinitesimal operators of the semi-group (we also say "of X_T"). We intend to solve in $D'g$ the integral relation (46.1B)

$$E^x g(X_\tau) - g(x) = E^x \int_0^\tau D'g(X_t)\,dt$$

where τ is a Markov time of X_T with $E^x\tau < \infty$, and D' can be replaced by D.

If τ is an ordinary time h, then, dividing both sides by h and letting $h \to 0$, we fall back upon the definition of D'. Yet, this is a "ready-to-wear" approach, in the sense that sample properties of X_T are not taken into account. On the other hand, if we use a random time τ of X_T defined in terms of its sample properties, the approach is fitted to the process—it is "made to measure." The analogy with Riemann versus Lebesgue integration is visible but not adequate. For, τ must be a Markov time of X_T and the approach becomes restricted to strong Markov processes. Furthermore, loosely speaking, the tailored τ would be the time X_T spends in smaller and smaller neighborhoods of x. To be precise, we take for τ the time τ_U that X_T takes to hit the complement of an open neighborhood U of x and let its diameter $|U| \to 0$. According to 43.4**c**, τ_U is a time of X_T when X_T is sample rightcontinuous. Thus from now on, *X_T is strongly Markovian and sample right continuous*.

The integral relation also requires that $E^x\tau_U$ be finite. The simple conditions below will suffice.

Let U, with or without subscripts, denote open sets and, given U_0, set $m(y) = E^y\tau_{U_0}$, $y \in \mathfrak{X}$.

a. *If there exist $s > 0$, $\delta > 0$ such that $P^x[\tau_{U_0} \leq s] > \delta$ for all $x \in U_0$, then $m(x) = E^x\tau_{U_0} < s/\delta < \infty$ for all $x \in U_0$. If $x \in U \subset U_0$ and $m(x) < \infty$, then $E^x\tau_U < \infty$ and $m(x) = E^x\tau_U + E^x m(X_{\tau_U})$.*

Proof. Let $A_t = [\tau_{U_0} > t]$. Since $A_{t+s} = A_t(A_s)_t$, where $(A_s)_t$ is the translate by t of A_s, and $X_t(\omega) \in U$ for $\omega \in A_t$, the first hypothesis yields

$$P^x A_{t+s} = E^x(I_{A_t} P^{X_t} A_s) < (1 - \delta) P^x A_t.$$

It follows, by induction, that $P^x A_{ns} < (1 - \delta)^n$ and the first assertion is proved by

$$E^x\tau_{U_0} \leq \sum_{n=0}^{\infty} \int_{ns}^{(n+1)s} P^x[\tau_{U_0} > t]\,dt < s \sum_{n=0}^{\infty} P^x A_{ns} < s/\delta.$$

Since $U \subset U_0$, it follows that $\tau_U \leqq \tau_{U_0}$ and the translate by τ_U of τ_{U_0} is $\tau_{U_0} - \tau_U$. Therefore, by the second hypothesis, $E^x\tau_U \leqq E^x\tau_{U_0} = m(x) < \infty$ and

$$E^x(\tau_{U_0} - \tau_U) = E^x(E^{X_{\tau_U}}\tau_{U_0}) = E^x m(X_{\tau_U}),$$

hence

$$m(x) = E^x\tau_U + E^x(\tau_{U_0} - \tau_U) = E^x\tau_U + E^x m(X_{\tau_U}),$$

The second assertion is proved, and the proof is terminated.

Let U denote open neighborhoods of x, and set

$$\tilde{D}g(x) = \lim_{|U|\to 0} \frac{E^x g(X_{\tau_U}) - g(x)}{E^x\tau_U} \text{ if } E^x\tau_U < \infty \text{ for some } U,$$

$$\tilde{D}g(x) = 0 \text{ if } E^x\tau_U = \infty \text{ for all } U;$$

it suffices to use the first form with the convention that the ratio therein is 0 when $E^x\tau_U = \infty$.

Denote by $\tilde{G}_d$ the set of those functions $g \in G$ for which $\tilde{D}g$ exists and belongs to G. We say that $\tilde{D}$ on $\tilde{G}_d$ is the *extended infinitesimal operator* of X_T or of its semi-group. Note that by **a**, setting $m(z) = E^z\tau_{U_0}$, $U \subset U_0$, and $P_{\tau_U}(x, S) = P^x[X_{\tau_U} \in S]$ hence $T_{\tau_U}g(x) = \int P_{\tau_U}(x, dy)g(y)$, we have

$$\frac{E^x g(X_{\tau_U}) - g(x)}{E^x\tau_U} = \frac{T_{\tau_U}g(x) - g(x)}{E^x\tau_U} = -\frac{\int P_{\tau_U}(x, dy)\{g(y) - g(x)\}}{\int P_{\tau_U}(x, dy)\{m(y) - m(x)\}}$$

with the same convention as above.

According to the integral relation, when some $E^x\tau_U < \infty$ then

$$\text{(C)} \qquad \tilde{D}g(x) = \lim_{|U|\to 0} \frac{1}{E^x\tau_U} E^x \int_0^{\tau_U} D'g(X_t)\, dt, \quad g \in G'_d,$$

in the sense that if either of the sides exists so does the other and both are equal. Thus, if the right side is $D'g(x)$ then $\tilde{D}g(x) = D'g(x)$ and our problem may be enlarged into a search for conditions under which the last equality holds, whether or not $E^x\tau_U$ are finite.

The equality is trivially true when x is an *absorbing* state, that is, almost all sample functions which start at x stay there forever. For then, $E^x\tau_U = \infty$ for all U and $\tilde{D}g(x) = 0$ by definition, while $P^x[X_t = x] = 1$ for all t implies that $T_t g(x) = g(x)$ for all t and $D'g(x) = 0$ by definition.

In fact, we may expect the equality to hold when almost all sample functions stay at x for some positive time. Similarly when $D'g$ is continuous at x provided $E^x\tau_U < \infty$ for some U. For then, given $\epsilon > 0$, $|D'g(y) - D'g(x)| < \epsilon$ for all $y \in U$ with $|U|$ sufficiently small, hence $|D'g(X_t) - D'g(x)| <$ for $t < \tau_U$ and

$$\left| E^x \int_0^{\tau_U} (D'g(X_t) - D'g(x))\, dt \right| < \epsilon E^x \tau_U.$$

However, we are concerned with the process and not with some of its r.f.'s such as those which start at some specified state. Then the required continuity of $D'g$ leads to considering "stable processes" and the required time interval of constancy leads to considering "jump processes." We intend to show that in either case, the term "extended infinitesimal operator" is justified, in the sense that $\tilde{D} \supset D'$, that is, the domain of $\tilde{D}$ under consideration contains that of D'; since always $D' \supset D$, we shall have $\tilde{D} \supset D' \supset D$.

We say that a stationary Markovian process X_T is a *jump process* if, for every $\omega \in \Omega$ and $t \geqq 0$, there exists an $h_0 > 0$ such that $X_t(\omega) = X_{t+h}(\omega)$ for $0 \leqq h < h_0$, equivalently, if X_T is sample rightcontinuous in the discrete topology (in the state space). Since our concern here is with stationary Markov processes, we reserve the term "jump process" for such processes. It follows at once from the definition that for a jump process X_T

$$P_h(x, \{x\}) \to 1, \quad P_h(x, S) \to I(x, S),$$

$$P_{t+h}(x, S) \to P_t(x, S), \quad T_{t+h}g(x) \to T_t g(x),$$

and X_T and its tr.pr. are Borelian. Furthermore, according to 46.1**A**(2°), X_T is strongly Markovian.

Let τ_1 be the first *jump time* of X_T: $\tau_1(\omega)$ is the time $X_t(\omega)$ first hits the complement of the singleton $\{X_0(\omega)\}$—the smallest open neighborhood of the state $X_0(\omega)$ in the discrete topology. The time τ_1 of the stationary strongly Markovian X_T is its stationary Markov time and $A_{t+s} = A_s(A_t)_s$ where $A_t = [\tau_1 > t]$. Since $X_s(\omega) = X_0(\omega)$ for $\omega \in A_s$, it follows that

$$P^x A_{t+s} = E^x(I_{A_s} P^{X_s} A_t) = P^x A_t P^x A_s,$$

and the nonnegative bounded nonincreasing function $p^x(t) = P^x A_t$ obeys the classical functional equation $p^x(t + s) = p^x(t)p^x(s)$, $t, s \geqq 0$. Therefore, $P^x[\tau_1 > t] = e^{-q(x)t}$, $q(x) \geqq 0$, and $q(x) < \infty$ since $q(x) = \infty \Leftrightarrow P^x[\tau_1 > 0] = 0$ (x is *instantaneous*)—contrary to the definition of a jump process. Thus $E^x\tau_1 = 1/q(x) > 0$ and $E^x\tau_1 = \infty \Leftrightarrow q(x) = 0$ $\Leftrightarrow x$ is absorbing. To summarize

b. *If X_T is a jump process and τ_1 is its first jump time, then*

$$P^x[\tau_1 > t] = e^{-q(x)t}, \quad 0 \leqq q(x) < \infty, \quad x \in \mathfrak{X}, \quad t \geqq 0,$$

and $E^x\tau_1 = 1/q(x) > 0$ is infinite or finite according as x is absorbing or is not.

In the discrete topology, with metric d defined by $d(x, y) = 0$ or 1 according as $x = y$ or $x \neq y$, every set $U \ni x$ is an open neighborhood of x and its diameter $|U| = 0$ or 1 according as U reduces to the singleton $\{x\}$ or has other points besides x. Therefore, we have to set $\tau_U = \tau_1$ in the definition of $\tilde{D}$ and, using **b**, it becomes

$$\tilde{D}g(x) = \frac{E^x g(X_{\tau_1}) - g(x)}{E^x\tau_1} = q(x)\int P_{\tau_1}(x, dy)\{g(y) - g(x)\}.$$

On the other hand, since $X_t = X_0$ for $t < \tau_1$,

$$q(x)E^x\int_0^{\tau_1} D'g(X_t)\, dt = D'g(x), \quad g \in G'_d.$$

Thus

$$\tilde{D}g(x) = D'g(x), \quad g \in G'_d$$

(whether x is absorbing or not), and

A. JUMP PROCESSES EXTENSION THEOREM. *If X_T is a jump process then $\tilde{D} \supset D' \supset D$ and, for every $x \in \mathfrak{X}$,*

$$\tilde{D}g(x) = q(x)\int P_{\tau_1}(x, dy)\{g(y) - g(x)\}, \quad g \in \tilde{G}_d \supset G'_d \supset G_d.$$

Note that for jump processes $\tilde{D}$ is an integral operator, and if the function $q(\cdot)$ is bounded then $\tilde{G}_d = G$.

Stable stationary Markov processes have been defined and investigated in 45.3: Their semi-groups $(T_t, t \geqq 0)$ leave the space C invariant, that is, transform bounded continuous functions g into bounded continuous functions T_tg, for every t. It suffices to consider these semi-groups on the invariant subspace C. Thus, the domains of D, D', $\tilde{D}$ are to be replaced by their intersections C_d, C'_d, $\tilde{C}_d$ with C. In defining $\tilde{D}$, we restricted ourselves to stationary Markov processes with Borel tr.pr. and sample rightcontinuity, for short to *rightcontinuous* processes. We also required strong Markov property, but (as in the case of jump processes) this requirement is superfluous in the case of stable processes;

according to 41.1**A**(1°), rightcontinuous stable processes are strongly Markovian. Since the domains of the corresponding semi-groups are restricted to C, hence all $D'g$ are continuous, the discussion which follows the definition of $\tilde{D}$ shows that then $\tilde{D} \supset D' \supset D$, provided $E^x\tau_U < \infty$ for nonabsorbing states $x \in U$. In fact

c. *Let X_T be stable. If $P^x[X_s \in U] = \delta > 0$ for some $s > 0$ and open U, then $P^y[X_s \in U] > \delta/2$ for some open $U_0 \ni x$ and all $y \in U_0$.*

Let X_T be stable rightcontinuous: If x is nonabsorbing, then $m(y) = E^y\tau_{U_0} < \infty$ for some open $U_0 \ni x$ and all $y \in U_0$.

Proof. Set $V_n = [z: d(z, U^c) \geq 1/n]$, so that $V_n \uparrow U$ and $P^x[X_s \in V_n] \to P^x[X_s \in U] = \delta$. Thus, for $\delta > 0$ there is an n such that $P^x[X_s \in V_n] > 3\delta/4$. For this n, define the bounded continuous function g by $g(y) = nd(y, U^c)$ or 1 according as $y \notin V_n$ or $y \in V_n$. If X_T is stable, then $g' = T_s g$ is also bounded and continuous and there exists $U_0 \ni x$ such that $g'(y) > g'(x) - \delta/4$ for $y \in U_0$. Then, from

$$g'(x) = \int P_s(x, dy)g(y)$$

and the definition of g, it follows that for $y \in U_0$.

$$P^y[X_s \in U] > g'(y) > g'(x) - \delta/4 > P^x[X_s \in V_n] - \delta/4 > \delta/2.$$

The first assertion is proved.

Let X_T be stable right continuous. If x is nonabsorbing, then there exists an open U with $d(x, U) > 0$ and $P^x[X_s \in U] = \delta > 0$ for some $s > 0$ and, by the first assertion, there exists an open $U_0 \ni x$ that we can take disjoint from U, such that $P^y[X_s \in U] > \delta/2$ for all $y \in U_0$. Therefore, for all $y \in U_0$,

$$P^y[\tau_{U_0} > s] < P^y[X_s \in U_0] < 1 - \delta/2$$

and, by **a**, $m(y) = E^y\tau_{U_0} < \infty$. The second assertion is proved, and the proof is terminated.

Let $m(z) = E^z\tau_{U_0}$, $U \subset U_0$, and recall that $P_{\tau_U}(x, S) = P^x[X_{\tau_U} \in S]$.

B. Rightcontinuous stable processes extension theorem. *If X_T is stable right continuous, then $\tilde{D} \supset D' \supset D$ and*

$$\tilde{D}g(x) = -\lim_{|U| \to 0} \frac{\displaystyle\int P_{\tau_U}(x, dy)\{g(y) - g(x)\}}{\displaystyle\int P_{\tau_U}(x, dy)\{m(y) - m(x)\}}, \quad g \in \tilde{C}_d,$$

or $\tilde{D}g(x) = 0$ *according as* x *is nonabsorbing or absorbing; if the state space is compact, then* $\tilde{D} = D' = D$.

Proof. The inclusion and limit assertions result from the discussion which follows the definition of $\tilde{D}$ and the restriction of the semi-group on C.

To prove the last relation, note that, by 40.3**B** and **d**, $D' = D$ and $C'_c = C_c = C$. Since $\tilde{D} \supset D'$, it suffices to prove that $\tilde{D} \subset D'$, that is, $\tilde{C}_d \subset C'_d$. Let $g \in \tilde{C}_d$ and set

$$\tilde{g} = (I - \tilde{D})g, \quad g' = R_1\tilde{g} = \int_0^\infty e^{-t}T_t\tilde{g}\,dt.$$

Since $\tilde{g} \in C = C'_c$, it follows that $g' \in C'_d$ and $\tilde{g} = (I - D')g'$. Thus, $\tilde{g} = (I - \tilde{D})g'$ because of $\tilde{D} \supset D'$. Therefore,

$$(I - \tilde{D})f = 0, \quad f = g - g'.$$

$\mathfrak{X}$ being compact, f attains its supremum on $\mathfrak{X}$ for some x_0 and the defining relation for $\tilde{D}$ implies that $\tilde{D}f(x_0) \leqq 0$, hence $f \leqq 0$; similarly, $-f \leqq 0$. Thus, $f = 0$, that is, $g = g' \in C'_d$. The proof is terminated.

We say that X_T is a *continuous stable* process if it is stable rightcontinuous and also sample leftcontinuous. The sample functions of a continuous stable process being continuous, the time τ_U is the time X_T first reaches the closed set U^c. Therefore, X_{τ_U} belongs to the boundary U' of U and the integrals in **B** can be taken over U' only:

$$\tilde{D}g(x) = -\lim_{|U| \to 0} \frac{\int_{U'} P_{\tau_U}(x, dy)\{g(y) - g(x)\}}{\int_{U'} P_{\tau_U}(x, dy)\{m(y) - m(x)\}}.$$

This expression is similar to the one which gives the ordinary Laplacian operator in terms of averages on spherical surfaces, except that there the averaging is with respect to a uniformly distributed measure. This leads to considering the foregoing operator as a generalized elliptic differential operator of second order or, in physical terms, a (*general*) *diffusion operator*, to be denoted by $\mathfrak{D}$. According to the convention made, we have, for every $x \in \mathfrak{X}$ and $g \in \tilde{C}_d$,

$$\mathfrak{D}g(x) = \lim_{|U| \to 0} \frac{1}{E^x\tau_U} \int_{U'} P_{\tau_U}(x, dy)\{g(y) - g(x)\}.$$

It follows that if $g = g'$ in a neighborhood of x then $\mathfrak{D}g(x) = \mathfrak{D}g'(x)$, and if g attains a relative minimum at x, then $\mathfrak{D}g(x) \geqq 0$. The foregoing terminology is further justified because, on sufficiently smooth functions, $\mathfrak{D}$ can be written as an ordinary elliptic differential operator, as follows.

C. DIFFUSION OPERATORS THEOREM. *Let $\mathfrak{D}$ be a diffusion operator and let $g_k, g_jg_k, j = 1, \cdots, n$, belong to the domain $\tilde{C}_d$ of $\mathfrak{D}$, in some neighborhood of a state x.*

If $f(y_1, \cdots, y_n)$ is twice continuously differentiable in a neighborhood of $(g_1(x), \cdots, g_n(x))$, then $\mathfrak{D}f(g_1(x), \cdots, g_n(x))$ exists and equals

$$\sum_{k=1}^{n} a_k \frac{\partial f}{\partial g_k} + \sum_{j,k=1}^{n} b_{jk} \frac{\partial^2 f}{\partial g_j \partial g_k}$$

where the derivatives are taken at $(g_1(x), \cdots, g_n(x))$,

$$a_k = \mathfrak{D}(g_k - g_k(x))(x), \quad b_{jk} = \mathfrak{D}(g_j - g_j(x))(g_k - g_k(x))(x)$$

and the b_{jk} form a nonnegative type matrix.

Proof. We have

$$f(g_1, \cdots, g_n) - f(g_1(x)), \cdots, g_n(x))$$
$$= \sum_k \frac{\partial f}{\partial g_k} h_k + \sum_{j,k} \frac{\partial^2 f}{\partial g_j \partial g_k} (1 + \delta_{jk}) h_j h_k$$

where the derivatives are taken at $(g_1(x), \cdots, g_n(x))$, $h_k = g_k - g_k(x)$ and $\delta_{jk}(x') = \delta_{jk}(g_1(x'), \cdots, g_n(x')) \to 0$ as the $h_k \to 0$, hence as $x' \to x$. Thus, the ratio in the defining expression of $\mathfrak{D}$ can be written as a sum of three terms: As $|U| \to 0$, the first term

$$\sum_k \frac{\partial f}{\partial g_k} \frac{1}{E^x \tau_U} \int_{U'} P_{\tau_U}(x, dy) h_k(y) \to \sum_k \frac{\partial f}{\partial g_k} a_k,$$

the second term

$$\sum_{j,k} \frac{\partial^2 f}{\partial g_j \partial g_k} \frac{1}{E^x \tau_U} \int_{U'} P_{\tau_U}(x, dy) h_j(y) h_k(y) \to \sum_{j,k} \frac{\partial^2 f}{\partial g_j \partial g_k} b_{jk},$$

and the squares of the summands of the third term

$$\left| \frac{1}{E^x \tau_U} \int_{U'} P_{\tau_U}(x, dy) h_j(y) h_k(y) \delta_{jk}(y) \right|^2$$

$$\leqq \max_{y \in U'} |\delta_{jk}(y)|^2 \left| \frac{1}{E^x \tau_U} \int_{U'} P_{\tau_U}(x, dy) h^2{}_j(y) \right| \left| \frac{1}{E^x \tau_U} \int_{U'} P_{\tau_U}(x, dy) h^2{}_k(y) \right|$$

converge to 0, since the first factor converges to zero and the two others are bounded. Finally

$$\sum_{j,k} b_{jk}\lambda_j\lambda_k = \mathfrak{D}(\sum_k \lambda_k h_k(x))^2 \geqq 0,$$

since the function on which $\mathfrak{D}$ operates attains a relative minimum at x. The proof is terminated.

To conclude this general discussion of the extended infinitesimal operator, let us mention that whenever it is a true extension of the weak infinitesimal operator, the last one is obtained by supplying further information. This information usually takes the form of boundary conditions. Thus, loosely speaking, the extended infinitesimal operator describes the behavior of the process *before* it reaches the boundary of the domain on which it is considered.

46.3. One-dimensional diffusion operator. In the one-dimensional case, the diffusion operator takes a specific differential form, and we proceed to establish this fundamental result of Feller, following Dynkin.

Let X_T be continuous stable with a one-dimensional interval state space $\mathfrak{X} = [\alpha, \beta]$ and the σ-field $\mathcal{S}$ of topological Borel sets in it. We take $U = (x_1, x_2) \ni x$; its boundary consists of the two endpoints to which correspond the pr.'s $p_{x_1} = P^x[X_{\tau(x_1,x_2)} = x_1]$ and $p_{x_2} = P^x[X_{\tau(x_1,x_2)} = x_2]$. Let $g \in \tilde{C}_d$. We know that

I. *If x is absorbing then* $\mathfrak{D}\,g(x) = 0$.

If x_0 is not absorbing, we can select $(x_1', x_2') \ni x_0$ so that, for all $x \in (x_1', x_2')$, $m(x) = E^x\tau_{(x_1',x_2')}$ is finite and

$$(\mathfrak{D}_1) \qquad \mathfrak{D}g(x) = -\lim_{\substack{x_1 \to x-0 \\ x_2 \to x+0}} \frac{p_{x_1}(g(x_1) - g(x)) + p_{x_2}(g(x_2) - g(x))}{p_{x_1}(m(x_1) - m(x)) + p_{x_2}(m(x_2) - m(x))}.$$

Formally, upon applying L'Hospital's rule, we obtain

$$\mathfrak{D}g(x) = -\frac{p(x)}{p(x)m''(x) + 2p'(x)m'(x)}\,g''(x) - \frac{2p'(x)}{p(x)m''(x) + 2p'(x)m'(x)}\,g'(x)$$

so that $\mathfrak{D}$ appears as a differential operator of second order (which may degenerate). In fact, we shall establish that $\mathfrak{D}$ is a generalized differential operator. For this purpose, we classify the states as follows:

x is a *right passage point* or a *left passage point* if $P^x[X_t > x] > 0$ or

$P^x[X_t < x] > 0$ for some t, and x is a *passage point* if it is a right and a left passage point. Note that if x is neither a right nor a left passage point, then it is absorbing.

II. *At one-sided passage points, $\mathfrak{D}$ is a generalized differential operator of first order.*

If x is a right but not a left passage point then $p_{x_1} = 0$ and

$$\mathfrak{D}g(x) = -\lim_{x_2 \to x+0} \frac{g(x_2) - g(x)}{m(x_2) - m(x)} = -D_m{}^+g(x).$$

If x is a left but not a right passage point then $p_{x_2} = 0$ and

$$\mathfrak{D}g(x) = -\lim_{x_1 \to x-0} \frac{g(x_1) - g(x)}{m(x_1) - m(x)} = -D_m{}^-g(x).$$

To find the differential form in the case of (two-sided) passage points, we require more information about the properties of such points. The arguments to be used are similar to those which preceded the one-dimensional case and, therefore, will be shortened.

Let $\tau_y(\omega) = \inf\,[t\colon X_t(\omega) = y]$ if this set is not empty and $\tau_y(\omega) = \infty$ otherwise. Thus, $\tau_y(\omega) = \tau_{[\alpha,y)}$ or $\tau_{(y,\beta]}$ according as $X_0(\omega) < y$ or $X_0(\omega) > y$; note that the subscript sets are open in $\mathfrak{X} = [\alpha, \beta]$. Denote by $p(x, y) = P^x[\tau_y < \infty]$ the pr. starting at x to reach y in a finite time, and denote by $p(x, y, z) = P^x[\tau_y < \tau_z]$ the pr. starting at x to reach y before reaching z. Note that $p(x, y, z) + p(x, z, y) \leqq 1$ and that x is not a right passage point if and only if $p(x, y) = 0$ for all $y > x$.

a. (i) *If $a < x < y < b$ or $a > x > y > b$, then*

$$p(y, a) = p(y, x)p(x, a), \quad p(y, a, b) = p(y, x, b)p(x, a, b),$$
$$p(x, a, b) = 0 \Rightarrow p(x, a) = 0.$$

(ii) *If x is a right passage point, then there exists $(x', x'') \ni x$ such that $p(x', x'') > 0$, and if all $x \in [a, b]$ are right passage points then $p(a, b) > 0$.*

Proof. 1° Let, say, $a < x < y < b$. Since

$$[X_0 = y, \tau_a < \infty] = [X_0 = y, \tau_x < \infty] \cap [\tau_a < \infty]_{\tau_x}$$

and $X_{\tau_x} = x$, it follows that

$$P^y[\tau_a < \infty] = E^y(I_{[\tau_x < \infty]}P^{X_{\tau_x}}[\tau_a < \infty]) = P^x[\tau_a < \infty]P^y[\tau_x < \infty].$$

This proves the first equality in (i), and similarly for the second equality.

Let $a < x' < x < x'' < b$. Let S' be the set of states $z' < x'$ and let

S'' be the set of states $z'' > x''$. We say that $X_T(\omega)$ crosses exactly n times from S'' into S' in time t if there exist n and only n pairs (t''_k, t'_k) with $0 \leqq t''_1 < t'_1 < t''_2 < \cdots < t'_n \leqq t$ such that $X_{t'_k} \in S'$ and $X_{t''_k} \in S''$. Let $\omega \in A \Leftrightarrow X_0(\omega) = x$ and $\tau_a(\omega) < \infty$, and let $\omega \in A_n \Leftrightarrow \omega \in A$ and $X_T(\omega)$ crosses exactly n times from S'' into S' in time $\tau_a(\omega)$, so that $A = \bigcup_{n=0}^{\infty} A_n$. Let $\tau_n(\omega) = \inf[t\colon X_t(\omega) = x$ and $X_T(\omega)$ crosses exactly n times from S'' into S' in time $t]$ if this set is not empty and $\tau_n(\omega) = \infty$ otherwise, so that $A_n \subset (A_0)_{\tau_n}$. If $p(x, a, b) = P^x A_0 = 0$, then, from $X_{\tau_n} = x$, it follows that $P^x(A_0)_{\tau_n} = 0$, hence $P^x A_n = 0$ and $P^x A = 0$. This proves the last assertion in (i).

2° If x is a right passage point, that is, $P^x[X_t > x] > 0$ for some t, then there exists an $x'' > x$ such that $P^x[X_t > x''] > 0$, hence there exists a neighborhood $(x - \epsilon, x + \epsilon)$ such that $P^y[X_t > x''] > 0$ for all $y \in (x - \epsilon, x + \epsilon)$. Therefore, $p(x', x'') > 0$ for any $x' \in (x - \epsilon, x)$. This proves the first assertion in (ii). If all $x \in [a, b]$ are right passage points then, by what precedes, there exist intervals $(x', x'') \ni x$ such that $p(x', x'') > 0$. These open intervals cover $[a, b]$, hence a finite number of them (x'_k, x''_k), $k \leqq n$, covers $[a, b]$. It follows from the first equality in (i) that $p(a, b) > 0$ and the second assertion in (ii) is proved. The proof is terminated.

b. *Let $p(a, b) > 0$. If $a < b$ then all $x \in [a, b)$ are right passage points, all $E^x\tau_{(a,b)}$ are finite, and $p(x, b, a) \to 1$ as $x \to b - 0$; if $a > b$ then $p(x, b, a) \to 1$ as $x \to b + 0$.*

Proof. Let $x \in [a, b)$. If x is not a right passage point then, by **a**, $p(a, b) = p(a, x)p(x, b) = 0$ contrary to the hypothesis. The first assertion is proved.

Since for $X_0 = y$ with $y \in [a, x]$

$$[\tau_b > t]_{\tau_x} \subset [\tau_b > t], \quad X_{\tau_x} = x, \quad \tau_{(a,b)} = \min(\tau_a, \tau_b) \leqq \tau_b,$$

it follows that

$$P^x[\tau_{(a,b)} > t] \leqq P^x[\tau_b > t] = E^a(P^{X_{\tau_x}}[\tau_b > t]) \leqq P^a[\tau_b > t],$$

where, letting $t \to \infty$,

$$P^a[\tau_b > t] \to P^a[\tau_b = \infty] = 1 - p(a, b) < 1.$$

Therefore, for some t,

$$P^x[\tau_{(a,b)} > t] \leqq P^a[\tau_b > t] = 1 - \delta < 1$$

and, by 41.2**a**, $E^x\tau_{(a,b)} < \infty$. This proves the second assertion.

Let $a < x < x_1 < \cdots < x_n \uparrow b$ and $X_0 = x$, hence $\tau_{x_1} < \cdots < \tau_{x_n} \uparrow \tau \leqq \tau_b$. Set $A = [\tau_b < \tau_a]$ and $A_n = [\tau_{x_n} < \tau_a]$ so that $A \subset \bigcap A_n$. Assume there is an $\omega \in \bigcap A_n - A$. Either $\tau(\omega) < \infty$ or $\tau(\omega) = \infty$. But, by sample continuity, on $[\tau < \infty]$, $X_{\tau_b} = b \leftarrow x_n = X_{\tau_n} \rightarrow X_\tau$, hence $\tau = \tau_b$. Therefore $\tau(\omega) < \infty$ implies $\omega \in A$ which contradicts the above assumption. Thus $\tau(\omega) = \infty$ so that $\tau_a(\omega) \geqq \tau_b(\omega) \geqq \tau(\omega) = \infty$, since $\omega \notin A$. It follows that $\tau_{(a,b)} = \infty$ while, by **a**, $p(a, b) > 0$ implies $E^x\tau_{(a,b)} < \infty$. Therefore $P^x(\bigcap A_n - A) = 0$,

$$p(x, x_n, a) = P^x A_n \rightarrow P^x A = p(x, b, a) = p(x, x_n, a)p(x_n, b, a)$$

and $p(x, b, a) > 0$, since otherwise $p(x, b) = 0$ and $p(a, b) = p(a, x)p(x, b) = 0$. It follows that $p(x_n, b, a) \rightarrow 1$, and similarly when $b < a$. This proves the last assertion. The proof is terminated.

c. *Let $p(a, b)p(b, a) > 0$ and let x vary over $[a, b]$.*

(i) *The function $p(x) = p(x, b, a)$ is continuous and increasing with $p(x) \rightarrow 0$ as $x \rightarrow a + 0$ and $p(x) \rightarrow 1$ as $x \rightarrow b - 0$, and*

$$p(x, x_1, x_2) = (p(x_2) - p(x))/(p(x_2) - p(x_1)), \quad a < x_1 < x < x_2 < b.$$

(ii) *The function $-m(x) = -E^x\tau_{(a,b)}$ is continuous and, with respect to the function $p(x)$, it is convex and has an increasing left continuous left derivative $s^-(x) = -D_p{}^-m(x)$ and an increasing right continuous right derivative $s^+(x) = -D_p{}^+m(x)$ which coincide on their continuity set $C(s)$.*

Note that, if the hypothesis holds then all points of (a, b) are passage points and if all points of $[a, b]$ are passage points then the hypothesis holds.

Proof. We use **a** without further comment.

1° Let $a < x < y < b$, so that

$$p(x) = p(x, b, a) = p(x, y, a)p(y, b, a) = p(x, y, a)p(y),$$

hence $p(x) \leqq p(y)$. If $p(x) = p(y)$ then either $p(x) = 0$ or $p(x, a, y) = 1 - p(x, y, a) = 0$. In the last case, $p(x, a) = 0$ and $p(b, a) = p(b, x)p(x, a) = 0$ contrary to the hypothesis. In the first case, $p(x, b) = 0$ and $p(a, b) = p(a, x)p(x, b) = 0$ contrary to the hypothesis. Thus $p(x) < p(y)$, and the first assertion in (i) is proved.

The limit assertions in (i) result from the corresponding ones in **b**, upon using for the first one the relation $p(x, a, b) = 1 - p(x)$. Similarly, as $x \rightarrow y - 0$, $p(x, y, a) \rightarrow 1$, hence

$$p(x) = p(x, y, a)p(y) \rightarrow p(y),$$

the function $p(x)$ is left continuous and, analogously, the function $1 - p(x) = p(x, a, b)$ is rightcontinuous. This proves the continuity assertion in (i). Since

$$p(x, x_1, x_2) = p(x, a, x_2)/p(x_1, a, x_2) = (1 - p(x, x_2, a))/(1 - p(x_1, x_2, a))$$

and

$$p(x, x_2, a) = p(x)/p(x_2), \quad p(x_1, x_2, a) = p(x_1)/p(x_2),$$

the last assertion in (i) follows.

2° If $a \leqq x_1 < x < x_2 \leqq b$ then

$$m(x) - p(x, x_1, x_2)m(x_1) - p(x, x_2, x_1)m(x_2) = E^x\tau_{(x_1,x_2)} > 0$$

so that, by (i),

$$m(x) > \frac{p(x_2) - p(x)}{p(x_2) - p(x_1)} m(x_1) + \frac{p(x) - p(x_1)}{p(x_2) - p(x_1)} m(x_2)$$

and the function $-m(x)$ is convex with respect to the function $p(x)$. The remaining assertions of (ii) result from this convexity, upon transforming it into ordinary convexity as follows. By (i), the function $p(x)$ has an inverse function $q(y)$. Set $n(y) = -m(q(y))$ and note that if $0 \leqq y_1 < y < y_2 \leqq 1$ then $a \leqq x_1 = q(y_1) < x < x_2 = q(y_2) \leqq b$. Thus

$$n(y) < \frac{y_2 - y}{y_2 - y_1} n(y_1) + \frac{y - y_1}{y_2 - y_1} n(y_2)$$

and the function $n(y)$ bounded from above (by 0) is convex. Since

$$s^-(x) = -D_p^- m(x) = D_y^- n(y), \quad s^+(x) = -D_p^+ m(x) = D_y^+ n(y),$$

the assertions follow from the corresponding properties of the function $n(x)$. The proof is terminated.

A. ONE-DIMENSIONAL DIFFUSION OPERATOR THEOREM. *Let X_T be continuous stable and let all $x \in [a, b]$ be passage points. Let $g \in \tilde{C}_d$ and*

$$m(x) = E^x\tau_{(a,b)}, \quad s^-(x) = -D_p^- m(x), \quad s^+(x) = -D_p^+ m(x).$$

On (a, b), the function $D_p^- g(x)$ exists and is left continuous, the function $D_p^+ g(x)$ exists and is right continuous, they coincide on $C(s)$, and

$$\mathfrak{D}g(x) = D_{s^-}^+ D_p^- g(x) = D_{s^+}^- D_p^+ g(x).$$

Proof. We apply **c** without further comment. Set

$$g(x, y) = \frac{g(y) - g(x)}{p(y) - p(x)}, \quad m(x, y) = \frac{m(y) - m(x)}{p(y) - p(x)}.$$

Relation $(\mathfrak{D}_1)$ becomes

$$\text{(1)} \qquad \mathfrak{D}g(x) = -\lim_{\substack{x_1 \to x-0 \\ x_2 \to x+0}} \frac{g(x, x_2) - g(x, x_1)}{m(x, x_2) - m(x, x_1)}.$$

Since

$$\lim_{\substack{x_1 \to x-0 \\ x_2 \to x+0}} (m(x, x_2) - m(x, x_1)) = \lim_{x_2 \to x+0} m(x, x_2) - \lim_{x_1 \to x-0} m(x, x_1) = s^-(x) - s^+(x)$$

it follows that

$$\lim_{x_1 \to x-0} g(x, x_1) - \lim_{x_2 \to x+0} g(x, x_2) = D_p{}^- g(x) - D_p{}^+ g(x)$$

exists,

$$\text{(2)} \qquad (s^+(x) - s^-(x))\mathfrak{D}g(x) = D_p{}^+ g(x) - D_p{}^- g(x)$$

and the last difference reduces to zero on $C(s)$. Set

$$\begin{aligned} G(y) &= g(y) - g(x) - D_p{}^- g(x)(p(y) - p(x)) \\ &= (p(y) - p(x))(g(x, y) - D_p{}^- g(x)), \\ M(y) &= -m(y) + m(x) - s^-(x)(p(y) - p(x)) \\ &= (p(y) - p(x))(-m(x, y) - s^-(x)). \end{aligned}$$

Note that $M(x) = 0$, for $y > x$

$$D_p{}^- M(y) = s^-(y) - s^-(x) > 0, \quad D_p{}^+ M(y) = s^+(y) - s^-(x) > 0,$$

and, upon letting $x_1 \to x - 0$ in (1),

$$\mathfrak{D}g(x) = \lim_{y \to x+0} \frac{G(y)}{M(y)}.$$

To obtain the asserted form of $\mathfrak{D}g(x)$, it will suffice to prove that L'Hospital's rule applies, as follows. First, we show that the limits of $h^+(y)$ and $h^-(y)$ as $y \to x + 0$ exist and are the same, upon setting

$$h^+(y) = D_p{}^+ G(y)/D_p{}^+ M(y), \quad h^-(y) = D_p{}^- G(y)/D_p{}^- M(y).$$

Suppose that $\liminf_{y \to x+0} h^+(y) < \limsup_{y \to x+0} h^+(y)$ so that there exist distinct c', c'' which lie between these limits. If c is either c' or c'' and $f(y) = G(y) - cM(y)$, then $D_p{}^+ f = D_p{}^+ G - cD_p{}^+ M$ changes signs in (x, y) for any $y > x$. Thus, there exist sequences $y_n, y'_n \to x + 0$ such that f attains relative maxima at the y_n and relative minima at the y'_n.

Therefore, $\mathfrak{D}f(y_n) \leqq 0$, $\mathfrak{D}f(y'_n) \geqq 0$ and, from $\mathfrak{D}f = \mathfrak{D}g - c$ it follows that $\mathfrak{D}g(y_n) \leqq +c$, $\mathfrak{D}g(y'_n) \geqq +c$, hence $\mathfrak{D}g(x) = +c$. Since c is either of two distinct numbers c', c'', this is impossible. Thus,

$$\lim_{y \to x+0} h^+(y) = \lim_{y \to x+0} \frac{D_p{}^+g(y) - D_p{}^-g(x)}{s^+(y) - s^-(x)}$$

exists, and similarly

$$\lim_{y \to x+0} h^-(y) = \lim_{y \to x+0} \frac{D_p{}^-g(y) - D_p{}^-g(x)}{s^-(y) - s^-(x)}$$

exists. Since $D_p{}^-g(y) = D_p{}^+g(y)$ on the everywhere dense set $C(s)$, the two limits coincide. Set now

$$F(z) = G(z)M(y) - G(y)M(z), \quad z \in [x, y]$$

and note that $F(x) = F(y) = 0$ so that F attains either its maximum or its minimum at some $z' \in (x, y)$. In the first case, $D_p{}^-F(z') \geqq 0 \geqq D_p{}^+F(z')$, hence

$$h^+(z') \leqq \frac{G(y)}{M(y)} \leqq h^-(z')$$

and in the second case, these inequalities are reversed. As $y \to x + 0$, hence $z' \to x + 0$, the extreme terms converge to the same limit while the middle term converges to $\mathfrak{D}g(x)$. Therefore,

$$\begin{aligned}\mathfrak{D}g(x) &= \lim_{y \to x+0} \frac{D_p{}^+g(y) - D_p{}^-g(x)}{s^+(y) - s^-(x)} \\ &= \lim_{y \to x+0} \frac{D_p{}^-g(y) - D_p{}^-g(x)}{s^-(y) - s^-(x)} = D_{s^-}{}^+D_p{}^-g(x).\end{aligned}$$

The function $s^+(x)$ being right continuous, it follows that

$$D_p{}^+g(x + 0) - D_p{}^-g(x) = (s^+(x) - s^-(x))\mathfrak{D}g(x)$$

and, taking into account (2), $D_p{}^+g(x + 0) = D_p{}^+g(x)$, that is, $D_p{}^+g$ is right continuous. Similarly for the remaining assertions. The proof is terminated.

Note that the passage points form an open set $\sum_j (a_j, b_j)$ and what precedes applies to any $[a, b] \subset (a_j, b_j)$. Furthermore, the foregoing form of $\mathfrak{D}g$ can be rewritten as follows:

III. *If the* $x \in [a, b]$ *are passage points, then*

$$\mathfrak{D}g(x) = D_s D_p g(x), \quad x \in C(s)$$

$$\mathfrak{D}g(x) = \frac{D_p g(x+0) - D_p g(x-0)}{s(x+0) - s(x-0)}, \quad x \notin C(s).$$

Together, I, II, and III determine the one-dimensional diffusion operator.

COMPLEMENTS AND DETAILS

Unless otherwise stated, the state space $(\mathfrak{X}, \mathcal{S})$ *is a separable locally compact metric space,* $\mathcal{S}$ *is its* σ*-field of topological Borel sets, and* $P_t(x, S)$, $x \in \mathfrak{X}$, $S \in \mathcal{S}$, $t > 0$, *is a stationary tr.pr. with* $P_t(x, \mathfrak{X}) \leqq 1$.

1. $P_t(x, \mathfrak{X})$ is nonincreasing as $t > 0$ increases.

2. Let $\tau \geqq 0$ be a r.v. not necessarily a.s. finite. Let X_t be a r.v. defined on $[t < \tau]$. $X_T = (X_t, t \geqq 0)$ is a r.f. of "lifetime τ."

If $P_t(x, \mathfrak{X}) \equiv 1$, then there exists a Markov r.f. of infinite lifetime with the given tr.pr. and an arbitrary initial distribution on $\mathcal{S}$. If $P_t(x, \mathfrak{X}) \leqq 1$ and $P_h(x, \mathfrak{X}) \to 1$ as $h \to 0$ for every x, then there exists a Markov r.f. X_T of possibly finite lifetime with positive pr., with the given tr.pr. and arbitrary initial distribution on $\mathcal{S}$:

Add an isolated point at infinite "∞" and determine a tr.pr. $P'_t(x', S')$, where the sets S' are sets S and sets $S + \{\infty\}$, so that it coincides with $P_t(x, S)$ for $x' = x$ and $S' = S$ and $P'_t(x', \{\infty\}) = 1$ or $1 - P_t(x, \mathfrak{X})$ according as $x' = \infty$ or $x' \neq \infty$. Construct X_T as above. Complete $\mathcal{B}(X_s, s \leqq t)$. Define $\tau(\omega)$ as the supremum of all rational r for which $X_r(\omega) \neq \infty$. "Curtail" X_T to have lifetime τ upon replacing the domain of X_t by $[t < \tau]$ and then replace $X_t(\omega) = \infty$ by an arbitrary $x_0 \in \mathfrak{X}$.

3. For all (x, S), if $P_h(x, S) \to I(x, S)$ as $h \to 0$, then $\{1 - P_h(x, \mathfrak{X})\}/h$ converges to a finite limit: Introduce the point at infinity and $P'_t(x', S')$ and apply section 39.1.

4. Apply section 39 to Markov processes with a finite number of states under the continuity condition and to Markov processes with a countable number of states under the uniform continuity condition.

5. Let $P_t(x, S)$, $t > 0$, be $\mathcal{S} \times \mathfrak{J}$-measurable ($\mathfrak{J}$ is the σ-field of Lebesgue sets in $(0, \infty)$).

a) $P_t(x, S)$ is continuous in t if and only if there exists a finite measure μ on $\mathcal{S}$ such that $P_t(x, S)$ is μ-continuous in S:

If $P_t(x, S)$ is continuous in t, take $\mu S = \int_0^\infty e^{-t} P_t(x, S)\, dt$. If μ exists, given t and S note that for $0 < \epsilon < t < t'$ and $|h_n| < \epsilon$

$$\iint_\epsilon^t |P_{s+h_n}(x, S) - P_s(x, S)|\, ds\mu(dx) = \int (dx) \int_\epsilon^t |P_{s+h_n}(x, S) - P_s(x, S)|\, ds,$$

where the inner integral converges to zero as $h_n \to \infty$ and, for some subsequence h'_n, $P_{s+h'_n}(x, S) \to P_s(x, S)$ for some $s \in (\epsilon, t)$ a.e. in μ, hence a.e. in $P_{t-s}(x, \cdot)$.

Thus,

$$P_{t+h'_n}(x, S) = \int P_{t-s}(x, dy) P_{s+h'_n}(y, S) \to P_t(x, S).$$

What if the state space is the Borel line and $P_t(x, S) = 1$ or 0 according as $x + t$ belongs or not to S?

b) Let $P_t(x, S)$ be right continuous in t. Then it is continuous in t: Note that to find μ finite such that $P_t(x, S)$ is μ-continuous in S, it suffices to know that $P_t(x, S)$ is t-measurable and that if $P_t(x, S) = 0$ for a.e. t, then $P_t(x, S) \equiv 0$.

If $P_t(x, S) = \int_S P_t(x, y)\mu(dy)$ where μ is σ-finite and $p_t(x, y)$ is (x, y)-measurable, then $P_t(x, S)$ is continuous in t.

c) If $\lim_{h \downarrow 0} P_h(x, S)$ exists for all (x, S), then $P_t^+(x, S) = \lim_{h \downarrow 0} P_{t+h}(x, S)$ exists, is a tr.pr. continuous in t, and coincides with $P_t(x, S)$ except for countably many values of t: Note that $P_t^+(x, S)$ is right continuous in t and that $P_t(x, S)$ has at most countably many discontinuity points in t.

6. Let X_T be a Poisson r.f. with the sample functions selected to be left continuous, and let $\tau(\omega)$ be the infimum of those t for which $X_t(\omega) = 1$. Then the Markov r.f. X_T is stable and τ is a time of X_T but is not its Markov time.

7. Let $\alpha_n, \beta_n > 0$, $p_n = \alpha_n/(\alpha_n + \beta_n)$, $q_n = \beta_n/(\alpha_n + \beta_n)$. Let $X_n = (X_n(t), t \geqq 0)$, $n = 1, 2, \cdots$, be independent Markov r.f.'s with two states 0 and 1, $X_n(0) = 0$, and

$$P(X_n(t + h) = 1 \mid X_n(t) = 0) = \alpha_n h + o(h),$$

$$P(X_n(t + h) = 0 \mid X_n(t) = 1) = \beta_n h + o(h).$$

a) $P(X_n(t) = 0 \mid X_n(0) = 0) \geqq q_n$, $P(X_n(s) = 0, t \leqq s \leqq t + h \mid X_n(t) = 0) = e^{-\alpha_n h}$. If $\prod q_n > 0$, that is, $\sum p_n < \infty$ then, at every time t, a.s. $X_n(t) = 0$ for almost all n.

b) If $\sum \lambda_n = \infty$, then $P(X_n(s) = 0, t \leqq s \leqq t + h$, for all $n \geqq m \mid X_n(t) = 0) = 0$ for every m.

Let $X = (X(t), t \geqq 0)$ be the joint r.f. $(X_1(t), X_2(t), \cdots, t \geqq 0)$. X is a Markov r.f. If $\sum p_n < \infty$ then X has only a countable number of states. If, moreover, $\sum \alpha_n = \infty$ then all these states are instantaneous.

c) Analytically, let the state space consist of sequences $x = (x_1, x_2, \cdots)$, $y = (y_1, y_2, \cdots)$, $\cdots$, of 0's and 1's with finitely many 1's. Set $p_t^{(n)}(0, 0) = q_n + p_n e^{-(\alpha_n+\beta_n)t}$, $p_t^{(n)}(1, 1) = p_n + q_n e^{-(\alpha_n+\beta_n)t}$, $p_t^{(n)}(0, 1) = 1 - p_t^{(n)}(0, 0)$, $p_t^{(n)}(1, 0) = 1 - p_t^{(n)}(1, 1)$. The function $P_t(x, y) = \prod_n p_t^{(n)}(x_n, y_n)$ is a tr.pr. which obeys the continuity condition, and ${}_0\dot{P}_0(x, x) = -\infty$.

8. Let P, Q, with or without affixes, be (pr.) distributions on $\mathfrak{X}$ and set

$$d(P, P') = \operatorname{Var}(P - P') = \int | P(dx) - P'(dx) |.$$

a) d is a metric and the space of distributions is complete in this metric.

Let X_T, $T = [0, \infty)$, be stationary Markovian with tr.pr. $P_t(x, S)$, $P_t(x, \mathfrak{X}) \equiv 1$, and distributions P_t, P'_t (of X_t) corresponding to initial distributions P_0, P'_0 (of X_0).

b) $d(P_t, P'_t)$ does not increase as t increases, whatever be P_0 and P'_0.

c) If $d(P_t, \bar{P}) \to 0$ as $t \to \infty$ for some P_0, then $\bar{P}$ is invariant (that is, $\bar{P}_t = \bar{P}$ for all t).

Suppose that (H): for every $\epsilon > 0$ there exist $C \in \mathfrak{J}$, a distribution Q, positive numbers a, b, t_1, and there exists t_0 for every P_0 such that

(i) $aQ(S) \leqq P_{t_1}(x, S)$ for all $x \in C$, $S \subset C$, (ii) $P_tC \geqq 1 - \epsilon$ for all $t \geqq t_0$, (iii) $P_tS \leqq bQ(S) + \epsilon$ for all $S \subset C$ and $t \geqq t_0$.

Then (C): there exists a unique invariant distribution $\bar{P}$ which is "ergodic" (that is, such that $d(P_t, \bar{P}) \to 0$ whatever be P_0):

d) Under (H), $d(P_t, P'_t) \to 0$ as $t \to \infty$ whatever be P_0, P'_0. Thus, there exists at most one invariant distribution and when it exists, it is ergodic.

e) Under (H), $d(P_{t_m}, P_{t_n}) \to 0$ as $t_m, t_n \to \infty$ whatever be P_0 and $t_n \mapsto \infty$.

f) Are conditions (H) necessary for (C) to hold?

9. Quasileftcontinuity (Blumenthal, Hunt, Meyer). Let $T = [0, \infty)$, X_t be $\mathcal{B}_t$-measurable, $\mathcal{B}_t \uparrow$, τ_n, τ be times of $\mathcal{B}_T$ and g be nonnegative measurable on $\mathfrak{X}$.

a) *$(X_T, \mathcal{B}_T)$ is Markovian with tr. pr. $P_t(x, S) \Leftrightarrow (P_{t-s}(X_s, S), \mathcal{B}_t, 0 \leqq s \leq t)$ are martingales.* What about adding "strongly"?

Let $(X_T, \mathcal{B}_T)$ be Markovian with semi-group $(T_t, t \geqq 0)$. Then $(g(X_t), t \geqq 0)$ is a supermartingale $\Leftrightarrow$ g is supermedian, i.e., $g \geqq T_tg$.

We say that g is *excessive* (uniformly) if it is supermedian and $T_hg \to g$ ($T_hg \to g$ uniformly and g is bounded); g (or X_T) is *quasileftcontinuous* (*qlc*) if $\tau_n \uparrow \tau$ a.s. $\Rightarrow g(X_{\tau_n}) \to g(X_\tau)$ (or $X_{\tau_n} \to X_\tau$) a.e. on $[\tau < \infty]$.

b) Let $(X_T, \mathcal{B}_T)$ be strongly Markovian with almost all sample functions rightcontinuous with left limits; we can take $\mathcal{B}_t = \mathcal{B}_{t+}$.

Lemma. Let g be bounded with $g(X_{\tau_n}) \to Y$ a.e. on $[\tau < \infty]$ as $\tau_n \uparrow \tau$ a.s. If $T_hg \to g$ uniformly, then $Y = E(X_\tau \mid \mathcal{B}_\tau^-)$ where $\mathcal{B}_\tau^-$ is the σ-field over the $\mathcal{B}_{\tau_n}$:

Reduce to bounded τ (replacing τ_n by $\inf(\tau_n, t)$ with arbitrary $t \in T$). Let $\sigma_n = \sup(\tau_n + h, \tau)$ so that, as $n \to \infty$, $P[\sigma_n \neq \tau_n + h] \to 0$ and $\Delta_2 = g(X_{\tau_n+h}) - g(X_{\sigma_n}) \neq 0$ with pr. $\to 0$. $\Delta_1 = g(X_{\tau_n}) - g(X_{\tau_n+h}) = T_hg(X_{\tau_n})$ and $\Delta_3 = g(X_{\sigma_n}) - g(X_\tau) = T_{\sigma_n-\tau}g(X_\tau)$ with $\sigma_n - \tau \leqq h$ converge vers zero uniformly in n, as $h \to 0$. Thus, given $\epsilon > 0$, $B \in \mathcal{B}_{\tau_k}$, then fixing h sufficiently small, for all n sufficiently large, $\left|\int_B g(X_{\tau_n}) - \int_B g(X_\tau)\right| < \epsilon$. Letting $n \to \infty$ then $\epsilon \to 0$, $\int_B Y = \int_B g(X_{\tau_n})$; and this equality extends on $\mathcal{B}_\tau^-$.

Uniformly excessive g on qlc X_T are qlc: For, lemma applies to bounded supermartingale $g(X_{\tau_n})$ and X_τ is $\mathcal{B}_\tau^-$-measurable by qlc of X_T.

If the semi-group $(T_t, t \geqq 0)$ is strongly continuous on invariant C_0, then X_T is *qlc:*

Reduce to compact $\mathfrak{X}$ (proceeding as in 40.3**d**). For $g \in C$, Y exists by existence of left limits and lemma applies. If $P[Y \neq X_\tau] > 0$, there is a sphere $V_X(r)$ with $PA = P[Y \in V_X(r), X_\tau \neq V_X(2r)] > 0$. With $g \in C$, $g \leqq 1$, $g = 1$ on $V_X(r)$, $g = 0$ on $V_X{}^c(2r)$, integrating $Y = E(g(X_\tau) \mid \mathcal{B}_\tau^-)$ on $[Y \in V_X(r)]$ yields $PA \leqq 0$-contradiction.

c) Qlc strongly Markovian processes with almost all sample functions rightcontinuous with left limits are *standard* (*Markovian*) in applications to potential theory. Combine with 40.3 and 41.1 to find conditions for existence of equivalent standard Markov processes.

BIBLIOGRAPHY

Titles of articles the results of which are included in cited books by the same author are omitted from the bibliography. Roman numerals designate books, and arabic numbers designate articles.

The references below may pertain to more than one part or chapter of this book.

PART FOUR: DEPENDENCE

I. Blanc-Lapierre and Fortet. *Theorie des fonctions aléatoires* (1953).

II. Cramér and Leadbetter. *Stationary and related stochastic processes* (1967).

III. Kaczmarz and Steinhaus. *Orthogonalreichen* (1935).

IV. Hopf, E. *Ergodentheorie* (1937).

V. Lévy, P. *Theorie de l'addition des variables aléatoires* (1954).

IV. Neveu, J. *Martingales à temps discret* (1972).

VII. Ville, J. *Etude critique de la notion du collectif* (1939).

VIII. Wold, H. *A study in the analysis of stationary times series* (1938).

CHAPTER VIII. CONDITIONING

1. Blackwell, D. Idempotent Markov chains. *Ann. of Math.* **43** (1942).
2. Cogburn, R. Asymptotic properties of stationary sequences. *Univ. of Calif. Publ. Stat.* **3** (1960).
3. Doblin, W. Sur les propriétés asymptotiques de mouvement régi par certains types de chaines simples. *Bull. Math. Soc. Roum. Sc.* **39** (1937).
4. Doblin, W. Eléments d'une théorie générale des chaines simples constantes de Markov. *Ann. Sc. Ecole Norm. Sup.* **57** (1940).
5. Halmos, P. and Savage, L. J. Application of the Radon-Nikodym theorem to the theory of sufficient statistics. *Ann. Math. Stat.* **20** (1949).
6. Nagaev, C. B. Some limit theorems for stationary Markov chains. *Teorya Veroyatn.* **2** (1957).
7. Neyman, J. Su un teorema concernente le cosiddette statistiche sufficienti. *Ist. Ital. Attuari.* **6** (1935).
8. Ueno, T. Some limit theorems for temporally discrete Markov processes. *J. Univ. Tokyo* **7** (1957).

CHAPTER IX. FROM INDEPENDENCE TO DEPENDENCE

9. Andersen, E. S. and Jessen, B. Some limit theorems on set functions. *Danske Vid. Selsk. Mat-Fys. Medd.* **25** (1948).
10. Blackwell, D. and Dubins, L. Merging opinions with increasing information. *Ann. Math. Stat.* **33** (1962).

11. Loève, M. Etude asymptotique des sommes de variables aléatoires liées. *J. de Math.* **24** (1945).
12. Loève, M. On sets of probability laws and their limit elements. *Univ. of Calif. Publ. Stat.* **1** (1950).
13. Robbins, H. The asymptotic distribution of the sum of a random number of random variables. *Bull. Am. Math. Soc.* **54** (1948).
14. Snell, J. L. Applications of martingale system theorems. *Trans. Am. Math. Soc.* **73** (1952).

CHAPTER X. ERGODIC THEOREMS

15. Birkhoff, G. D. Proof of the ergodic theorem. *Proc. Nat. Ac. Sc. U.S.A.* **17** (1931).
16. Blum, J. R. and Hanson, D. L. On invariant probability measures. *Pacific J. of Math.* **10** (1960).
17. Doob, J. L. Asymptotic properties of Markov transition probabilities. *Trans. Am. Math. Soc.* **63** (1948).
18. Dunford, N. Spectral theory. *Trans. Am. Math. Soc.* **54** (1943).
19. Dunford, N. and Miller, D. S. On the ergodic theorem. *Trans. Am. Math. Soc.* **60** (1946).
20. Dowker, Y. Invariant measure and the ergodic theorem. *Duke Math. J.* **14** (1947).
21. Halmos, P. Approximation theories for measure preserving transformations. *Trans. Am. Math. Soc.* **55** (1944).
22. Halmos, P. An ergodic theorem. *Proc. Nat. Ac. Sc. U.S.A.* **32** (1946).
23. Hartman, S., Marczewski, E. and Ryll-Nardzewski, C. Théorèmes ergodiques et leurs applications. *Coll. Math.* **2** (1951).
24. Hurewicz, N. Ergodic theorem without invariant measure. *Ann. of Math.* **44** (1944).
25. Khintchine, A. Zu Birkhoffs Lösung des Ergodenproblems. *Math. Ann.* **107** (1933).
26. Loève, M. On almost sure convergence. *Proc. Sec. Berkeley Symp. on Stat. and Prob.* (1951).
27. von Neumann, J. Proof of the quasi-ergodic hypothesis *Proc. Nat. Ac. Sc. U.S.A.* **18** (1932).
28. Oxtoby, J. C. On the ergodic theorem of Hurewicz. *Ann. of Math.* **49** (1948).
29. Ryll-Nardzewski, C. On the ergodic theorems. *Studia Math.* **12** (1951, 1952).
30. Riesz, F. Sur la théorie ergodique. *Comm. Math. Helv.* **17** (1945), **19** (1947).
31. Yosida, K. and Kakutani, S. Operator-theoretical treatment of Markov's process and mean ergodic theorem. *Ann. of Math.* **42** (1941).

CHAPTER XI. SECOND ORDER RANDOM PROPERTIES

32. Cramér, H. On the theory of stationary random processes. *Ann. Math.* **41** (1940).
33. Cramér, H. On harmonic analysis in certain functional spaces. *Ark. Mat. Astr. Fys.* **28** (1942).

34. Cramér, H. A contribution to the theory of stochastic processes. *Second Berkeley Symp.* (1951).
35. Karhunen, K. Über lineare Methoden in der Wahrscheinlichkeitsrechnung. *Ann. Acad. Sci. Fenn.* **37** (1947).
36. Khintchine, A. Korrelationstheorie der stationäre stochastischen Prozesse. *Math. Ann.* **109** (1934).
37. Kolmogorov, A. Stationary sequences in Hilbert space—in Russian. *Bull. Math. Univ. Moscow* **2** (1941).
38. Loève, M. Fonctions aléatoires de second ordre. *C. R. Acad. Sci.* **220** (1945), **222** (1946); *Rev. Sci.* **83** (1945), **84** (1946).
39. Maruyama, G. The harmonic analysis of stationary stochastic processes. *Mem. Fac. Sci. Kyusyu Univ.* **4** (1949).
40. Slutsky, E. Sur les fonctions aléatoires presque periodiques et sur la decomposition des fonctions aléatoires stationaires en composantes. *Act. Sci. Ind.* **738** (1938).

PART FIVE: ELEMENTS OF RANDOM ANALYSIS

I. Billingsley, P. *Convergence of probability measures* (1968).
II. Breiman, L. *Probability* (1968).
III. Doob, J. L. *Stochastic Processes* (1953).
IV. Dynkin, E. B. *Markov Processes* I, II (1965).
V. Friedman, D. *Brownian motion and diffusion* (1973).
VI. Gihman and Skorohod. *The theory of stochastic processes* I (1974), II (1975).
VII. Ito and McKean. *Diffusion processes and their sample paths* (1965).
VIII. Hille and Phillips. *Functional Analysis and Semigroups* (1957).
IX. Karlin and Taylor. *First course in stochastic processes* (1975). (1954).
X. Lévy, P. *Processes stochastiques et mouvement brownien* (1965).
XI. Meyer, P. A. *Potentials and Probability* (1966).
XII. Nelson, E. *Dynamical theories of Brownian Motion* (1967).
XIII. Neveu, J. *Mathematical Foundations of Calculus of Probability* (1965).
XIV. Skorohod, A. V. *Studies in the theory of random processes* (1965).
XV. Wax, N. (editor). *Selected papers on Noise and Stochastic processes* (1954).

CHAPTER XII. FOUNDATIONS; MARTINGALES AND DECOMPOSABILITY

1. Dobrushin, R. L. On the Poisson law for distribution of particles in space—in Russian, *Ukrain. Math. J.* **8** (1956).
2. Dobrushin, R. L. The continuity condition for sample martingale functions—in Russian, *Teorya Veroyatn.* **3** (1958).
3. Dynkin, E. B. Criteria of continuity and of absence of discontinuities of the second kind for trajectories of a Markov random process—in Russian, *Izv. Ak. Nauk SSSR* **16** (1952).
4. Ito, K. On stochastic processes I (Infinitely divisible laws of probability), *Jap. J. Math.* **18** (1942).

5. Kallenberg, O. Conditions for continuity of random processes . . . *Ann. Prob* **1** (1973).
6. Slutsky, E. Alcune proposizioni sulla teoria delle funzioni aleatorie, *Giorn. Ist. Ital. Attuari* **8** (1937).

CHAPTER XIII. BROWNIAN MOTION AND LIMIT DISTRIBUTIONS

7. Araujo and Le Cam. Remarks on a theorem of Donsker and Kolmogorov. *Ann. Prob*—to be published.
8. Bachélier, L. Théorie de la spéculation. *Ann. Ec. norm. sup.* **17** (1900).
9. Bachélier, L. Probabilitiés des oscillations maxima. *C. R. Acad Sc.* **212** (1944).
10. Blumenthal, R. M. An extended Markov property. *Trans AMS* **85** (1950).
11. Cogburn and Tucker. A limit theorem for a function of the increments of a decomposable process. *Trans. AMS* **99** (1961).
12. Donsker, M. D. An invariance principle for certain probability limit theorems. Mem. AMS **6** (1951).
13. Dvoretzky, Erdös, and Kakutani. Nonincrease everywhere of the Brownian motion process. *Proc. 4th Berk. Symp.* **2** (1960).
14. Einstein, A. Die vonder der molekularkinetischetheorie der wärmer geforderte bewegung van in rehenden slüssigkeiten suspendierten teilchen. Ann. d. Phys. 11 (1905).
15. Hartman and Winter. On the law of the iterated logarithm *Amer. J. Math.* **63** (1941).
16. Hunt, G. A. Some theorems concerning Brownian motion. *Trans. AMS* **81** (1956).
17. Khintchine, A. Ein Satz der Wahrscheinlichkeitsrechung. *Fund. Math.* **6** (1924).
18. Kolmogorov, A. Über die Grenzwertsätze der Wahrscheinlichkeitsrechnung. Izv. Ak. Nauk (1933).
19. Lamperti, J. On convergence of stochastic processes. Trans. AMS **101** (1962).
20. Le Cam, L. Convergence in distribution of stochastic processes. *Univ. Calif. Publ. Stat.* **2** (1957).
21. Paley, Wiener, and Zygmund. Note on random functions. *Math. Zeit.* (1933).
22. Prohorov, Y. V. Convergence of random processes and limit theorems in probability theory, Teoria Veroyt **1** (1956).
23. Strassen, V. An invariance principle for the law of the iterated logarithm. *Zeit. Wahrsch.* **3** (1964).
24. Strisower, E. Sanitas Leonis probabilati mathematicae auxiliaris. *Misc. arcana* (1977).
25. Wiener, N. Differential space. *J. Math. Phys.* **2** (1923).

CHAPTER XIV. MARKOV PROCESSES

26. Austin, Blumental, and Chacon. On continuity of transition functions, *Duke Math. J.* **25** (1958).
27. Blackwell, D. Another countable Markov process with only instantaneous states, *Ann. Math. Stat.* **29** (1958).

28. Blumental, R. An extended Markov property, *Trans. Am. Math. Soc.* **85** (1957).
29. Doblin, V. Sur certains mouvements aléatoires discontinus, *Skand. Akt.* **22** (1939).
30. Doob, J. L. Topics in the theory of Markov chains, *Trans. Am. Math. Soc.* **52** (1942).
31. Doob, J. L. Markov chains—denumerable case, *Trans. Am. Math. Soc.* **58** (1945).
32. Doob, J. L. Brownian motion on a Green space, *Teorya Veroyatn.* **2** (1957).
33. Dynkin, E. B. Markov processes and semi-groups of operators—in Russian, *Teorya Veroyatn.* **1** (1956).
34. Dynkin, E. B. Infinitesimal operators of Markov processes—in Russian, *Teorya Veroyatn.* **1** (1956).
35. Dynkin, E. B. One-dimensional continuous strong Markov processes—in Russian, *Teorya Veroyatn.* **4** (1959).
36. Dynkin and Yushkevitch. Strong Markov processes—in Russian, *Teorya Veroyatn.* **1** (1956).
37. Feller, W. Zur Theorie der Stochastischen Prozesse (Existenz und Eindeutigkeitssätze), *Math. Ann.* **113** (1936).
38. Feller, W. On the integro-differential equations of purely discontinuous Markov processes, *Trans. Am. Math. Soc.* **48** (1940); *Errata Ibid.* **58** (1945).
39. Feller, W. Semi-groups of transformations in general weak topologies, *Ann. of Math.* **57** (1953).
40. Feller, W. The general diffusion operator and positivity preserving semi-groups in one dimension, *Ann. of Math.* **60** (1954).
41. Feller, W. On second order differential operators, *Ann. of Math.* **61** (1955).
42. Fortet, R. Les fonctions aléatoires du type de Markov associées à certaines equations lineaires aux dérivées partielles du type parabolique, *J. Math. Pures Appl.* **22** (1943).
43. Hunt, G. A. Markov processes and potentials, *Ill. J. of Math.* **1** (1957), **2** (1958).
44. Hunt, G. A. Markov chains and Martin boundaries, *Ill. J. of Math.* **4** (1960).
45. Ito, K. On stochastic differential equations, *Mem. Am. Math. Soc.* **4** (1951).
46. Kendall, D. G. Some analytical properties of continuous stationary Markov transition functions, *Trans. Am. Math. Soc.* **78** (1955).
47. Kinney, J. R. Continuity properties of sample functions of Markov processes, *Trans. Am. Math. Soc.* **74** (1953).
48. Kolmogorov, A. N. Über die analytischen Methoden in der Wahrscheinlichkeitsrechnung, *Math. Ann.* **104** (1931).
49. Kolmogorov, A. N. On some probabilities concerning the differentiability of the transition problems in temporally homogeneous Markov processes having a denumerable set of states—in Russian, *Uch. zapiski MΓY*, **148** (1951).

50. Lévy, P. Systèmes markoviens et stationaires: cas dénombrable, *Ann. Ec. Norm.* **68** (1951), 69 (1952).
51. Maruyama, G. Continuous Markov processes and stochastic equations, *Rend. Circ. Mat. Palermo* **4** (1955).
52. Maruyama, G. On the strong Markov property, *Mem. Kyushu Univ.* **13** (1959).
53. Maruyama and Tanaka. Some properties of one-dimensional diffusion processes, *Mem. Kyushu Univ.* **11** (1957).
54. Meyer, P. A. Fonctionnelles multiplicatives et additives de Markov, *Ann. Inst. Fourier* **12** (1962).
55. Neveu, J. Théorie des semi-groupes de Markov, *Univ. of Calif. Publ. Stat.* **2** (1958).
56. Ray, D. Resolvents, transition functions and strongly Markovian processes, *Ann. of Math.* **70** (1959).
57. Sevastianov, B. A. An ergodic theorem for Markov processes and its application to telephone systems with refusals—in Russian, *Teorya Veroyatn.* **2** (1957).
58. Yushkevitch, A. A. On strong Markov processes, *Teorya Veroyatn.* **2** (1957).

INDEX

Page numbers in *italic* type refer to Volume I; those in roman type to Volume II

Graduate Texts in Mathematics

Soft and hard cover editions are available for each volume up to vol. 14, hard cover only from Vol. 15

1 TAKEUTI/ZARING. Introduction to Axiomatic Set Theory. vii, 250 pages. 1971.
2 OXTOBY. Measure and Category. viii, 95 pages. 1971.
3 SCHAEFFER. Topological Vector Spaces. xi, 294 pages. 1971.
4 HILTON/STAMMBACH. A Course in Homological Algebra. ix, 338 pages. 1971. (Hard cover edition only)
5 MACLANE. Categories for the Working Mathematician. ix, 262 pages. 1972.
6 HUGHES/PIPER. Projective Planes. xii, 291 pages. 1973.
7 SERRE. A Course in Arithmetic. x, 115 pages. 1973.
8 TAKEUTI/ZARING. Axiomatic Set Theory. viii, 238 pages. 1973.
9 HUMPHREYS. Introduction to Lie Algebras and Representation Theory. xiv, 169 pages. 1972.
10 COHEN. A Course in Simple Homotopy Theory. xii, 114 pages. 1973.
11 CONWAY. Functions of One Complex Variable. 2nd corrected reprint. xiii, 313 pages. 1975. (Hard cover edition only.)
12 BEALS. Advanced Mathematical Analysis. xi, 230 pages. 1973.
13 ANDERSON/FULLER. Rings and Categories of Modules. ix, 339 pages. 1974.
14 GOLUBITSKY/GUILLEMIN. Stable Mappings and Their Singularities. x, 211 pages. 1974.
15 BERBERIAN. Lectures in Functional Analysis and Operator Theory. x, 356 pages. 1974.
16 WINTER. The Structure of Fields. xiii, 205 pages. 1974.
17 ROSENBLATT. Random Processes. 2nd ed. x, 228 pages. 1974.
18 HALMOS. Measure Theory. xi, 304 pages. 1974.
19 HALMOS. A Hilbert Space Problem Book. xvii, 365 pages. 1974.
20 HUSEMOLLER. Fibre Bundles. 2nd ed. xvi, 344 pages. 1975.
21 HUMPHREYS. Linear Algebraic Groups. xiv. 272 pages. 1975.
22 BARNES/MACK. An Algebraic Introduction to Mathematical Logic. x, 137 pages. 1975.
23 GREUB. Linear Algebra. 4th ed. xvii, 451 pages. 1975.
24 HOLMES. Geometric Functional Analysis and Its Applications. x, 246 pages. 1975.
25 HEWITT/STROMBERG. Real and Abstract Analysis. 3rd printing. viii, 476 pages. 1975.
26 MANES. Algebraic Theories. x, 356 pages. 1976.

27 KELLEY. General Topology. xiv, 298 pages. 1975.
28 ZARISKI/SAMUEL. Commutative Algebra I. xi, 329 pages. 1975.
29 ZARISKI/SAMUEL. Commutative Algebra II. x, 414 pages. 1976.
30 JACOBSON. Lectures in Abstract Alegbra I: Basic Concepts. xii, 205 pages. 1976.
31 JACOBSON. Lectures in Abstract Algebra II: Linear Algebra. xii, 280 pages. 1975.
32 JACOBSON. Lectures in Abstract Algebra III: Theory of Fields and Galois Theory. ix, 324 pages. 1976.
33 HIRSCH. Differential Topology. x, 222 pages. 1976.
34 SPITZER. Principles of Random Walk. 2nd ed. xiii, 408 pages. 1976.
35 WERMER. Banach Algebras and Several Complex Variables. 2nd ed. xiv, 162 pages. 1976.
36 KELLEY/NAMIOKA. Linear Topological Spaces. xv, 256 pages. 1976.
37 MONK. Mathematical Logic. x, 531 pages. 1976.
38 GRAUERT/FRITZSCHE. Several Complex Variables. viii, 207 pages. 1976
39 ARVESON. An Invitation to C^*-Algebras. x, 106 pages. 1976.
40 KEMENY/SNELL/KNAPP. Denumerable Markov Chains. 2nd ed. xii, 484 pages. 1976.
41 APOSTOL. Modular Functions and Dirichlet Series in Number Theory. x, 198 pages. 1976.
42 SERRE. Linear Representations of Finite Groups. 176 pages. 1977.
43 GILLMAN/JERISON. Rings of Continuous Functions. xiii, 300 pages. 1976.
44 KENDIG. Elementary Algebraic Geometry. viii, 309 pages. 1977.
45 LOEVE. Probability Theory. 4th ed. Vol. 1. xvii, 425 pages. 1977.
46 LOEVE. Probability Theory. 4th ed. Vol. 2. approx. 350 pages. 1977.
47 MOISE. Geometric Topology in Dimensions 2 and 3. x, 262 pages. 1977.
48 SACHS/WU. General Relativity for Mathematicians. xii, 291 pages. 1977.
49 GRUENBERG/WEIR. Linear Geometry. 2nd ed. x, 198 pages. 1977.
50 EDWARDS. Fermat's Last Theorem. xv, 410 pages. 1977.
51 KLINGENBERG. A Course in Differential Geometry. xii, 192 pages. 1978.
52 HARTSHORNE. Algebraic Geometry. xvi, 496 pages. 1977.
53 MANIN. A Course in Mathematical Logic. xiii, 286 pages. 1977.
54 GRAVER/WATKINS. Combinatorics with Emphasis on the Theory of Graphs. xv, 368 pages. 1977.
55 BROWN/PEARCY. Introduction to Operator Theory. Vol. 1: Elements of Functional Analysis. xiv, 474 pages. 1977.
56 MASSEY. Algebraic Topology: An Introduction. xxi, 261 pages. 1977.
57 CROWELL/FOX. Introduction to Knot Theory. x, 182 pages. 1977.
58 KOBLITZ. p-adic Numbers, p-adic Analysis, and Zeta-Functions. x, 122 pages. 1977.